VOYAGER TALES
Personal Views of the Grand Tour

David W. Swift
University of Hawaii
Manoa, Hawaii

American Institute of Aeronautics and Astronautics, Inc.
1801 Alexander Bell Drive
Reston, Virginia, 20191

Publishers since 1930

American Institute of Aeronautics and Astronautics, Inc., Reston, Virginia

Library of Congress Cataloging-in-Publication Data

Swift, David W., 1927–
  Voyager tales: personal views of the grand tour / David Swift.
    p. cm.
  Includes bibliographical references (p.).
  1. Voyager project. 2. Planets—Exploration. 3. Astronautics in astronomy. 4. Jet Propulsion Laboratory (U.S.)—Employees—Interviews. 5. Astronomers—United States—Interviews. 6. Aerospace engineers—United States—Interviews. I. Title.
QB601.595 1997 629.43'54'0922—dc21   97-20179
ISBN 1-56347-252-X   CIP

Copyright © 1997 by the American Institute of Aeronautics and Astronautics, Inc. All rights reserved. Printed in the United States of America. No part of this publication may be reproduced, distributed, or transmitted, in any form or by any means, or stored in a database or retrieval system, without the prior written permission of the publisher.

To Earth—
Whose uniqueness was made more evident by these Voyagers

Where one Englishman traveled in the reign of the first two Georges, ten now go on a grand tour .... There is scarce a citizen of large fortune but takes a flying view of France, Italy, and Germany.

[1772]

The grand tour was conceived of as a comprehensive, once-in-a-lifetime venture .... (p. 159)
Not a brief spree but a peregrination lasting years, the grand tour marked the young men who undertook it with a cosmopolitan outlook that persisted for a lifetime. (p. 156)

John A. Quadrado
Architectural Digest, April 1995

FOREWORD

In the *Odyssey*, Homer recounted the journey of a fictional character—described as "a man of many wiles who wandered far and wide"—who took 20 years to traverse a mythical universe of a few hundred miles. In a comparable 20 years, the actual adventures of the two Voyager spacecraft have taken them billions of miles across the solar system and out into interstellar space on a mission that will continue until the end of time. If Homer were still with us, he no doubt would be impressed.

My experience with Voyager was a very personal one. I was an executive with Martin Marietta, the maker of the Titan line of launch vehicles, which was responsible for giving each spacecraft its initial boost off the Earth's surface. On August 20, 1977, I drove my son, Greg, to Cape Canaveral to watch what promised to be a spectacular launch. Despite his young age, Greg shared with me a fascination with all space missions, manned and unmanned. Many times we camped out under the open sky and counted satellites passing overhead. This was to be Greg's first Titan launch, and he was excited in the way that only 13-year-olds can be.

As we waited during the last minutes of the countdown in mission control, just a few miles from the launch pad, a thought crystallized in my mind about the future of the spacecraft and the future of my son. Like anyone connected with a space mission, I was reluctant to say anything aloud, for fear of "jinxing" the mission—a strange behavior for hard-nosed engineers, but a common one! Finally, the Titan Centaur lifted off amid a tremendous explosion of flame and steam and sound. I remember how my heart nearly stopped as I worried about every sequence the Titan was to perform and felt sorry for my friends from General Dynamics who built the Centaur upper stage and would have to sweat, not just for minutes, but for hours.

As the majestic rocket rolled into its exit trajectory, I shared with my son my insight. "Greg," I said, "do you realize that when Voyager approaches its first destination—Jupiter—two years from now, you will be in high school?" He looked at me with awe. I continued, "And when it reaches Saturn, you'll be in your junior year. When it passes Uranus, you'll have a college degree. And when it catches up with its last stop in the solar system—Neptune—12 years from now, you might be designing spacecraft of your own." Over the years, as Voyager reached its milestones, Greg and I have exchanged calls and postcards, reminding each other of that day and noting his progress and that of the spacecraft whose launch we witnessed. And, yes, today my son is helping to design spacecraft.

Now it is two decades since the beginning of the Voyagers' adventures. Having given us never-before-seen views of the cold volcanoes of Io, the tortured terrain of Miranda, the dazzling rings of Saturn, and the icy grandeur of Triton, the Voyagers have left the solar system and are streaking away from their birthplace at some 35,000 miles per hour. Our understanding of the universe has flourished, as Neil Armstrong noted when he said, "Voyager [is a] modern Odysseus, [which] has uncovered the unimagined intricacy and awesome spectacle of the giant planets of our home solar system."

All of us in the aerospace community stand in awe of the accomplishments of these intrepid spacecraft, whose performance reflects the extraordinary planning and execution of thousands of engineers and hundreds of creative and highly skilled scientists. This book recounts the hopes and dreams and prayers of many of these participants, capturing their thoughts in an oral history that reminds us again of the enormous technical challenges that faced the Voyager team. How fortunate we are that they succeeded so brilliantly.

For in the final analysis the Voyagers not only sent back enormous amounts of data and breathtaking views of our strange and wondrous planetary neighbors, they also made us more keenly aware of the treasures of our own Earth. Perhaps the legacy of the Voyagers is best expressed in the words of the poet T. S. Eliot:

> We shall not cease from exploration
> And the end of all our exploring
> Will be to arrive where we started
> And know the place for the first time.

Norman R. Augustine
Chairman and CEO
Lockheed Martin Corporation

CONTENTS

PREFACE xi

ACKNOWLEDGMENTS xv

INTRODUCTION 1
   The Voyager Project 3
   Views of the Grand Tour 5
   History of Robotic Exploration 25
   References 31

INTERVIEWS 33
   Overview 35
      Edward Stone   Project Scientist 37

   Pathfinders 59
      Gary Flandro   Graduate Student, Advanced Projects Group ... 61
      Roger Bourke   Supervisor, Advanced Projects Group 75
      Charles Kohlhase   Voyager Mission Design Manager 83

   Project Managers 101
      Bud Schurmeier   Development Phase 103
      John Casani   Prelaunch Phase 113
      Robert Parks   Cruise Phase 121
      Peter Lyman   Jupiter 1 133
      Raymond Heacock   Jupiter 2 and Saturn 1 145
      Esker Davis   Saturn 2 159
      Richard Laeser   Uranus 173
      Norman Haynes   Neptune 181
      George Textor   Interstellar Mission 189

   Laboratory Directors 197
      William Pickering   Director 1954–1976 199
      Bruce Murray   Director 1976–1982 209

   Engineers 221
      William Shipley   Space Development Manager 223
      Henry Cox   Tracking Data Systems Manager 239
      Thomas Gavin   Project Assurance Manager 249
      Glen Southworth   Television Recording Engineer 257
      Robert Cesarone   Navigation Team Task Leader 261
      Howard Marderness Voyager Spacecraft Team Chief 275
      Charles Avis   Deputy Chief, Image Science Data Team 281

   Scientists 289
      Arden Albee   Chief Scientist 291
      Moustafa Chahine   Chief Scientist, Jet Propulsion Laboratory 299

Scientists (continued)
- Ellis Miner — Assistant Project Scientist 307
- Candice Hansen — Voyager Imaging Experiment Representative .. 315

Support Services 327
- Robert Post — Photo Laboratory Supervisor 329
- Peter Wells — Head Dispatcher, Transportation 339
- Anita Sohus — Documentation Representative 345
- John Brickett — Plant Protection Chief 357
- Bobbie Fishman — Visitor Control Greeter 363
- Marty Greer — Automotive Technician Leadman 371
- Judith McGavin — Project Secretary 375
- Deborah Gag — Telephone Operator 381
- Jack Blanco — Senior Guard, Plant Protection Department 385
- Catherine Swift — Undergraduate Research Fellow 389
- Jurrie van der Woude — Public Information Officer 391

CONCLUSIONS 407

APPENDICES 415
- A. Voyager Chronology 417
- B. Voyager Phases Discussed 419
- C. Science Investigations at Saturn 421
- D. JPL Grand Tour Memo 1966 423

INDEX 425

PREFACE

The Voyager project was an impressive technological and scientific achievement, made possible by a highly effective organization. Therefore I prepared this book of original interviews with participants in the project, discussing their views on Voyager's origin, development, obstacles, crucial decisions, accomplishments, and broader implications.

This book differs from others about Voyager, and space exploration in general, in at least two ways: topic and presentation.

Topic

It is concerned with subjective perceptions rather than objective "facts".

The book looks behind the scientific results, to the people and processes involved in attaining them. My focus is the Jet Propulsion Laboratory (JPL), which designed, built, and flew the spacecraft. I present a cross section of this unusual organization, interviewing a broad spectrum of JPL personnel, from the director and chief scientist, down through engineers and technicians, to the level of secretaries and security guards.

They offer insights into the operation of this highly successful mission, insights that are worth consideration by other projects struggling for survival in the increasingly competitive global economy. What did JPL do right? What things should be avoided?

Along with their thoughts on management, technology, and science, I've also inquired about a range of personal and philosophical implications of the project. What impact did it have on their daily lives? Do their experiences with Voyager suggest insights into the future of human interaction with complex, sophisticated machines?

To find out, I began talking with Voyager personnel at JPL in 1981, during the Voyager 2 encounter with Saturn, and continued through the subsequent encounters during the rest of the 1980s. Twenty-four of the three dozen interviews in this book were conducted in 1990; they were expanded and the rest were added through 1995.

These people are a sample of the thousands who made the "Grand Tour" possible. The project resembled a pyramid. Personnel at the top were obvious, well known, but further down more people were involved and the picture was less clear. Individual contributors were obscured by numbers, so I reached into their ranks almost at random, to pick representatives who give some indication of the contributions, however indirect, provided by themselves and their colleagues.

The focus is on JPL personnel but I include one outsider to indicate the enthusiasm generated by Voyager and to suggest the contributions made by the private sector. Glen Southworth of Colorado Video tells why he donated his expertise to transfer pictures of Earth to Voyager's golden record.

Few studies have documented a major development in science and technology while it was actually happening. Fewer still have presented this emergence from the participants' perspectives, rather than as interpreted by someone else.

Presentation

The second unusual feature of this book is the presentation of entire interviews rather than fragments or paraphrases. The transcripts are printed almost verbatim, edited to eliminate repetitions or to smooth the occasionally awkward grammar of spontaneous conversation.

This approach has two advantages. The respondents' own words provide a sense of individuality and immediacy often missing in standard, third person accounts. This is not a digest of their ideas; it is an album of verbal portraits, depicting each individual's style of thinking and speaking, thereby highlighting the people behind the project.

Furthermore, allowing people to speak for themselves reduces the danger that they will be misinterpreted by the journalist or professor writing a more typical book. I am not evaluating their statements, or trying to tell you what they think.

This can become significant when people differ about a particular event: for example, who first conceived the idea of the Grand Tour? Instead of my being the judge, imposing my opinions and biases, you can go directly to Voyager personnel and decide for yourself what actually happened.

Of course, deciding what actually happened may sometimes be difficult because more than 3000 people were involved with Voyager over nearly two decades. This book presents barely one percent of them, so it is likely that not all points of view have been included, and it is almost certain that important contributions of many deserving individuals have been unintentionally overlooked. This becomes even more evident when we consider that 11,000 work years had already been devoted to Voyager by the time of the Neptune encounter in 1989, and the project is expected to continue well into the 21st century.*

Although this report was partially funded by JPL, nobody attempted to influence my selection of people or topics. I asked for advice but the choices were entirely my own, directed by my desire to discover what enabled this project to succeed. Any complex human endeavor will experience problems, stresses, and mistakes, but the ultimate demonstration of Voyager's overall competence is its remarkable achievements.

Author's Background

Why is a sociologist writing about robots in outer space? The connection is my concern for competence in organizations. This concern originated from my personal observations during the early days of space exploration.

I originally aspired to be an aeronautical designer. During my childhood in San Francisco the dirigible *Macon* rumbled overhead, and in 1935 I watched the first Martin Clipper fly through the still unbridged Golden Gate on its maiden voyage to Hawaii. I built models, drew planes, collected photographs, studied Buck Rogers, and read avidly about aviation and astronomy. Upon entering the University of California at Berkeley I majored in engineering. After a year I turned 18, was drafted into the Army Air Force in February 1946 and sent to Wright Field, Dayton, Ohio. It was the experimental center for testing the latest military aircraft, and the mecca of aviation enthusiasts. I looked forward to the experience.

As a technical illustrator I worked with blueprints for German rocket weapons, and a V-2 stood next to my barracks. I soon realized that the technical problems of spaceflight were well on the way to solution, but a very different set of problems was also evident at Wright Field—problems involving people and organizations.

I was one of the few soldiers in the 4000th AAFBU Intelligence Division with less than a bachelor degree. It was a unit of engineers, scientists, and other highly trained specialists, many with advanced degrees. During the war they had been employed at Douglas, Boeing, Lockheed, North American, and other technically oriented companies, but when the war ended they were reclassified 1A and drafted into the armed forces. They were placed at the very bottom of the military hierarchy and were assigned tasks that made little use of their abilities.

Their time and training were wasted on traditional military activities, including marching, drills, guard duty, cleaning latrines, picking up litter, working long shifts in the kitchen, and so forth. Even the engineers and scientists who eventually gained exemptions from the more onerous tasks were still required to live in barracks on the base, with all that entailed. To

---

* *Voyager Neptune Travel Guide,* NASA JPL 89-24, June 1, 1989, pp. 135, 275.

leave the base, even when off duty, they had to request permission, which was sometimes denied for no apparent reason other than the whim of the sergeant in charge.

Their treatment differed sharply from that accorded German scientists and engineers. After the Third Reich collapsed, 100 were brought to Wright Field, where they had more personal freedom than their American counterparts. They were treated with deference by the U.S. Army and were not subjected to the same strictures controlling the American engineer and scientist soldiers. They wore civilian clothes, could leave the labs and the base when they pleased, and were not required to perform menial tasks.

Why were skills being wasted? Why were professionals who had demonstrated their ability to manage their own lives by completing rigorous college curricula, and who had been considered essential to the nation's survival, now subjected to control over every aspect of their existence? And why were former enemies being treated better than their American counterparts?

The situation was especially perplexing because it occurred despite official, high-level policy. A February 1946 War Department circular had ordered that American engineers and scientists be placed in a "critically needed specialist" category. After basic training they were supposed to be assigned only to duties that would make full use of their scientific and technical abilities.

A close witness to this, I became intrigued by competence in organizations. This interest continued while teaching in public schools, and into my career as a sociologist studying organizations. Voyager provided an opportunity to examine a project that achieved far more than its originally stated objectives, and it allowed me to return to my early interests in aviation and space.

Preview

Following the introduction, which includes a brief history of space exploration, the main part of the book presents interviews with 37 participants in the project. Each interview is preceded by a biographical sketch and the person's photograph. The book closes with a conclusion, appendices, and an index.

This book is not intended to be read straight through at one sitting. Instead it is a series of conversations with project personnel, a source book for researchers as well as for general readers who want more information about these persons as people. It provides vicarious contact with some of the most competent engineers, scientists, managers, and technicians of our time.

ACKNOWLEDGMENTS

Like the Voyager project, this book is the product of many people. Dale Cruikshank revived my early interest in space, which had been pushed aside for decades by more mundane matters. Kenneth Burtness called my attention to the unusual Voyager mission which, among other things, carried sights and sounds of Earth on a journey far beyond our solar system. And I met Glen Southworth, the imaginative engineer who transferred those sights to the golden recordings carried on the two spacecraft.

Jurrie van der Woude has been a prime mover in writing this book. Despite his being besieged by requests from eminent scientists and famous newscasters for photographs, he took time to point a sociology professor toward interesting people and topics. Edward Stone found funds to partially support the preparation of this book even though it does not deal directly with science.

In the hectic months approaching this manuscript's encounter with the publisher, Charley Kohlhase was a tireless source of ideas, information, and constructive criticism; similarly helpful were Ellis Miner and Anita Sohus. Among other JPL people who assisted along the way are Jim Doyle, Michael Hooks, Laurie Lincoln, William McLaughlin, Ed McMahon, Mary Beth Murril, and Jim Wilson.

Outside JPL, valuable aid was provided by John Calderone, Sue Cowing, Elliot Cutting, Grania Davis, Ben Finney, Mark Kahn, William T. Larkins, Kevin Polk, Donald Tarter, Rodger Williams, Dawn Yocom, and Paul Zarchan.

In Hawaii, Carol McCord has long given general encouragement and specific help on my various projects. Eric Johnson, too, has been valuable in many areas. At home, Lois Swift has contributed far beyond the quiet support traditionally attributed to spouses: she is a very good editor whose assistance ranged from detecting unneeded verbiage to helping me choose between several equally attractive ideas and wordings. Similar aid was provided by Catherine Swift.

The American Institute of Aeronautics and Astronautics (AIAA) is the professional society representing 40,000 aerospace engineers, managers, scientists, and technicians. A number of them are included here and several have been honored with AIAA awards. In 1971, AIAA published a report, co-authored by an engineer interviewed in this book, that is highly relevant to Voyager—AIAA Paper 71-187, "Design of Grand Tour Missions."

Finally I appreciate the part played by *Astronomy* editor Richard Berry, who years ago asked a most provocative question.

INTRODUCTION

THE VOYAGER PROJECT

The Voyager missions to the outer planets revolutionized our knowledge of astronomy. The two spacecraft, launched in 1977, greatly increased our information about Jupiter and Saturn; provided the first closeup views of Uranus and Neptune, the outermost major planets in the solar system; and took a final series of photographs, a "family portrait" of the sun and its retinue of tiny dots, one of which is us—Earth.

Then, having virtually completed the initial reconnaissance of our solar system, the Voyagers flew on toward interstellar space. There is nothing more for their cameras to see, but they will send us other kinds of data well into the 21st century. And, after we lose contact with them, they will continue their silent journeys for untold millennia, carrying sounds and pictures of the planet from which they came.

Significance

The Voyagers' 11 science instruments produced more data about the outer solar system than had been gained in all previous centuries.

Although the Apollo program and the space shuttles received more attention because humans were onboard, Voyager's scientific returns surpass those from the manned projects, and were obtained for a fraction of the cost in life and money. Not burdened with the complex demands of life support, Voyager could focus more directly and efficiently on obtaining scientific data.

Another contribution of the Voyager project was psychological and philosophical. Voyager broadened our concept of our place in the cosmos. We now have a deeper awareness of the solar system, the setting in which our planet exists. Several factors are involved: the size of Earth, the number of planets and satellites in the solar system, the distances between them, and the apparent absence of life.

Apollo depicted Earth as a huge ball; Voyager showed us as a tiny dot. Apollo presented clouds, oceans, and continents in such detail that we could almost see where we live. But from the edge of the solar system, all this is compressed into a pinpoint of light, barely discernible against the blackness of space.

The Voyagers discovered many moons around the outer planets, increasing to 60 the number of known worlds in the solar system. The immense distances separating them were dramatized by the time it took the spacecraft to travel to them. Radio communication, requiring several hours, further emphasized the vastness of space.

Voyagers' demonstrations of size, number, and distance may at first glance seem to diminish Earth's importance. However, awareness that Earth is only one small world among many, separated from others by vast distances, may increase humanity's appreciation of Earth's uniqueness and value.

This appreciation may be enhanced by Voyager's findings that, despite the great diversity of the many other worlds, they appear inhospitable to life as we know it. The Voyager explorations virtually extinguished lingering beliefs that advanced life forms might now exist elsewhere in the solar system.

Another aspect of Voyager was that it was the first to reach the outermost planets. In centuries to come there may be many more visits to these remote bodies, but there can never be another *first* approach. "Firsts" require more boldness, more risk, more energy than do subsequent, repeat, followup actions.

Voyagers' travels have been likened to those of Columbus, but they have far surpassed his in time expended, distance traveled, and number of new worlds discovered. And unlike historic terrestrial journeys into unknown realms, the Voyagers have no foreseeable end to their travels, and may wander for eons among the stars.

Although the spacecraft will have ceased long since to communicate with us, they are capable of communicating with intelligent beings in some distant part of the galaxy who might intercept them and play the recordings containing sights and sounds of Earth.

Such thoughts, stimulated by the Voyagers, are bound to influence our conception of ourselves and our place in the universe.

Difficulties

The enormous distances to the outer planets presented a formidable obstacle to visiting them. Even radio waves, traveling at the speed of light, take hours to reach them, and no manmade object is able to attain even a small fraction of light's velocity. Craft of the early space age would require dozens of years to traverse the billions of miles to Uranus and Neptune.

But JPL engineers noticed that a rare pathway to the outer planets would open briefly in the late 1970s. The stately motion of these planets in their immense orbits was such that once in about every two centuries they would be in a special alignment that would allow a spacecraft to fly past all four of the distant giants, getting a boost from the gravity of the nearest planet, and hurling it on toward the next one, farther out. The power needed to accomplish this without gravity assist was beyond the capability of 20th century technology.

The opening of this pathway could be neither hastened nor postponed. The timing was determined primarily by the orbit of Neptune. This outermost of the major planets, 30 times farther from the sun than Earth, takes 165 years to complete one circuit around the sun. So the orbits of these planets formed a gigantic lock, opened once every 176 years by a cosmic key to provide the stepping stones to the edge of the solar system.

The last time this alignment occurred was at the beginning of the 1800s. Jefferson was president of the 13 United States, Beethoven was just beginning to compose symphonies, and Napoleon was marching across Europe. George Washington had died only two years earlier and Abraham Lincoln had not yet been born. There were no railroads, telephones, automobiles, radios, airplanes, or television.

But even if spaceflight technology had been available back then, the significance of this alignment would not have been fully appreciated; Uranus had been discovered only 20 years earlier, and the existence of Neptune was not yet known.

Effective Organization

How was the Grand Tour accomplished? The opening of the pathway to Neptune was not sufficient in itself; it had to be used effectively. Success required extremely precise navigation, arriving at a planet within a few miles of the aim point after traveling billions of miles. Passing too close to the planet would bend the trajectory too sharply, while passing too far would not bend it enough. It was like hitting a golf ball 3000 miles across North America and sinking a hole in one.

Midcourse corrections were made, but were complicated by the immense distances that greatly stretched out reaction time. The delay was especially critical approaching an encounter, as the spacecraft plunged toward the planet many times faster than a bullet. It took hours for data on the Voyager's precise direction and location to reach Earth, and more hours to radio course adjustments back to the speeding craft, still more hours to verify that the correction had been received and executed, and still more to verify that the correction had the desired effect on the craft's trajectory.

An error anywhere in this complex sequence could result in failing to obtain scientific measurements, or straying too far off course to proceed to the next planet. Successfully handling such challenges demanded competence—competence of individuals and of the organization.

Obviously there were capable people within JPL who could figure out what needed to be done to accomplish this. But success required more than individual expertise; it also required that JPL's structure encouraged personnel to exercise their skills cooperatively. Therefore a major objective of this study was to identify the characteristics of JPL that fostered success.

Great Voyager Images

Of the thousands of images taken by the Voyagers, the following represents five major events in their flights: encounters with the four planets and the Family Portrait mosaic.

Jupiter's Great Red Spot.

Saturn and moons from 8,079,000 miles. Tethys is at the upper left. Dione is closer to Saturn and its shadow is on Saturn at the lower left.

Uranus's moon Miranda.

Neptune and Triton three days after flyby. Triton, the smaller crescent, is closer to the viewer.

Earth as seen from Voyager 1, 3.7 billion miles away, on February 14, 1990. This pale blue dot, tiny as it is, has been magnified approximately six times to make it visible here. In the original photograph Earth was only 0.12 pixel (picture element) in size.

VIEWS OF THE GRAND TOUR

This section provides a visual introduction to JPL and the Voyager project. Following a view of JPL as it appeared in 1995, we go back to its origins in the 1930s and then proceed roughly in chronological order, looking at highlights and also at random glimpses into the lab's routine daily activities: from early concepts, the planning and development stages, through launches and encounters to the results, the landmark closeup images of the outer planets and their satellites. And finally, the "picture of the century," the family portrait of our sun and its retinue of tiny dots, one of which is us, as seen from almost 4 billion miles away.*

These photos, diagrams, clippings, and excerpts from technical manuals only begin to depict the diversity and magnitude of effort that went into this remarkable mission—of which the solar system phase was just the beginning.

---

* The closeup images of the outer planets and the Family Portrait are included in a separate color section.

Opposite, top: Jet Propulsion Laboratory, Arroyo Seco, Pasadena, California, 1995.

Opposite, bottom: On October 31, 1936, in the Arroyo Seco riverbed, students and co-workers of Dr. Theodore von Kármán from the Guggenheim Aeronautical Laboratory, California Institute of Technology (GALCIT) fired their first rocket motor. Left to right are Rudolph Scott, Apollo M. O. Smith, Frank Malina (cofounder and director of JPL 1944–1946), Edward S. Forman, and John Parsons.

This page: Impressed by the GALCIT experiments, the Army Air Corps, in 1938, requested research on a number of projects, including rocket-assisted takeoff. The MIT representative chose something he considered more serious—airplane windshield de-icing—and added, "Kármán can take the Buck Rogers job." Von Kármán did, and in August 1941 jet-assisted takeoff rockets turned this Ercoupe into the first U.S. rocket-assisted airplane, cutting takeoff time and distance almost in half.

Top: On January 31, 1958, JPL Director William Pickering (left), James van Allen, and Werner von Braun display a model of Explorer 1 at a news conference after confirming that it had become the first American satellite to attain orbit.

Bottom: Voyager trajectories. Both spacecraft have passed beyond the outermost known planets. Voyager 1 was deflected upward above the ecliptic, the plane of the planets, by its pass beneath Saturn. Voyager 2 was deflected below the ecliptic by its dive over Neptune.

Opposite page: Opening page of Flandro's *Journal of Spacecraft* article describing the swingby technique and solar electric propulsion for the Grand Tour.

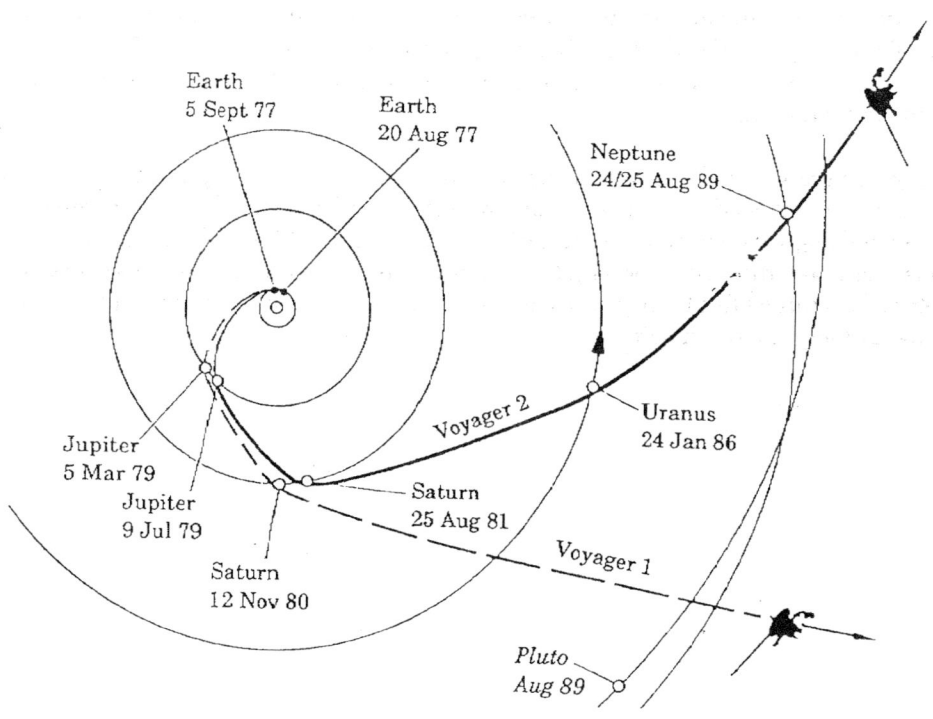

Solar Electric Low-Thrust Missions to Jupiter with Swingby Continuation to the Outer Planets

G. A. FLANDRO*
University of Utah, Salt Lake City, Utah

The advantages of combining the Jupiter swingby technique with optimized solar electric propulsion for missions to the Jovian planets are evaluated. Significant payload gains are achieved without sacrificing the reduced trip time which results from energy gained in the Jupiter encounter. Of particular interest is the "grand tour" to Jupiter, Saturn, Uranus, and Neptune which is best launched in 1977. Use of low-thrust propulsion on the Earth-Jupiter leg of the trajectory increases the payload by a factor of 3 for the standard Atlas/Centaur launch vehicle.

Introduction

THE gain in heliocentric energy available from close passage of the massive planet Jupiter can greatly decrease required trip time to the outer planets.[1-3] Although launch energy is also reduced, rather large launch vehicles still are required to accommodate adequate payloads. Other studies[4,5] have shown that the application of low-thrust electric propulsion in Jupiter flyby missions can significantly increase payload, and solar electric systems appear to be developing at a rate which should make them available for flight early in the next decade.[6] This paper considers the combination of the Jupiter swingby technique with solar-electric propulsion, which, in effect, opens the entire solar system to unmanned scientific exploration utilizing launch vehicles already in advanced stages of development. Basic mission analyses for Earth-Jupiter-Saturn, Earth-Jupiter-Uranus, Earth-Jupiter-Neptune, and Earth-Jupiter-Pluto flights are presented. Of special interest is the possibility of a "grand tour" mission involving close passage of Jupiter, Saturn, Uranus, and Neptune in a single flight. Optimum launch dates are established for each of the aforementioned missions, and performance is evaluated in terms of the tradeoff between payload and required trip time.

Because of the complexity introduced into the mission analysis process by incorporation of low-thrust propulsion,[4] only two launch vehicles are evaluated in detail: 1) Atlas SLV-3C/Centaur and 2) Atlas SLV-3X/Centaur. This results mainly from the inseparability of the escape and interplanetary phases of powered flight in low-thrust analyses. The spacecraft itself must be regarded as part of the launch vehicle, and the system must be optimized over the entire trajectory. The performance results are based on essentially current electric propulsion state-of-the-art (powerplant specific mass of 30 kg/kw). Significantly better performance may be available before the suggested launch dates. The advantages of the combination of low thrust and the swingby method are verified by comparison with an appropriate ballistic vehicle utilizing the Atlas/Centaur/Burner II launch vehicle system.

Utilization of Solar Electric Power in Jupiter Swingby Trajectories

A space vehicle closely passing a massive body moving relative to the inertial heliocentric coordinate system is perturbed in such a way that its kinetic energy is changed. The energy gained (or lost) by the spacecraft is balanced by a loss (or gain) of energy by the planet. The magnitude of the energy change depends on the following: 1) distance of closest approach, 2) hyperbolic approach speed of the spacecraft, 3) velocity of the planet relative to the sun, 4) mass of the planet, and 5) direction of approach of the spacecraft. Figure 1 illustrates the encounter hyperbola. The energy change during the encounter is approximately

$$\Delta E = \mathbf{V}_p \cdot (\mathbf{V'}_o - \mathbf{V'}_i) \quad (1)$$

where $\mathbf{V}_p$ is the velocity vector of the planet, and $\mathbf{V'}_i$ and $\mathbf{V'}_o$ are the relative incoming and outgoing asymptotic velocity vectors. The maximum energy gain corresponds to a trajectory which grazes a forbidden sphere concentric with the planet and with radius equal to the sum of the equatorial radius of the planet, the maximum atmospheric depth, and an allowance for guidance uncertainty.[1] Figure 2 shows the maximum possible energy increase plotted vs the incoming hyperbolic excess speed for each of the major planets. Jupiter swingby produces the greatest heliocentric energy change for a given approach velocity because of the massiveness of this planet coupled with its higher heliocentric speed as compared to the other major planets.

The energy increment is conveniently expressed in the form

$$\Delta E = fE^* \quad (2)$$

Fig. 1 Encounter hyperbola.

Presented as Paper 68-117 at the AIAA 6th Aerospace Sciences Meeting, New York, January 22–24, 1968; submitted January 29, 1968; revision received May 27, 1968. This work presents the results of one phase of research carried out in part by the Systems Analysis Research Section of the Jet Propulsion Laboratory, California Institute of Technology under Contract NAS 7-100 sponsored by NASA. The author wishes to express his thanks to T. A. Barber for innumerable suggestions and to C. G. Sauer and A. Joseph of the Jet Propulsion Laboratory for assistance in generating the trajectory data used in this study.

* Assistant Professor of Mechanical Engineering; formerly Member of the Technical Staff, Systems Analysis Research Section, Jet Propulsion Laboratory, California Institute of Technology, Pasadena Calif.

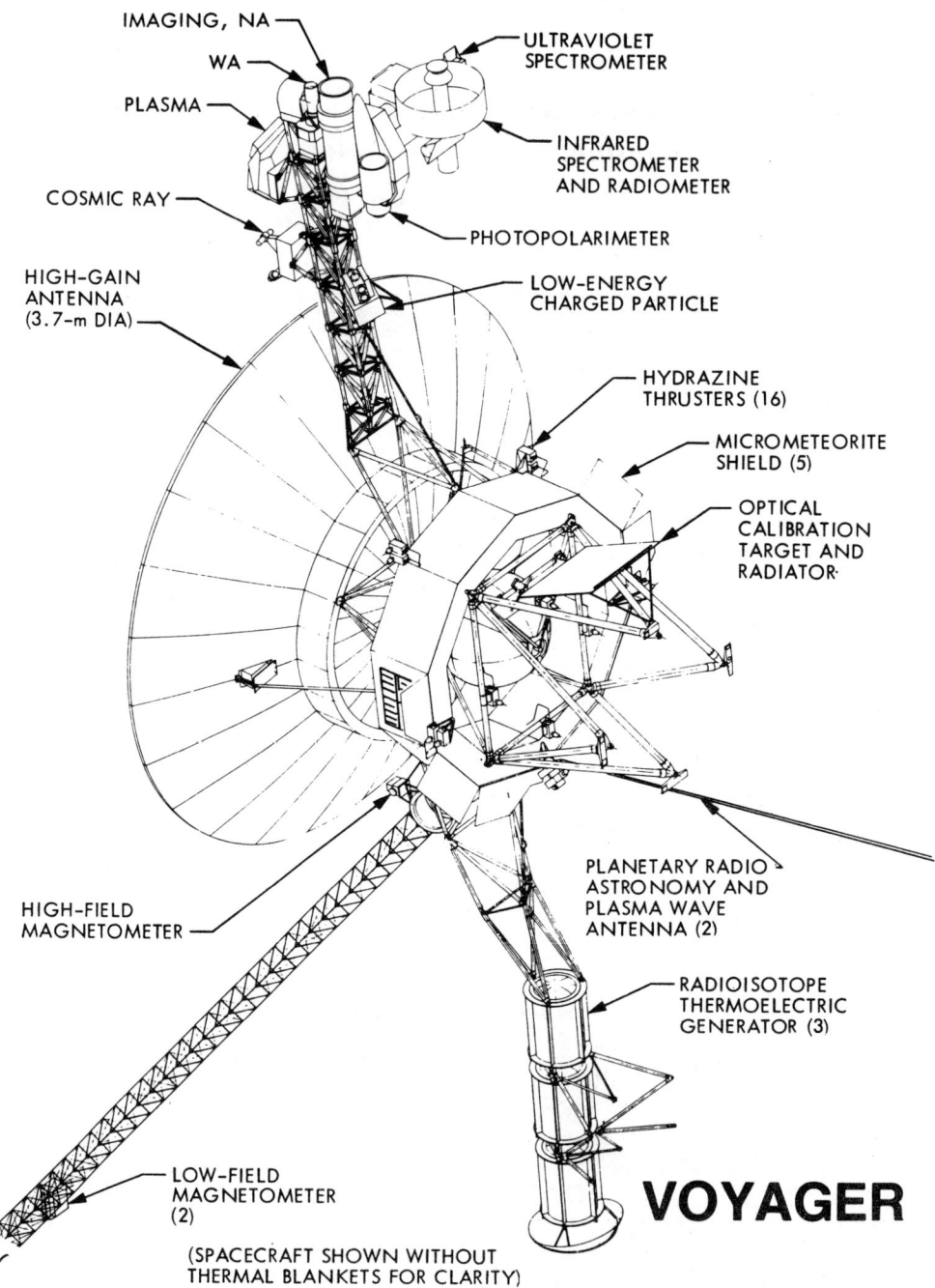

Voyager diagram.

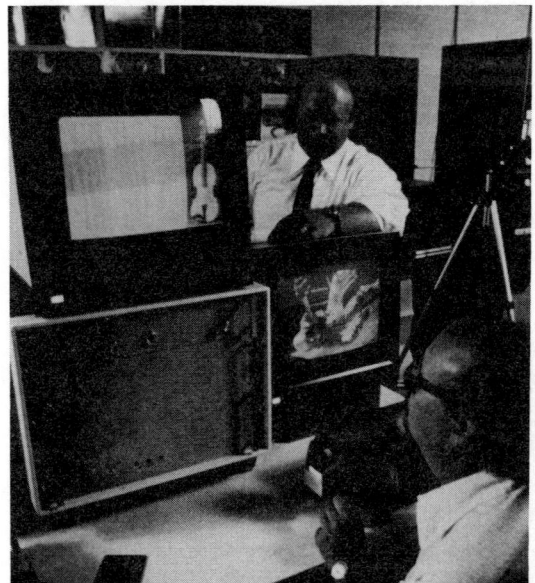

Left: A few weeks before launch, Glen Southworth of Colorado Video, Inc., transfers pictures of Earth to the recordings carried into space by the Voyagers.

Sounds and sights of Earth: Both Voyagers carry a gold-coated copper record as a message to possible extraterrestrial civilizations that might encounter the spacecraft in some distant time and space.

Each record contains 118 photographs of our planet, ourselves, and our civilization; 90 minutes of classical, jazz, and folk music; sounds of Earth including frogs, surf, birds, a rocket launch, and a kiss; and greetings in 54 languages—including that of whales. The record, together with a cartridge and needle, is fastened to the spacecraft in a gold-anodized aluminum case that also illustrates how the record is to be played.

Voyager 1 was launched September 5, 1977, from Cape Canaveral, Florida, on a Titan/Centaur rocket. Although Voyager 2 had been launched 16 days earlier, Voyager 1's shorter, faster trajectory enabled it to reach Jupiter four months ahead of Voyager 2.

The 230-ft antenna of JPL's Deep Space Network at Goldstone, California, in the Mojave Desert. Two similar stations are located near Madrid, Spain, and Canberra, Australia.

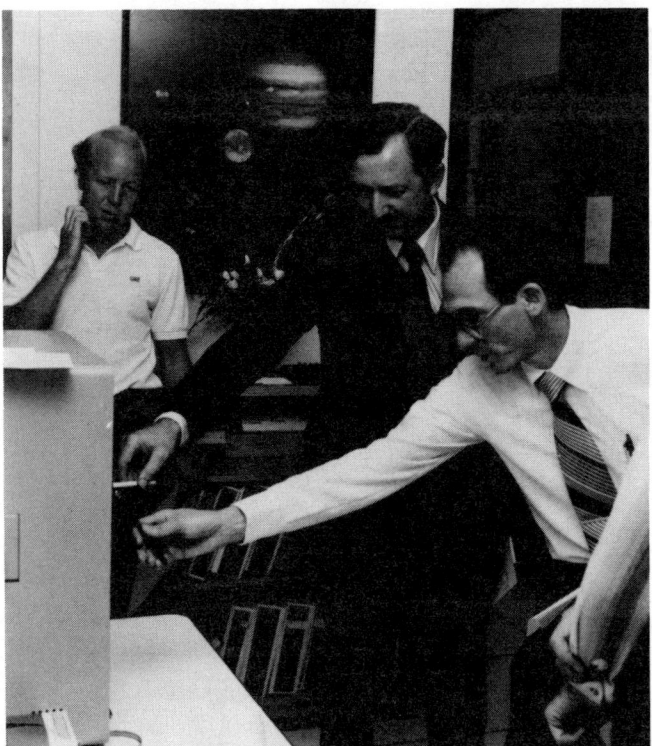

Left: During the first Saturn encounter, November 1980, Project Scientist Edward Stone and Deputy Project Scientist Ellis Miner confirm the correct operation of one of the science instruments onboard Voyager 1.

During the Voyager 1 Saturn encounter, JPL Director Bruce Murray organized a discussion of "Saturn and the Mind of Man." Left to right are Philip Morrison, MIT; Carl Sagan, Cornell; Walter Sullivan, *New York Times*; Murray; and author Ray Bradbury.

Spacecraft subsystems area in Building 264 where technicians and engineers monitor the signal strength of the data stream coming back from the instruments on Voyager 2 during its encounter with Saturn.

During planetary encounters several scientists would summarize their recent discoveries at 10:00 a.m. press conferences. At 2:00 p.m. one scientist would provide a more detailed report. Here Harold Masursky of the Imaging Team analyzes Voyager 2 data from Saturn.

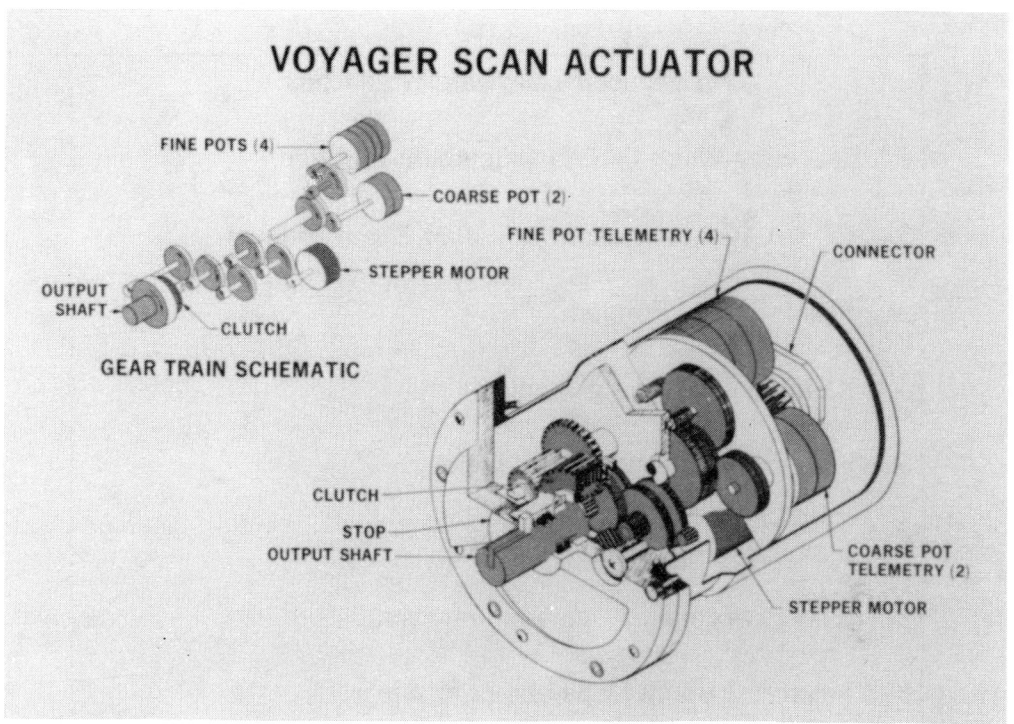

Scan platform actuator. The electric motor that points the scan platform containing Voyager's cameras and three other instruments.

Problem plaque. When Raymond Heacock (seated) left the Voyager project in January 1981, his colleagues gave him a plaque bearing 14 small engraved plates representing hardware problems that occurred in the Voyagers' development and flights. They challenged him to remember the specifics of each. He did, and he then recalled more that they had overlooked. Also pictured, left to right, are Ed McKinley, Charles Kohlhase, Albert Nakata, Tony Hagar, Richard Laeser, George Textor, and Esker Davis.

Heacock's Hardware Headaches

1. Traveling Wave Tube Development Problem.
2, CCS/AACS Processor Mulitlayer Board Problem.
3. Power Relays Problem.
4. Radiation Hardening Problem.
5. Electrostatic Discharge Problem.
6. 77-2 AACS Memory 1 Failure.
7. RTG Performance Degradation.
8. Voyager 2 Boom Deployment Problem.
9. Voyager 2 RCVR 1 Power Supply Failure.
10. Voyager 2 RCVR 2 Tracking Loop Capacitor Failure.
11. Voyager 1 Scan Platform Actuator Problem.
12. Voyager 1 Canopus Tracker Cone Angle Problem.
13. Solid-State S-Band Power Amplifier Degradation.
14. Voyager 1 Photopolarimeter Filter Wheel Drive Problem.

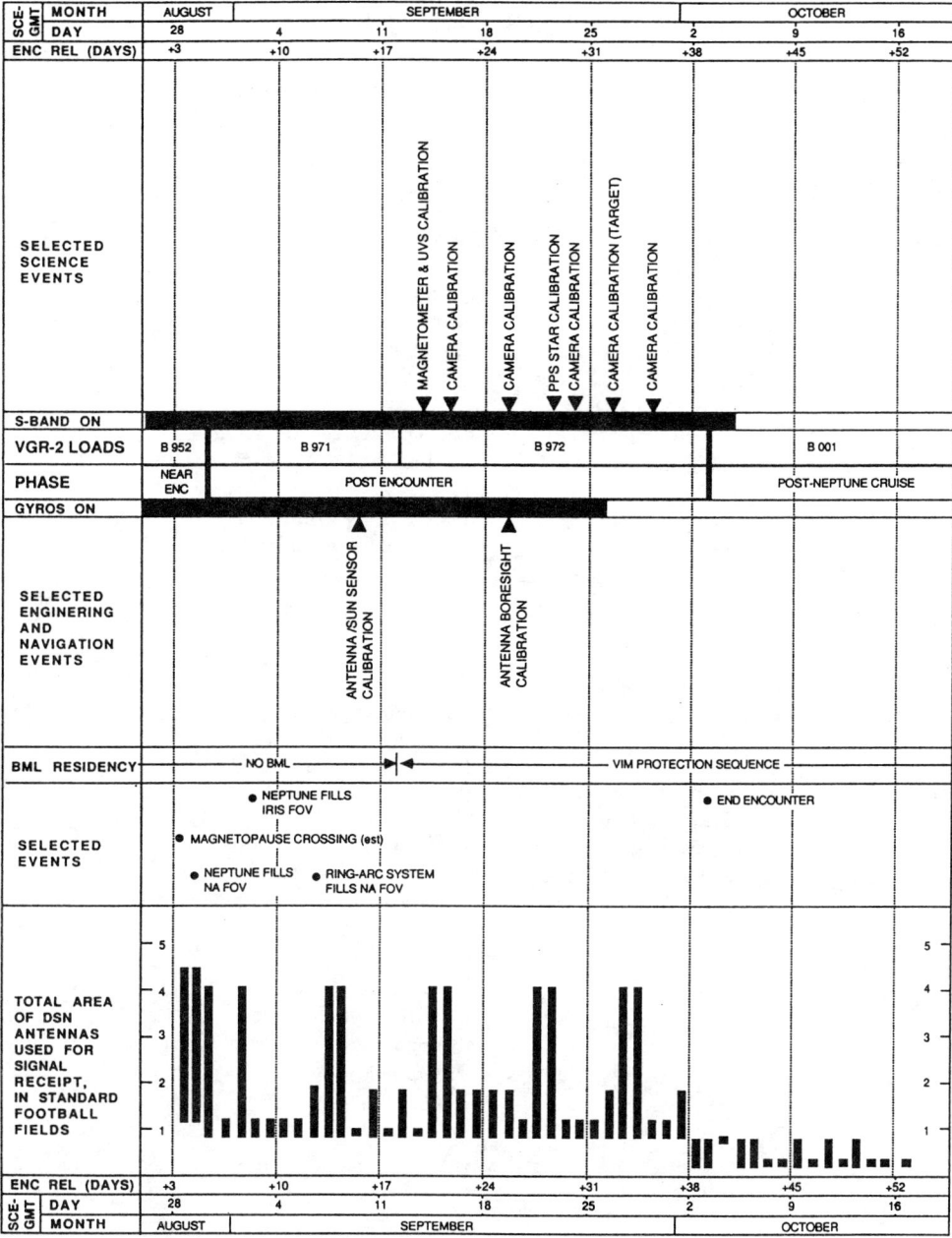

Voyager 2 Neptune encounter overview time line for the postencounter phase and early post-Neptune cruise.

URANUS ENCOUNTER TIME LINE

Jan. 24, 1986 (DOY 024)

PST START	PST STOP	EVENT	COMMENTS	RANGE TO TARGET
12:46p	12:47p	Radio-astronomy observation of Uranus. Data recorded for playback at 7:35a on DOY 027 and 11:44a on DOY 028.	Search for lightning or electrostatic discharges near closest approach.	82,000 km 51,000 mi N/A
12:47p	12:54p	Spacecraft maneuver.	Roll +65 degrees to Canopus and lock	82,000 km 51,000 mi
	12:48p	Plasma-wave observation of Uranus. Data recorded for playback at 7:48a on DOY 027 and 11:57a on DOY 028.	Search for lightning and plasma wave signals at closest approach.	325,000 km 202,000 mi
12:47p		CLOSEST APPROACH TO UMBRIEL.	This is the time at the spacecraft; arrival of signals on Earth will occur 2 hours, 45 minutes later.	86,000 km 53,000 mi
12:54p				
12:58p	1:09p	Ultraviolet observation of Uranus with one picture. Data recorded for playback at 8:02a on DOY 027 and 12:11p on DOY 028.	Study atmospheric constituents and structure by observing the star Gamma Pegasi as it exits occultation by the dark pole of Uranus. Complements illuminated pole (Gamma Pegasi) and equatorial (Sun) measurements. Includes photopolarimeter study of upper atmosphere.	

VIEWS OF THE GRAND TOUR

Charles Kohlhase (left), Jim Blinn (standing), and Patricia Cole at their special workstation where the Voyager planetary flyby animations were made as well as the special effects for several computer graphics sequences used in the PBS COSMOS series hosted by Carl Sagan.

Transportation Dispatcher Peter Wells assigns an urgent request to move equipment for the Neptune encounter.

Hawaii audiences view Uranus image relayed directly from JPL over a slow scan link arranged by the author.

During the Uranus encounter Torrence Johnson (left) and Lawrence Soderblom (center) of the Imaging Team and Photo Lab Supervisor Robert Post select images of Uranus's moons for a press conference.

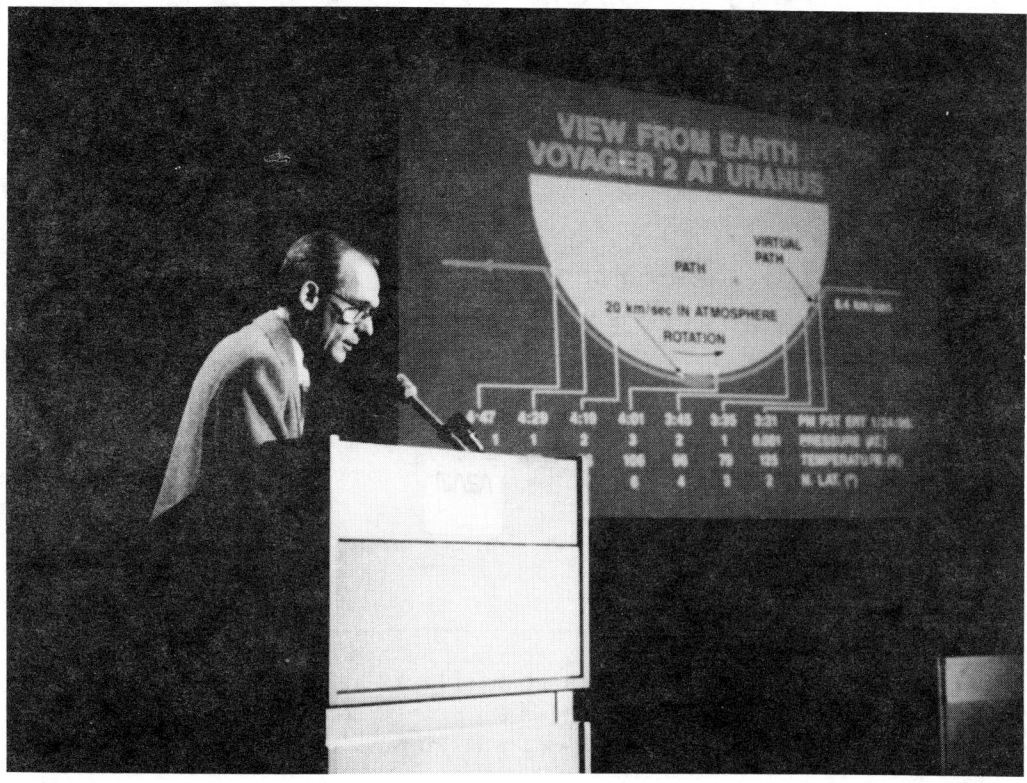

Edward Stone at a 10:00 a.m. press conference in JPL's von Kármán auditorium, announcing discoveries made by Voyager 2 at Uranus.

As Voyager 2 approaches Neptune, members of the Navigation Team gather for an update. Clockwise from left are Mark Ryne, Chris Potts, Robert Cesarone, Karl Francis, Duane Roth, and Bob Jacobson.

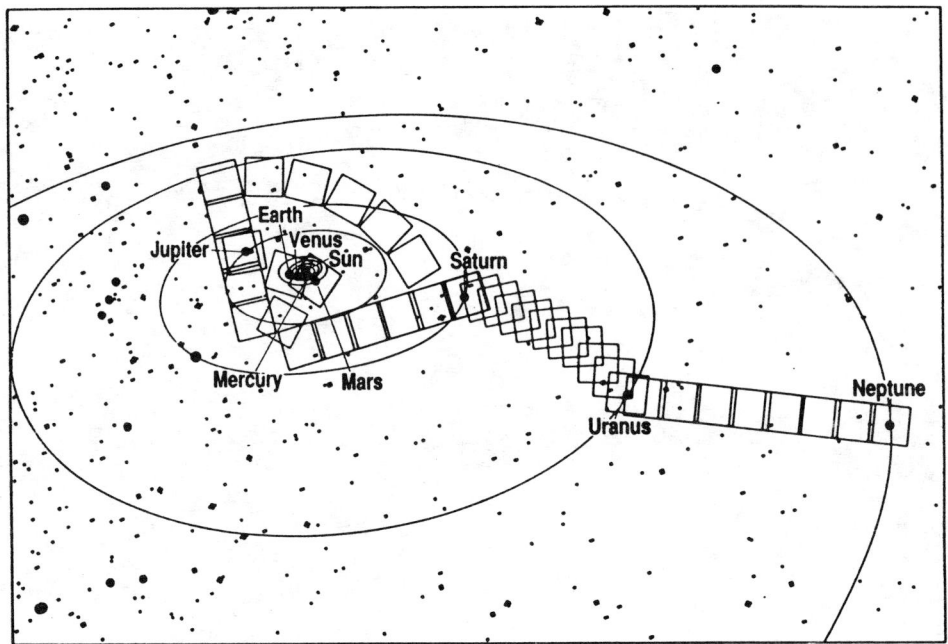

Diagram showing the 60 photos—39 wide angle and 21 narrow angle—taken by Voyager 1 on February 14, 1990, for the "family portrait" of our solar system. These were the last of the approximately 67,000 images taken by the two spacecraft since the 1977 launches.

HISTORY OF ROBOTIC EXPLORATION

The history of space exploration presents a paradox. Although the idea of space travel is at least 2000 years old, the concept of unmanned exploration is very recent, hardly mentioned until after World War II. Yet the initial exploration of the solar system was done by unmanned robots, and was completed in barely 30 years.

Furthermore, rockets, which made this possible, did not become the favored method of propulsion until around the 1920s, although they, too, had existed for centuries.

Why is unmanned exploration such a recent concept?

The Concept of Space Travel

Space travel is an ancient idea. A story of a trip to the moon was told in *Of the Wonderful Things Beyond Thule*, by Antonius Diogenes, who lived sometime between 356 B.C. and 160 A.D.

The first novel about interplanetary travel was Lukian of Samosata's 160 A.D. *Vera Historia*. His hero's ship is caught in a whirlwind and borne to the moon. Another of Lukian's stories, *Icaromenippus*, describes a flight to the moon using wings taken from an eagle and a vulture.

But further thoughts about space travel were inhibited for more than 1000 years by the prevailing concept of the solar system. The sun and the moon were presumed to be the only material bodies besides the Earth, and even the nature of the moon was debated. Was it really a body like the Earth or was it more ethereal; a perfect, unblemished disk or sphere? God was incapable of creating anything less.

It remained for Galileo in 1610 to demonstrate that the moon and planets were actual places, and therefore potential destinations that could be visited. His telescope revealed that the moon was not a perfect, pristine sphere; and Jupiter, Venus, Saturn, and Mars were more than mere pinpoints of light, they were real worlds. Consequently, to write seriously about journeying to them by vague mystical methods no longer sufficed. The method of travel had to be thought through more carefully.

Charles Sorel in 1622 was among the first to suggest space travel by mechanical means: "Engins", which he did not describe. But in 1648 a detailed discussion of flying chariots to take men to the moon was presented by Bishop John Wilkins. He was one of the first to consider space travel seriously, and to approach it with scientific thinking. His enthusiasm and his faith in the eventual conquest of space exerted an influence that reached into the future as far as the time of Jules Verne.

The first interplanetary novel in English literature appeared in 1638, Francis Godwin's *The Man in the Moone*. Twenty-one years later a retelling of this story became the first German work on space travel. In 1666 Margaret Cavendish wrote *The Blazing World*. Her heroine, perhaps the first female space traveler, visits the moon and other planets.

The next two centuries saw increasing attention, serious and satiric, given to space travel and the means of accomplishing it. Edgar Allen Poe, Alexander Dumas, and even millionaire John Jacob Astor were among the many who wrote novels about travel to the moon.

Jules Verne's 1865 classic, *From the Earth to the Moon*, historian Ron Miller noted, "has had more influence on the history of astronautics than any other work of fiction. There is powerful evidence that the entire history of rocketry would have been retarded had this book not been written." Miller refers to its impact upon Tsiolkovskii, Goddard, Oberth, and von Braun.

Interest in spaceflight reached a higher level of intensity when three influential groups formed: the Society for Spaceship Travel (VfR) in Germany 1927; followed shortly by the

American Rocket Society in 1930; and the British Interplanetary Society in 1933. The emergence of these organizations was a turning point in the history of spaceflight. Interest no longer relied only on isolated individuals. Now there were groups of enthusiasts, supporting, and stimulating each other.

But despite growing interest in spaceflight, doubts about it persisted right up until Sputnik. In 1945 A. V. Madge of the Royal Astronomical Society expressed doubts that interplanetary travel would ever become possible, because of the vast distances, the difficulty in aiming, and collisions with meteors. For Jerome Meyer in *Readers' Scope*, the deleterious physiological effects of weightlessness made interplanetary travel impossible. Astronomer Fred Hoyle wrote in 1950, "One has to be an optimist to believe that a trip to the moon and back will be safely accomplished within the next 100 years." And in 1956 Britain's Astronomer Royal, Sir Richard Wooley, declared space travel to be "utter bilge".[1]

Various factors contributed to this pessimism in the years before Sputnik, but the most basic problem was propulsion. Spaceflight would remain a dream until this problem was solved. How could a real craft actually rise off the ground and travel through space? The answer lay in 1000-year-old technology.

Rocket Propulsion

Though rockets had been known for a very long time, they were not considered the favored method of propulsion until the 1920s. Previously the most frequently suggested method had been some form of antigravity.

The use of gunpowder goes back at least to 850 A.D., when the Chinese were creating fireworks for religious festivals. In Europe, by 1242 Roger Bacon had developed a precise formula for gunpowder, and had also created potassium nitrate, saltpeter.

By this time, rockets were being used in warfare. The Chinese employed them in 1232 to repel the Mongols who were besieging the town of Kai-Fung-Fu. Arabs fired rockets in the 1249 siege of Damietta, and in Italy rockets were used in the 1379 assault on Chioggia.

Italian military engineer Joanes de Fontana proposed in his 1420 *Book of War Machines* many rocket designs for combat use, including a rocket-powered torpedo. He included the first rocket car for use as a battering ram against walls and gates, but he did not intend his rocket vehicles to carry passengers. It was Conrad Haas (1529–1569) who first suggested using rockets for manned flight. He envisioned a two story cylinder with windows and a door, perched atop a large stick.

In 1680 Czar Peter the Great established a factory in Moscow to make rockets for signaling and illumination by the Russian army. Military use of William Congreve's rockets is remembered in the American national anthem, referring to their use against Fort McHenry in the War of 1812.

The conceptual basis for rocketry had been provided by Isaac Newton in 1686. His Third Law of Motion, published in *Principia*, explained how rockets worked: "To every action there is always opposed an equal reaction." This was a landmark in the history of space travel, although its implications were not fully understood until the 20th century.

The first accurate description of rockets for space travel appeared in the 1852 novel *Gulliver Joi*, but the author Elbert Perce did not take the method seriously. Jules Verne did. He correctly understood how rockets operated in the vacuum of space, and was the first to suggest their use, in steering his manned projectile toward the moon. However, Verne's understanding exceeded that of many scientists, with the result that other methods of propulsion were favored over rockets until well into this century. For example, antigravity powder was suggested by famed inventor Thomas Alva Edison in his 1910 movie *Trip to Mars*.

Nevertheless Tsiolkovskii, Oberth, and others were developing the conceptual foundations of rocketry, and in 1916 Robert Goddard demonstrated by experiment that rockets would not only work in a vacuum, but also their performance would be improved. Ten years later Goddard achieved the first flight of a liquid fueled rocket: it rose 41 feet and landed 181 feet away.

The first airplane powered completely by rockets, the Heinkel 176, flew in 1939. The effectiveness of rocket propulsion was finally settled by the deadly demonstrations of the V-2 in World War II. In 1949 a WAC Corporal, carried aloft atop a V-2, soared to an altitude of 244 miles, becoming the first manmade object to enter empty space.

On October 4, 1957, Sputnik, the first artificial satellite, settled into orbit, circling the globe every 96 minutes. A month later, Laika the dog became the first living creature to orbit the Earth. On April 12, 1961, Yuri Gagarin became the first human to fly in space, making one orbit of the Earth in a flight lasting 108 minutes. Apollo 8 was the first manned craft to orbit the moon, in December 1968, followed by the Apollo 11 moon landing July 16, 1969.

Apollo was the culmination of 2000 years of thinking about space travel—*manned* space travel. But a different mode of space exploration was also developing: unmanned robots.

Robotic Spacecraft

Robotic flight did not receive much attention until after World War II. Earlier proposals had included Goddard's 1920 plan to hit the moon with a rocket charged with flash powder. Another lunar probe was envisioned in 1926 by Arelsky, who described an instrumented, unmanned rocket 30 meters long.

In 1945 Claussen suggested that sale of photographs of the moon's "dark side", taken by an unmanned rocket, would produce enough revenue to finance a later manned rocket to the moon. He proposed forming a stock company for the project. Two years later a five-stage unmanned moon rocket weighing 25 tons was described by Martin Summerfield of the California Institute of Technology.

Such projects, and unmanned spacecraft in general, would require new technologies, some of which were hardly even imagined in previous centuries. Radio, television, and computers were major requirements, along with other advances such as miniaturization, tape recorders, and lighter, more durable materials.

The first radio message across the Atlantic, transmitted by Marconi in 1901, began a new era. The first broadcasting stations began in 1920: KDKA Pittsburgh and WWJ Detroit. In 1926 radio communication with passengers on a moon voyage was suggested by A. Platonov. Twenty years later G. Edward Pendray proposed unmanned, instrument-carrying rockets that would land on the moon and telemeter information back to Earth about lunar conditions, as preparation for later manned missions and eventual moon colonization.

The first experimental television station, W2XBS New York City, went on the air in 1930, and the first commercially licensed stations began in 1941. While the ability to view distant events would ultimately play a vital role in space exploration, serious rocketeers during this period were struggling with more immediate problems. For example, in June 1943 at a special demonstration for Nazi Gestapo chief Heinrich Himmler, a prototype V-2 crashed seconds after launch into a nearby German airfield destroying three parked planes and blasting a crater in the ground 100 feet across, prompting Himmler to comment that it was a very effective close combat weapon.[2]

In later decades, as the basic problems of spaceflight were solved, communication with distant craft was accomplished with arrays of huge antennas. NASA's Deep Space Network (DSN), operated by JPL, was established in 1958 to track the first U.S. satellite, Explorer 1. The DSN has three stations: Goldstone, California; Canberra, Australia; and Madrid, Spain. By the time of the Neptune encounter each station had acquired two 112-ft-diameter antennas and one that is 230 ft in diameter.

To receive Voyager 2's extremely faint signals from Neptune, 2.8 billion miles away, the DSN antennas were augmented by cooperation with other installations: the National Science

Foundation's Very Large Array of 27 82-ft antennas near Socorro, New Mexico; Australia's Parkes 210-ft antenna 200 miles from the DSN's Canberra complex; and the 210-ft antenna near Tokyo operated by the Japanese Institute of Aeronautical Sciences.

These facilities were able to pick up Voyager's transmissions from the Neptune encounter, even though they reached Earth 20 billion times weaker than the power needed to operate an ordinary digital wristwatch.

But not even this highly sensitive technology could overcome another kind of problem: time. Commands from Earth, traveling at the speed of light, took 4.1 hours to reach Voyager at Neptune, and acknowledgment of their receipt was not received back on Earth until 8.2 hours after they were sent. Data telemetered from Voyager at Neptune was 4.1 hours old the instant it was received.

For distant spacecraft, not even radio messages traveling at the speed of light would suffice. Information from the craft, perhaps indicating a need for rapid response, a call for help, would take too long to reach Earth, and our reply would take the same amount of time to return to the distant craft. So sending and receiving signals from the edge of the solar system was not in itself the problem. The problem was—and is—the time required for radio messages to travel the immense distances between planets. Thus, while unmanned spacecraft did not need life support systems, they did need something else, perhaps even more complicated: some form of intelligence, some capacity to make decisions about guiding the spacecraft and gathering data or carrying out whatever else the objective of the mission might be.

The answer to this need was the computer, perhaps the least expected, most surprising technological advance in spaceflight. While its ancestry also extends back into the previous century, a useful working model did not materialize until the gigantic, room-sized ENIAC of World War II. Within a couple of decades, tubes had been replaced by transistors and integrated circuits, permitting a device weighing only a few pounds to process vastly more data than its huge predecessors.

These technological advances gave spacecraft two new capabilities. They could remain in contact with the Earth, sending and receiving messages over vast distances, many times farther than that between the Earth and the sun. And they could store enough information and capability to be, to a considerable extent, self-directing and self-correcting.

Telecommunications and onboard computers meant that a human pilot was not needed on the spacecraft, and so, freed from the complex requirements of providing life support, attention could be focused more directly on scientific and other concerns, and the distance and time of the journey could be greatly expanded. In fact it was not even necessary to bring these robots back to Earth. The stage was set for explorations of a magnitude undreamed of a few decades earlier.

Technology was not the only factor delaying the advent of robotic flight. The concept simply was not there to stimulate thought. The idea of using intelligent machines to operate a vessel and obtain information about distant places seldom occurred in earlier times.

It was more natural to imagine people doing things rather than robots. After all, it was humans who were exploring the oceans and the continents: the Columbuses, the Marco Polos, and their intrepid colleagues. It was just taken for granted that people would be aboard a spacecraft; so there was no need for substitute nonhuman forms of intelligence.

This presumption could continue as long as space travel was only a dream. So in 1939 Eando Binder could still write about a "grand tour" in which a manned spaceship visited every planet in the solar system in less than 10 days. But when the dreams started to come true, when people were on the verge of actually going into space, realities of the vastness of space and the very long time required to traverse it could no longer be ignored.

The initial manned ventures into space provided a yardstick with which to measure what lay ahead. Communication with manned vehicles, operating no farther away than the moon, would take less than two seconds, but would require 32 minutes to reach Jupiter, and four hours, at the speed of light, to reach Neptune. Objects such as spacecraft could actually attain only a miniscule fraction of the speed of light. So visions for manned visits to the planets, other than our neighbor Mars, gave way to missions by robots.

Robotic spaceflight required the development and convergence of concepts as well as of technologies. When this convergence finally occurred, the results were explosively rapid. Thirty two years after the first artificial satellite had orbited the Earth, robots had visited all the major planets and were heading out of the solar system into interplanetary space, but humans had not traveled beyond the moon.

Planetary Exploration

Five years after Sputnik, Mariner 2 became the first manmade object to reach another planet, passing Venus on December 14, 1962, at a distance of 21,594 miles. Mars was visited for the first time on July 14, 1965, when Mariner 4 flew within 6118 miles of the planet and sent back data revealing craters on the surface.

The first actual landings on the two planets were made by Soviet craft. On December 15, 1970, the Venera 7 capsule transmitted data from Venus's surface for 23 minutes. The following year the Mars 2 spacecraft entered orbit around Mars on November 27, surveyed the surface, and ejected a lander capsule. Though it crashed, it was the first manmade object to reach the Martian surface. Mars 3 entered orbit on December 2, 1971. Its lander descended successfully and began transmitting a television picture, but stopped 20 seconds later.

Mercury was first visited on March 29, 1974, when Mariner 10 flew by at a distance of 431 miles.

This completed the initial reconnaissance of the inner planets, Earth's immediate neighbors. But what about the outer planets, the gas giants beyond Mars and the asteroid belt? These distant worlds posed greater challenges, described by David Morrison of the Voyager Imaging Team:

> Several barriers stood in the way of a successful exploration of the outer solar system. The most difficult to overcome was the vast distance which necessitated larger rockets and much more reliable, automatic spacecraft that could spend years unattended in space and still provide flawless operation as they reached a remote world. Other barriers included the difficulties of communicating over such vast distances and the lack of a satisfactory solar power source so far from the sun. It was clear that a new generation of spacecraft employing fundamental advances in electronics, communications, and reliability would be required to reach to Jupiter, Saturn, and beyond.[3]

Distance could be overcome by using gravity assist to take advantage of the rare alignment of the outer planets in the late 1970s and 1980s. Flying by a planet at just the right distance bends the spacecraft's trajectory and flings it on, at higher speed, to the next planet.

Jupiter, the nearest of the outer planets, is 10 times farther away from Earth than Mars or Venus. But there were other challenges besides distance. In addition, to reach it required crossing the asteroid belt, consisting of tens of thousands of minor planets a mile or more in diameter and even larger amounts of debris, from boulders down to microscopic dust. Collision with a pebble-sized object could destroy a spacecraft. A second threat was the intense radiation in Jupiter's magnetosphere that could damage the spacecraft's many electronic components.

To explore beyond Mars and assess the dangers of the asteroid belt and Jupiter's radiation, the Pioneer Jupiter mission was established in 1969. Designed for reliability and simplicity, the craft were patterned after successful Earth orbiting probes. The craft weighed 568 lb and transmitted data to Earth with only 8 watts of power. Pioneer 10 was launched March 2, 1972, passed Jupiter on December 3, 1973, and became the first manmade object to explore beyond Jupiter and eventually to leave the solar system.

Pioneer 11 was launched April 5, 1973. Passing within 27,000 miles of Jupiter on December 2, 1974, its trajectory was bent so sharply that it dropped back toward the sun and flew to the other side of the solar system, reaching Saturn on September 1, 1979. Flying just 21,000 miles above Saturn's clouds at 71,000 miles per hour, it then continued on out of the solar system in the opposite direction from Pioneer 10.

The Pioneers had demonstrated that the asteroid belt and Jupiter's radiation could be surmounted, thus paving the way for the Voyagers.

Voyager had been conceived in the late 1960s as the Outer Planets Grand Tour, and by 1970 NASA developed plans envisioning four spacecraft. The first two would be sent to Jupiter, Saturn, and Pluto. These would be launched in 1976 and 1977. They would be followed two years later by two more spacecraft, aimed for Jupiter, Uranus, and Neptune. The cost of the project would be $750 million, but NASA's budget was slashed and the Grand Tour was canceled. A more modest mission, costing only a third of the original budget, was approved and began officially on July 1, 1972.

This new project used only two spacecraft and would just go to Jupiter and Saturn; the more distant planets were not included. Based on the proven Mariner design, the mission was initially named Mariner Jupiter-Saturn. The name was changed to Voyager in 1977, and in August and September of that year the two 1800-lb spacecraft were launched.

They were sent on differing, complementary trajectories, because no one path could explore the varied aspects of Jupiter and Saturn and their many moons. Voyager 1 was actually launched a few days after its twin, Voyager 2, but arrived at Jupiter several months earlier on March 5, 1979, because it flew a shorter, faster trajectory. Voyager 2's longer route was chosen to permit examination of other aspects of the Jupiter and Saturn systems and subsequently to enable continuation to Uranus in 1986 and Neptune in 1989.

After Voyager 2 completed its Neptune encounter, Voyager 1 looked back on the solar system and on February 14, 1990, took a series of photos for a mosaic of the sun's family. Both craft are continuing on into interstellar space and are expected to send data back to Earth until well into the 21st century.

Voyager 1 is heading for the constellation Camelopardalis, and Voyager 2 is flying toward Andromeda. Although they are traveling about 35,000 miles per hour, it will take them a million years to reach the dozen stars nearest our sun—and their voyages will have barely begun.

REFERENCES

1. Chicago Tribune Press Service, London, January 5, 1956.

2. Erik Bergaust, *Wernher von Braun*, Washington, DC: National Space Institute, 1976, p. 70.

3. David Morrison, *Voyages to Saturn*, Washington, DC: National Aeronautics and Space Administration, Scientific and Technical Information Branch, NASA SP-451, p. 14.

* * * *

Background on the period before Sputnik became the first artificial Earth-orbiting satellite on October 4, 1957, was obtained primarily from Ron Miller, *The Dream Machines*, Malabar, FL: Krieger Publishing Co., 1993.

Since Sputnik, general sources are Kenneth Gatland, *The Illustrated Encyclopedia of Space Technology*, New York, NY: Harmony Books, a division of Crown Publishers, 1981.
David Baker, *The History of Manned Spaceflight*, New York, NY: Crown Publishers, 1982.
Ian Redpath, *The Illustrated Encyclopedia of Astronomy and Space, Revised Edition*, New York, NY: Thomas Y. Crowell, Publishers, 1979.

For robotic exploration a good overview is Bruce Murray's *Journey into Space*, New York, NY: W. W. Norton and Company, 1989.

Details of the planning that led up to the Voyager project are presented in Craig B. Waff's *Jovian Odyssey: A History of NASA's Project Galileo*, working draft, Nov. 8, 1988, Chap. 3.

Basic references for Pioneer and Voyager are David Morrison and Jane Samz, *Voyage to Jupiter*, NASA SP-439, 1989, and Morrison, *Voyages to Saturn*, NASA SP-451, 1982.

Later stages of the Voyager mission have been described in three types of JPL publications: *The Voyager Uranus Travel Guide*, JPL D-2580, Aug. 15, 1985; and *Voyager Neptune Travel Guide*, JPL 89-24, June 1, 1989, both edited by Charles Kohlhase.
Press releases, especially the Press Kits distributed during encounters.
The 99 *Voyager Bulletins* prepared by Anita Sohus.

Many other JPL documents of a more technical nature, primarily for in-house use, are exemplified by the *Science and Mission Systems Handbook*, JPL D-498, Feb. 1, 1983; and by the *Voyager 2 Saturn Encounter Timeline (Aug. 21 to Aug. 28, 1981)*.

William Sims Bainbridge, *The Spaceflight Revolution: A Sociological Study*, New York, NY: Wiley, 1976, includes a detailed analysis of the German rocket program.

INTERVIEWS

Overview

Edward Stone has been Voyager's Project Scientist since its official beginning in July 1972. An active researcher in cosmic ray physics, he is deeply committed to educating the public about science, and uses data from planetary encounters to explain what science is all about.

He brought extensive scientific expertise to management, and in the process transformed the administration of project science. Voyager at times involved contradictory interests: engineers' concerns for spacecraft safety versus scientists' desire for more data. The 11 scientific experiments often had conflicting requirements, especially during the precious minutes when the spacecraft were hurtling past a never-before visited planet at tens of thousands of miles per hour—which instruments should gather data, and for how long?

His effectiveness in handling such issues led one project manager to observe that it was amazing to see him interact with the scientists. "It almost appeared that he knew as much or more about their experiments than they did."

He also mediated effectively among various demands with NASA Headquarters, or within the Jet Propulsion Laboratory, where there were other projects, as well as other concerns not directly related to science.

High regard for Stone's competence and fairness resulted in his appointment in 1991 to be director of the entire Jet Propulsion Laboratory.

EDWARD C. STONE
Project Scientist
Born January 23, 1936   Knoxville, Iowa

After attending Burlington Junior College, Edward Stone studied physics at the University of Chicago, earned a master of science in 1959 and a doctorate in 1964. He then joined the Caltech as a research fellow.

He rose through the academic ranks to professor of physics in 1976, chairman of Caltech's Division of Physics, Mathematics and Astronomy in 1983, and vice president for Astronomical Facilities in 1988. That same year he also was appointed chairman of the California Association for Research in Astronomy, which operates the Keck Observatory in Hawaii.

Since 1961, Stone has been a principal investigator on nine NASA spacecraft missions and a co-investigator on five other NASA missions for which he developed high-resolution instruments for measuring the composition of energetic cosmic-ray nuclei.

He has served as project scientist for the Voyager mission since 1972, participating in both mission operations and hardware development, and has coordinated the efforts of 11 teams of scientists in their studies of the outer planets. In 1991 he was appointed director of JPL.

In addition to five NASA medals, he has received the National Medal of Science, the Leroy Randle Grumman Medal, the American Philosophical Society Magellanic Award, the Association for Unmanned Vehicle Systems National Award for Operations, the International von Kármán Wings Award, and the Dryden Medal and Space Science Award of AIAA.

Among his many memberships are the National Academy of Sciences, the American Philosophical Society, and the International Academy of Astronautics. He is a Fellow of the American Physical Society, the American Geophysical Union, and AIAA and a member of the American Astronautical Society.

His work is his hobby. His main recreation is reading a daily newspaper.

As members of the Voyager flight team, we were privileged to participate in a journey that revealed dozens of new worlds of unexpected diversity and gave us a new view of the solar system. This journey greatly expanded my appreciation of the importance of scientific cooperation and teamwork in learning the most we could from the observations of so many distinctive phenomena. Even as our knowledge and understanding rapidly evolved with each planetary encounter, we became humbler about our ability to anticipate all that we would find and more strongly committed to the need for further exploration. There is also a broader interest in exploring and understanding the unknown, and this journey afforded us a unique opportunity to describe what we were learning in a way that would engage a much wider audience in the scientific enterprise and its role in shaping the future.

—Edward C. Stone

EDWARD STONE

I was raised in Burlington, Iowa, on the Mississippi River. I was interested in science as a youngster. That was a result of reading magazines like *Popular Science* or the *Book of Knowledge*, which showed how things worked. Eventually, I started building crystal radio receivers and then super heterodyne radio receivers as the experimental aspect of doing science. Then when I got to junior high school I could actually start taking science courses, so my interest in science developed steadily over the years.

Then you turned up at the University of Chicago.

Right. After having gone two years to Burlington Junior College, my physics instructor, Wilfred White, recommended that I consider the University of Chicago, so I applied and was accepted. I entered the master's program; that was 1956.

In October 1957 the space age began with the launch of Sputnik and the following summer I started working in Professor John Simpson's group. He had been flying cosmic ray detectors on balloons and was very interested in moving into space to measure cosmic radiation. He had as a colleague Professor Peter Meyer, who was also doing balloon work with a student, Rochus Vogt.

In Vogt's thesis experiment, a balloon experiment, they discovered a flux of electrons in the cosmic radiation. Nucleons, such as protons and alpha particles, had been detected previously, but no electrons. So they were codiscoverers, along with Jim Earl, of the electron component of the cosmic rays arriving at Earth. That discovery led to Vogt's coming to Caltech as an assistant professor in 1962 to set up a group in space physics, and subsequently he invited me to join him when I finished my doctorate thesis.

My thesis experiment to measure cosmic rays and solar energetic particles was launched on an Air Force satellite, Discoverer 36, in December 1961. I finished the analysis of those measurements and others from Discoverer 31 and arrived at Caltech in January 1964. We launched our first Caltech instrument five years later on OGO-6.

In 1972 I was invited to become the Voyager project scientist by Bud Schurmeier, who was then project manager. This was a part-time position, so I continued my research activities on campus and was at JPL about 30 percent of the time, at least initially.

How does Voyager differ from other projects, before or after?

The major difference is that Voyager revealed so many distinctive worlds, having such unexpected diversity. No other mission has explored so many different worlds. The number of discoveries is really quite unusual and is unlikely to happen again because, after all, you visit a place for the first time only once.

The Pioneer 10 and 11 spacecraft had been to Jupiter and Saturn first, but their capability was limited by design and so left a great deal for Voyager to discover when it followed along some years later. Even at Jupiter and Saturn, Voyager still had a lot to discover, especially about the moons and the weather systems. I think the principal differences from previous projects are in terms of the scientific impact and also in terms of public interest.

Speaking of public interest, one of the things about the Voyager project that strikes an outsider is how well the project and the project leader have been able to enlist laymen's interest in it. Let's look at you, starting out life as a physicist; how did you personally, and JPL, develop such effective communication skills?

I think it's partly a matter of practice. It was also a matter of interest in understanding nature. That is what science is all about, understanding things. I just happened to be interested in understanding a rather broad range of subjects and that was one of the appeals that the mission had for me when I agreed to become its chief scientist—the project scientist—in 1972. I saw it as an opportunity to really become involved in understanding a much broader range of endeavors than my own particular area of research.

One thing that you learn when you teach freshman physics is that you explain things best when you're teaching yourself. In that sense, by teaching myself what we were learning and what we were supposed to be learning, it was possible for me to explain it to others. I think any good teacher recognizes that as one of the essential elements of teaching: teaching yourself, trying to understand in a simpler, more fundamental way what it is you're trying to communicate to your students.

Talking about Voyager to the general public was much the same. The things that Voyager discovered, although they were diverse and distinctly different from things here on Earth, were familiar enough in form that you could communicate about them—geysers, for example. Although on Triton geysers are driven by the evaporation of frozen nitrogen and here on Earth it's evaporation of boiling water, they're still geysers. And so there is a common visual framework that you can use to try to convey this exotic new form of geysering in terms that a nonexpert can understand.

Scientists tend to do that—reduce the complexity of nature to something that you've already understood and then understand how it's different. That process helps you simplify the amount of information you have to carry around in your head. I think that is the basic process that goes on, and I find it interesting to understand things in that way, and it's important to communicate those understandings.

Early on, I became quite aware that planetary exploration and astronomy are accessible to the public. There are discoveries that can be described in nontechnical terms. In some areas of science, the discoveries are more technical and more difficult to communicate, but that is not the case for planetary exploration, solar system exploration, or astronomy, although some aspects of these are very technical.

I went to the Pioneer-10 flyby of Jupiter in December of 1973 as a guest, because I was the Voyager project scientist and we were going to be there next. I was able to observe the whole discovery process and the communication of what was being discovered during this first flyby of Jupiter with its intense radiation environment and immense magnetic field.

There was an audience of maybe 200 reporters wanting to know what was being discovered. And I realized that there were very few opportunities where scientists are being requested—not offering, but being *asked*—to please tell us what you're learning, please tell us what it means. I realized this was such a valuable opportunity for teaching, for communicating what science was all about, that we really had to take it seriously. We had to organize ourselves to do as good a job as we could, given the amount of effort and thinking time we had, to communicate clearly what we were finding.

After that experience with Pioneer, I realized that the Voyager encounters were going to be a marvelous opportunity for communicating, not only what we were learning about the planets, but communicating what science was all about in a more general fashion—the whole process of science, from the initial observation to the eventual understanding of what was happening. Voyager would be a very good tool for communicating why scientists do what they do, and why it's interesting to be a scientist.

I wanted to communicate to the public that it is indeed fascinating to discover things. That is the reason scientists work the way they do: because they enjoy understanding something about nature.

Were there some specific organizational steps that you took to further that?

We had to be concerned about accurately reporting "instant science": that is, reporting, within a day or two of seeing something, what you think you understand, and what you've learned. You have to be careful that you are doing science and not just guessing, that this is true scientific process. To provide some of the usual kinds of scientific checks and balances, I put into place a series of meetings that would lead up to the press conference itself. Usually we'd have a scheduled press conference at 10:00 in the morning each day for a period of about 10 days.

First thing each morning, at 8:00, we would have a brief report on the plan for that day and any recent observations. This was a project-wide interaction with all the teams reporting. That afternoon we would have a science discussion, a form of peer review, which is the normal way scientific review takes place. (You write a paper, you submit it, it is reviewed by your peers. After that, if the paper is of value, it is published.)

We compressed all of this into a period of 24 to 48 hours. Whoever felt they had an important new observation would report it to the entire scientific group, which could involve 100 to 150 people during an encounter. There were questions and answers, debates about whether or not the interpretation was appropriate, or whether the observation was unambiguous.

We would decide whether or not it was ready; that is, if we understood it well enough to announce it as a discovery, or whether we felt we would be in a better position with another day's worth of data. If that happened, we would examine it again the next day, and as soon as we felt confident as a group in what had been discovered, we would schedule it as part of the press conference.

Once we decided we knew enough to announce a discovery, the items for the press conference the next day would be given to our overnight graphics support. Images are one thing—they're the most spectacular things, they're fairly easy to talk about, and seemingly the easiest things to understand, but there were 10 other scientific investigations that did not normally produce images as such. We needed to convey their less-accessible results in a graphic or cartoon form to make explanations of these other areas of scientific discovery more accessible.

An overnight graphics service would make cartoons of the results during the night, so they would be available for the press conference the next day.

There are a couple of other aspects to this. We were discovering all sorts of things, and of course there was a great deal more discovery than any of us had anticipated. It turns out nature is much more diverse than we could have imagined. That is fortunate because, both from a scientific point of view and from a public point of view, you learn more when you discover something totally different from what you'd seen before.

Progress is most rapid, however, if what you are seeing is not too different from what you've seen before, so you have at least some relevant insight or experience. If it is too bizarre, you may take a very long time to understand it. We've seen many diverse objects in the outer solar system, but they are not so diverse that you can't begin to understand or describe them. So it was just about the optimum for rapid understanding.

The other interesting thing about the whole process of preparing for these briefings is that it's really a highly compressed form of the process that science goes through in the laboratory all the time. Normally you feel that if you understand or discover one or two things a year you have a reasonably successful research program. Very typically the pace is that you'll make an observation, you'll gain some understanding, and that will lead to another observation, and perhaps in another six months you'll have another new observation. Slowly but surely, step by step over months or years, you will expand your base of understanding of nature.

But in the case of these encounters it's all compressed into, literally, a few days. You make an observation one day, and you know the next day that you're going to have a better observation. Rather than having months in which to digest and replan, you have only a few hours before you have a new set of data that will be better than the last. So the whole process is remarkably compressed. In retrospect this compression seems obvious, but I had not really anticipated it.

Compression makes it a very good medium for communicating because one can, within a period of days, go from essentially no knowledge to quite a bit of knowledge, and in some cases even to understanding. Normally the scientific process is much more drawn out and consequently there is less immediate understanding than is offered by these planetary encounters.

You mentioned retrospect. What other things came up that you had not anticipated?

I certainly had not anticipated there'd be so many new things. We all knew that we were on a mission that was going to make discoveries, because any time you look at nature the first time you almost surely are going to discover something you hadn't anticipated. But I don't think any of us expected such a degree and such a diverse set of discoveries.

The thing that developed along the way, which I had not really anticipated, was that over the years the Voyager science teams worked very well as a complete team rather than as 11 distinct investigations. In the past, individual scientific investigations often remained focused on their own data and did not necessarily interact much with the other scientific investigations, at least in real time. But, in the process of preparing for press conferences and in planning the sequence of observations for each encounter, the science team became much more integrated across the 11 instrument-oriented investigations.

Several things contributed to this. Sequence planning can be looked at as a matrix. You have all the columns of the matrix; each column is a particular investigation that was selected by NASA. There were 11 of them. Imaging was an investigation, and it had actually three subdivisions in it: the planet, the rings, and the moons. Also there were the infrared investigation, the ultraviolet investigation, the photopolarimeter, the radio science, magnetometer, plasma, low energy charged-particle, cosmic ray, plasma wave, and planetary radio astronomy investigations. Those were the 11 columns of the matrix.

Early on, in trying to decide which observations we should make, it occurred to me that we really needed—besides the individual investigations representing their own specific interest—rows in the matrix that were complete areas of study: for instance, the Atmosphere Working Group, where several of the experiments or investigations would contribute to the study of the atmosphere of the planet. Some would contribute to the study of rings; other scientific studies might be the satellites, or the magnetosphere.

So I organized working groups with members selected from the relevant scientific teams, which would then organize themselves as discipline-oriented teams. These groups provided interaction and cohesion among the individual teams that greatly contributed to the overall rapidity of discovery.

There was another thing that I had not really anticipated, although in retrospect it is quite obvious. The way the science decisions are made on a project like Voyager is that each of these scientific teams has a leader called the principal investigator or team leader, appointed by NASA. They form what is called a science steering group, which the project scientist chairs. The science steering group made the scientific decisions for the project.

This group attempts to achieve consensus, but in some cases there clearly was no way that we could do everything that was desired. Therefore I soon realized that a project scientist ultimately has the responsibility, if there is not a consensus, if everyone cannot agree on what to do, then the project scientist has to make the decision.

What that really means is the project scientist decides who gets to make discoveries. And when you put it that way, you realize the impact it has on the individuals involved, because that is what it's all about: making discoveries. Somehow you have to decide. You won't know what's out there until you look, but before encounter you must make some judgment as to which thing to try to observe when you can't observe everything and, therefore, which things to try to discover.

And if you decide to observe this rather than that, it may be Team A that makes a discovery rather than Team B. Different individuals will get to make a discovery, depending on which way the decision is made. That is another very interesting part of the process that I hadn't really thought about. These decisions were very personal ones for those involved, because the decisions affected what they were going to get to do, what they would observe, and what they would discover.

And their subsequent careers?

Well, none of these discoveries were of such a magnitude that someone could say, "There goes my Nobel Prize." Still it is true that it would affect what they could study and what we would learn about the planet. There were a number of such decisions. Most decisions of this sort were worked out among the team members themselves, as part of the disciplinary working groups.

But there were some key observations where it was just not possible to resolve the issue that way because you could not do two things at the same time, and they were both important. That is where I had to educate myself and make sure that the other team leaders were equally well-educated. I had to be sure that there was really a tradeoff that had to be made, that both observations were desirable, and that it wasn't obvious that one or the other should have priority. In such instances, both had valid and strong cases, but nevertheless we had to make a choice.

In retrospect I think that most of the decisions that had to be made that way were probably the correct ones. So the process worked reasonably well, and I think that all the investigators involved would agree, even though each one of them occasionally did not get an observation they really wanted. If they look at the total of what they observed and the total of what Voyager accomplished, I suspect that by and large they would agree that it was a reasonable process. One could argue about specific things here and there, but if you look at the sum total, we succeeded—and I mean "we" the whole group—in arriving at a reasonably optimum scenario for using Voyager to its limit.

Another aspect that I hadn't realized when I started is that the project scientist serves another very important role as an impedance match between the scientific desires and the engineering capability. There has to be a delicate balance, what I call dynamic tension. The scientists are always pressing to do as much as can be done, which is correct, but the engineers always want to be sure that when we try to do something we can really deliver, that it can be done, and that it is also correct. You want to do as much as you can, but you don't want to attempt more than you actually can do because then you won't succeed and you will achieve less than if you had tried for less.

You have to try to do just the right amount. You shouldn't try to do too little. That would be very conservative and safe, but would deny scientific discoveries that should have been made. At the same time, you must not try to do more than your team—not only the science team but also the engineering team—can actually accomplish, because this has to be done on a certain time line. The planet was not going to wait. We couldn't say, "Whoops, let's take another six months and get ready." We knew years in advance that we would arrive at a known, predetermined time. We had to be ready, so we had to set a cutoff point for planning and outline the right amount of work that we knew we could do without ever having done it before.

There was the delicate balance that had to be achieved, and the project scientist was really in the middle, trying to understand how strong the various science drivers were. Of the many science drivers, some are stronger than others. At the same time, we had to consider which of the many engineering constraints or concerns are the strongest. Where might there be some flexibility to provide an impedance match to reach this optimum point?

As I said at one of the press conferences, it's very much like an athlete in a marathon run. What's important is to have exactly the right pace to get there first. You can fail two ways. You can fail because you burned yourself out before you get there, or you can fail because you didn't run fast enough to stay in a position to win. The winner is the one who does it exactly right. I think that the process optimized very well the amount of science we received from Voyager.

It was a very complex process, involving a lot of interaction between the scientists first of all and then between the science and the engineering capabilities. Our capability kept growing, encounter after encounter, both from an engineering point of view and from a scientific point of view, but we always managed to let it grow just the right amount. There are very few things where in retrospect we could have said, "We really could have done that and we were foolish not to have taken it on."

I don't think there were any major things that we could point to, with the exception of running the scan platform until it jammed at Saturn, which we didn't know would happen. We overused it, but that wasn't because we had knowingly pressed any limit; we had no idea that there was a lubrication problem in the actuator that moved the platform. We learned the hard way that for Uranus and Neptune we should run the scan platform more slowly, which we did. With the exception of a few such things, we didn't overstress the spacecraft.

That was a very interesting process. I had no idea how interesting and important it would be when I started, because I'd never been involved with such a complex program that had so many opportunities and such a capable spacecraft to take advantage of those opportunities. Another thing that I hadn't really appreciated was the importance of system design in the spacecraft itself. This is the reason we could extend the Voyager mission so far beyond its original design, which was a four-year mission to Saturn, to a 12-year mission, and now for even longer.

Let's talk about getting to Neptune, which is three times further from the sun than Saturn and three times longer in-flight time.

The reason that it was possible to do all that was because the engineers were clever enough in designing the various subsystems in the spacecraft and in connecting them together to provide extra capability. That is what system design is all about. How do you make these various subsystems on the spacecraft? How do you optimize them so that you have a robust and capable spacecraft design? They succeeded, well beyond what could have been expected.

We were doing things for which nobody had ever said, "Well, if we put this in, we can do such and so five years from now." But there was enough intelligence about the basic system design that even though a specific technique or application was not thought of in 1973 through 1975 when it was being designed, the design was flexible enough so it was possible for subsequent generations of engineers to do unplanned-for things. So a lot of what we accomplished at Uranus and Neptune goes right back to the system design work.

That was the first time I had been involved in such a process, and I now appreciate how important the systems engineering was in deciding what capability to build into Voyager. We would have had a much different mission if that capability had been more conservative. It was a very important part of making Voyager the success it was.

How did this idea get started, of matching the impedance?

I realized it early in the process because it was clear that those were the issues we were facing, and we had to face them right at the time we were doing the system design. We had to make decisions about how much capability was enough. Just as in the case of the encounters, you had to pick the right amount of observing to do, planning for not more than you could do and not less than you should do. The same was true in the hardware area.

You want to build as capable a spacecraft as you can afford in the time you have, but you also have to understand, as you press the limits of this design, what you're gaining. That is, you want to be careful that you don't spend your time optimizing some aspect of the spacecraft capability that is less essential scientifically than another aspect. There is a limited amount of time and attention you can spend on pressing the boundaries of capability before you freeze the design and build it.

Consequently, there was a detailed interaction as to what the scientific gain would be for a given capability, and that had to interact with the difficulty of gaining those capabilities. Again, there was this impedance matching between scientific desires and engineering capabilities, right at the very beginning of the system design itself.

It soon became clear to me that it was an important role for the project scientist, and to exercise this role you had to understand the scientific objectives and capabilities of the instruments. At the same time you also had to have some understanding of the engineering aspects, so that you could appreciate how hard the limits were, because in the engineering area as in the science area, there are different levels of difficulty.

You have to understand from the engineering side when something is a lot more difficult and when it's just a little more difficult, because in one case you might say, "We really can't. It doesn't make sense to push harder there, but it might make sense to push harder here where it's less difficult, but there is still a reasonable science return." It's that understanding that I enjoyed developing, and I realized it was contributing to the overall mission design.

What would be an example of a particular instance in which the team was considering the tradeoff between engineering capability and scientific possibility?

One recent one was at Neptune. The tradeoff was in how close to fly by both Neptune and Triton. There was only one way to fly by both, which was to fly over Neptune's north pole, deflecting the spacecraft down to an encounter with Triton. The closer you fly to Neptune the sharper the spacecraft trajectory is bent and the closer it then comes to Triton. So there is a connection between how close you must fly to Neptune to how close you are able to fly to Triton.

For some scientific purposes, for instance, imaging of Triton, closer is better. For other investigations a somewhat more distant encounter is better. For example, we can measure the density of an atmosphere by observing its effects on the radio beam from the spacecraft as it propagates from the spacecraft to Earth. This requires pointing the antenna at the top of Neptune's atmosphere as the spacecraft disappears behind the planet. If the spacecraft flies too close to the planet, timing uncertainties would prevent us from pointing the antenna in the right direction at the right time.

So we had a real tradeoff. If we flew too close, we might lose radio science data on Neptune's atmosphere, but we'd get better images of Triton. However, if we backed off far enough to be safe for the radio science occultation of Neptune, we'd lose high-resolution images of Triton. It turned out the navigators provided a solution.

We could pick a trajectory that would allow us to fly within 40,000 kilometers of Triton if we could predict when we would arrive at the north pole of Neptune to within a second. An uncertainty of five seconds was too much because we wouldn't know when to start pointing the radio antenna at the top of Neptune's atmosphere. The spacecraft was moving so fast and so close to Neptune that, if you're off by five seconds where you're pointing the spacecraft, you're pointing in the wrong direction.

We had to know exactly when and where to point the spacecraft to within a second. We had never delivered a spacecraft to within a second before, but the navigators felt that they could do it. So the question was how close to Neptune could we get? We kept pushing closer, and they kept finding new, inventive ways to improve on the timing, including a new way to do the trajectory correction maneuver.

It turned out that the process worked very well. We arrived within one second of the prediction, and we successfully executed both the close flyby of Triton and the radio occultation.

Those were engineering innovations that were in response to high-priority scientific objectives. We had to decide, based on the engineers' estimate, that the timing uncertainty would be small enough to take that option. We made that decision although that was not a conservative thing to do. If we wanted to be conservative we would have chosen one or the other. We would've given up on the radio science and said, "We'll just go for Triton, close." Or we would have given up on getting so close to Triton and said, "We want to really be sure we do radio science 99.9 percent guaranteed."

Well, either would not have been the optimum. That would have guaranteed one or the other, but it would not have provided the optimum science. The optimum science is picking the closest you think you can get, but using judgment—using the right judgement, getting close, but not too close.

We spent about a year on that process because the engineers were very responsive in developing and in refining their tools. They had a lot of experience with previous encounters. We'd had five other encounters, and so they were more confident that they could reduce the uncertainty to one second.

That is an indication of the kind of tradeoffs that we often would spend a year studying before deciding what to do. We didn't want to be too conservative and unnecessarily give up science, but we did not want to be overly optimistic and not get the science either. That is a recent example of many such iterations that have occurred.

What would be an example of a really difficult decision that the project scientist had to make?

One I certainly remember had to do with choice of the flyby trajectories at Jupiter. We had to choose these before launch, because you have to launch the spacecraft in the right way to get to the planet at the right time. We had decided that Voyager 1 would arrive at Jupiter first and would fly past very close to the moon Io, which we knew was going to be interesting, although we had no idea it had active volcanoes.

That meant going into a distance of about five radii from the center of Jupiter, where the radiation environment is very intense. There was a trajectory discovered, which I think Geoff Briggs found, that took us close not only to Io but to Ganymede and Callisto—three moons from closeup on this one flyby.

We had decided for reasons of safety to fly Voyager 2 at 10 radii from the planet, where the radiation environment was less intense, so that if Voyager 1 did have a problem, Voyager 2 would be safer. That meant we couldn't get near Io, which orbits at six radii, and the tradeoff we had to make was one of getting close to Europa, which Voyager 1 could not do, or flying behind Ganymede so that we could use a solar occultation to search for a possible atmosphere of Ganymede. These were two different things. One was imaging the surface of the fourth Galilean satellite, Europa. The other was trying to verify a suggestion that Ganymede had an atmosphere. That was potentially quite exciting, to find that a moon has an atmosphere.

The way we study thin atmospheres, such as was suspected on Ganymede, is to watch the "sunset" with the ultraviolet spectrometer. The absorption of the sunlight by the atmosphere tells us how much gas is there. Or, if there is enough atmosphere, we can use the radio beam from the spacecraft to watch the atmosphere refract, or bend, the radio beam sent back to Earth. But you have to fly behind the object to do that.

There was no trajectory to both image Europa and examine Ganymede's atmosphere. We had to make a choice, and it involved two different science teams who were going to be potentially making discoveries. So this was one of the cases I mentioned earlier, where the decision on trajectory was also a decision on which individuals were going to be making discoveries.

We decided to go with Europa. If you want to understand the evolution of the four major moons around Jupiter, there would be a serious gap if you did not have the comparable information on Europa that you had on the other three. That was the correct decision. Subsequent analysis of ground-based data indicated there was no atmosphere on Ganymede. So in that case we knew even before getting there that that was the right decision to have made.

That was a decision which, once it was made and once we were launched, there was no way to change. We could not say, "Well, we want something new." We couldn't. We were on a trajectory that was going to pass Europa and not behind Ganymede. There were other issues like that, affecting trajectories, where once you make the decision you've decided what kinds of discoveries you're going to emphasize.

Were some of those decisions relatively easy to make?

Oh yes. For many it was straightforward, and we had a process in place. But there also were some where there were two incompatible objectives and the decision had to be made, one way or the other. It was an apples and oranges decision: you were trying to ask, "Is this apple better than that orange?" It's a very difficult balance to weigh, because they're two different objects: an atmosphere of a moon and another moon. They're just not the same thing.

How many times did you, as project scientist, have to make the final decision on such matters?

It was not a large number; perhaps 10 or 20. There were many smaller decisions, but I'm talking about major decisions. They occur one at a time, and after 15 or 20 years you forget. That is the reason I have 38 volumes of notebooks in which I've recorded such decisions because I knew I wouldn't remember them all.

What would be another example?

Another decision we had to make was about the encounter with Titan at Saturn. There was a desire to have an encounter with Titan before an encounter with Saturn itself; in other words, to encounter Titan when it was out in front of Saturn. That's because when you fly by Saturn you might run into particles in the ring plane, damage the spacecraft, and then not be operating afterward when you arrive at Titan. This was called the "Titan before" option: you encounter Titan, learn what kind of atmosphere Titan has, and then you encounter Saturn.

The other option was to do a "Titan after," which had the advantage of providing a better occultation of Saturn's rings. We had several key observations we wanted to make. One was to measure the atmosphere of Titan. That means we had to go behind it at encounter so we could watch the sun set with the ultraviolet and watch the refraction of the radio beam.

But we also wanted to go behind Saturn's rings for exactly the same reason. We wanted to use the radio beam from the spacecraft, coming through the rings and being scattered, as a way of determining the distribution of sizes of particles that make up Saturn's rings. To do that you have to go behind the rings in a particular way to be most effective. We'd like to have a passage in which, as seen from Earth, the spacecraft goes diametrically right behind the rings. It turns out that "Titan after" provided the best ring occultation geometry, but it did have the risk of having to fly by Saturn first.

The radio science team looked very carefully at the possibilities, and their desire was "Titan after." But it was clear to me that "Titan before" made a lot of sense for safety reasons and because on approach you could image it and then fly behind it for the occultations. If Titan is on the other side of Saturn, and you fly behind Titan, you won't see the lighted side of Titan close up. As it turned out, Titan was hazed over so there was not a lot to see anyway, but we couldn't be sure of that until we arrived there.

In any event there were several reasons why, scientifically, we wanted to do Titan before, besides the safety issue. And so the question was, could we find a geometry for the "Titan before" trajectory that would give us a satisfactory ring occultation, even if not the optimum? That meant asking the question: "Okay, I understand what you would like ideally, but what else would do the job, even if it wasn't the ideal?"

It turned out that Len Tyler, the radio science principal investigator, found a particular geometry of the flyby that would allow us to analyze one half of the rings in an almost ideal way. Instead of cutting behind the rings on both sides of the planet, we cut behind the rings on only one side, and that basically did the job. We selected "Titan before", and it worked very well.

I don't know that at Uranus we had any major trajectory issues because our flyby was so constrained by having to go on to Neptune that our flyby distance was tightly defined. The only questions were when we would arrive there and how close to Miranda we'd come. But I don't recall that as a major tradeoff discussion.

At Neptune though, another issue involved the occultation of Triton. We wanted Triton to block Voyager's view of the sun so we could study Triton's atmosphere with the ultraviolet spectrometer, and we wanted it also to block the spacecraft's view of the Earth to do an occultation study of Triton's atmosphere with the radio beam. However, the sun and the Earth are in different places, and so the optimal locations behind Triton for viewing the two occultations also differed, by about the radius of Triton.

Both occultations were important. They were both atmospheric occultations, so this was not a case of apples and oranges; it was really two apples. There were two ways of determining the atmosphere—different parts of the atmosphere, so they were not identical. I became convinced that we needed them both because we didn't know how much atmosphere Triton would have, and the radio science could detect a denser atmosphere while the ultraviolet measurements could detect a much thinner one.

Obviously each team, if given their choice, would have optimized their occultation for the center of Triton. But I felt that was not the right answer, provided we had enough confidence in our aiming that, if we put them equally off-center, we'd get them both. That puts one of the two at more risk than if it had been at the center, and puts the other one at less risk than if it had been at the edge.

Then we had a question of assessment and navigation: How accurately could we deliver the spacecraft? Could we define those two tracks precisely enough, and guide Voyager between them so that Triton would interrupt both the radio beam on its way to Earth, and also interrupt the sunlight on its way to the spacecraft?

About seven days before closest approach, I became convinced that the navigators were, with 95 percent confidence, going to be able to get them both, so I decided we would go for both rather than optimizing one or the other. This was a hard decision because it was possible that one of them might be lost that way. But I thought the probability was high enough, and the importance of both was high enough to try it.

So you aimed halfway in between?

Yes, and it turned out to be the correct decision. The navigators delivered us on the right path behind Triton. Those were very important observations, and we needed them both. The atmosphere was too dense for ultraviolet alone to have measured it, so we needed the radio science too.

That was another decision that was based on an integration of scientific importance and assessment of what the engineers and navigators were telling us they could do. That is judgment, because you don't know until it's all over whether it's really going to work. There have been a number of things like that along the way.

Let's go back to what persuaded you to take this chance to be the project scientist. You mentioned the idea that there was a lot to learn. What might have been some of the other variables, pro and con?

One of the questions was whether or not it was feasible to be an effective project scientist while committing only 30 percent of my time to that role. I was a professor, and that was already a major commitment. I was also concerned whether or not from a scientific point of view it made sense to spend 30 percent of my time in what has often been viewed as a service role. It was not clear whether having an active scientist, one who's doing his own research, in that role was really all that important; whether it would make that much difference. And if it didn't make that much difference, then it was probably not a good use of my time at that stage in my career. Thirty percent is a lot of one's time.

But weighed against those uncertainties, which there was no way to assess because it was untried, was the possibility—because this was a very important mission with a lot of discovery and with a very broad range of science on it—that it was an opportunity both to learn a lot and to contribute, depending on exactly how the role was structured, how well it worked, and how it evolved. I was strongly encouraged to do it by the Caltech administration, and their strong encouragement finally led me to say, "All right, I will. It sounds like a reasonable thing to do."

We started on it, and it worked much better than I had any reason to expect. It was an experiment; it turned out to be very interesting, and I felt that I was really contributing; it wasn't just a pro forma sort of thing. I could see that my experience was helpful to the scientists and the engineers, and they felt it was helpful too. That all meant that it worked very well, because what the project scientist was trying to do was to help the science teams do what they were selected to do. That is really what the job's all about, because it's the science teams that really do it in the end. My job was to help make that happen.

Was this the first time this concept had been used?

As far as I know, this was the first time a university professor had such an active role in a major planetary project. It was certainly the first time I'd ever been a project scientist, and it would be the only time. It's the sort of thing you do only once. At the time, it was very unusual to have a non-NASA person being the project scientist. Normally it was a NASA employee, or, in the case of JPL, a JPL employee who was the project scientist, because it's considered to be a project job, which at JPL had often been full-time or at least a half-time activity.

Did it actually work out to be 30 percent of your time?

I would guess it was on the order of a third of my time on the average, but during encounters it was full time. As I gained experience and as the teams gained experience, I managed to drop to about 20 percent.

I would have guessed that the demands on your time would increase as the spacecraft grew older and traveled farther from the Earth. Didn't that increase the complexity of the mission and therefore the demands on you?

No. Fortunately I've become smarter and so has everybody else, so things are done much more efficiently now. The sixth encounter was a lot easier than the first, even though it was more complicated, because everyone was more confident. At the first encounter no one knew what to expect. We were walking off into the unknown in terms of capability. By Neptune we had done it many times, and although we were trying new things that we'd never done before, we were confident of the basic approach. So that allowed us to take on these new tasks without having them overwhelm us.

What would be the biggest surprises that you've encountered on the project?

There were a lot of surprises. If I had to pick one, I guess I would pick the volcanoes on Io as the first major surprise. That really told us that the outer solar system was not a bland reproduction of the same phenomena over and over again. Each of these worlds was distinct and different. From there on we expected to be surprised.

Even so, we were still surprised because just expecting to be surprised doesn't mean you don't get surprised, but we were no longer blasé about the fact that nature was quite diverse and interesting. The Io volcanoes really drove that point home. After that, we were all much more conscious that we were going to be seeing things that we hadn't expected, and that was indeed the case, time after time.

What were some others that you'd put in this category of major surprises?

Certainly at Saturn a major surprise was the complexity of Saturn's rings: discovery of spokes and all of the structure in the rings was a surprise. The fact that the F-ring is kinked was a big surprise. We knew from Pioneer 11 that it was a narrow ring. We thought it was being shepherded by two moons, and indeed that is what we found. That was not a surprise, but it was a surprise to find it kinked.

Titan itself was not the same surprise because there were already different models, one of which described the kind of atmosphere we found. It was more a matter of learning what was there, already knowing what some possibilities were.

Perhaps the most interesting surprise was one that is an inference; that Titan may have lakes of liquid ethane on its surface. Some of the pre-Voyager models suggested that there might be lakes of liquid nitrogen. It turns out it's not cold enough for that, but instead there may be liquid ethane. This is similar to what we'd expected—liquids on the surface—so it's not in the same class as things we hadn't expected at all, as in the case of Saturn's rings.

At Uranus there were many surprises, but I'll pick two. One is that the poles of the magnetic field were down near the equator on that planet. The magnetic field is extremely tilted, and the center of the magnetic field is offset from the center of the planet by almost a third of the radius of the planet.

The other surprise, in the sense that there was no anticipation of it at all, was that Miranda has one of the most complex geologic surfaces of any body, and yet it's a tiny moon, only about 300 miles in diameter. That was really quite astounding—and that was after having already seen moons that had a lot of tectonic activity on them. We were still surprised to find that this tiny moon had been that active.

At Neptune the Great Dark Spot was a surprise. And Triton—we knew it was going to be different, so you might say nothing was a surprise because we expected it to be unlike anything we'd seen before, and it was. But it was still a surprise to actually see these big ice calderas, craters from ice volcanoes, and geysers—nitrogen geysers. It was still a surprise, even though we had told ourselves it wasn't going to be anything like we'd seen before.

Coming back to Saturn, another surprise was the jet stream at Saturn's equator, which was 1100 miles per hour. This was just the first of many weather surprises. As we went further out in the solar system, we found higher speed winds rather than lower speed winds. There is less solar energy input, yet the winds are faster. Neptune is the end-member of that set, where the winds may be as much as 1300 or 1400 miles per hour, even though there is a trivial amount of sunlight and internal energy to drive the winds compared to Jupiter—one-twentieth of what there is at Jupiter.

At Jupiter we expected that the Great Red Spot would be a hurricane, but the real surprise was the widespread turbulence. The atmosphere is violently turbulent, and it's riddled with smaller hurricanes; "smaller" meaning "only" Earth-sized hurricanes rather than something three Earths across. There were literally dozens of these storms. That degree of activity had not been anticipated at Jupiter.

The discovery of the Io torus—where one to two tons of sulphur and oxygen a second are being stripped off the moon Io, forming a doughnut-shaped region of material circulating around Jupiter and creating a torus that glows in the ultraviolet—was a big surprise, even though there were ground-based clues and indications from the prior Pioneer flybys that there was something interesting there. I think no one had really imagined that there would be such a massive amount of material, creating a torus which then inflated Jupiter's magnetosphere to two to three times its normal size.

The giant size of the magnetosphere had been discovered by Pioneer, but it was not understood why it was so much larger than it should have been. We now know that it is inflated by the centrifugal forces of a disk of sulphur and oxygen spinning outward and inflating Jupiter's magnetic field, all coming from Io. So that was a big surprise.

At Neptune the magnetic field was a big surprise: it looked like Uranus. Many of us had thought Uranus was a peculiar planet, an anomaly, but Neptune is very much the same way. So we now have a real puzzle: why does the flow of fluids inside these two planets generate a magnetic field having its poles down near the equator?

What were your biggest disappointments in the Voyager program?

I can't think of anything major. I can't even think of any minor disappointments, to be quite honest. Everybody keeps asking me at the press conferences, "How would you rate the success?" And I would say, "200 percent." We always discovered so much more than we expected; so many more answers than questions we've asked. And the questions we'd asked were simple compared to the answers we got. How can we be disappointed when we learned so much that we weren't even smart enough to know we were going to learn?

What frustrations were there?

I guess I'm not a person who gets frustrated easily, so I tend to be optimistic that somehow we're going to figure out what to do.

This project has been so successful, and yet there were many problems after launch?

Many problems, yes.

How, then, do you explain the success?

I think that most of the difficulties we had after launch were learning problems. We had one critical hardware failure, and that was the command receiver on Voyager 2, but everything else were learning problems. These are very complex spacecraft, and it takes awhile to learn how to run them. Because of budgetary constraints, we had not staffed up to the level we needed in order to avoid making mistakes.

The spacecraft had been designed very carefully, so that it would protect itself against anomalies. That is, if the spacecraft found itself spinning when it shouldn't be, it would stop itself and change its thruster system because it assumed that maybe a thruster was stuck open, or it would change the control electronics. It would do all this automatically whenever it sensed something that was outside of what had been programmed into it as normal performance.

The problem is you don't know exactly what normal performance is until it's launched, because you cannot test the spacecraft on Earth with all of its booms deployed, to know exactly how rapidly it's going to rotate. So you calculate how rapidly it *should* rotate, and if your calculation is off by only one or two percent, that's a good calculation.

But one or two percent off in a day's worth of rotations turns out to be many minutes, and so the spacecraft would say, "Whoops! Took too long to complete that roll; must be a problem," and would safe itself. Then we would have to spend several days unsafing the spacecraft when nothing had been wrong in the first place, except that we didn't know, to the last few percent, how rapidly the spacecraft was going to roll.

These are learning problems. You learn by flying the spacecraft, and we did not have enough staff so we were getting behind. That is, every time the spacecraft saved itself, it would require effort to understand why the roll rate was different from what had been assumed ahead of time. That meant that you weren't doing the planning for the normal things that had to be done, so eventually it became clear that we needed additional staff, and within six, seven, eight months we were back in operation. We had our first encounter in January 1979, which was less than 18 months after launch, so it was a reasonably rapid learning process.

With the Hubble Space Telescope one sees the same learning process going on. It's a very complicated machine, and you cannot fully exercise it in any way until it's in zero G. That is when you first find out how it works, and as it is being done remotely, you can't easily make adjustments. You have to tell the spacecraft to do something and watch it for a day and then tell it to do something again. With Voyager it was even more difficult because it was receding from Earth.

I never felt that we had any serious, fundamental problems. We just had to learn how to use this complicated, capable machine. It reminds me of video recorder stories: people joke about how hard a VCR is to use. Well, the more capable an instrument is, the smarter people have to be to use it. This was a very capable spacecraft, and even smart people have to learn how to use these complicated machines.

Would you agree with my impression that the Voyager project did much better than many projects in avoiding cost overruns?

Yes, it was certainly a well-managed project in terms of cost. But we were quite fortunate that when we started the program in 1972 it had a large enough budget so that there was a built-in reserve, and we could solve problems without having a budget crisis whenever a new problem developed. When Pioneer 10 flew by Jupiter in December 1973, it found a radiation environment 1000 times more intense than had been assumed. That was a year after we had started designing Voyager, so we had to restart design work because the spacecraft we had envisioned would not have survived that harsh environment.

We had to revisit the circuit designs. Nothing had been built yet, but a lot of the design work had been done. Parts had been procured, but we had to get new parts. We had to install spot shielding in many places. We spent a year, all of 1974, redesigning, hardening the spacecraft against a much more intense radiation environment. That extra cost could be covered by Voyager's budget.

Today, projects often start without an adequate reserve. Then the normal development problems, which you always have when you're doing something for the first time, tend to surface as cost overruns, when it's really a matter of not having a sufficient reserve in the first place, which you need when you design something you've never designed before. So with Voyager we were quite fortunate. We had a budget that allowed the project manager to address problems and not have a budget crisis. That certainly is an important factor.

Voyager also built on JPL experience with the previous Mariner series. This was a more elaborate Mariner than there'd ever been before, but it's basically a Mariner. It was called Mariner Jupiter-Saturn '77 (MJS 77) when we started because it was really evolution of that series.

That very strong technical heritage allowed the engineers to take on new tasks with the confidence that they were do-able tasks. That's easiest to do when you have an experience base where you've already managed to digest a few things, and you know how much you can bite off next time around. You learn by doing. There had been a lot of learning going on with the previous Mariners and the Viking Orbiter, and that all contributed to capabilities that were built into Voyager, and to the fact that Voyager was well-managed, was launched on time, and was within budget.

Many of the problems or concerns you had as project scientist were scientific. What other challenges did you encounter? Personnel, for example—stresses here or at home?

These things are all done by people, so that's just part of doing the job. You have to learn to work with people, and everybody's different. But I don't consider those anything out of the ordinary.

Any time you go into space you are clearly taking on a significant challenge, and once you commit yourself to that challenge you find that occasionally it does require long hours to achieve the thing you set out to achieve.

On the other hand, the compensation for that is getting to see things for the first time that humankind hasn't seen before. So that's the trade: between making the commitment, and the effort that one has to be willing to follow up with, having made the commitment.

As John Kennedy said about going to the moon, we're not doing it because it's easy, we're doing it because it's hard. It is hard work. Going into space is indeed a challenge. Even today, 30 some years after the first planetary flyby, it is still challenging, and there are *still* risks in planetary exploration. That will be the case for decades to come.

Is this different from other occupations, as far as personal involvement?

I think there is a difference in space research. As a scientist I think so, and it must also be true for engineers. The thing you're building leaves Earth, and you can't bring it back to fix it—we're talking about planetary exploration now. There are a couple of characteristics that are important in being willing to undertake space research, basic planetary exploration.

You have to be patient, because it takes a while to get the result, and you have to be optimistic, because you have to depend on so many things working right in order to get that result. But in return for the patience and optimism, most of the time you achieve things which haven't been achieved before.

When you look at the 30 percent time you started out with, how much of that 30 percent would have been science and engineering, how much would have been fiscal, how much would have been personnel, and other things we haven't mentioned?

I didn't spend a lot of time on personnel supervision because there was a science manager. That was one way of making this part-time job possible: to have a full-time science manager who was responsible for managing the science activities. I had a series of good science managers working with me who were responsible for the smooth operation of the science office. I provided overall policy and guidance in those areas and interacted with the science manager, but it really did not require a lot of my time. My time was mainly spent on understanding what the issues were and working with others to help them understand what the issues were.

Then you didn't get much involved in hiring, firing—decisions of this sort?

No. In hiring I would be involved because I had the final decision if we wanted to bring someone on board in the science area. And in any firing, if there were issues that I needed to be aware of, but I did not have to deal with the routine part of that. I couldn't have; there wouldn't have been enough time.

Fiscal concerns? To what extent were those a major area of your duties?

Certainly budgetary issues at the top level; that is, the cost of the mission. Our initial mission was just through Saturn. We had to repropose to NASA for the Uranus mission, then repropose for the Neptune mission, and then repropose now for the interstellar mission.

I would be involved with each of those stages of budgetary proposals in terms of establishing the overall science part of the budget, as well as working with the project manager in talking with NASA Headquarters about the overall project budget. I had top-level involvement with budgetary matters, but I didn't have a daily or monthly level of involvement.

Is it correct to say that you were doing mainly scientific and technical decisions?

Well, it's not just decisions; a lot goes into making a decision. A lot of routine meetings are needed in order to make sure homework is being done before a decision is made. But in the end my main effort has been focused toward making the right decisions; making sure that I was educated enough and others were educated enough so the right decisions were being made. During the two-week period of the most intense activity during each encounter, a majority of my time was devoted to the process of understanding the scientific results and interacting with the large number of science reporters and writers that had gathered at JPL.

What lessons does Voyager offer for other projects, not just in space, but perhaps also for other kinds of challenges?

I think that the main thing is to try to understand what's going on. That's an obvious recommendation. Scientific decisions should not be made by majority vote. Somebody ought to take the responsibility for understanding enough of the issues to be able to make a decision. The project manager has that responsibility ultimately for the entire mission, but the project scientist can and should take that responsibility for a lot of the issues that are science tradeoff issues.

It can be delegated and broken up, and that's another way to do it. But I think it's not as effective or satisfactory as if there is *one* person who ultimately has responsibility for understanding enough of the issues so that the right issues are addressed, the homework's done, and the right decisions are made.

Even when I wasn't making decisions myself, I wanted to be sure that the decisions were being made on the basis of homework and evaluation, and not just on the basis of political pressure, where some group or other is just much more active in making their case. Giving it to whichever team is strongest in making their case is another way to let decisions be made, but it's not the optimum way. I think the optimal way is for someone to probe and understand and in the process help everybody else understand what the issues are.

Whenever you have a limited resource—and any time you have a spacecraft flying by a planet or even one in orbit, you have a limited resource because you can't do everything you could imagine that you would like to do—the challenge is to prioritize the various options that are available and to do so in a way that can be accepted by the individuals involved.

This is probably not an uncommon problem in any situation where there are limits. It comes back to the general discussions about how one makes such decisions: trying to understand what the issues are, being sure that the individuals involved recognize that you have understood the issues, that the individuals understand that there really is a tradeoff to be made, and that the competing options are worthy ones. These are all part, at least in a scientific milieu, of reaching a decision that can be accepted.

Do you feel that most of the Voyager mission is over?

If you want to talk about the number of discoveries, I'd have to say Voyager has now made the majority of its discoveries. On the other hand, there are still some major discoveries left. These have to do with the interaction of the sun with interstellar space. The space between stars is filled with a dilute gas, ionized gas, a plasma, called the interstellar medium, that has come from the explosion of other stars, supernovae and so on. When stars explode, most of the mass of the star is ejected into interstellar space, mixing with gas and plasma ejected from other stars.

That gas and plasma eventually collapse to make a new star. Our sun formed out of such gas, the interstellar medium, four and a half billion years ago. Our sun now has a wind blowing radially outward from it at about a million miles per hour. It's called the solar wind. That wind blows a bubble in the interstellar medium. The bubble is called the heliosphere. It holds out the gas from the other stars. Every star has such a bubble around it.

We don't know how far it is out to the boundary of the heliosphere, the so-called heliopause. It may be 10 billion miles; that is, 100 astronomical units or so. (The Earth's at one astronomical unit, and Neptune is at 30.) No one really knows. It could be that close. It could be 11, 12 billion miles or more.

In any case, if nothing breaks on Voyagers 1 and 2, we should be able to track them for another 25 years, at which point Voyager 1 will be 130 times as far from the sun as the Earth. It's possible that by then Voyager will have reached the interstellar medium, returning data from that region of space for the first time. That would be an important discovery: that is, to leave our solar cavity, our heliosphere, behind and cross over into interstellar space.

I don't think that will happen before the year 2000, but it's not unreasonable to expect that it might happen sometime between 2000 and 2015. Nobody knows, and that's what makes it so exciting. So that's an example of a discovery that lies ahead. There are other things we'll find out there that we *won't* know about until we get there.

In the case of the solar wind, there will be a sonic shock as the supersonic wind approaches the heliopause, just like a shock—the sonic boom—from a supersonic jet aircraft. It may be only 70 to 80 times as far from the sun as the Earth is; in other words, two times as far out as the spacecraft is now.

We might find that shock—where the solar wind slows down abruptly from a million miles per hour to 250,000 miles per hour—sometime in the next 10 years, possibly before the year 2000. That would be our first indication that the solar wind is approaching the heliopause: when it suddenly abruptly slows down, then we'll know. We'll have our first measure of the size of this bubble. We won't be out of it yet, but we'll know that we may have only another 10 years to go before we do get out of it, before we finally reach interstellar space.

There are these intermediate milestones that will tell us something that we don't know. We do not know today how big the heliosphere is. That's the reason I'm anxious that the Voyager Interstellar Mission, VIM as it's called, continue as long as at least one of the two spacecraft is functioning.

What are some of the factors involved in this continuing?

Continued funding.

What does it look like at the moment?

It's in NASA's five year plan and the Voyager Interstellar Mission has started. I'm sure every year the plan will be updated, but I have no reason to believe that NASA will discontinue tracking Voyager. This is a unique opportunity. Besides the two Voyagers, there are two Pioneers, 10 and 11, which flew by Jupiter, and Pioneer 11 also flew by Saturn. They, too, are on their way out of the solar system. This fall, 1990, Ulysses will be launched, which will be the first spacecraft to go out of the plane of the planets and over the solar pole. Ulysses will arrive over the south solar pole in 1994.

This coming decade we will have five spacecraft in deep space. We're unlikely to have that again in the foreseeable future, because most future planetary spacecraft will go into orbit around planets or land on their surfaces. They won't be escaping from the sun. This is both our first opportunity to survey the heliosphere and may be our last for a very long time. That's another reason I'm quite anxious that we keep these spacecraft alive.

In the perennial competition for limited space exploration funds, where you run up against the space shuttle and the space station, etc., how confident are you about continuing funding for Voyager?

I'm optimistic. I'm basically optimistic anyway and think that given such a unique opportunity, and short of a disaster for the country as a whole or for NASA, it would be astounding if we didn't track Voyager for as long as it was operating.

You mentioned that there was a time when the project was understaffed?

We were definitely understaffed for the first six to eight months after launch. Since that time our staffing level has been reasonable. Now, for the Voyager Interstellar Mission, we are shrinking to a smaller staff than we've ever had before, and we'll see if this new, very much reduced staff is adequate. We believe it is, but we've never flown these complicated spacecraft with so few people before, and that's going to be a new experience for us in another year.

Was this staff reduction voluntary?

It was necessary to continue getting the support we needed. Once there were no more planets to visit it was clear that the level of support had to be reduced. It was a question of what the new level would be; what number NASA felt sustainable for a long period of time. I would much rather have a sustained level of support for many years than have a lot of money for a few years and then nothing. That would be the wrong answer. The right answer is a reasonable, adequate level of support that can be sustained for the next 10, 20, or 30 years. It's in everybody's interest to make do with the minimum in the long haul we have ahead of us.

What were some of the other tasks and key people that played a part in Voyager yet might be overlooked? For example, the Deep Space Network.

The Deep Space Network has played an essential role in the success of the Voyager Mission. For example, by linking together several antennas, including 27 that make up the Very Large Array in New Mexico, it was possible to considerably increase the rate at which images and other scientific data could be transmitted from Neptune. In fact, without the DSN enhancements and new programs on board the spacecraft, our ability to receive images from Neptune would have been marginal at best.

Did you have overall responsibility, even though they might not have been directly under you, for doing these types of things?

No, not for the Deep Space Network. That's not part of the Voyager project. That's a separate activity at JPL. It has an interface with the project, but it's a separate institution. There are interface agreements between a project and the Deep Space Network. We indicate what our requirements are, and they agree to a certain amount of tracking at a certain time. There's a very elaborate procedure for making sure that our requirements for tracking times agree. The flight operations office handles that interaction between the Voyager project and the Deep Space Network. I'm somewhat removed from that.

Flight operations certainly would be an area to look at.
Yes. Now it's being scaled down, but it was organized into several different offices. There was a Flight Science Office that had a science manager. There was the Flight Engineering Office with an engineering manager and a Flight Operations Office with the operations manager. There was a Mission Planning Office, of which Charley Kohlhase was the manager. Those were the key implementing offices, which then directly reported to the mission director, who recently was George Textor who is now the project manager. Those individuals have all played key roles.

They've changed, many of them, because that's the nature of this kind of activity. Over a long period individuals move on in their careers. It's only the scientists that tend to stay on and on and on, because we get to see new things every time. For the engineers, career objectives are probably better met by gaining experience, learning how to do this level of activity, and then moving on to another project where their experience can be applied. That's proper career growth for a project manager: after they've run Voyager for awhile they move on to assume further responsibilities. Some are now assistant laboratory directors, for example.

So you tend to find many project managers. George Textor is the eighth project manager even though there has only been one project scientist. As project scientist you work with many, many individuals who've been involved over the years in various substantive ways.

On the organizational chart are you, as project scientist, at the very top?
No. The project manager is. He is ultimately responsible for what happens, what goes right and wrong. As project scientist, I reported to him and was responsible for making recommendations about the scientific aspects of the mission, but in the end it was up to the project manager to decide what to do. He has final say on all decisions.

Theoretically he could ignore you?
Yes. He could have decided, "That's nice, but for other reasons we're going to do it *this* way."

As I said, there have been eight project managers, starting with Bud Schurmeier, who was the implementing, beginning project manager. I consider him to be the father of this whole thing.

If Bud Schurmeier is the father, how would you describe your own role?
I don't know. I'm not sure that's a good analogy to be extended. I think of Schurmeier as the father because he was the project manager who first studied the Grand Tour mission. The possibility of a grand tour of the outer planets had been known for several years, but Bud was given the responsibility in the late 1960s to develop the Grand Tour as a study project. When that turned out to be too expensive, he quickly reacted and came up with MJS 77 that NASA decided was affordable.

He was the project manager who oversaw all the critical system design effort that I mentioned to you, through about 1976, when John Casani took over. So many of the capabilities—the system design, the philosophy, which have made all this possible—were built in when Bud was the project manager. That is the reason I consider him the father of the Voyager project.

Charley Kohlhase is another key player because he has been the mission designer, not from quite the beginning, but from '74, '75 anyway, well before launch. He developed all the various scenarios, which we could then trade off, like this trajectory versus that one, and so on. That has been a key role in providing the framework in which decisions could be made. That is called mission design, and that's been a critical role as well.

You came on in 1972?

Yes, in the latter part of the year. The project started in July of '72. The experiments, the 11 teams, had not yet been selected when I agreed to become project scientist in the fall of 1972. Our first meeting of the Science Steering Group was in December 1972.

Do you intend to stay on until 2020?

Sure, but not to 2020—I will be retired before that, but I certainly see no reason to stop.

Do you see Voyager as an inanimate object, a living thing, or what?

Oh no. I don't think of Voyager as a living thing. I think of Voyager as a tool, as an extension. It's like using a telescope; it's the same. It's a very capable tool that we've designed, and we have learned how to use, but I don't personalize it. Others may, but I don't.

What impact has it had on you?

It's clear that Voyager's had a very large impact on me personally. I have a much broader perspective of science than I would ever have had otherwise. By being put in the position of having to understand things, I have understood a lot more than I would have otherwise, and I've enjoyed it. Otherwise, I probably wouldn't have done it. The fact that I was in a situation where I needed to learn meant that I did. So it's really been a valuable experience and I would highly recommend it.

When you think of Voyager, what words would you use for it?

"Journey of a lifetime" is one phrase I've been using. I think it really has been a journey of a lifetime; using the term "journey" in the broadest sense.

Pathfinders

These were some of the engineers in the early days who discovered the rare planetary alignment which made the Grand Tour possible, and who developed the trajectories to utilize this alignment. This involved more than geometry: could a flight path that initially looked promising actually be navigated without consuming too much of the precious fuel that controlled spacecraft maneuvers?

After finding trajectories that were navigable, still further questions involved science, the purpose of the mission in the first place. To what extent would a trajectory facilitate all of the important scientific observations? Would the spacecraft be in the right place to conduct experiments? How could the full capabilities of the spacecraft and the ground system be used to reliably return the largest amount of scientific data?

There were so many factors to be considered that ultimately, as mission designer Kohlhase recalled, "We searched through 10,000 different flight path possibilities to pick the best ones."

GARY A. FLANDRO
Graduate Student, Advanced Projects Group

Born March 30, 1934   Salt Lake City, Utah

After receiving his bachelor of science in mechanical engineering at the University of Utah in 1957, Gary Flandro worked at JPL as a Sperry employee on Sergeant missile trajectories and aerodynamics. While studying for a master of science (1960) and doctorate degree (1967) in aeronautics from California Institute of Technology, he worked part-time for JPL.

His academic career began in 1967 as assistant professor of mechanical engineering at the University of Utah. In 1985 he became a professor at the School of Aerospace Engineering, Georgia Institute of Technology. He was awarded the Chair of Excellence in Advanced Space Propulsion at the University of Tennessee (UTSI) in 1990. He has supervised over 80 masters and 20 doctoral students.

In addition to university teaching and research in such areas as aerodynamics, combustion, acoustics, and astrodynamics, Flandro has worked as a consultant for 20 aerospace corporations. He has more than 70 publications and is co-author of textbooks on aerodynamics and rocket motor combustion instability.

He is an Associate Fellow of AIAA. In 1970 Flandro received the M. N. Golovine Award of the British Interplanetary Society in recognition of the discovery of the Voyager Grand Tour multiple outer planet mission opportunity. He received the Edward M. Glass Award for outstanding research for the U.S. Air Force in 1981.

He holds a commercial glider pilot rating and the FAI Diamond soaring badge. He has built two sailplanes and is completing the design and construction of a three-quarter scale replica of the Supermarine Spitfire. Other interests include amateur astronomy, model airplanes, hiking, and cross-country skiing. He is currently collaborating with a Russian group at the University of St. Petersburg on innovative design of solar sail space missions.

> *I believe that the most precious natural resource in the world is its young people. I have passed up many opportunities for financial and professional gain in order to maintain a position in which I can help young people to understand the creative application of the methods of scientific research. I have attempted to convey to my students the great sense of fulfillment that can be attained in such a search for truth. I feel great sorrow for the manner in which this most important resource is now being wasted.*
>
> —*Gary A. Flando*

GARY FLANDRO

How did you happen to be at JPL?

I had worked there previously just after finishing my bachelor of science degree at the University of Utah. I was hired by Sperry Univac to work in a liaison role at JPL on development of the Sergeant missile trajectories and aerodynamics. So my association with JPL began at a quite tender age in the fall of 1957.

I believe that I hold the record for the largest number of goings and comings to JPL as I carried out my graduate studies, spent a short time with Sperry, and taught math and physics in the Tonga Islands for a year. After a two-year stint as an engineering instructor at the University of Utah, I continued on at Caltech studying for a doctorate degree in aeronautics. During this period I worked part-time at JPL, with the full expectation that I would seek a permanent position there when I finished my thesis research in rocket combustion instability. It was during the spring of 1965 that I first began work on the outer planet missions.

Up to that time only preliminary mission analyses had been done on outer planet exploration. This was at a time when space flight hardware, guidance, and communications technologies were stretched to the limits by the requirements of unmanned lunar, Mars, and Venus exploration. No one was seriously contemplating flights to the outer planets, since they are much more difficult to reach in terms of required launch energy and flight duration.

The task of carrying out detailed outer planet mission studies was assigned to me in March 1965 by my supervisor, Elliott (Joe) Cutting, who was at the time the head of the advanced projects group. This was to be a serious, in-depth study to determine if there were any truly useful outer planet mission opportunities; the practical type of unmanned mission that could be accomplished with launch systems, guidance techniques, and spacecraft mechanical and electronic technology available at that time.

Cutting and Fran Sturms were studying the application of intermediate planet gravity assist in missions to Mercury via close passage of Venus (that work led to the highly successful Mariner 10 Venus-Mercury flyby launched in 1973). Based on this experience, Joe suggested that a similar trajectory shaping method might be applicable in outer planet exploration.

It was widely believed at the time that employing an intermediate planet encounter would always lead to a longer trip time than could be achieved with the best available direct trajectories. This appeared to be true in the Venus/Mercury mission designs and the inner solar system round trips (e.g., Earth-Mars-Venus-Earth) first proposed by Hohmann in 1928, studied in detail by G. A. Crocco in the mid-1950s, and later by Minovitch. It seemed to me that careful application of the energy gained in a properly configured "swingby" could be used to advantage in shortening trip times to the outer planets.

I was already familiar with the idea of using intermediate planet orbit perturbations in the design of interplanetary trajectories. Most of my practical knowledge of the technique came from study of the works of Krafft Ehricke. His textbook, *Space Flight (Vol. II, Dynamics)* published in 1962, presented a comprehensive discussion of the application of the energy gained in a close gravitational encounter with an intermediate planet to either gain or lose speed relative to the sun. I used Ehricke's works in all of my mission analysis efforts at JPL and was greatly inspired by them as I began my quest for trajectories to Jupiter and beyond.

The mechanics of orbital energy modification in a planetary encounter, the "gravity-assist", or "swingby" effect were presented in Ehricke's text. He wrote, "If at all possible, maneuvers for changing the heliocentric orbital elements should be carried out during the hyperbolic encounter with a planet, rather than in heliocentric space. The greater the planet's mass, the greater the energy saving." In the preface of *Space Flight* he notes that "... hyperbolic encounter (with an intermediate planet) represents a potentially important means of central force field orbit change without propellant expenditure." He obviously saw with great clarity the benefit of using the massive planet Jupiter as an energy source in the exploration of the outer planets.

I was of course aware of Michael Minovitch's efforts, and it was my impression that he had been concentrating on the inner planets—Mars and Venus round trips. He had just published a JPL technical report that described solar system escape, out-of-ecliptic, and close solar probe missions utilizing Jupiter swingbys. In fact these had already been proposed by Ehricke. On pages 1107–1109 of *Space Flight, Vol. II* are detailed diagrams showing the trajectory designs just mentioned.

Anyone interested in the early history of solar system exploration should examine Ehricke's marvelous astrodynamics texts. They contain not only a vast array of astrodynamics tools, mission analysis techniques, and applications of the type described, but also a comprehensive history of space technology.

Dr. Krafft Ehricke had been one of von Braun's colleagues at Peenemünde before coming to this country under Operation Paperclip. He was described by Grayson Merrill [editor of the Van Nostrand series on Guided Missile Design in which Ehricke's books appeared] as "... a beloved humanist as well as a fine scientist." Interestingly, during the time these books were written, Ehricke was working at General Dynamics/Astronautics as the program director of the Centaur upper stage rocket, which was to play a key role in launching the Voyagers on their way to the outer solar system.

Minovitch later claimed he had invented the concept of gravity assist, or what he termed "gravity thrust", but in fact, I believe Ehricke should be given careful consideration if an inventor is to be identified. He clearly understood rather completely the application of energy gained or lost in a hyperbolic planetary encounter in modifying the orbit of a spacecraft. I've endeavored to motivate historians to pay more attention to Ehricke's early contributions in this regard.

In reality, the basic ideas behind gravity assist were known as far back as the 1800s. Astronomers (Leverrier, the discoverer of Neptune among them) interested in comet motions were quite familiar with the gravitational modification of the orbit of a comet or asteroid when passing close to a massive planet like Jupiter. The mathematical formulation of the orbital energy change that these astronomers developed for estimating the amount of energy gain or loss and the resulting modification of the orbit were the same ones that we used later in our outer planet mission analyses. They were already over 100 years old when we put them to work in space flight applications.

So that is where I started. I was looking for ways to reach the outer solar system efficiently, making practical missions possible. The essence of the problem was that a mission design approach was required that both shortened the flight duration and provided for a payload of significant size using available launch vehicles. I began with the idea of employing gravity assist as a means for supplementing the energy imparted to the spacecraft by the rocket booster at its departure from the Earth. With that in mind, the key element was to examine the geometry of the solar system to see if practical flight paths would be available at not too distant future times.

Thus began the search for actual time periods during which outer planet flights would be feasible. The first step was to graph the heliocentric positions of all the planets versus time (Fig. 1).

I noticed that the traces for all four of the major (outer) planets crossed in the 1980s. I was intrigued by the conjunction of Saturn, Uranus, and Neptune and found by computation that this planetary configuration would not occur again for about 175 years. This number later appeared in many places, including infamous discussions of the so-called "Jupiter Effect".

It therefore became quite evident that, if you are looking for trajectories first passing Jupiter on the way outward, this was the correct time period on which to focus. I was struck by the fact that all of the outer planets, Jupiter, Saturn, Uranus, and Neptune, were on the same

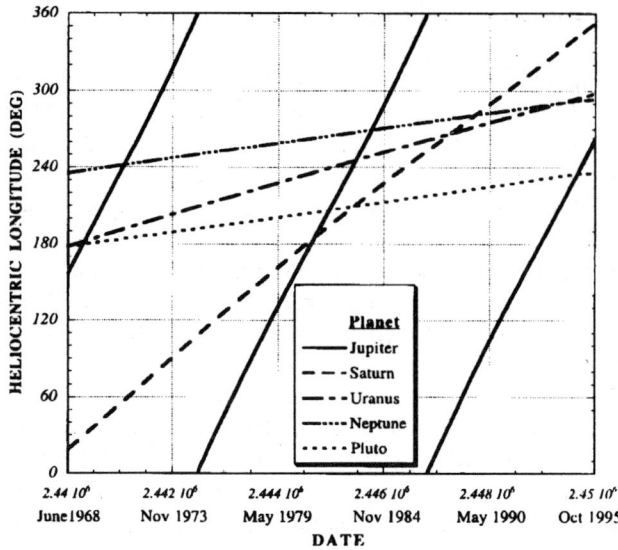

Fig. 1 Heliocentric positions of the outer planets vs time.

side of the sun (within a range of about 50 deg). So, why not look for a *single trajectory* that would pass each planet with the shortest possible trip time between?

Study of the planetary position diagram indicated that a spacecraft launched from the Earth in the mid-1970s on a two-year trajectory to Jupiter would enable access to all of the major outer planets (except Pluto) with a single vehicle. It was also clear that trajectories including Jupiter, Saturn, and Pluto could be designed, but I was more interested in Uranus and Neptune as final targets.

This realization occurred in the spring of 1965, while I was working in my office on the third floor of building 180 at JPL. I distinctly remember the feeling of awe as it first dawned on me that this mission was available at just the right time, leaving about 10 years to market the mission concept and to design and build a spacecraft. The next such opportunity would not appear until about 175 years later! I then hurriedly sketched the Earth-Jupiter-Saturn-Uranus-Neptune trajectory configuration and excitedly described the potential mission to my office mate, but as I recall his only response was a condescending smile. Joe Cutting, however, was very interested in these preliminary findings and encouraged a detailed study of the opportunity.

Since there was no guarantee that such flight paths were actually available until it was demonstrated that passage distances at each target were close enough for scientific measurements but not too close, I began a careful search for the actual trajectory profiles. I studied these at first, using so-called "conic" trajectory approximations. I did all of my trajectory work with a computer program that had been written for JPL by the Lockheed Missile and Space Company in northern California. It was a simple interplanetary ballistic trajectory program that approximated the flight paths with conics (ellipses and hyperbolas).

I did not set up the computer runs myself because in those days JPL provided a closed-shop computer system. I would work each day, setting up the launch dates and flight times for a number of runs, and I would then take these to Don Snyder, the cognizant programmer, to be run overnight. The next morning, if all went well, I would have a stack of computer printouts representing my next increment of data to analyze. Helen Ling and her outstanding group of "human computers" sometimes aided me in plotting and cross-plotting results.

One of the most difficult parts of the computation was the "matching" of trajectories at an intermediate planet. I would run a set of, say, Jupiter-Saturn trajectories and then match these with continuation Saturn-Uranus orbits. These had to jibe properly on both the incoming and outgoing legs.

All of the detailed mission analysis for this first four-planet study was accomplished by graphical methods; I did not use a computational method such as the one that Mike Minovitch developed. I did not feel that a blind, large-scale searching technique would give me the physical insight I was looking for. After all, the determination of the correct time period had already been accomplished. All that remained was to carry out the detailed calculations to prove the concept and accurately portray the best launch dates and flight times.

Mine turned out to be a sufficiently accurate and yet quite practical multiplanet trajectory design method given my self-imposed time constraint: the work had to be done by the end of the summer. By July of 1965 I had a quite complete idea of the best launch dates. I then began writing a paper describing the findings, which appeared later in *Astronautica Acta* (Vol. 12, 1966). This was the first publication outside of my internal JPL memoranda describing what would later evolve into the four outer planet Voyager missions. I'm sure it would have been academically wiser to publish the results in one of the more widely read technical journals of the time. I later found that very few readers in my target audience paid much attention to *Astronautica Acta*.

My Ph.D. advisor at Caltech, Professor Frank Marble, had been approached by the editor-in-chief, Martin Summerfield, in search of some appropriate material for publication. I had described my findings in a graduate seminar at Caltech, showing some of the multi-outer-planet trajectory results I was getting. After observing that this was somewhat removed from the work I was supposed to be doing for my thesis, Frank suggested that I should publish in *Astronautica Acta*.

Were you working on this all by yourself at this time?

There was no one else at the Lab other than a few advanced-mission specialists much interested in such matters. The focus at that time was mainly on implementing the Mars flights (Mariner and Viking) and soft lunar landings (Surveyor). An overwhelming portion of the Lab's resources was devoted to carrying out the exploration of the moon, Mars, and Venus. Just a few of us were looking at some of the more distant opportunities. To my knowledge I was the only person at that particular time doing outer planet work.

How did it come about? Did your supervisor say, "Gary, why don't you look at this?"

No, not really. Although he would have allowed me to select any research topic I thought worthwhile, Cutting felt that gravity assist in general showed great promise in making the outer solar system accessible. I was just given the general theme of "outer planet missions," so I took it upon myself then to search out the practical means for accomplishing them. The Grand Tour was the outcome.

At what point in your thinking did you have the idea "Aha! Maybe I can get all four planets"?

That came right at the very beginning of the study, in the spring of 1965, when I had just made the plot showing the heliocentric longitudes of the outer planets. I could see that the paths all crossed in the late 1980s. That is, the outer planets would all be on the same side of the sun. I knew at that point "Yes! Here is the way to do it!" All that remained was to prove by means of detailed trajectory calculations that a practical mission could be designed.

When you were first looking at Jupiter and what you could do to go on to Saturn, was that thought of getting out to Neptune in the back of your mind?

Absolutely! That was there in the thinking, along with Uranus, right from the outset. I could see that doing flights with a final target of Saturn (or in fact any of the other outer planets including Pluto) by means of Jupiter gravity assist would be quite interesting. Given the economics and expected tight budget constraints, it seemed more likely that a mission proposal would succeed if we could encounter several outer planets in a single flight. So that was the goal right from the beginning.

I carried out detailed trajectory calculations for Earth-Jupiter-Pluto, Earth-Jupiter-Uranus, Earth-Jupiter-Neptune, and Earth-Jupiter-Saturn as well. I felt that these were necessary, since it seemed likely that the Grand Tour should be backed up by some less ambitious exploratory opportunities that would serve as a fall-back position.

While I was still at JPL, there was much concern that missions as ambitious as the Grand Tour could not actually be accomplished; I received considerable negative response at first. One of the activities I engaged in was the "marketing" of the multi-outer planet mission concept to those who might later be involved in the required guidance, communications, and spacecraft mechanical and electrical design. Most felt that it would not be possible to build a machine with the longevity for such a mission. I think it is important to note here that there was at this time considerable skepticism regarding the potential of the Grand Tour as a practical mission undertaking for JPL. Many openly scoffed at the idea.

What time are we talking about here?

The 1965–66 period. At that time I was seeking a really fast trajectory that would take less than eight years to travel all the way to Neptune via Jupiter, Saturn, and Uranus. It should be pointed out that in conducting a mission study, one does not seek just a single trajectory solution. There must be an entire set of solutions to yield an appropriate "launch window" during which the mission can be initiated. In fact, one endeavors to define sufficiently long launch windows in several successive years to provide a contingency for any problems in readying the spacecraft and its launch vehicle. I found that the best launch windows for the Grand Tour were in the fall of 1977 and '78.

Since my *Astronautica Acta* paper presented a rather broad picture of outer planet exploration, including a discussion of the mechanics of gravity assist and no fewer than six different classes of missions including the Pluto opportunities, there was not room for very extensive plots of the Grand Tour trajectory results. I chose to display the launch date/arrival date (at Neptune) contours only for the 1978 mission opportunity (Fig. 2).

I had, however, done all the detailed calculations for the 1976 and 1977 opportunities as well. This involved hundreds of trajectory profiles for each of the launch years. It is important to see that there is not just a single trajectory, but a very large family of them. It was necessary to investigate the entire family of trajectories in order to determine the following items: 1) the launch dates yielding the shortest times of flight to the final target planet; and 2) the sensitivity of the trajectory to the launch energy from the Earth.

The latter requirement was met by constructing drawings of the launch date vs arrival date as contours of constant launch energy, traditionally called the "C3" by JPL engineers. In simple terms, C3 is the kinetic energy of the spacecraft relative to the sun as it leaves the gravitational field of the Earth per unit spacecraft mass. So the contours show exactly when you must launch to achieve the lowest required launch energy, or in other words which trajectory yields the highest potential spacecraft mass for a given mission duration.

On this basis, the Voyager mission designers later selected the Titan Centaur launch vehicle because it could impart the necessary launch energy to a payload of appropriate mass, about that of a Volkswagen "Beetle." Ehricke's contributions must be mentioned again here, since he led the team at General Dynamics that developed the Centaur liquid hydrogen/oxygen upper stage that made the mission possible.

If you examine Table 1 you will see that I had identified 1977 as the best year for the Grand Tour launch. This is the launch opportunity that was chosen for the actual mission.

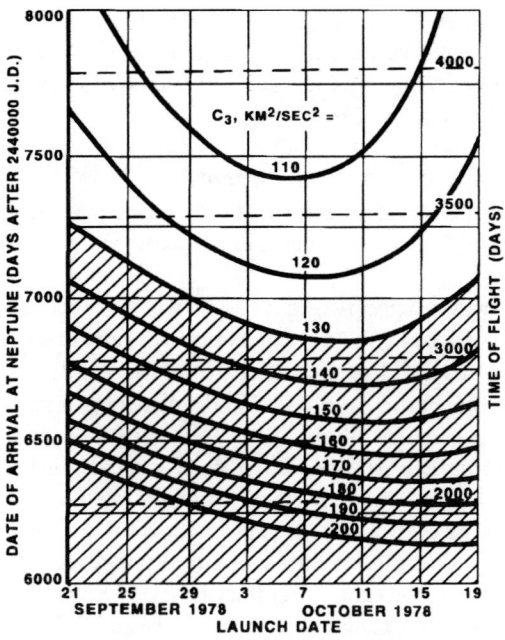

Fig. 2 Neptune launch and arrival dates during the 1978 launch window opportunity.

To make a fast flight (seven to eight years), a very close passage of Saturn was required, with the trajectory passing between the planet upper cloud layers and the rings to achieve the extra gravity assist boost. That particular possibility was rejected later on because there was concern that there would be ring material in that zone, the almost invisible inner rings, that would be a hazard to the spacecraft. Therefore, the future Voyager mission designers opted for a considerably slower trajectory that went outside the rings. Looking back on things, this choice seems ironic in view of the concern then that even the shorter flight times could not be accommodated.

I studied a whole range of trajectories and flight times, but my favorite was that one that took you out to Neptune in about eight years. This represented a remarkable time savings. A direct optimal flight to Neptune would require a flight duration of about 30 years! But when I suggested such things, many people at JPL would say, "This is ridiculous, we can't build a spacecraft that will last over two years, let alone eight years!"

Table 1 Multiple-planet trajectories to the outer solar system

Mission	Launch years				
Earth-Jupiter-Saturn-Escape		1976,	1977,	1978[1]	
Earth-Jupiter-Uranus-Escape	1977,	1978,	1979,[1]	1980,	1981
Earth-Jupiter-Neptune-Escape	1977,	1978,	1979,[1]	1980,	1981
Earth-Jupiter-Pluto-Escape	1975,	1976,	1977,[1]	1978,	1979
Earth-Jupiter-Saturn-Uranus-Neptune		1976,	1977,[1]	1978	

[1]Optimum launch year.

There was much concern over the practicality of the mission from the standpoint of the longevity of the required electronics and mechanical components in the interplanetary environment. Even an eight-year trajectory was considered to be a very difficult task for the mechanisms and electronics that were available then.

So all of this wonderful work was as a student?

Yes, that's correct. I was at the final stage of my course work in aeronautics at Caltech, had completed the two required foreign language exams, and had begun serious thesis research for the doctorate degree.

I must admit that I was disappointed that I got very little recognition for my work in light of later events, but I guess that sort of thing is natural. I did receive some recognition, but not from sources within the United States. The British Interplanetary Society (BIS) awarded me its annual Golovine Award in 1970, in recognition for the outer planet work. The British were, I think, more likely to be following things in the *Astronautica Acta*. Based on suggestions from Elliott (Joe) Cutting, Dr. William Pickering, the JPL director at the time, wrote a letter to BIS recommending that the Golovine award be made to me in recognition of my outer planet mission research. This was not just for the Grand Tour, but for the entire set of outer planet missions I had proposed.

I did not manage to raise the funds necessary to travel to England to accept the award. So the Dean of Engineering at the University of Utah, Max Williams, who was in England on business at the time, attended the award ceremony for me and reportedly enjoyed the whole thing immensely. I was pleased that someone had noticed my work. All of this took place, of course, before any of the actual planning for the Voyager mission had begun at JPL. I know that Arthur C. Clarke, a longtime member of the British Interplanetary Society was aware of my work; he used the idea of the Jupiter-Saturn gravity assist in the original screenplay for *2001, A Space Odyssey*. Apparently the final script had to be shortened so all the important final events took place at Jupiter instead of Saturn as he had originally written it.

No recognition from NASA?

Most certainly nothing from NASA; nor anything from JPL other than Dr. Pickering's kind recommendation for the Golovine Award made long before the mission had become a reality.

Several of the people who have written things have done an excellent job of portraying the events as they happened. There's a fine book, *Planets Beyond* by Mark Littman, that received several literary awards. This book describes both the discovery and the later exploration by spacecraft of the outer planets.

Littman gave you some credit by name?

Yes. In fact I was invited by him to contribute a vignette on the history. You should also consult Joe Cutting and the other members of the original Mission Analysis Group for their views on the history. I think because I left the Lab before the Voyager project was underway that many JPL people have forgotten the true origins of the mission. Their minds were elsewhere at the time. Part of the problem was the 10 year time span between my discovery of the mission opportunity and the actual implementation of the concept.

Charley Kohlhase, who worked on the standard trajectories, appeared not to have remembered who did the original work until I reminded him around the time of the Neptune encounter. In his intense concentration on the trajectory design, he lost sight of the fact that had I not discovered the mission when I did, it is most likely that he would never have had the chance to do his part. The timing was critical. That entire 10-year interregnum before he started work in 1975 was needed for the turbulent political and technological stew to settle into something like a realistic approach to the mission.

Many myths have arisen about the origins of the Voyager mission. Some of these were the result of the political nature of the mission during the early days of its marketing by JPL to NASA and by NASA to the federal government. Since I was pretty low on the totem pole and had already left JPL, no one felt much need to mention my connection with the discovery of the mission. I accepted these misconceptions, but it was sometimes difficult. In looking back, I can see that what happened is a lesson on human behavior. I have found it interesting that there was little contention until the Voyager mission became a reality. Then the trouble really began.

I left JPL as a rather naive young fellow, I could not conceive of the possibility that I would not in time be acknowledged in some fashion for the work I had done. I thought this would happen automatically, since I had properly documented my work and presented it to many people both at the Caltech campus, at JPL, and in various technical meetings.

Those at JPL who brought everything together certainly deserve major credit for the magnificent job that they did.

What was the origin of the term "Grand Tour"?

About mid-1965, Joe Cutting recognized that we had indeed identified a practical mission opportunity. Therefore he set up an appointment with the JPL chief scientist, Homer Joe Stewart, to describe the possibilities to him. I showed Dr. Stewart slides for a presentation I made earlier to the Los Angeles AIAA section. This was a detailed description of the available mission profiles including flights to Pluto.

By that time I had been able to determine the view angles and passage distances at each of the target planets, demonstrating that useful photography and scientific measurements could be made at each planetary encounter. It also had become clear that the mission would be politically correct. That is, each leg of the trajectory required less than four years; therefore we could probably count on the support of highly placed politicians who, of course, are always on a four-year schedule.

During this meeting Homer Joe recalled the earlier work on multiplanet trajectories carried out in the 1950s in Italy by G. A. Crocco (father of Professor Luigi Crocco of Princeton University fame). Crocco had proposed an Earth-Mars-Venus round trip gravity-assist mission, which he had called the "Grand Tour". This name was of course a perfect description of our new mission proposal, and it was at this moment that the name "Grand Tour" was adopted to describe the four-planet trajectory.

I have been amused over the years as many other people have claimed that they were responsible for attaching this name to the mission. Dr. Stewart gets the credit for this one. He issued a press release within a day or two after our meeting, and the first steps in publicizing and marketing the Grand Tour mission had begun.

How were the final trajectories chosen?

Later, the analysts who designed the standard trajectories for the Voyager chose a longer flight duration of about 12 years. This was largely predicated on the need to achieve the best view angles, passage distances, and lighting for photography at as many of the satellites of the outer planets as possible. In fact, the Voyager 1 trajectory was designed with the objective of acquiring pictures of Saturn's large moon, Titan. To accomplish this, it was necessary that the trajectory was deflected upward out of the ecliptic plane and off the flight path needed to continue on to Uranus and Neptune (Fig. 3).

What other concerns were there?

There was much concern about "how do you get any useful data back from those distances?" Communications engineers were at the time still grappling with the data transmission problems

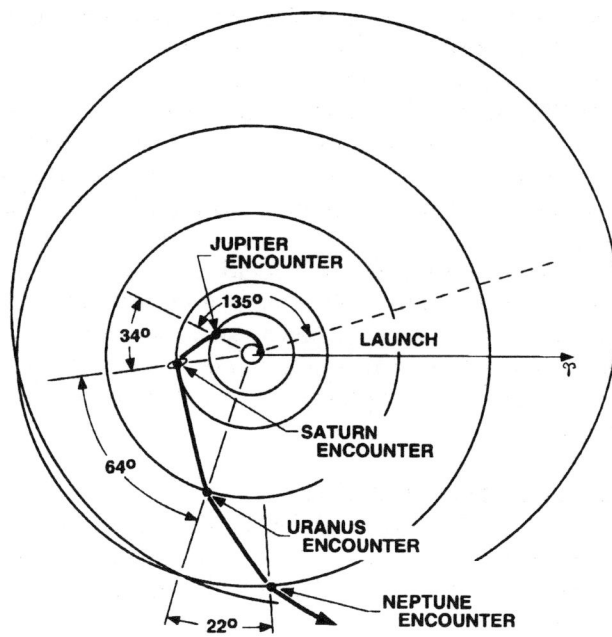

Fig. 3   Later version (1966) of a Grand Tour slide shown to Dr. Stewart.

from Mars. I recall Bruce Murray's "Martian Horror Story," describing the grave limitations imposed by constraints on the rate of data transfer to the Earth. It looked quite difficult even from that distance, so in going to Jupiter and beyond, it appeared that the data transmission and handling would be extraordinarily difficult.

However, about that time the deep-space communication net was becoming a reality, and that was an important ingredient in making long-range planetary exploration feasible; the large Earth-fixed antennas made it possible to implement practical data transmission rates. There were also, of course, rapid improvements in electronics, magnetic data storage, and so on.

Another worry was how to accurately guide the spacecraft at extreme distances. An oft-expressed concern was that we were using simple conic trajectory programs in the mission design. How well did these represent the actual mechanics of the solar system? However, those fears proved groundless. It was quite practical using Earth-based radio guidance and midcourse propulsion maneuver strategies developed by JPL to carry out the necessary accurate guidance—as the final Voyager trajectories so elegantly demonstrated.

There was also concern that Jupiter's magnetic field and radiation belts were so powerful that the spacecraft electronics could not survive a close passage with the planet, but that worry was based on some misconceptions about how such fields affect the spacecraft systems. It turned out not to be an insurmountable problem.

Yet another imagined hazard was the asteroid field between Mars and Jupiter that must be traversed by outer planet vehicles. It seemed that this might be a major threat to the spacecraft. It turned out not to be. All that was needed was to carefully design the final standard trajectories so that any large bodies were avoided.

When did the atmosphere at JPL shift from skeptical to supportive?

That happened after I left JPL to accept an academic position at the University of Utah in the mechanical engineering department in September 1967. This part of the story can best be told by those who later continued the Grand Tour work at the Lab. I will attempt only to briefly describe important events as I watched them develop from afar.

In the same time period my mission analysis interests had shifted to application of solar electric low thrust and solar sails. I did trajectory studies that employed solar electric propulsion on the Earth-Jupiter leg of missions like the four-planet Grand Tour. This approach greatly enhanced the payload, but required application of a promising new space flight propulsion technology, electric propulsion, that has not yet even as we speak reached the stage of practical application.

Before leaving JPL, I enjoyed working briefly with Al Joseph, in the design of a computer program that could be used to accomplish the analysis of multiplanet missions in a more automated fashion. Al produced an excellent program that enabled the complex searching procedures to be carried out without the tedious handwork that I had used in the first Grand Tour study. Later versions of Al Joseph's program were used, I believe, in the detailed Voyager trajectory design work. Such efforts as these in the late 1960s demonstrated to me that there would soon be an explosion of interest in outer planet exploration at JPL.

A study to implement outer planet exploration started in June of 1969 and was based on an extension of the successful Mariner technology. In 1970 James van Allen, who had by then become a devotee of the proposed outer planet missions, chaired a NASA team examining the scientific opportunities that could arise from implementing the Grand Tour. Successful Mars exploration had been accomplished, and I suppose everyone saw that the time had come to reach for more distant targets and for even more ambitious missions. So began a serious effort to bring about outer planet unmanned exploration. Bruce Murray, the Caltech planetary scientist, worked hard to get the planners to devote attention to scientific observations of the moons of the outer planets as well as the planets themselves.

At first, NASA enthusiastically supported the Grand Tour, and a rather grandiose scheme was envisioned in which several spacecraft would fly the mission and would drop off atmospheric entry instrument packages and orbiters at each of the target planets. These were going to be heavy spacecraft, to be carried on the expensive Saturn V of Apollo fame.

The original plan even earned the endorsement of President Nixon, who expressed his full support. There was an extensive publicity effort underway. Werner von Braun displayed his public backing in various articles in the popular press despite his passionate devotion to manned spaceflight. Homer Joe Stewart received much credit in the media during this time as the originator of the Grand Tour, which I suppose was appropriate since he was the chief scientist at JPL when I discovered the mission opportunity.

At this point, early 1972, ominous clouds appeared on the horizon. The first was the announcement of the space shuttle development. This was followed immediately by notification to JPL that the Grand Tour project was cancelled because all the money in the shrinking NASA budget was needed for the space shuttle and continued emphasis on manned spaceflight. There was a perceived need to preserve the manned space program at any cost. There were several highly placed individuals in NASA who attempted to bring all the outer planet work to an immediate halt along with other plans for unmanned exploration of the solar system.

Even when scaled-back versions such as the MJS (Mariner Jupiter-Saturn) were proposed by JPL, they received scant support at NASA Headquarters. By this time such proposals were also not being received favorably by the Nixon White House because of the perception that the public was no longer interested in space exploration. The novelty of the Apollo moon landings had apparently worn off. Bruce Murray aptly described the situation, "JPL was on everybody's hit-list awaiting the next big NASA cut." It seemed that JPL had no useful role in the shuttle-dominated NASA of the future. There was much resistance to the whole idea of outer planet exploration, since this required JPL to carry out the work.

On what basis?

I suppose it just took money and attention away from the manned program, and especially the shuttle space transportation system. Also, it doesn't take too much imagination to perceive the petty jealousies (probably spawned by JPL's spectacular successes with the Mariners and other unmanned missions) displayed in some of the actions of the NASA management at that time.

The MJS survived somehow and finally evolved into the Voyager project. In about 1975, former colleagues of mine at JPL began the arduous task of designing the standard trajectories to implement the Grand Tour mission opportunity. Charley Kohlhase, who later managed the Voyager flight operations with Paul Penzo and Joe Beerer, picked up where I left off and accomplished the next stage of the job in great style. This involved running thousands of trajectories in search of the ones that would yield the best photography and scientific measurements at not only the four planetary targets, but also their moons. I believe Charley deserves much credit for setting the search criteria that led to the final mission design.

In response to an RFP for scientific payloads for the MJS, Dr. Don Groom, a physics professor at the University of Utah and I prepared a proposal for a clever energetic particles telescope. We weren't selected as experimenters. However, the team of investigators finally selected came up with some wonderfully innovative techniques to get the best possible information from the unmanned exploration, as later events proved.

Meanwhile, NASA had *prohibited* the Lab from mounting anything beyond a Jupiter-Saturn mission. So when the project finally reached the hardware development stage, the Lab had only received funding to do a Jupiter-Saturn flight. But in spite of all that, the magnificent JPL team once again demonstrated extraordinary creativity and undertook to build a robust spacecraft that could accomplish the entire four-planet mission. Fortunately, later on, when it was too late for NASA to stop it without looking petty, the Voyager team was able to carry out the full Grand Tour mission (the Voyager 2 spacecraft), despite the politics. All of this on a two-planet budget!

How did NASA make its disapproval known?

They were directly opposed to it. NASA Deputy Administrator Hans Mark saw absolutely no value in scientific exploration of the outer solar system. He, in fact, was quite serious about ridding NASA of the pesky JPL entirely.

Did they say "Hey you guys, we understand you are thinking about wandering beyond Saturn; don't do it"?

Not exactly, but there is no question in my mind, that had they fully realized what was going on in the spacecraft assembly building at JPL, they would have been most displeased.

This was in the late 1960s?

No, this would be at a later time, during development of the Voyager spacecraft and the early stages of flight in the late 1970s. A fascinating 1991 book by William E. Burrows, *Exploring Space*, presents a very penetrating analysis of the politics of the time.

And your own observations support what Burrows said?

Pretty much. All of this recalls some other words by Bruce Murray, the JPL director during the Voyager years, to the effect that, "While the Voyagers functioned smoothly in space, circumstances in their terrestrial birthplace were not so harmonious."

It also brings to mind William Burrows' fascinating description of some of the events during those marvelous hours at JPL during the Voyager 2 Neptune encounter in August 1989. I was there with my teenage son Troy to witness the encounter. While I was being interviewed by a British journalist, I noticed Hans Mark and some other dignitaries on the other side of the von Kármán auditorium.

Burrows describes the response of Hans Mark as George Alexander, NASA Office of Public Information head, showed to him the Voyager spacecraft full-scale model on display in the von Kármán auditorium. As Burrows tells it:

> The fact that Voyager 2 had reached Neptune after such an arduous long-duration flight was a tribute to the quality of U.S. engineering, Mark told Alexander as he continued to appraise the spacecraft without visible emotion. He might have added that the feat was no less a tribute to its team's and the larger community's resilience, to their ability to withstand and ultimately prevail over duplicitous bureaucrats concerned only with what was politically expedient on the home planet as well. He might have. But he didn't. Instead, Hans Mark finally turned and purposefully retraced his steps to the rear of the auditorium, once more being careful to stay clear of the cables that were carrying the triumph at Neptune around the world. He did not smile.*

Do you think attitudes about space exploration have changed since then?

The present reality is somewhat frightening. Perhaps it is a demonstration of some of the societal trends in America that have taken us so far back from the great promise of space exploration. This is especially evident in the current lack of interest in science in the young people and in the rise of pseudoscience as a substitute for reality. I guess I can understand why typical high school students think that it is much smarter to put their efforts into things that "make money" instead of the arduous work required to prepare for a career in science or engineering.

Over the years I have felt the obligation to convey to students the great excitement and fulfillment I experienced in my early connection with outer planet exploration. Close to home, my own students, both graduate and undergraduate, have received much of value from this experience. I'm not so sure I did much good with some others. During the late 1960s there was considerable interest and apparent understanding among groups of high school students and astronomy clubs I talked to about outer planet exploration.

There has been a noticeable decline of interest and understanding with the passage of time. My most recent and most disappointing experience came in a talk I gave to a group of high school students at a rural town in middle Tennessee. I noticed during the presentation that there was considerable agitation and lack of attention. To them I was clearly just another nerd. I finished by showing some of the spectacular pictures taken by Voyager 2 at Jupiter, Saturn, Uranus, and Neptune. The first question that was asked was, "How did they fake those pictures? They aren't real are they?" This was followed by: "Do you believe in UFOs?" The media and our poorly trained and unmotivated high school math and science teachers are part of the problem.

I must take this opportunity to thank my wonderful teachers in grade school and high school who maintained my interest in science and technology. They are the ones who deserve recognition.

Speaking of pictures, do you have any photos, charts or other illustrations from the early days of the Grand Tour?

I have several of the original graphs I described earlier. I have the drawing I told you about illustrating the heliocentric longitudes of the outer planets. It was just a pencil drawing made for my own use. I had two 11-in. × 17-in. pieces of graph paper taped together so I could expand out the plots to see the details more easily. I also have copies of the original flight trajectory sketches.

While I'm talking to Roger Bourke, what particular points should I pursue with him?

I think the historical matters regarding the earlier discovery of the gravity assist by the astronomers is one of the aspects with which he is intimately familiar. That again is a fact that

---

* William Burrows, *Exploring Space*, Random House, New York, 1991, p. 412.

is not very well known, and it is only right that those who did the work should get the credit. That's why I'm so adamant that some credit be given to Krafft Ehricke. I tried hard to get the Smithsonian Air and Space writers to mention these matters in their recent article on the origins of the Voyager missions; apparently they weren't listening. Some of us who were involved in the early work have grown weary of trying to straighten out the historical record as it was presented in that and other articles over the years. I hope this will be the last time I have to attempt to do so.

I think you'll find Roger a fascinating chap. He was at JPL during all events I have described. Please ask him what his perception was of how this mission began. It's my impression that he was aware of most of the happenings I have described to you.

Going back even before the times we've talked about so far, when did you first become interested in science or space?

Oh golly, that takes me right back to my earliest memories. I received a beautiful book on astronomy from my mother when I was about six years old. In the frontispiece appeared an illustration of the solar system, showing all the planets. Jupiter, Saturn, Uranus, and Neptune were placed in what I guess the artist considered a typical configuration. There they all were, lined up just as they needed to be for the Grand Tour! I must have been impressed by that because I drew similar diagrams in my later work on the mission.

I remember thinking as a youngster how great it would be to fly through space. There was in my book an artist's conception of a space vehicle of the sort you might expect in a vintage 1940s book, and I was most intrigued by it. I thought how neat it would be to go all the way through the solar system and pass each one of those outer planets shown all lined up just the right way. I suppose that might have sown the seed that later grew. What an amazing piece of good fortune to find the planets lining up in an ideal pattern at just the right moment.

I've been interested in space flight, astronomy, and engineering in all of its forms right from childhood. I built telescopes when I was a youngster. I bought my first telescope kit (for one dollar) when I was in the first grade. It was supposed to reproduce the performance of the 'scope used by Galileo. That is, it was not a very good telescope, but I built it with some help from my father and it worked. This was the beginning of a lifelong interest in amateur planetary astronomy. In my teens I spent many fruitless hours looking for evidence of the canals on Mars. I enjoyed watching the motion of Jupiter's moons and studying the cloud configurations and the Great Red Spot. However, my favorite viewing object was and will always be Saturn and its magnificent rings.

Looking back over the Voyager project, were there any lessons from it that could be applied to other projects, not necessarily just in space, but generally to any large organization?

Oh yes, absolutely. In fact, one of the things I learned was that you have to be very careful not to be discouraged by experts. I was being told at JPL that I was wasting my time with outer planet exploration. The experts would say "you can't do this, that, or the other." This important principle was expressed most elegantly by the science fiction writer Arthur Clarke in the form of Clarke's Law: "If a highly respected and well-established authority tells you something is possible, then he is probably right; if he tells you something is impossible, he is probably wrong."

I think this principle is absolutely correct, and I make all my students learn it so they won't get discouraged by being "experted". When they have an idea they should use their own mental resources and creativity to dig right to the roots. Good things almost always come from such an effort. I attribute the success of the Voyager mission to that attitude on the part of the JPL team. They followed what they saw was the correct path. They simply ignored all the opposition to the greatest possible extent.

ROGER D. BOURKE
Supervisor, Advanced Projects Group
Born June 23, 1938   San Francisco, California

Roger Bourke attended Stanford University from 1956 to 1964 and was awarded a bachelor of science in engineering, a master of science in engineering mechanics, and a doctorate in aeronautics and astronautics.

Upon graduation he joined JPL as a member of the Advanced Projects Group in the Systems Division. In that capacity he studied a number of potential future planetary missions from a mission analysis and design perspective. In 1966 he collaborated with Gary Flandro on an early comprehensive analysis of the Outer Planet Grand Tour opportunity of the late 1970s and brought in the first considerations of the technical and implementation implications of this mission, considered at the time a "mathematical curiosity".

He became supervisor of the Advanced Projects Group and in the early 1970s led mission design for the Grand Tour, which became Mariner Jupiter-Saturn '77 (MJS 77).

Subsequently he held various positions at JPL including deputy, then manager of the Systems Analysis Section. Later he managed the JPL Advanced Studies Program; the Advanced Programs Branch of the NASA Headquarters Solar System Exploration Division; Mission Design, Analysis and Operations office of the Mars Rover Sample Return project; the Mission and Systems Engineering element of the Exploration Initiative Office; and other positions.

He was also an adjunct associate professor of management at the University of Southern California. Currently he is in charge of the international programs part of the Mars Exploration program.

Having flown for 30 years, he holds a commercial pilot certificate with instrument rating and over 4000 hours of pilot-in-command experience. He has been president of the Mooney Aircraft Pilots Safety Foundation and is active in aviation circles. An avid mountain sportsman, he enjoys backpacking, camping, mountaineering, and skiing with his family. His current challenge is to build a house in the mountains.

> *Knowledge is the key to a better future for all humankind. At rare points in history, humans suddenly and decisively expand their knowledge about a particular phenomena. The results, and the impact of those knowledge explosions, cannot be forecast. Even today, we can't fully appreciate the implication of Voyager's journey into space. In a moment in history that will never be repeated, we have made a giant step forward in knowledge; I was incredibly lucky to be there.*
>
> —Roger D. Bourke

ROGER BOURKE

When did you get interested in science?

I graduated from Stanford in '64 in aeronautics and astronautics and came right to work here at JPL. I came into the Future Missions development area. My job was to start working on things that were beyond what we were doing right now. So it was natural that any advanced project would be something that would be in my area. I was fresh out of school, 26 years old with a doctorate when I went to work here—education, no experience.

I began working here two days after the beings of this planet had launched their first successful mission to Mars and the second successful mission to another planet. I think it is important to set the context of where we were in 1966, which is the earliest records I can find of my personal involvement mucking around with the Grand Tour.

There was a book written after Mariner Venus '62, called Mariner 2, which was titled something like "109 days to Venus." The reason it had that title was because it was so extraordinary for a spacecraft to last that long and go that far and function and do its thing when it got there. Two years later two flights got launched off to Mars. One failed as a result of a launch problem. The second one, Mariner 4, was a success; it was nine months to Mars.

The people involved in that held their breath for nine months, because there was a constant concern that something would go wrong, that the spacecraft would fail. It was so far away that the data rate coming back was 8.3 bits per second. It took days to return a couple of dozen pictures. It was absolutely at the state-of-the-art at that time.

Now here we were, a year after that successful arrival at Mars, working on a mission that was going to go more than 15 times longer flight time and more than 15 times greater distance, out to Neptune. At that time it was an absolutely audacious idea that probably only young kids like Gary Flandro and me, who didn't know any better, would take on.

There was a whole confluence of good fortune. One was that the atmosphere, the culture within JPL let young kids such as Gary and I take on such audacious ideas when all the old salts knew it wouldn't work. That, plus the horizon of space was blossoming out in front of us here. We were at the leading edge of planetary exploration. People were very ambitious. The Cold War was on, driving all of this stuff. There was the race with the Russians to do better and better in space. We were in the midst of the Apollo development activity in '66.

And this confluence of the planets also emerged in which this very unusual alignment came at just the right time. You could get one spacecraft to four outer planets in one opportunity in 1977 when the technology had just gotten up to the point that you could actually do that. It was a lucky break.

Historian Craig Waff reminded me of some of this. He found this memo authored by Gary and me in 1966 which says, "some comments on a possible study of a mission to Jupiter, Saturn, Uranus, and Neptune." The first point was that somewhere between Gary and Mike Minovitch they discovered that the ballistics were possible, but that's a long way from actually making a mission possible, so a few of us—and I'm sure we were stimulated by our bosses to do this—wanted to move it from a geometric curiosity to something that might be a contender for an actual flight. I think I was tied for the earliest person to be in that phase of it, and that was 1966 or thereabouts.

It subsequently became legitimate three or four years later and attracted a lot of money. There was a big development program that took place called TOPS (Thermoelectric Outer Planet Spacecraft). We worked on navigation. I recently searched my personal library and found some old papers, including "A Grand Tour of the Outer Planets." Also there was an article in the Caltech magazine *Engineering and Science* in May 1970. There's an AIAA paper in '71 "Design of Grand Tour Missions." And in '72 one on Mariner Jupiter-Saturn (MJS).

I think of my involvement in two rather distinct stages. One was when it was a curiosity. It moved from being a curiosity to being a possibility in '66, '67, '68. A couple of things happened. One is that planetary missions started to become more successful. We launched a mission to Venus in '67, two more to Mars in '69. All those worked. People started to gain confidence in it.

Secondly, one of the real issues about this idea of flying to all these outer planets was whether you could actually do the navigation well enough. The ballistic trajectory relies on your going through a single well-defined point by each intermediate planet so you get on to the next. And one question was, since you cannot get the precise point, you can only get some vicinity around it, is that vicinity good enough to make it work?

I think the key to that was unlocked by two guys here, Fran Sturms and Joe Cutting, who did some of the analysis, not for the Grand Tour but for the Venus-Mercury opportunity in 1970. They showed in a paper that our navigation capability was sufficient that you could fly Earth to Venus to Mercury and do it with propulsion that was reasonable, and tracking that was reasonable, and sensors that were reasonable, and all the kinds of things that looked like you could actually make it work. There was an inference, the correct one, that because the Venus-Mercury flight was navigationally feasible, that the Grand Tour was also navigationally feasible.

That was a very key point in moving the Grand Tour out of the curiosity category and into the possibility category. I can't say that one morning people woke up and started taking it seriously; it evolved towards being something that would be seriously considered. Of course it was a challenge for the spacecraft people. The idea of building a spacecraft for which you couldn't use solar cells; you had to use nuclear power on board, something that would communicate over vast distances, one that could last for such a long time—all these were big, big challenges for some of the folks on the engineering side.

This was an activity, a project, a moment in human history that was conceived, born, and raised at JPL. What a fortunate experience to have had a personal hand in it! I'm forever grateful that some accident of life took me to that point and from there to here.

So this thing moved from a curiosity to a possibility in the late 1960s, and as it became a possibility it started to attract more attention. Particularly the challenge of building a spacecraft became fairly concentrated and resulted in a funded advanced spacecraft development project, which Bill Shipley led. He's a good guy to talk about that phase of it. NASA started to formally recognize this, and there were plans for flights in '77 to all four planets and one in '79 to Jupiter, Saturn, and Uranus—there was a variety of plans for hitting the outer planets. At one point there was a Jupiter-Saturn-Pluto 1977 launch, and a Jupiter-Uranus-Neptune 1979 launch.

Then along came an event in about 1972 when NASA pulled in their horns, for financial reasons. They concluded that the Grand Tour was simply too expensive, and they wanted us to cut it back to something that was more feasible financially. One of the big cost drivers was trying to build a spacecraft that could operate reliably for this enormously long flight time, so the obvious thing to do was to make the flight time less, and you couldn't get to the outer planets with a short flight time.

So we cut the proposed mission off at Saturn, and it metamorphosed over the course of a weekend from Jupiter-Saturn-Uranus-Neptune, under the Grand Tour, to MJS. Because the cutback was all wound up in the budget discussions in Washington, our plans for Uranus and Neptune were kept quasi-secret, and we couldn't tell anybody what we were doing.

When the budget is in its more critical phases, there's kind of a lid put on discussion. This was the rationale for the secrecy: NASA was about to propose something in the next fiscal year budget, namely MJS 77. NASA Headquarters had to give it some technical direction, so if they were going to change this mission into something different, there had to be some technical basis for it.

I remember one weekend we came in and were working on a terminal which was far less advanced than the one on my desk right now. It was the early days of graphics terminals, and we were drawing pictures of this planet with rings around it. People came by and asked us what we were doing. Since MJS 77 was embargoed by the budget process, we weren't allowed to tell them anything about this new mission we were creating, but it was so obvious it was Saturn because it's got rings. No other planet was known at that time to have rings, and we were designing missions that would go to Saturn in a comically secret way.

But something came out of that, which was another key event in the history of Voyager. The bureaucratic and financial forces which came into play said okay, we're going to sell this mission that only goes to Jupiter and Saturn—we'll call it MJS, and it will be appreciably less costly than the whole Grand Tour. Our task was to redesign the mission so it would be scientifically interesting and rewarding within that particular context.

One of the most fascinating elements of going to Saturn was the opportunity to investigate its moon Titan, because at the time it was the only satellite known to have an atmosphere; big satellite, interesting spot. So we came up with a mission which we called JST—Jupiter-Saturn-Titan—meaning that in the process of flying by Saturn it would also have some sort of close approach to Titan, and it would allow us to get a lot more information on Titan than we would otherwise get during the Grand Tour in which we didn't necessarily come close to Titan at all. So it was clear that for the first flight we would concentrate on getting Jupiter science, Saturn science, Titan science.

But then we said to ourselves, "Assume that the first flight is successful; what are we going to do with the second one?" It's already going by Jupiter and on its way to Saturn, to arrive at Saturn three months after the first one. And we said, "What if we got as much as we could reasonably expect to get from Titan; might we do something else with the second one?"

Of course what we engineers wanted to do was to target the second one such that it would go by Saturn and on to Uranus. So we called the second one JSX, where X was a variable depending on what the outcome of the first was. Had the first one failed, then X would have been T, but our hope was that the first one would be successful and X would become U—and sure enough, it worked.

I think that was a marvelous invention, which was one of many things that enabled the Grand Tour that eventually came into being.

What specifically are you referring to as the invention?

The invention was to design from the outset the second mission to be adaptive, such that if the first was successful, there would be no preclusion for the second to go on to Uranus and hence to Neptune. Once you let these options close, they are very hard to open again. We purposely made that option without a lot of noise about it because we were afraid that if Washington saw what we were doing they might squelch it because, "Hey, you're not following the guidelines. We are only going to Jupiter and Saturn; there's no money to go to Uranus and Neptune." Our idea was that once the thing is pointed at Neptune, maybe the money will somehow materialize, which is what eventually happened. So I think of that as one small triumph of the engineers over the bureaucracy, for the everlasting benefit for all mankind.

Let's focus on your personal perspective about that fascinating period, as opposed to somebody else writing about what "objectively" happened. What do you remember from that period about what you were experiencing: your frustrations, satisfactions, disappointments with respect to Voyager?

I was a mission designer at the time. By the time this transition from the Grand Tour to MJS came along, I was the supervisor of the group, which I was in a few years earlier when I was at the very birth of it. I had a lot of opportunity to communicate directly with the program manager in NASA Headquarters, Warren Keller, whose job was to sell this mission.

Warren kept wanting me and the project to commit to more and more in terms of what the return would be. That is, how many satellites can you fly by? Not just how many planets, but how many satellites? How many pictures? He wanted quantitative measures that he could go wave and say, "Look how many planet-sized bodies we're getting on this project!"

I wanted to be fairly conservative about what we could realistically do, because we'd worked out ways in which we could fly by Jupiter and come fairly close to two or three satellites. I thought this was pretty damned innovative, to be able to not only get information about Jupiter and also come close to these bodies that are the size of Mercury, but it seemed like there was a never-ending appetite on the Washington end. Once we said, "I think we can squeeze out two satellites on one flyby and three satellites on the other," they said, "Okay, we'll take all that, and give us more."

On one hand, I felt we were being dragged into more than was prudent and even reasonable to commit to. On the other hand, looking back, no one at the time could have possibly imagined that rich cornucopia of scientific information that came out of these missions. It boggles one's mind to see how much Voyager produced. In some sense, the guys in Washington were trying to drag us more to what turned out to be reality, and I in particular was trying to say, "I'm not sure we can do all of that." So there's one feeling from that period.

Another was a strong sense that this really was a once-in-a-lifetime opportunity. It was a once-in-many-lifetimes opportunity to fly this thing to these far outer planets. It was exciting to be part of it, and the people with whom I worked were excited to be part of it. It really was on the frontier of space, the honest-to-god frontier. As a guy said when I was describing some of the events in my life, he said, "You didn't watch the space program on television." And I said, "That's right, I didn't—I was there!" That sense of creating these ideas in my own mind, that we were able to turn into this event in human history, was heady.

What specifically were you doing?

I was doing mission design. How you arrange trajectories such that you can maximize the scientific observations you want to take. There were tricks. You had a few variables you could play with, having to do with when you launch, when you arrive, what altitude you fly by, when you arrive at the next planet. What we were trying to do was to take those variables and manipulate them to get some excellent observations of the satellites as well as the planets. We had to figure out such things as how to go by Saturn in such a way that you are not damaged by the rings, and then can go on to Neptune or go to Titan.

The intellectual contribution that I was a part of was the design of the mission that allowed those kinds of things; to be able to not come too close to Jupiter where the radiation environment was severe, get close to as many satellites as you can, don't bang into the rings of Saturn, go inside the rings if you need to, or outside the rings, or through the Cassini division, get closeup data of Titan, and so forth.

How did you do that?

We'd run trajectory programs. Initially, in the 1966 era, we'd have stacks of computer printouts, consisting of line after line of numbers. You'd give them to somebody to plot things, or you'd try to plot them yourself, and from that you'd try to infer things about which trajectory would be most effective for what you're after.

As we moved into the 1970s, we were able to write some software that would make this output graphical, so you could put on the screen in front of you a picture of what the planet looked like and what the trajectory looked like that went by this planet. Then we would manipulate the variables at our disposal—flyby distances and arrival times and that sort of thing—such that we'd get a new picture on the screen that was better, in the sense that we were thinking about "better", than the last one. So I messed around at this screen with my colleagues, pecking away at the keyboard, and got up new pictures. Eventually we'd say, "Let's use that one."

Pictured in the AIAA 1972 paper is a spacecraft coming by Titan and then passing by Saturn and going through the Earth's shadow and on. Another diagram shows it going behind Titan as seen from the Earth. That's important because if you can get the spacecraft to go behind a planet, you're able to observe the radio signal passing through the atmosphere of the planet and infer stuff about the atmosphere that's very revealing. So some of the things we were trying to do was to arrange these flybys such that they will go both close to Titan and behind Titan as seen from the Earth, and close to Saturn and behind Saturn as seen from the Earth.

So one of my jobs was designing this stuff. It was a team effort. There is a paper written by five of us. All five of us are still at JPL, 25 years later, mucking around in this area.

What was your relation to Gary?

In 1966 he and I were both in this future projects or advanced projects group, and so we were colleagues. I don't think I was ever his boss; organizations change, and I became the leader of that group, but by that time Gary had left. He had been working here and going to Caltech and back and forth, and eventually left Caltech and went on to be a faculty member at the University of Utah. We were professional colleagues and friends.

How did your work differ from what he was doing?

Gary was more working on details of the trajectory, and my work was more thinking about the entire mission, which included the science, the spacecraft, the communication, and all that stuff. The trajectory was fundamental, but there's a lot of things added on. So my perspective was broader and much less detailed.

How does Mike Minovitch enter the picture?

Early on I wasn't very aware of Mike Minovitch. I didn't see much of him. In the era in which he was really part of this discovery of this multiplanet scheme, my contact with him was pretty minimal. Later in his career I turned into his boss, but it was after the Grand Tour stuff and not really associated with it. Gary was close to Mike and was able in some sense to derive from Mike some principles and concepts that led Gary to discover this four-planet opportunity. I think Gary is actually the discoverer of this four-planet mission.

When he noticed that the orbits of the four planets did line up, he said he was very excited. He had been wondering about going to the outer planets ever since childhood. "Aha, it really could happen!"

That remark about going to the outer planets since he was a kid reminds me of my own background. Before I came to JPL, somebody asked me what would you like to do if you went to work at JPL? I said, fantasizing, I'd like to work on seeing if you could fly something to Jupiter, which was very far away and very ambitious at the time.

So I came here with a propensity towards what would be considered far out kinds of planetary space travel. It wasn't a burning passion, but it certainly was a curiosity: could these things be done? The intrigue of the opportunity that Gary discovered captured me. Could it be done? Can you solve this puzzle? Well, let me work on it a little, see if I can. That was one of the attractions that I had for this kind of stuff.

Do you remember when you actually first became aware of this?

No. Was there a memorable conversation along those lines? Unfortunately not. In fact, if it were not for this 1966 memo, I would have lost a lot of memory of our initial investigation of the Grand Tour. Homer Joe Stewart was the boss of the future projects planning business at

JPL in that era, and I have some recollection of us giving him a presentation on this stuff, and I recall that he was awfully skeptical—as well he should have been, because what I described was a very audacious undertaking.

Now I wish I could free myself from all the constraints that have piled up on me in the subsequent 30 years, so I could think in that kind of fashion now. I doubt that I can. We tend to get ourselves more and more into channels as we go along.

I spoke of the Voyager mission as being conceived, born, raised and flourished all from JPL. A question one might ask is what was there about this place that caused that to happen? The openness with which we have functioned has been a key to it. The openness to ideas, and let the best idea win out—that is an intrinsic quality of this place, which is probably not something that every place enjoys.

The early period is so crucial; the standard questions I'd ask about it were your biggest disappointment, your greatest pleasant surprise, the toughest decision you had to make in relation to these events leading up to Voyager.

The most pleasant surprise was the one I've already talked about, which has to do with the navigation feasibility issue: switching from "probably can't be done" to "probably can be done." Because after you've got the trajectory patched together all the way out there, *the* greatest issue, I felt, was whether it could be realistically navigated. It never occurred to me that you couldn't build a spacecraft that would last for 12 years, or would run on nuclear power, and all those sorts of things. Those, I figured we could do. I just didn't know that it was possible to do the navigation, so a very pleasant and significant surprise was this navigation work on the Venus-Mercury mission, from which we could infer that the Grand Tour could be done.

The most unpleasant surprise was the lack of support in Washington, not necessarily by our immediate contacts in the program office. As "kids" in our 20s or so, it was inconceivable that there would be anybody who wouldn't fall in love with this idea.

What period would that be?

This was in '71, '72 when it switched from the Grand Tour to MJS. That was an emotional setback that was depressing, although we recovered well.

That was that weekend you mentioned?

Yeah, we went in and designed something for MJS, which did sell, and built into it this escape to Uranus and Neptune feature that was subsequently exercised. My personal contribution was being the first to come up with this adaptive quality of the two flights, such that we could eventually go on to Neptune. The covers of *Aviation Week* that have the great pictures of Uranus and Neptune are hanging on my office wall at home because I thought somehow I added one step toward getting those things to happen.

That was an emotional highpoint, too, being able to come up with a plan that met all the bureaucratic constraints and still got us what we were really seeking, or at least did not block us from what we were really seeking, which was to take this once-in-two-centuries event, this four-planet-grand-tour opportunity, and make something out of it. Jupiter-Saturn opportunities occur about every 13 years, so that wasn't a big deal.

What was the toughest decision you had to make? A typical kind of such decision is that either way you choose, somebody is going to be disappointed.

Being a mission designer at that time, I don't remember anything in which there was such a tough decision. I may have been a little too junior to actually make a decision that really adversely affected people. In fact the kinds of decisions I made were usually ones which got you a little bit more, and so everybody got slightly happier, as opposed to some people getting less happy and others getting more happy. At least that's how I remember it, 25 years ago.

What would be an example?

For each new presentation we would add one or two features to the project. Flying by Titan is a feature. Flying behind the rings is a feature. Flying behind Titan as well as by Titan is a feature. Being able to go by satellites of Jupiter and also go by Titan is a feature. With each increment we would try to keep all the nice features we had, and add another one in.

By working on this problem for a long time, we were pretty successful at doing that. That was rewarding. It was like a cornucopia; you could go in and figure out how to get one more piece of fruit out, and you kept doing it. Again, I don't think anyone really imagined the explosion of science that came out of this. The idea that we would go out and discover two or three times more satellites than were known to exist before we started this mission, was something that we just missed entirely. Active vulcanism on Io. All this fancy structure in the rings of Saturn. The terribly intriguing surface of Miranda. And then Triton with ice volcanoes.

Nobody I knew had enough imagination to come close to getting all that stuff in. So, exciting as it was then, it turned out to be an order of magnitude more exciting when it actually came true.

And you played a part in making it come true.

Yes, and I'm damned proud of that. As I say, it was a lot of accidents getting together at the right time here.

CHARLES E. KOHLHASE
Voyager Mission Design Manager

Born August 15, 1935   Knoxville, Tennessee

Charles Kohlhase earned a bachelor of science with honors in physics from Georgia Tech in 1957 and a special master's science degree in engineering from UCLA in 1968. He joined JPL in 1959 following two years in the U.S. Navy as a lieutenant (jg) aboard the carriers Essex and Independence.

He has 38 years of expertise in planetary exploration, with focus on trajectory and deep space navigation design, systems engineering, risk assessment, and overall mission design, planning and scientific return. While on the Viking project, he was selected in December 1974 to be the mission analysis and engineering manager of MJS 77, which later was renamed Voyager. After 15 years with Voyager, he was appointed in 1990 to be science and project engineering manager for the Cassini joint U.S./European mission to Saturn and its moon Titan.

He was twice awarded NASA Outstanding Leadership Medals for his contributions to Voyager mission design. He has written numerous professional papers and popular articles and given many talks and seminars to national and international audiences. His special effects produced for Carl Sagan's COSMOS educational television series brought him a recommendation for an Emmy award. *Spaceflight* magazine acclaimed him the world's leading designer of unmanned space missions.

His other activities include writing, photography, model building, wilderness exploration and preservation, golf, and software development for personal computers. His photographic and creative digital images have appeared in fine arts galleries in Southern California and in national magazines.

> *Rise early and seize each day, learn much and use this knowledge well, spend time with those you love, never abuse your pets, use logic to fight the irrational for it is everywhere, be ethical to the highest, treat your body with respect, defend the environment and its wildlife as a knight would protect King Arthur, meld mind and heart for greatest creativity, follow your dreams, and become all that you can be.*
>
> —Charles E. Kohlhase

CHARLES KOHLHASE

What steered you in the direction of JPL?

I suspect that the answer lies in the fact that when I was young my father was a perfectionist: everything he did, he did very well. He was critical of me; he was very hard on me. He was an excellent cabinet maker. If I worked on a small box or something, if I didn't do it exactly right he would criticize me, and he was constantly after me in that sense.

Well, the only things he did not understand, or was not proficient at, were things such as math, physics, the sciences. I was good in math and science in high school; I went to a military academy, and they had high standards. So when I went to Georgia Tech, my father wanted me to study the practical things that he liked, but somehow I had the courage to rebel. I switched my major from mechanical engineering, which he wanted me to take, to physics. He told me it was the greatest mistake of my life, but somehow I stuck with the decision. I did very well at it, I think because he could not compete with me in that area.

But then, when it came time to look for a job—well, I loved reading, I loved adventure stories, so here I had this ability in science, and what was I going to do? I had one offer to go to Oak Ridge and work in a laboratory on some nuclear thing, but none of that seemed as exciting an adventure as exploring space. When I checked around, in 1957, the only place in the country that was getting into space exploration was JPL. They were talking about the first mission to the moon, and so it was a natural for me. I sent them my resume, talked to them, got a job offer here. I got job offers at defense companies as well, but my dream of adventure won out over building tanks or making better bombs at Oak Ridge. So it was an easy choice to select JPL, even though their salary offer was lower than those of most other places.

So to answer your question, it was because I was good in science and my father wasn't, and I read a lot and loved adventure. That was the bottom line.

How did you get involved in the Voyager project?

I was very fortunate. I came aboard in December of 1974. A couple of years earlier I had been working on the Viking program. At that time Viking was managed by Langley Research Center, but the chief contractor was Martin Marietta. As a result of working with JPL on the Viking program, Martin Marietta had occasion to meet people that they'd like to have leading some of their departments. So they had been making offers to steal away selected JPL people, and that was beginning to annoy Dr. Pickering, the JPL director.

I didn't know this had been going on. I got a call from Martin Marietta wanting to hire me to run a systems division in Denver, so I flew back to talk with them. Their offer to me just happened to be the straw that broke the camel's back. When Dr. Pickering learned that I had gotten an offer, I guess he decided, "Look, this has got to stop."

I suppose you could argue that in the democratic process anybody has a right to offer anyone something, but he felt they were using their intimate knowledge of JPL to pick the quote "best people" away from us. So he called them and said, "Lay off of Kohlhase." I hadn't been out looking for a job; I was happy here. I was flattered to think that the Lab would intervene on my behalf.

I remember talking to Bud Schurmeier or some manager that worked for Pickering, and he basically said, "We realize you have a right, obviously, to seek any jobs you want, but we would hate to lose your abilities and we'd be happy to keep you in mind for some opportunity in the future. What would you like?" And I said, "Well, the job I've always wanted was to be the mission analysis and engineering manager for a flight project." They said, "Okay, we'll keep that in mind."

Two years go by and the mission analysis and engineering manager for what was then called MJS 77 was a man named Ralph Miles. Ralph was having serious physical problems with muscle spasms in his back, and he had to give up the job. I applied for it and was selected.

Whether my selection over the other candidates was a result of this intervening that Pickering did to discourage my being hired by Martin, and then management's sort of unofficial promise or wink of the eye to say, "Okay, we'll take care of you some day," I'll never know. But I can tell you how thrilled I was in December of 1974 to be selected to be the mission analysis and engineering manager of the MJS 77 project.

We had a contest just before launch, when John Casani was the project manager. All of us got tired of saying this mouthful: Mariner Jupiter-Saturn 1977. Let's give it a real name. And so people submitted names, like Nomad, Pilgrim, and Pioneer. There were some stars—Antares maybe—and Voyager was in the list, too. We all voted, and Voyager came out the winner. We liked that name better than any other, although there was a little bit of superstition, little bit of negative vibes, because the Viking program had originally been called Voyager.

There used to be this previous space mission called Voyager that was to go to Mars. There were going to be two orbiter-lander combinations flying piggy back on the top of a Saturn C5 launch vehicle. Because that project became costly, it was finally scrapped and in its ashes arose what became Viking, which was to send orbiter-landers to Mars, but on separate launch vehicles.

We all knew that there had been this prior Voyager project and it had met with its demise, so we said, "Wait a minute. Are we bestowing bad luck on this MJS 77 project?" And people said, "No, come on; engineers and scientists aren't superstitious. It's a great name, let's go with it." So the name Voyager was chosen.

Basically how I got on this project was that it's the sort of thing I always wanted to do, because my forte is the ability to take a complex problem and simplify it to its parts and then make those parts fit together. I'm like a mission architect. I'm knowledgeable enough about each of the systems and subsystems that make it up that, although I might not be an expert in spacecraft attitude control or nuclear power sources, for example, I understand all of those subjects very well so that I know how they should fit together.

So it was the kind of job that I was tailored to do nicely; wanted to do if only given the opportunity. And the opportunity came in the form of the person that first had the job, Ralph Miles, contracting a malady that he and the USC school of medicine couldn't figure out how to treat for the longest time.

Tell me about your early days on the Voyager project.

Those were very exciting times. Paul Penzo, Joe Beerer, and I searched through 10,000 different gravity-assist flight path options from Earth to the outer planets, varying the launch and arrival dates over a large period. We eventually could only target about 100 of these, and of course only two were finally flown. In order to filter the multitude down to the best cases, we had to apply many criteria to each case. We naturally preferred arrival dates for which not only preferred satellites—Io at Jupiter and Titan at Saturn—but as many of the other satellites would be as close to the swingby corridor as possible. Because of the large number of Jovian and Saturnian satellites, we spent a lot of time picking and choosing the best arrival conditions at each planet.

We had to worry about whether the promising options could be navigated with the relatively small amount of corrective propellants aboard the spacecraft. For example, when each spacecraft flew past Jupiter, the huge planet's mass turned the trajectory by nearly 90 degrees. If the swingby should be too close or too far off the perfect path, then we would be deflected too much or too little, causing us to expend precious propellant to get back on course for Saturn.

I remember the first time a few of us did the analysis to try to estimate how much fuel would be required to correct the swingby, the errors in missing this idealized swingby corridor. We sort of knew how big the navigation errors would be, flying past Jupiter and—using physics which we'd had in school, and celestial mechanics and all those things—we could map those errors into dispersions leaving Jupiter, and we could calculate how much fuel it would take to correct those errors. But we also knew that the trajectory flew by some of the large Galilean satellites, and their gravities would have effects on the swingby trajectory as well.

So we tried to calculate all these things. And we came up with a figure that said, "We've got to set aside this much fuel to correct for the swingby errors at Jupiter." But it didn't look like that much and everybody said, "Oh come on, it's got to be more than that." We rechecked the calculations and we couldn't find anything wrong so we said, "Okay, it looks like all we need is whatever it was, 15 meters per second—some amount."

Then I had a few bad dreams at night, in which I dreamed that somehow we did something wrong and that we'd missed some terms, and the actual demands were bigger and the spacecraft just couldn't get to Saturn. And I thought, "Oh God, how would I like to be responsible for overlooking something that ...." But as it turned out, it worked just as predicted.

We took actions to minimize any possible environmental threats, such as the near-Jupiter radiation levels and the near-Saturn ring particle hazards. We avoided solar conjunction, knowing that communications with the spacecraft would be disrupted if the sun was between the craft and the Earth. We even considered small arrival time adjustments to cause key science events to occur over preferred radio telescope view periods. So the final Voyager trajectories were very carefully chosen, with many more conditions applied than done by prior researchers whose primary goals were to identify a feasible space of multiplanet opportunities.

The beauty in this mission was to be able to plan it ahead of time, think it all through, work all the details out, and then watch it happen—and not really be surprised. Of course, there were some surprises in the spacecraft's own behavior as a machine. This was a complicated beast with five million equivalent electronic parts, and it had its own fault protection logic. If things didn't happen the way they were supposed to, the onboard computer appeared to take its own action in some cases.

And because we'd never flown Voyager in space before, after the first one was launched and the second one was launched, there were a lot of little things that happened that shook people up. In hindsight those were mainly just the consequences of setting the reaction levels of the spacecraft too tightly. We had the threshold levels to which it would respond too tight. And just general inexperience in ever flying it in space. But as the months went by, we figured out how to do it and made the spacecraft work very well.

There were surprises in the form of occasional failures. We lost one of the radios on Voyager 2. It was in the process of switching from the primary receiver to the backup receiver, which it wouldn't have done had we sent this command to it; that's another whole story. But the spacecraft was supposed to be able to switch okay. It made a switch from one receiver to the other, and when it came back to the primary receiver, it had shorted out somehow. Nobody knows why. We can't figure out the failure mechanism.

So there were some surprises in terms of that kind of a sudden failure, but basically we succeeded in the planning of the mission and the execution of the sequences, and the taking of the pictures, and sending the data back, and dealing with the increased distances.

When we decided to fly on to Uranus and Neptune, we knew that we'd be going a lot further away, and so the data rates would be lower. The increasing telecommunications distance would mean we couldn't get as much data back unless we compressed pictures onboard or added more antennas to the listening capability of the Earth. Of course we did all those things.

Also the light levels were fainter, so the cameras would have to shutter for longer and longer exposure times. Therefore, in order to not smear images of these satellites as we raced by them, holding the shutter open a long time, we learned how to do target motion compensation. We reprogrammed the spacecraft in-flight and did some things on the ground to make the ground systems better too. We were able to do many things like that because basically we were working with a well-designed machine that people had put together years earlier.

Anyway, that's how I came to join Voyager and I've enjoyed the challenge and particularly the outcome. When two spacecraft travel over four billion miles each and in 12 years see 50 different worlds, what can you say? There's one of mankind's ventures that will stay in the history books for a long time.

Following along with this, what are some of your strongest, most vivid impressions of the project?

Well, my impressions come in two forms. There are global impressions and local impressions. The global impression, of course, is that there was a unique opportunity known as the Grand Tour. It's only available for three consecutive years once every 176 years. In other words, we could have have launched in 1976, '77, or '78. Those are like 30-day launch periods, roughly 13 months apart, because in a year the Earth comes back to its same position, but Jupiter has gone one-twelfth of its way around the sun, so you need another month to catch up again. So these little launch periods were at about 13-month intervals.

But what happened in 1976? The inner planets move around faster than the outer ones, and so as the years are going by, Jupiter is still gaining on Saturn. Well, in 1976 it was so far behind Saturn that the gravity-assist corridor at Jupiter needed to go on to Saturn was real close. The spacecraft would have to get a tremendous amount of deflection. In fact, it was unsafely close. In 1977 it was just perfect. That is, it flew right through the orbits of all the Galilean satellites. And we chose '77.

In 1978 it was still possible, but Jupiter now had overtaken Saturn enough that we wanted very little deflection. So we would have to fly so far from Jupiter with the 1978 launch that it did not come close to the moons at Jupiter. So '77 was the most favorable opportunity of this little three-year set. That three-year set will start again 176 years from 1976, and there'll be three more years and the middle year will probably still be the best one, and so forth.

Well, globally, U.S. technology happened to be properly poised at a place back in the early 1970s when this mission was approved, and we knew about the gravity assist that offered the Grand Tour opportunity. The technology was ready, and we got the funding to use it. That's all remarkable in a way, because that opportunity won't occur again 'til the middle of the 22nd century. And we not only got permission to do it and plan it, but then it was executed flawlessly.

That came about because of, in large measure, a lot of the people that were on the program, particularly in those crucial early years—1974, 5, 6, and 7—were so devoted to designing the spacecraft and the mission as well as they could possibly do. And these people were waking up in the middle of the night or taking showers or driving to work and thinking, "Oh, we ought to do this a little differently or that a little differently." There was a great deal of personal motivation to do it well. So that's a kind of a global good feeling I have about it.

The next memory is that the results were not a disappointment to people. I mean, you saw the tremendous variety of worlds in the solar system. You saw volcanoes on Io, and thousands and thousands of ringlets in the Saturn rings, and unusual features on small, icy moons; just everything you could possibly want to see: geysers on Triton. Can you imagine capping things off with the Neptune flyby, looking at its large moon Triton that's 400 degrees below zero Fahrenheit and actually finding active geysers on it, going up in this very thin atmosphere and being blown by winds and deposited?

Let's face it: our moon is not that exciting a place, right? It's just a dry, airless (no atmosphere), cratered, somewhat dead body. What if we had seen 50 moons like ours? We'd have been a little bit disappointed. But we didn't. We saw Jupiter with its turbulent atmosphere with all the colors in it, from the trace amounts of phosphorous and sulfur and organic compounds and so forth. We watched the wind speeds of hundreds to over 1000 miles per hour. We saw a lot of wonders, I guess, is the way to put that. The solar system had more diversity than people had imagined. So those are probably the two global points.

Now locally, there were all kinds of things that happened. We almost lost Voyager 1 during the launch. Most people wouldn't know about that. The Titan rocket center liquid stage had a malfunction, a small one, that caused the fuel and oxidizer to not mix at the right ratio. And when it shut down, it had failed to give the Centaur the amount of velocity it needed, so the Centaur had to expend 1200 lb extra of propellant to complete its mission. And as you may or may not know, when the Centaur engines finally shut down for Voyager 1 ....

You had 3.4 seconds left?
Yeah, so you did know about that.

Yes, but I'm glad to get it in your words.
There were lots of little high points that had to do with scares or tremendously exciting things. Like that famous narrow angle image taken in March of 1979, 13 million miles from Jupiter, that showed Io and Europa in front of the planet. And you could begin to see for the first time that they had detail on them. You could just make out that Io had this golden color and seemed to have unusual markings, and Europa seemed to look different. We knew every day these would become bigger and bigger, and that finally we'd go by and see them with tremendous detail.

I remember feeling like a kid again, like Tom Sawyer, sailing to an island that he had never been on before and wondering if there's wildlife there? Are deer there? Is there something else there? Well, suddenly seeing these worlds for the first time, with their own special makeup, and knowing that each day that makeup would be revealed more and more—oh, it had us all running into work early and looking at the pictures and everything else. And crossing our fingers that the spacecraft would keep working, because it is, after all, a piece of hardware. Your car radio can fail. Your washing machine can fail. But we didn't want the spacecraft to fail.

And of course, there was the glamour of Voyager. But there's also the hard work, which puts family situations under stress. When the man or woman in the family is coming here, putting in 10, 12 hour days, it's taking its toll on the family. And you're somehow hoping that they will understand that this is important, and it's not going to last forever, and there'll be better times in the future. A lot of us expected the people we loved to somehow understand, and many of them did and many didn't. Many homes broke up because too little time was devoted to your wife or your husband or what have you.

This is the kind of thing that people should be aware of—that these glorious pictures didn't show up on the evening news automatically; people really worked hard and paid a price for them.
They did, they did, and that even happened in my own case some. I didn't spend enough time with my children. They respected me for what I was doing technically, but they missed not having me be more a part of their lives. And that's not good. It's a rare person that can understand and accept that. You should interview Candy Hansen, for example. She'll tell you of people that work on the project that actually remembered when their babies were born relative to the encounters. You see, if you have an encounter that starts in 1979 and one that is in 1989, that span of 10 years—you're going to have a number of children. You're going to have children born, you're going to have people dying of cancer.

In fact, there's a bizarre story: We lost two Voyager science managers. They had the same job. They had the same office. When one died the next one moved into that office. They both died of cancer of the pancreas, which is a relatively rare disease. If you look at the odds like, oh, one chance in 100,000 per decade or so, you're going to get cancer of the pancreas. As a matter of fact, the financial planning assistant in a neighboring office died of cancer of the liver. I had cancer of the thyroid gland, but I trace that back to radiation that my parents okayed a dermatologist to give me for acne many years ago.

But this, the cancer of the pancreas and liver thing that happened a few years ago, began to frighten some of the people that worked in the area. They said, "What's going on around here? Are there cleaning solvents somewhere in the closets, or radiations coming in we can't see?" In fact, I did a quick analysis on it, wrote it up, and we brought some people in to examine the area, and we never could find any cause for it. But nevertheless, people often wondered about that. But the point is, people die during the course of the project. People were born. People married; people divorced. All these things happened. Life is still going on during this time, as it is with all occupations.

There were exciting times. I appeared at a Star Trek convention one time, speaking to 4500 people in one audience. I was scheduled to appear between William Shatner and DeForest Kelley. That was a thrill. Appeared on *Saturday Night Live* one time. Squired Angie Dickenson around a few times. So you have everything. You have the lows when you're working Saturdays and Sundays and things aren't going well. You have the glamour when things are at the top, and you're on top of the world. It's all there.

But the skills that I think proved to be the most valuable were exhibited by those people that had what I would call rational imaginations. You obviously have to have technical training to simply know what to do, but you also have to have the imagination to know what you want to do. Or what would not only be exciting, some worthwhile goal which you can imagine, but you must know technically whether that goal is reasonable, that you can expect to achieve it. A lot of people that worked on Voyager were constantly thinking, "Let's see, what can we do to make this better?" And "Is that going to work, from our knowledge of whatever it is—orbital mechanics, spacecraft design, how fast the cameras can slew, where we might find a new body, whatever?" "How can we make this mission better?"

And there was constantly a core of such people, though obviously not all the people that worked on the project, when you realize at one time there were as many as 3000 people in the program a couple years before launch. Whereas, during the graveyard cruise between the planets, I think we got down to 100 at one point in time. And the Neptune encounter had maybe a couple hundred people onboard. Maybe more, maybe 250; can't remember those numbers exactly. Obviously not all those people were equally devoted to the project, but there was always a core of people that energized the group, and that core was very devoted.

That's probably the case with most projects. If you're designing a new Ferrari body in Italy, there will be the ones that are energized with ideas and how to do it. Then there are the others that convert those into blueprints or what have you, and they're just doing their jobs. Voyager was no different, but the energized group was really energized, and the other people then tended to support that group.

I'll always look back on Voyager well. It was a great endeavor. Didn't hurt anyone. I suppose, other than the launches, it didn't pollute the Earth either. Didn't start a war. It gave the United States a good image; that is, the national prestige of having pictures of the planets on the covers of everything from *National Geographic* to French and Egyptian magazines. I've seen Voyager results written up in about every language you can imagine. The world got excited about the results coming back from the outer solar system. I recall the fervor with which they published pictures and articles and stories in magazines in countries all over the world. People liked it.

And it showed, in fact, that you could carry out complicated challenges if you wanted to. There are a lot of things we could do on the Earth to make it better that we aren't doing. It's not because we don't have the ability to; it's probably because the politicians aren't spending the funds and endorsing it, or perhaps big business doesn't see a profit in it. We could clean up acid rain if somebody really wanted to. They just seem to want to study it, that's all. That's the problem.

The difficulty I have is in saying things that don't sound trite. I feel it's inappropriate for me to be patting our backs, and yet I know that what happened was a historic event; it really was. And the only thing I can say is a lot of people worked really hard, took some personal losses, but made it all happen. And there's certainly an application. If there was ever an argument for the value in education, this is one.

You can sell cars and real estate with a minimum of training. Don't quote me on this, but real estate agents and title insurance companies are overpaid. I used to joke years ago and say that if you wanted to determine whether one person's salary was in the right relationship to another's or not, you took the ratio of the logarithms of the amount of time it took each person to do the other person's job. How long would it take for me to become a good bank teller? I could probably learn how to do it in two or three days. For sure in a month, I'd be a pretty good one.

How long would it take a bank teller to be able to design a Grand Tour mission, assuming they could ever do it? Well, they'd have to at least understand math and physics and astronomy and celestial mechanics. And so it would probably take them four or five years just to get the technical skills to go and apply them. Five years is what, 60 months. The ratio? Now, I shouldn't be making 60 times the salary of the teller, but maybe the logarithm or cube root of 60. I should be making three or four times their salary. So when people say, "Oh come on, that poor teller is only making $15,000 a year and you're making $60,000 a year, how do you explain that?" Well, I just gave you an explanation.

I don't know why I strayed on to that.

We were talking about education.

You read a lot that the educational competency of particularly the high schools, the schools before you get to college, has really declined in the United States relative to the other countries. Now we're ranked way down. And I often thought, "Oh come on, it can't be that bad."

But I was a judge at a science fair about two months ago, and if I had not been told the grade level of the students that were submitting the exhibits, I would've said fourth grade. They were eighth grade. I was horrified. When they wrote on their cardboard placards what they were doing and what they expected to show, I don't think a single one that I reviewed had gotten through anything they'd written without some grammar or misspelling errors. Their English was rotten. Their logic often was flawed. It was appalling to see what had happened.

It has been argued that one of the benefits of the space program is that missions like Voyager excite the young mind—the young kids 10, 11, 12 years old—and may provide just enough spark that they are the ones who somehow get through high school and go on and do better. So one of the benefits of a technical achievement that is also exciting to the mind, to the imagination, is that it helps slow by a teeny bit our decline in educational levels.

It also says that there are some occupations where you don't go to school just to have the credential and get through with it, but that you can really use the training; it could be applied. I think that's true with Voyager.

I have certainly had a great time working on the program, and I'm sorry to see it end. I feel like I'm back in the trenches now on a new project and will have to kill myself for five years until we launch it. And kill myself for another five years 'til it gets to the destination. And there's no guarantee it'll work as well as Voyager did. But at least, having been on Voyager, as I told you earlier, I had that part of my needs satisfied. There are other people working on the new project that I'm working on now that have never been through a successful project. Their hopes are all applied on this next one. Mine are too, but it's not as easy for me to become as excited about it, having just come through Voyager.

What were some of the surprises of the Voyager project, the unexpected events both pleasant and unpleasant?

Okay, one of the unpleasant ones would be the Titan rocket anomaly during the Voyager 1 launch that had us all chewing our fingernails. I told you that the Centaur didn't get enough energy from the Titan, but it had a 45-minute coast around the Earth before it fired up again. So we all had 45 minutes to try to guess whether there was enough fuel left on Centaur to make it. There was, with a few seconds to spare.

I was down at the command center at the Cape with a headset on, and I could hear the people at General Dynamics in San Diego who make the Centaur—the launch vehicle contractor—talking and making calculations and basically saying, "We think it's got enough." And John Casani and I were sitting there, saying, "Well, we hope you're right." And then having the Centaur have a normal guidance shutdown as opposed to burning to fuel depletion. See, if it had run the tanks dry, that would have meant it had not quite achieved the velocity it needed, but fortunately it shut down on a normal guidance command. That was obviously a scary moment.

The loss of the Voyager 2 primary radio receiver was bad. To deal with that, we designed a backup mission load and put it in the spacecraft's main computer and just let it hibernate there, so that if we lost the one surviving receiver, this little backup mission load would activate when we needed it. And when we got 50 days out from Saturn, it would sequence the spacecraft and send back scientific data.

We didn't like to find ourselves in a position where a single-point failure would do us in. As long as we still had two of something, we didn't worry so much. If we lost one, well, we had a backup. But one receiver was already dead, and if we lost the second receiver the spacecraft would never hear from us again. It wouldn't know what to do and basically its mission would then be cut short.

What else? There weren't many bad things that happened. We missed some pictures at Saturn on Voyager 2 when the scan platform stuck. Right as it stuck, we misguessed, misdiagnosed what the problem was, and it took us weeks to find out what the real problem was. See, we had had a previous scan platform slow down—I don't remember which spacecraft it was on—from a teflon screw dropping into the gear teeth of the drive chain on the scan platform. And the solution to that was just to run the gears until we had smashed it out.

Well, when the Voyager 2 scan platform stuck, people said, "Oh, probably more teflon. Let's send it a bunch of commands and just work it out." But we couldn't get it to move, because that was not the cause of the problem. We had been running the platform at high rate, and a lubricant had migrated away from this little gear-to-shaft interface. Imagine a little shaft about the thickness of a toothpick with a little gear spinning around on it. That thing evidently warmed up enough that the lubricant migrated away. Then the surfaces became drier, which led to galling and a certain amount of friction, and the surfaces began to expand and catch. So the worst thing you could do would be to run the unit at high rate.

We then restricted the rate at which we could slew Voyager 2's platform. We moved it at only low rate. And once we finally let the thing alone, let it rest, the lubricant migrated back in again, and we found if we ran it at slower rate, it didn't hang up anymore. So we did that. But when it first stuck at Saturn, during a sequence to look at one of the moons, that was a shock. Everybody said, "Oh no, what's going wrong now?" That was one of the low points.

The high points. There are far more high points than low points. The discovery of the volcanoes on Io was a high point. And you know the history behind that discovery.

Please repeat it for the record.

Well, the optical navigation people were having some difficulty. The way they navigate is they look at the picture of one of the natural satellites against the background of stars. They try to measure where the center of the satellite is relative to these known stars, and they use that as a data type to solve for where the spacecraft is, on its flight path coming into the planet. Because if the spacecraft is off-course, then the satellite's relative position to the star background will shift a little bit and they can tell where they are. And they were having trouble fitting the limb of Io, because it had a bulge on it.

This young lady, Linda Morabito, asked, "What's going on here? What is that lump there?" And the more she looked at it, she said, "That's not a blemish or a flaw. That must be a volcano or something." She was not a member of the imaging team. She then called someone on the imaging team and said, "Look! What do you think this is?" And they said, "Oh my gosh, we think it's a volcano!"

Then they began to look in other pictures of Io and discovered other volcanoes. And they found eight active volcanoes when Voyager 1 flew by and looked at Io, and a ninth one when Voyager 2 flew by. But all nine weren't active at the same time. The largest volcano, Pele, erupts to a height 30 times the height of Mt. Everest, and the fallout zone covers an area the size of France. If you think Mt. St. Helens was impressive, it's just a little pipsqueak compared to Pele. So the volcanoes of Io were exciting.

Voyager responded satisfactorily to these situations because of your concerns, your efforts, before launch.

Yes, but I was also concerned during the actual flight of the spacecraft. With the small staff that I had, we did worry about mission risk and mission contingencies. For example, when we were going to Neptune we knew many months before we got there that Neptune had rings around it, or at least ring arcs—it turned out later they were complete rings. We felt that Neptune probably had a magnetic field and trapped radiation belts like Jupiter, Saturn, and Uranus had. We didn't know how strong they were. They might have been a problem.

We also wanted to fly very close to Neptune to get enough gravitational deflection to reach Triton five hours after the Neptune closest approach, which would mean we would be coming fairly close to the upper atmospheric fringes of Neptune itself. And so we tried to calculate where we could safely aim with respect to all these concerns and try to assure people that the risk we were taking was small, and that it was worth the risk.

We also asked ourselves questions like, as we're approaching Neptune, will any of our measurements tell us something about the environment so that, if we have to, we can maneuver the spacecraft? So all of that was part of dealing with environmental risk.

But there are still other contingencies. The issue of contingency planning goes far beyond just an unknown environment out there somewhere. We also had to worry about things like, what if there's a major earthquake in Southern California and it knocks out JPL and the tracking station at Goldstone? What are we going to do? Actually that's an environmental risk, only it's here; it's not there. As you may know, the probability of a major quake in Southern California is about two percent per year, and we tended to worry about threats that were at the one percent or greater level.

You can't worry about everything. You can't worry about something that has a one chance in a million of happening. I don't think a herd of elephants are going to run through the laboratory and smash all the computers that send the commands up. But what we said is, "Let's think of everything that has a one percent or greater chance of happening to us in some period of time, and then decide how would we try to bullet-proof ourself to that."

In the case of earthquakes, you might say there's nothing you can do about them. Well, if we were counting on generating an important load of instructions that were going to be transmitted from the Laboratory or from Goldstone, you could argue that we should generate that tape ahead of time and be able to send it from one of the overseas stations, if our facility here was knocked out of action. So we did things like that.

We also worried about spacecraft hardware. Obviously it can fail. Voyager seemed to work very well, but that would not be surprising if you went through all the reliability analyses that had been done before launch, where data existed on parts and circuits and things, that had been produced and operated for certain periods of time. There were test programs at different places that would be able to tell you that if you leave on this high-voltage photomultiplier tube for more than 6000 hours or whatever, you have some chance of losing it.

We took all of that data, and we knew that several spacecraft subsystems had more than a one percent chance of failing in a period of a year. The CCS—the Command Control Subsystem—did, the flight data subsystem did. The tape recorder? There is only one tape recorder aboard each spacecraft, and so we didn't have redundancy. We couldn't lose one tape recorder and then just go to the other. But the tape recorder turned out, ironically, to be a very reliable device, even though with your own tape recorder at home, you're often running back to Circuit City or wherever to get the thing fixed.

Was that made in-house here or was it an off-the-shelf manufactured one?

It was not made at JPL; it was made at a company called Odetics. They do really quality work, and they delivered us some exceptional equipment. The tape never broke; the heads didn't wear out; all worked fine.

Anyway, we looked at different parts of the spacecraft and we asked, "What if we lose half of the central computer memory? What are we going to do?" You always make sure you sequence your most valuable observations. Just because the computer program that is sequencing the spacecraft has been cut in half, doesn't mean the science value return is cut in half. Because, if that program is running from a day before closest approach to a day after closest approach, doing a lot of things, if I tell you the most important Triton observations all happen between three hours after closest approach and six hours after closest approach, you can bet I'm going to make sure those are in my short program.

So it often is not that painful to do contingency planning. The scientific community might tell you it's really painful, but you can do a certain amount of planning without harming the science return very much. So we did that.

We obviously had to worry about the X-band frequency we were transmitting all the data back over, because it's vulnerable to rain. So if there's a thunderstorm over the Spanish antenna right as you're beaming down your most important data, you have a harder time picking the signal out of the noise, because as the X-band radio wave passes through water molecules, they oscillate and it adds noise to the signal. Well, you might say, "What in the world can you do if there's rain in Spain? Aren't you just out of it?"

No, not necessarily. If you put some of your most important data on the tape recorder at the same time you're sending it down in real time, if you have an outage on the ground, then you can always play back the tape recorder later and get it back. In fact, that strategy is actually kind of clever for both a tape recorder failure and a ground failure, because if you've got some real important data and you're sending some of it down in real time and some of it back later from the tape recorder, you're bound to get it sooner or later.

We rarely did contingency planning for multiple failures. In other words, it's just too improbable that two one-percent events would happen at the same time. The way probability works that would be 0.01^2, which is only one chance in 10,000. And so we did not feel we had to cover double rare events.

Typical tasks you were involved in?

Again, before launch, lots of design team meetings, lots of time spent developing requirements on the different elements that comprised the project. Lot of time selecting the flight paths, lot of time making sure we could navigate them. Things like that.

After launch, the time was spent putting together these mission rules. We're on the way now, and we're going to have to navigate from time to time to control the flight path. But as we approach each of these planetary systems, we need to have previously developed all these computer loads that will sequence the craft, and point it, point the platform, take the pictures, send the data back.

My office—the mission planning office—got the earliest start in developing these sequences because we had to put out the rules or constraints within which they were to be developed. There's a boundary or envelope within which you've got the resources to implement this certain job. The science community may want to do 10 times more things than we can afford. And on the sequencing team, someone who's worried about making sure all the bits are right in the spacecraft computer doesn't necessarily have a global view of whether the fuel's being used up too fast because he or she is sequencing a lot of roll maneuvers. But using fuel too fast means that we may look great at Jupiter but we won't have enough fuel at Saturn or whatever.

So the mission planning office looked at the whole mission and then allocated what we could afford to do at each of the planetary encounters along the way, so that we could still get to the last place. And then we took what we'd allocated for Jupiter and broke that down into a set of rules that said, "Okay, you can develop 10 flight sequences. They're to execute between 80 days before closest approach and 30 days after. You shall not use any more fuel than such-and-such an amount, or you can't use the tape recorder any more than such-and-such an amount, or even the television cameras." We found out that the total on-time for the cathodes in the cameras led to a degradation such that you just couldn't leave the cameras on forever.

Now, we here on the ground had to adapt, too. As the spacecraft had failures of certain kinds, then our rules might change. For example, we lost one of Voyager 2's radio receivers, remember, so now it only has one radio receiver left, which means it has a single point of failure. So my office said, "That worries us. We could lose this mission; if we were to lose that other receiver, it'd all be over." So we put out guidelines saying a backup mission load will be developed, and it will occupy 750 computer words inside this computer memory that's sequencing the spacecraft.

In other words, we took that much space away from the nominal mission. Where the science and sequence people would like to have stuffed everything in the computer memory they could, we said, "Look, you can't have all the space. We're going to take a little space over here. We want to develop this other load and put it in there, and it's just going to stay there, forever, so if we lose that other receiver, it will at least get us some return from the next planet. Now, it may be costing you a little bit on your nominal mission, but it isn't costing you that much. Maybe your science return is 95 percent what it could have been. But if we lose that receiver, you're going to lose the next planet altogether without this backup load."

The receiver that survived was not totally healthy. It had a kind of hearing problem. The range of frequencies it would respond to was too narrow. If it had its normal bandwidth of 100,000 Hz or cycles per second, where it could lock up—then, if a command from the Earth came up and it was within several thousand hertz of the rest frequency of the receiver—it wouldn't have mattered, because it had a tracking loop that would just go over and grab it. But the tracking loop capacitor had failed, so that meant that if these commands didn't get in within 96 Hz of the rest frequency, the spacecraft didn't hear them.

To show you how little that is, the rest frequency of the receiver would change by 96 Hz if you just changed the temperature in the bay in which the receiver was mounted. If you changed it by half a degree Fahrenheit and didn't know it had changed by half a degree Fahrenheit, then you'd have the wrong frequency. Which meant that every time that we did maneuvers that pointed the side of the spacecraft at the sun and just warmed it a little bit, then it changed the rest frequency in that bay. Or if we turned things on and off that were just dissipating different amounts of watts in an adjoining bay, that could change the lock-up frequency.

So we had to develop rules for saying that if you do these kinds of activities, they are apt to change the best-lock frequency of the spacecraft. And we told the sequence design people, "Therefore don't schedule the next big load of instructions coming up from the ground 12 hours after the spacecraft has just done this trajectory correction maneuver, because the temperature in that bay will not have settled back that quickly. And when you go to load that next load, it won't get in, and then everything will be screwed up."

So we established rules known as "Command Moratoria". These were time periods ranging from 12 to 72 hours, where, depending upon what had been done to perturb that frequency, then no commands were allowed in a period of some number of hours following that. So when they built the sequences, they had to honor that. That's what I mean by a constraint.

It's like telling someone to build you a house. You have some carpenters and painters and you say, "Go build me a house, but, by the way, I've got a few constraints. It can't be over two stories high, it can't be over such-and-such an amount of square footage. The paint you use has to be resilient to acid rain because I'll live in an area that's near a foundry. I want all my wires grounded."

Well, the mission planning office basically said, "Here are a bunch of rules to follow, and as long as you follow these rules, you can design any sequences you want. But just watch out that you don't do too much of this, or whatever." And that was primarily what I did. Those rules came out first, then the science teams decided what observations they wanted to take. Then the sequence team looked at the mission constraints and the science observations, and they built a sequence that would do as much science as the science team wanted and not violate these rules.

These rules were called "Mission Design Guidelines and Constraints." They were in a big thick book. It was organized in chapters, and it covered the Jupiter phase, the cruise, Saturn encounter, and so forth. You could then open it and you would find things like when the encounter was to start and end, how much of this and that you could do, and so forth.

Then the sequence team had the nitty-gritty work of running these big ground computers to design the sequences, converting them into loads of instructions that went up to the spacecraft to be executed. But there was a checking process. As they would develop sequences, those sequences were reviewed by my office to make sure they had not violated mission design guidelines and constraints, and by the science office to make sure they were capturing more or less the amount of science they had in mind. So that's the essence of it.

It may be familiar to you, but it's mind boggling to me the amount of forethought and planning that went into it.

Well, we also try to honor little things like not having major spacecraft activities occurring on Thankgiving holidays and Christmas holidays. Suppose you have an emergency then, and half your people are off somewhere visiting relatives and having turkey. So some of the mission design rules imposed what were known as "quiet periods" on the design of the sequences, and these periods would include major holidays. For example, the time between Christmas and New Year is not the time to be running the spacecraft through its most hectic paces.

We would also tell people to not try to do too much with the spacecraft when the sun was in between the Earth and the spacecraft, because the radio signal picks up a lot of noise when it goes by the sun. If you get an emergency situation during that time, you may not be able to reliably command the spacecraft. So when you think about planning, you mustn't overlook any little thing like that or it'll get you, sooner or later.

That intrigues me. When you read mysteries or you see movies where a criminal is trying to plot the perfect crime and he's trying to think of absolutely *everything* so that he can pull this off without something going wrong—a mission planner's job is like that, only it doesn't have an evil intention. But you ask yourself, "Have I overlooked something?"

We would look to see if any other project had something important going on. Suppose there's a multimillion dollar spacecraft over here, and it's firing its engines to go into Mars orbit on a certain date. Then our guidelines would say, "Don't do any spacecraft maneuvers during this time period." Now, there were times when *our* times were so important, we would not give into anybody else. Obviously, if we are close to the Neptune encounter, I'm not going to be saying, "Don't do anything."

At such times we are issuing rules that are going out to the rest of the world. What we did not want to have happen is to have radio frequency interference occur right as we're going by one of the planets, beaming our data down, and the Deep Space Network is down here trying to receive it, and geez, there's another spacecraft in low Earth orbit, right in the way, operating near our frequency, and we lose the data. So we put out radio frequency interference requests to not communicate on certain frequencies at certain times.

These requests had to go through the State Department. I think that there was even an instance of a Soviet Earth-orbiting craft of some kind that turned off transmissions; the Cold War didn't seem to intrude in these scientific ventures. You might argue that we shouldn't publish the periods that are the most dangerous for us, because somebody will sabotage us, but that never happened.

So we're not always yielding to somebody else's high activity. There are times when we're saying, "Look, this is real important for us." In fact, during those times, we captured most of the large antennas on Earth. We said, "We want them for ourselves. We'll only let other people have tracks just to maintain survival tracking. That is, you'll get just enough coverage so that you can keep monitoring your craft, but we don't want you doing a lot of hot activities." So it went both ways. Right around encounter, Voyager really had a lot of influence over other people's activities.

In fact, months before we got to Uranus, we knew we were going to fly by Uranus on January 24, 1986. We designed it that way. Then I looked up the shuttle manifest and found out that a shuttle was due to launch on January 20, four days before we were to fly by Uranus. Well, since we would have been flying from 1977 to '86, eight and a half years—we've been trying to get this Voyager to Uranus for eight and a half years—it would be dumb if we launched a shuttle right at that instant. If the shuttle had an emergency, because of the presence of people onboard, it could take some of our tracking support away, right when we needed it.

So we tried to get the Challenger shuttle launch slipped so it wouldn't be on top of our Uranus encounter. We made this request through laboratory management, and they agreed that the shuttle should be slipped a few days. It went all the way to NASA Headquarters, and it finally went up to Beggs who was the administrator then. Everybody under him said, "Good idea. We ought to do it." But Beggs said, "Request denied." They asked him why. He was purported to have said, "The White House doesn't want the launch slipped."

That fact never came out in the press.

As a matter of fact, there was a plan that McAuliffe would cause a little Voyager command to be sent from one of our control centers here. Reagan could easily have said, "Yeah, I wanted to make a big spectacle in space, but I would never have compromised safety if I'd thought that putting pressure to launch the Challenger would've put anybody at risk. Obviously, I would never have done that." So he could have had an easy response. But anyway, our request was denied.

As it turned out, the Challenger launch slipped anyway, to January 28. It slipped just past our encounter and then, much to everyone's horror, when it launched it did fail. There were some hundred news reporters out here in von Kármán auditorium—three-quarters of them then ran and jumped on airplanes and flew to the Cape to cover the tragedy. So in a sense, not only was that tragic, but some of the glory from the Uranus encounter was diminished by the departure of the press to cover hotter news.

What were your most important decisions, toughest decisions?

Before launch, as I said earlier, one of the harder decisions was sizing the propellant tanks to navigate the spacecraft. Because it was a situation where you had to have complete faith in analytical methods—that is, nobody'd ever done this before—and you had to write down the equations and make the calculations that would say, "If I have a certain navigation error as I swing by Jupiter and get this tremendous deflection by this planet, what's it going to cost me to correct that?"

I had to decide and finally say, "This is how much you need." Remember, we can't keep asking for large amounts of propellant because we couldn't launch that big a payload. Again, it's a tradeoff. Putting my signature on something saying, "This is enough to get us there"— I worried about that. That's the primary thing that comes to mind before launch.

We played this on the safe side. Io held a high priority at Jupiter, because it had been observed to orbit within a torus of sodium atoms and a bunch of interesting things that we don't need to go into. But in order to fly by Io, the first swing by Jupiter would have to be in as close as five Jupiter radii. And we knew that the radiation dose that Voyager 1 would get would be pretty hefty. It might knock out some hardware.

So we said we better design the Voyager 2 flight path for a safer flight by Jupiter. We targeted it to go by at 10 Jupiter radii instead of five and still also have the property that it could go on to Uranus and Neptune if circumstances permitted. That wasn't a hard decision; it seemed like a smart decision. But it did mean that if Voyager 1 failed before it flew close to Io, Voyager 2 could not have a close Io encounter. It would basically be outside of Io's orbit when it went by, whereas Voyager 1 went right inside its orbit.

But that guaranteed that if the radiation belts of Jupiter were a real problem—and Voyager 1, let's say, got damaged enough that it could not last the additional years to reach Saturn—that Voyager 2 would probably survive the Jupiter encounter and would get to Saturn. So there's a case where we really put Saturn's interests ahead of Io's interests, rather than repeat another Io flyby. Even if we'd tried to do that, though, the two arrivals would have been too close to each other anyway. We'd have had operational problems.

After launch, that's another story. We had a big surprise. One of the things that went wrong, nobody realized what happened. When we sized the propellant tanks, we sized them to be able to produce a certain amount of velocity change with the spacecraft. In other words, what the navigators really do is to say, "If I get off course, I've got to change speed by so many feet per second, or meters per second, to get back on course." Now, how much propellant that corresponds to is a function of the specific impulse of the thrusters and the mass of the spacecraft, and what have you.

But our pure commodity was delta V. That delta is a mathematical symbol meaning change; V is for velocity. So our term delta V meant how much velocity change ability does the spacecraft need to make this trip. So we specified delta V, and it turns out we specified the right value. But that mapped into a certain tank size. We launched the Voyager 2 spacecraft first. We did our first maneuver to correct the launch vehicle injection errors about 10 days after we left Earth. And we consumed the amount of propellant we thought would be required to get the velocity change that our tracking said we needed to correct for these injection errors, but it wasn't enough.

People said, "What is going on?" They figured out that the spacecraft is this box-shaped thing with an antenna on top and some tanks back here. There are still some struts hanging on the back of the spacecraft where the spacecraft and kick stage attached to the Centaur. Now, the trajectory correction thrusters fire to make maneuvers, or to change the velocity of the spacecraft, and it was known that some of their exhaust would impinge upon these little struts and that you would lose a certain efficiency. It was calculated as a loss factor. If you were the thruster looking out at the world, you'd see these wrapped struts blocking some of your exhaust.

Everybody said, "We're going to lose seven percent from gas impingement on the struts." That is, one pound of fuel will give you an amount of velocity seven percent less than it would if you had nothing out there. Well, it turned out that the people who made those calculations—and they weren't my people, they were in another division somewhere—underestimated the effect. And in fact, the loss factor was 19 percent.

That is a big difference. If you've got a tank that's supposed to hold 230 lb of N2H4, which is hydrazine, and you think that produces a certain amount of velocity, and now you find out it's 12 percent shorter than you thought it was, we suddenly found out that we were marginal in having enough propellant to get to Saturn, and we were negative for getting to Uranus and Neptune. So we said, "What can we do? We're on the way."

Well, we thought of this clever thing to do. I mentioned that if you miss the navigation corridor you get these dispersions you've got to fix. We had originally planned to fix them up by doing a maneuver at Jupiter plus 70 days. We were going to track the spacecraft until we got out a little over two months past encounter and then correct the trajectory. Well, knowing something about the laws of physics, we knew that if you wanted to correct an energy error or a flight time error to Saturn, you can correct it a lot more efficiently when you're moving fast than when you're moving slowly.

So we said, "If we could do that maneuver earlier, near Jupiter closest approach, it won't cost as much delta V as it would 70 days later." In fact, it'd cost maybe one-third as much—big savings! But people said, "Wait a minute. If you do the maneuver back here, you're doing it right in the middle of all these science observations. And if you're turning the spacecraft, the antenna won't even be pointed at Earth anymore, and you'd have to put all that data on the tape recorder."

Well, we found out that basically you could stay Earth-pointed—it turns out the Earth is back in this direction—and not do any maneuvers to lose Earth lock, and fire the engines on Earth point, get this correction done, and not interfere too much with the science. We still worried about it a little, but it all came off well.

Then there was some concern about flying near the rings of Saturn. We flew outside the main rings but there was always a fear that you might get hit by a particle or something. But at Neptune—we were coming very, very close into Neptune. When we made all the calculations, it looked okay, but you asked what did I worry about? I worried a little bit that maybe we'd estimated the risk without sufficient conservatism, even though we felt pretty confident.

Two final topics. How Voyager differed from other projects, and lessons learned and significance of the whole thing.

Okay, the differences. Of course, each project the Lab does tends to get more capability than the last one. The payload's bigger, and there are more scientific instruments. It's more complicated. The data rates were much higher because we went to X-band frequency. We were transmitting 115,000 bits per second from Jupiter, while some of the early missions in the 1960s were sending back only 16 bits per second. There's just no comparison. The telecommunications performance is way up, the part count: five million equivalent electronic parts on each Voyager.

The way Voyager differs is it's a more sophisticated machine. Also, it differed a lot from the Pioneer spacecraft, which were spinners, because they could not store long sequences on board and then execute them later when the clock counted down. You pretty much had to just send real time commands up to the Pioneers and say, "Do this, do this, do this, do this." And so it was easier to cause Voyager to do more things than it was with Pioneer.

In terms of lessons learned, that's the funny thing. Normally, any time you do some activity that's at all complicated, in hindsight you say, "Ah boy, I would've done all these things differently," because hindsight's always better than foresight. The unique thing about Voyager is that even in hindsight, after all the years have gone by, there are very few things we'd do any differently. This is a good feeling. We did them well enough the first time around that there are very few regrets.

Now, you could say, "We shouldn't have lost that radio receiver by allowing ...." I didn't tell you the whole story on that, but the spacecraft has its own fault protection logic, like the HAL computer from 2001. And one of its fault protection routines does the following: if it hasn't heard from the Earth in some period of time—and you can set that; seven or eight days—it assumes that something's gone wrong on the spacecraft, like maybe its receiver has failed. So it decides to switch to its backup receiver. People on Earth are supposed to be talking to it all the time. If it doesn't hear from them, its onboard logic says, "The receiver that I'm listening on must not be good; I'd better go to this other receiver." And so it switched over.

Now, it was designed so that it *could* switch over and switch back; it shouldn't have failed. But there are still a lot of people kicking themselves. It's like turning a lightbulb on and off. It would take me too long to tell you why this isn't a good example, but if a lightbulb you're using is working and you got another lightbulb right next to it that is not on, you rarely, just for the heck of it, turn that first bulb off and turn the second one on and then go back to the first one, because you risk a chance that something might happen.

And so, although we had this fault protection logic onboard, we would prevent this automatic switchover by sending, every week, a little trivial command. And what that did is it reset this command-loss timer and it gave you another week, as long as you sent it. Well, we were preoccupied with a problem on Voyager 1 at the time the Voyager 2 command loss timer ran out without getting this little reset command, and then went through the switching, and then, whomp, we lost a receiver.

That didn't happen to be my responsibility; the operational crew was supposed to regularly once a week send this command up. Because Voyager 1 was having a problem, its scan platform was sticking from this teflon that was down in the gears. It had most people's attention, and they just forgot to send this little once-a-week command to Voyager 2.

Now, if the spacecraft had worked as designed, it wouldn't have mattered. It would've gone to the backup receiver. It wouldn't have failed, and then after some number of days, it would've gone back to the primary receiver and it would've kept working. But it is true that the failure coincided with this onboard switching back and forth, and therefore there are a number of people kicking themselves for not having sent up that simple command.

But in hindsight there were few things that we would've done differently. The basis for saying that is the excellent scientific return. The missions were very successful. There was almost no way to get any more information out of them. When the scan platform on Voyager 2 stuck 100 minutes after Saturn's closest approach, we lost a few pictures of Tethys. It wasn't like we didn't see it at all; we just didn't finish the complete mosaic.

Nobody foresaw this very subtle lubricant migration problem on the actuator's tiny gear shaft. You can say what we should've done differently is build a better motor so the scan platform would never have stuck. Well, they built the best one they could. But other than the loss of the receiver and the stuck scan platform—we lost an instrument, I think on Voyager 1, with the Jupiter radiation dose. But we knew that was a risk. We wanted to get in there and get Io, and we saw the volcanoes on Io. It was all worth it.

I tend to think that we stuffed our sequences too tightly, but you can't knock success in hindsight. People were complaining and working hard and extra hours to try to fit in every observation they could. But they made it all work.

Voyager was able to do a lot of special things and yet was not so complicated that we couldn't fly it. It took us a little while to learn how to fly, but it's not some machine with millions of words of computer memory, processing thousands of instructions and everything. It is far more sophisticated than early spacecraft, which lets it do a lot of things, but not so sophisticated that we humans couldn't still understand and keep up with it. It's just about right. Good balance. We have to be careful that some of the future spacecraft we're designing don't become too complex to fly.

How about lessons from Voyager that might be applied to other projects?

Sometimes it's hard to give advice without sounding trivial. The things that come to mind may sound so obvious.

The most important thing of all is choose a small team of highly motivated, creative, technically sharp people who all understand the goal, who will work together and make decisions fairly quickly, who will not be shackled by excessive bureaucracy, by lots of status reports to various and sundry management units. They have to be able to make decisions in a fairly timely way and not be required to run those decisions by half the world.

And they have to have a feel for the answer. Most of the ones I know, and that was true of the Voyager team, knew their fields technically, but beyond that, if they got a question they'd never heard before, they still had a feel for the answer, and that feel was usually right. Those people exist! You just have to find them. You don't need very many of them. You can have 10 of them, and 1000 people following, and that's okay.

There may be some people who don't want to hear me say that. In a complete democracy you let everyone make an input. In new management courses and seminars you involve everybody. But it's been my experience—at least on the Cassini project, where an awful lot of people have to say okay before anything gets done—that you can go too far. Voyager worked well because the people in charge were very competent, creative, motivated, friendly and didn't beat around the bush a lot; they just got on with the job.

That's the chief lesson to be learned. All the other details are just technical details; like you do parts selection early and make sure you allow enough time for testing. You could go through this whole litany of schedulistic steps that have to happen, but every project can draw those diagrams and can look at them on the wall, but to make the project or program succeed, you still need this core of decision makers who get on with the job. If you don't have those types of personnel, these schedules aren't going to save the day. They will just be guides, but they won't save that project relative to some problem that only the right kind of people can solve.

You don't need many, only on the order of a dozen or so highly motivated, capable people with a good sense of humor; you can't be getting into fights with each other. People that have not only the technical skills but also an intuitive feel for the answer. They just *know* what to do.

What about other problems, outside of aerospace?

That works everywhere: building a bridge or buying a piece of land in Montana or deciding where to invest one's money. That brings up the other aspect: you have to be a logical thinker. You have to be able to take a problem, no matter how complicated it is, and very quickly identify its essence, what really matters.

The typical researcher buries himself or herself in a laboratory for a long time—there may be a need for these people, but they tend to work problems in great depth, a lot of thoroughness. If the answer depends on 15 variables, they'll treat all 15 variables. The secret to doing a tough job quickly, though, is to be able to know that, out of those 15 variables, only two make any real difference, that two of them determine 97 percent of the answer and the other 13 only the last three percent.

You can make the right decision as long as you know what the dominant parameters are: I've got a lot of considerations here, but all that really matters is A, B, and C. Focus on those and look at their interplay, and then decide. That's an ability that some people have, but many don't have. Throw out all the branches that don't matter, and then go down the essential path, and do that efficiently, and that makes an enormous difference. You can get answers in a week instead of a year if you go that way, so that's crucial. The more complex the task at hand, the more important it is to have that skill. That applies to the people who influence or control the movement of the project day by day; they all must have that skill.

Voyager will no doubt be the greatest professional experience of my life. I would do it over and over again. I wouldn't choose anything in its place. And the great thing is that it's something that can never be taken away. I can be an old man one day, telling stories to my grandsons and point to Sirius, which is the brightest star in the heavens, (and in fact that's the general direction in which Voyager 2 is heading) and say, "Harry and Cammy, there's a spacecraft headed out that way, and it flew this remarkable Grand Tour mission back in the 1970s and 1980s, and here's what happened," and just watch their little eyes light up. And that makes me really feel good. That part of me feels complete.

Project Managers

The project manager has the ultimate responsibility for the mission. Other people, including the project scientist, may offer recommendations, but the project manager has the final say.

All of the Voyager project managers were engineers. The first was Bud Schurmeier, who came on in 1970 and guided the transformation of the Grand Tour from the original proposal, envisioning four newly designed spacecraft, to the final, cheaper mission that used only two, based on the successful Mariners. John Casani prepared the launches, and subsequently a different manager was appointed for almost every encounter.

After Neptune, George Textor took over the last phase, the interstellar mission, which will continue until the fuel for communicating is exhausted, sometime around the year 2017.

HARRIS M. (BUD) SCHURMEIER
Project Manager, Development Phase
Born July 4, 1924   St. Paul, Minnesota

Bud Schurmeier entered Caltech in 1942 and received a bachelor of science in mechanical engineering in June 1945. He was commissioned as a naval aviator in May 1947. He returned to Caltech graduate school and earned a master of science in 1948 and a professional degree in 1949, both in aeronautical engineering.

He started work at the JPL in the fall of 1949 as a research engineer working on the assembly and calibration of the new 20-in. supersonic wind tunnel. He subsequently managed the Wind Tunnel Section and then the Aerodynamics Division. He was deputy manager of the Sergeant guided missile program before taking on the task of creating and managing a Systems Division for the Laboratory. In 1962 he was appointed manager of the Ranger project and after completing that, he managed the Mariner Mars '69 project. In 1970 he was selected to manage the Grand Tour project that was subsequently scaled back and named the Mariner Jupiter-Saturn '77 Project (MJS 77) and later named the Voyager project. In 1976 he left the Voyager project to assume management of JPL's Civil Program activities. He served as associate laboratory director for Defense and Civil Programs from 1981 until his retirement in 1985.

NASA awarded him several medals for his management of the planetary projects, and he twice received the Astronautics Engineer Award from the National Space Club. He gave the AIAA von Kármán lecture in 1976.

He is a Fellow of AIAA and a member of the National Academy of Engineering. He was a member of the Apollo 13 Failure Investigation Team and the Hubble Space Telescope Repair Mission Review Board. He chaired the Galileo project and the W. M. Keck Observatory Project Review Boards for 10 years.

Since retirement, he has been an avocado grower and active in the affairs of the Planetary Society, the Soaring Society of America, and the Auxiliary-Powered Sailplane Association, as well as the City of Oceanside. He skis, surfs, and flies a sailplane for recreation.

He is married and has four children and seven grandchildren.

Education, hard work, honesty and integrity,
responsibility for ones own actions, consideration for others,
thinking before speaking or acting, and respect and support of family
are all essential but are not sufficient.
The human species must take better care
of our precious home, planet Earth.
The exponential population growth is causing
serious degradation in the environment and the quality of life.
We must find solutions other than continual growth increases
to solve problems.
—Harris M. (Bud) Schurmeier

HARRIS (BUD) SCHURMEIER

I went to grammar school and the first two years of high school in St. Paul. Then we moved to just outside of Chicago, and I finished high school in Winnetka, Illinois.

I've always been interested in airplanes. I built models as a kid: flying models, gasoline powered, rubber-powered, indoor models. I wanted to be an aeronautical engineer, although I also debated about being an airline pilot. I applied to Caltech and took a bunch of exams, and then a dean came out and interviewed me. He spent most of the time trying to discourage me, probably because I was not an excellent student. However, the last student from my high school they had turned down a few years earlier had then gone to MIT and was the head of his class. MIT and Caltech had always been rivals. I think they were afraid of doing that again, so I got accepted.

I came to Caltech, spent my first year there as a civilian. Then, because World War II had started and I either had to join one of the officer candidate programs or get drafted, I joined the Navy, since I also liked boats and sailing. I was very fortunate in that they set up a Navy V-12 program at Caltech, so I finished the last three years under the navy.

When I was about to graduate, the navy came around looking for pilots so I signed up for flight training and went to preflight at St. Mary's. I was there when the war ended. They gave us the option of getting discharged, but I wanted to fly so I stayed in. In '47 I finished flight training, qualified aboard a carrier and qualified in flying PBYs, and got my commission in the reserve. I left the service, went back to Caltech for two years of graduate school, got a professional engineering degree in '49, and got married.

At that time the one place I really wanted to work was Boeing. On my honeymoon I drove up north, visited Boeing, and interviewed with them. But it turned out that '49 was a very poor year for engineers, so I didn't get hired there. I came back to Pasadena and was sitting around thinking about what to do when a guy at JPL called me up and said, "Hey, we have a new supersonic wind tunnel just being delivered and we've got to put it together and calibrate it. Are you interested in coming to work here?"

I said sure, that sounds like fun; I'll go to work there for a couple of years while it's interesting and learn something. To make a long story short, every time I thought I'd learned about all, they found something else interesting for me to do, so I ended up staying there for my whole career.

I did aeronautical research. We designed and built the hypersonic wind tunnel. JPL was part of Army ordnance at that time, and converted the Corporal rocket into a guided missile, and then designed the Sergeant artillery rocket. I was asked to help manage that, and did that for awhile.

Then JPL was growing larger and had a number of projects. Bill Pickering, who was the director at that time, decided it needed a systems division. It had a bunch of technical divisions, and the way JPL did the system design was kind of by committee, which is okay as along as it is a small group of very closely knit people. But when it got large, on a diverse group of projects, we needed to have a group that worked only on the systems aspect. Bill asked me to develop and manage a system division.

What year would that have been?

About '59, right after JPL became part of NASA. I did that for a while. Then JPL got heavily involved in the space business: Explorer, then it designed the Ranger project to the moon. Then, after the first five Rangers failed, each for a different reason, Congress and NASA were getting unhappy and wanted a change in management. I was asked to head up the Ranger project. I took over that around '60–61.

We worked hard and we launched Ranger 6, which looked like it was going to be a highly successful mission. It got all the way to the moon but then the cameras didn't work. After we went back and looked at the data, we found that during the launch phase an ionized layer that developed around the missile shorted a couple of pins on the umbilical cord connector, which turn the cameras on. We have that so we can turn on the cameras on the pad and check to see that they are working before we launched. Unfortunately the power supplies would work at atmospheric pressure, and they'll work in vacuum, but in a partial pressure region they will arc over. So it arced over and burned out the power supplies. When we went to turn them on for good, they didn't work.

That really caused a stink. There were big problems with NASA, and there was a congressional investigation, and a NASA failure board came out and looked at it. We suffered through that and had to go back to Congress and testify. Those guys had their axes they were trying to grind.

If you go back and look at the history of the Ranger project, at the time it was started the space program was clearly in competition with the Russians. This was really before the Apollo program got started, so the basic concept guiding the Lab's design of the Ranger program was that it would be relatively low cost, a short schedule and reasonable risk. That was why there were identical launches for the first two, which were just to check out the parking orbit concept, and then three launches at trying to land a hard capsule on the moon. You have three of them because you're still trying to find your way.

Well, after the five failures NASA said that programatic concept is not acceptable; we've gotta have less risk. So we stopped and spent a great deal of time trying to increase the reliability of the whole system, which increased the cost considerably. We had that one problem with Ranger 6. We fixed that particular problem, reviewed and inspected the complete camera system. We launched Ranger 7, and that was highly successful, as were 8 and 9. In this period also the first mission to Venus was launched, Mariner 2, and then the first one to Mars, Mariner 4.

Then a big, grandiose mission was designed, called the Voyager—not the present Voyager—not the Grand Tour.

This was much earlier than that. We were going to launch two spacecraft, complete with orbiters and landers, on top of a Saturn rocket. But that got so big, so much risk, so much money all on one launch, that it was canceled. In place of that they initiated a Venus mission, Mariner 5, and then Mariner 6 and 7 was the second Mars mission. I managed the second Mars mission.

At that time, in the late 1960s, a bunch of studies had gotten started, looking at the concept and the particular trajectories that were available for gravity assist missions to the outer planets, and it turned out that the ideal opportunities were in the 1970s.

So an integrated advanced technology effort was started, which was called TOPS: Thermoelectric Outer Planet Spacecraft. That was to design the various systems that would be needed for a mission to the outer planets. It had a self test and repair computer that would be highly redundant and test itself and other subsystems, and would then switch to redundant subsystems if there was a failure. There were new communication systems, new attitude control elements, and radiation hardening of components, because we knew the radiation environment of Jupiter was quite severe.

Well, that went along, and then the Lab proposed the Grand Tour to NASA, and I got asked to head it up. It was going to be based on the TOPS spacecraft. The idea was to design a single spacecraft that could be used for a number of different missions. It would be quite versatile, but it was going to be expensive. The basic Grand Tour mission was to Jupiter, Saturn, and Pluto. Two spacecraft would do that, then a couple of years later another dual launch would go to Jupiter, Uranus, and Neptune. So there were JSP—Jupiter-Saturn-Pluto—and Jupiter-Uranus-Neptune (JUN), and they were the two missions of the Grand Tour.

When did you first become aware of the concept of the Grand Tour?

It was at the time that the TOPS advanced technology project was started, which was around '68, '69. The trajectory analysis work had been done, and we could see that those opportunities were coming up in the next decade, which then precipitated the TOPS project to design the spacecraft that would be needed.

Planning for the Grand Tour proceeded, but at the last minute, I think it was for the '71–72 budget, NASA decided, no, they would not propose it because there were negative views from a number of people in the scientific community—negative in a political sense. It was a commitment for such a long period of time, way beyond the four-year political horizon. The scientists were concerned that you design it now and then it would be 10 years later before it was launched and they would get the data. They were afraid that a lot of the instruments would be out of date. So, at the last minute it got canceled by NASA before it was proposed.

So the opposition included scientists, because it would take too long?

Yes. It was large and expensive, and the time scale was too long. Scientists are used to making an experiment and finding out and then changing it and doing something more, and since they were exploring a lot of unknowns, their desire was to do the thing more piecemeal, so they could use what they learned to help design the next thing.

I had assumed that the cancellation was the result of antispace budget concerns.

It was not just that; it was more the concern of scientists. When we were doing the Grand Tour, we went through a big effort to select a science steering group that worked with us in designing the instruments for it. Although there was not a formal selection at that time of the guys who would actually fly, it was clear to a lot of scientists that if they weren't part of that group, they weren't going to be part of it for years and years and years, so they wanted to have opportunities more often to allow broader scientist participation.

NASA and the Space Science Board asked the project to come up with something that's not so expensive; that's more limited in scope. Can you come up with something that's more reasonable and still take advantage of those trajectory opportunities in the late 1970s?

We went home and worked night and day for a couple of weeks, and came up with a design that, instead of being based on the TOPS spacecraft technology, would be based on the Mariner because we had a good legacy on Mariner spacecraft. We made a number of changes to the Mariner design for that particular mission, and proposed that there would be only two spacecraft and just go to Jupiter and Saturn. We called that the Mariner Jupiter-Saturn mission, MJS 77, because it was to be launched in '77. So that was the original name of Voyager.

NASA said if you can get the scientific community and the Space Science Board to support it, we'll propose it. We went back and got the Space Science Board to review it and approve it and support it, and therefore NASA proposed it. It got approved, and off we went.

When it became obvious that you would not get the Grand Tour, did you and your colleagues think, "Okay, we've only got approval to Jupiter and Saturn, but if we do things right we can perhaps push it to Uranus, possibly even to Neptune?"

No, we didn't have that concept at the time we started designing the spacecraft for Jupiter and Saturn. The idea was that we'd have a Jupiter and Saturn project, and we'd have a Jupiter-Uranus-Neptune project for '79, completely separate. But as we got involved and looked at that design of the basic spacecraft, we said wherever it's reasonable and cost effective, let's put in the capabilities that will be needed for the '79 mission, so we won't have to start from scratch to completely redesign the spacecraft; and so we did that as we went along. Clearly we thought at that time we'd get the JUN mission project approved, but we were wrong.

But just in case.

Well, once you start the design down that way, you really can't back out, because it wasn't until very late in the fabrication and testing of these spacecraft that it was clear that you weren't going to get those things, but you had already put that capability in there.

Were the four craft to be designed more or less the same, for 1977 and 1979?

That was the idea. The basic spacecraft and the basic computer to handle the communication system, the attitude control system, the programs and sequences would be identical.

The Mariner spacecraft had no radiation hardening; that had been a key cost driver in the original TOPS spacecraft. So MJS started out with no radiation hardening. Then the Pioneer 10 and 11 were launched. One of their prime purposes was to determine what the radiation environment was around Jupiter. Our plan was we'd fly by Jupiter and use the gravity to get out to Saturn; if we found from the Pioneer flyby that the radiation environment was a little higher than what we had assumed, we would just fly a little further away, and we'd still have that same mission. We thought we had enough flexibility in the mission design to accommodate changes.

Without doing the hardening?

Right, without doing the hardening. We didn't think the electron environment was severe. But we found, when the results from Pioneer 10 came back, the electron environment was about 1000 times stronger than what we had assumed, and it did not fall off as rapidly as you went away from Jupiter. The result was to fly the spacecraft far enough away to be sure that it would survive the radiation, which otherwise would essentially destroy the mission.

So, very late in the game, we had to embark on a radiation hardening program. We did that in three ways. First of all, if you could replace an electronic component with a radiation hardened one; that was the simplest thing to do. Second, if you couldn't do that, or only do that partially, could you change the circuit so that it could stand some degradation in the components and still function properly? If those two things weren't successful, then could you shield it?

Well, we used all three of those in varying degrees, in different combinations, to solve the radiation problem. Then we had to qualify the equipment to find out that it would stand it.

Since there would also be an opportunity in '79 to go to Jupiter, Uranus, and Neptune, we tried to design the spacecraft so it would work for all of them. You would just put some different instruments on it if NASA approved that second, JUN project. So we designed some versatility into it. We put in a different attitude control system, which was an upgrade from Mariner, which we decided was cost effective.

Late in the mission design, when it was clear that there was not going to be a JUN mission, there was an effort to evaluate the instruments and see about their capability for Uranus and Neptune, although they were designed basically for just Jupiter and Saturn.

The one that was most severely compromised was the infrared spectrometer of Rudy Hanel of the Goddard Space Flight Center. So there was a crash effort to design an upgraded instrument that would be put on one of the spacecraft. By then the idea had evolved that if the first spacecraft that went by Saturn was highly successful, then we would target the second one such that when it went by Saturn, it would then go out to Uranus. The idea was to put this upgraded instrument on that spacecraft. Unfortunately they just couldn't get it built and qualified, so we had to go with the two original instruments.

The versatility we put into that spacecraft allowed us to do a lot of the things needed to go to Uranus and Neptune. We put in a coding system that increased the communications capability, and that, coupled with arraying the ground antennas, allowed us to get significant data rates at the distance of Uranus and Neptune. It all worked out great.

By the time Voyager was launched, I was no longer the project manager. Bruce Murray was now the director of the laboratory. The United States had won the race to the moon, and the space program was going down some, and the Lab was embarking on nonspace activities: civilian applications of the technology, and some defense applications. Bruce asked me to head up that part of the activity at the Laboratory, so I was off in non-NASA space work, and John Casani became the project manager. He carried it through the launch and the difficulties it had. Subsequently everything worked out and it was a great success.

So the 1979 Voyager mission was disapproved before the first two were launched in 1977?

Yes, that's what precipitated the effort to design this upgraded infrared instrument. When the '79 mission wasn't approved, we thought we'd upgrade one of those spacecraft that would go out to Uranus and Neptune. There were a couple other alternatives we looked at, but that was all that was really tried for.

We designed the Voyager mission to use the Titan IV launch vehicle with a Centaur upper stage, but that was not quite enough energy to get us all the way, so we added a propulsion module. We just took a small standard solid rocket and attached it to the spacecraft. It had its own little attitude controls, stronger than the ones that were on the spacecraft, but it used all the electronics that were on the spacecraft, so we called it a propulsion module instead of a stage. And after it injected the spacecraft on its way to Jupiter, we ejected that spent propulsion module.

When did that redesign happen? Was it after the cancellation of the third and fourth Voyagers?

It was not really a cancellation because it was never approved; it was never a part of the program. The only thing that was approved were the two Voyagers in '77. We proposed the JUN mission, but they didn't approve it. But then we cut it back and said, "Let's make it one launch each year; make it a three-launch combination: one in '77, one in '78 and one in '79." We would target the subsequent launches depending on how successful the first one was. That didn't get approved, basically because NASA didn't want to buy any more Titan IVs. NASA decreed that all future launches would use the shuttle.

So you went from proposals for four, then down to three, and then the two that we finally got.

Yes. They went from the Grand Tour, ratcheted all the way down to the two that were just Jupiter and Saturn, MJS 77 with the idea that we would propose, as a separate project, the two in '79. Then that didn't get approved. Then we said how about a combination of three, and that didn't get approved, so it was left with just the two launched in '77. That then precipitated the design of an upgraded infrared instrument to go on one of them. It was that sequence, with the idea that if it's successful at the first planet, we'd go on to the next one.

It turned out to be that way. We were very fortunate, and the fact that we had put this flexibility and capability in the basic spacecraft allowed us to reprogram it and do the things necessary to send it on to the other planets.

How open or well known was this capability to go to the two outer planets at the time you were working on it? Was it some guys saying, "They won't let us do that, but I got here in my top drawer a little extra stuff we can put in, so, when the people at Headquarters aren't looking, we can slip this in?"

No, it really wasn't that way. To give you an example, at the time that the spacecraft was being designed there was some interest in standard subsystems—a standard transmitter system, a standard attitude control system—that could be used on various spacecraft, so that you didn't have the expense of starting from scratch each time, with everybody designing a new one. So they funded some money to develop standard subsystems.

We saw that and said, "We'll make a partnership deal with that program office at headquarters that's supporting the design of these subsystems, and the Mariner Jupiter-Saturn project, to jointly fund the development of a new attitude control system. It would be much more versatile. It would have programmable electronics instead of being hard wired for its limit cycles and all those things.

That's how the new attitude control system got in there; it was a joint effort. If they hadn't been able to fund part of it, would we have put that sophisticated one in? I don't know, because we had to look at how we could fit it within the budget and all those kinds of things. So there were a few things like that that helped us get these improvements in there.

At that time we believed they would approve the '79 mission; it seemed logical because, as I said earlier, one of the reasons that they didn't like the initial Grand Tour was that you had to make a design and select the instruments way back in the beginning, and then that was it. They wanted a more piecemeal approach: fly one mission, see what it was like, and then fly another one. We fully thought, as they continued to plan the program, they would approve that and get to those other two planets. But it didn't turn out that way.

What was behind the veto of the 1979 launch?

By the early stages of Voyager, which was then called MJS 77, NASA had made the decision to develop the shuttle, and for economic justification they said we will launch everything on the shuttle; therefore we won't have any other launch vehicles.

When NASA decided we wouldn't have any other vehicles, there wasn't a way to launch JUN unless they were to buy some more Titan IVs, and they didn't want to do that. I think that was one of the strong reasons.

There was also the concept, by a bunch of NASA types when they started on the shuttle, that we can't fly planetary missions; we'll just stop the planetary program for a few years, and wait until we get the shuttle going, and an upper stage.

So a major reason the 1979 mission fell through was because you ran out of Titans?

Yes, they did not want to buy any more Titans. They had to buy those from the Air Force.

Voyager 1 and 2 used the last two Titans?

Right. At that point they were the last two; we just sneaked in, in a sense. That's why there was no Titan available for Galileo or anything else, so when Challenger went, Galileo had to be delayed because all the other unmanned launch vehicles had been cut out. That was one of the big reasons why they cut off the '79 missions.

But Bruce Murray objected strongly. In a lot of ways he made himself very unpopular with NASA, but he fought like hell for the planetary program to try and keep it alive.

This perspective, as I mentioned earlier, is quite different from the one I had: the reason that the Grand Tour was scratched, the reasons that the Voyagers 3 and 4 were terminated, was supposedly because of shortsightedness in NASA Headquarters and budget queasy congressmen in a movement against space.

I think there was some of that in the background, because after we'd won the race to the moon there was an increasing view that we'd better start working on problems here on Earth, and why do we need all this space stuff? So that was certainly there, but it was not, I don't think, the big reason, because there was still a lot of enthusiasm in general for the space program, and there still is.

Would it be fair to say, then, that the cancellation of the 1979 missions was at least in part because of the concern for the forthcoming shuttle; would this have been similar to the cancellation of the space plane program of the 1950s and early 1960s?

I think one of the key reasons for that was the realization that it was going to be very difficult, on a manned spacecraft, to get the stability for the kind of measurements wanted. The big thing at that time was remote sensing of the Earth, with big imaging systems and all that. That's what the Air Force was planning: to put those instruments on a spacecraft that had men marching around on it. But then they decided it would be better to have unmanned satellites up there that can be very rigid and fixed and motionless.

So it wasn't competition for funds for shuttle development?

No, because I think the DynaSoar got canceled before shuttle came into being. But later, when the shuttle was proposed, there was a big push in the administration to say to the Air Force, "You guys are going to use that shuttle for your launches also," so there was a plan to phase out the Titans and to use the shuttle. The Air Force didn't like it, but that was the edict that came down, until the Challenger accident, and then they said "screw you; we gotta have some capability," so the Titan came back into the fore again after that.

Going back to your association with Voyager, what were the hardest decisions that you had to make; your most difficult choices?

There were a lot of programmatic decisions one would make, like should we embark on the upgrading of the attitude control system? It was a new system; it had to be designed and developed and qualified. Or should we stick with the old, reliable one that we had a lot of experience with? Well, that's a judgment call. I don't know that I'd call it a hard decision, but it's the kind of decision that came up all the time.

It's hard to reconstruct the basis on which we decided it then, but it just seemed like that was the right thing to do. It added a little more risk because it was a new system, but we also got a lot more capability. So those kind of programmatic, technical decisions went on.

Another category of decisions that came up was JPL is organized into a bunch of technical divisions and a systems division, and boundaries get fuzzed over. I can remember one that related to jurisdictional issues. There were two groups working on the approach guidance that we needed to target the planets and satellites and point the cameras and the other remote sensing instruments at the selected targets.

The approach guidance concept was to use the camera to take some pictures of the satellites and the planet against a star background over a period of time, analyze those, and out of that compute where the spacecraft was, where it was going, relative to the targets. Then you could decide when and how to point the cameras and how to make a given maneuver.

Well, that had been worked on by a couple of different groups, and they got to the point where there was duplication and conflict, and you'd say, "How are we going to resolve this?" One of the decisions was, "Let's put all of that in one division and have people transfer if they wanted to continue work on the approach guidance."

One of my key efforts in being the project manager was selecting the right people, because you can't do it yourself; you gotta have the people. I had to make the decisions—scheduling, programmatic cost—as were needed to keep the thing going, but the basic detail work was done by good competent strong engineers, and selecting the key managers and engineers was a key part of management.

There were certainly cases where several guys wanted the job; you had to select one, and the others would be disappointed, so there were those kinds of decisions. Some of them were hard to make. You got two guys that were very well qualified, and how do you decide? Then you get down to a gut level and say: "I think this guy has the better chance of doing the best job."

There was another decision, which I wouldn't say was a difficult one but it was interesting. When we were doing the Grand Tour, we had a science steering committee that worked with us in the design of the mission; they provided the science input. This was a wide group of scientists from around the country. The guy who chaired it was Robbie Vogt, a professor of physics at Caltech. He later became the first chief scientist at JPL. He was in the German army in World War II, and the trials and tribulations he went through were amazing and heart wrenching—a fascinating guy, very competent.

Anyhow he was the chairman of that first science steering group, and worked with us on the design of the Grand Tour. We had not selected a project scientist for the MJS 77 Voyager mission; the project had been approved, now where do we get a project scientist?

Robbie and I had lots of discussions, and he was the one who proposed Ed Stone; they worked together at Caltech. I talked with Ed, but he was very reluctant to become the project scientist. Science guys don't like to get into administration and management; they want to do their science. I talked with him a number of times and tried to set up a scheme where there would be an engineer who had a strong science bent who would work with us on details and live with us in the project, so that Ed did not have to spend all of his time, or even a great deal of his time, working as the project scientist. We thrashed around and finally got him convinced to become the project scientist on that basis. It was highly successful, probably the best arrangement that we'd had on any of the previous missions.

This was something new at JPL?

Yes, Robbie and I were the key guys, because Robbie was the chairman of the steering committee and he knew Ed. So between the two of us, with help from other people who Ed respected, we got Ed to be the project scientist; and you know what subsequently happened ....

Talk about crucial, key decisions!

As it turned out, Ed was the one guy out of the senior management of the project who was there through the whole thing. He was the project scientist on the whole mission and subsequently became director of the Lab. But as I said, he was very reluctant in the beginning to get involved in any of the management. I can understand that because there's a lot of routine stuff you gotta do, to keep things organized and going. We set it up with a guy who would help him to do that, so Ed could provide the intellectual, science background as well as make the many science tradeoffs that were required.

Ed had the respect of the outside scientists, which was *very* important. He participated enough with the spacecraft and mission design so that he understood things, so that he could make the decisions and convince the various scientists who didn't get all they wanted that that was the right decision. It would be very difficult for us, as engineers, to directly convince a scientist, when he didn't get all that he wanted. It was a great arrangement, and worked out fine.

What about your own workload; were you working a strict 40-hour week, so you could punch in at 8 a.m. and punch out at 4:30 p.m.?

No. For most of my career at JPL, I was working closer to 10 hours a day. You'd get involved in it, and if you had to go back Saturday you'd go back there, and when you had a mission and problems arose, like on Ranger when we had failures, we had to go back and redesign it.

I remember the failure of Ranger 6; it was on a Saturday evening. It looked like it was successful and going great, and we were just getting to the encounter and it would be Saturday night. I went out and bought a whole bunch of bottles of champagne, put 'em in a wash tub, and put them in my van. Then we had the failure and had a press conference. I drove home and got up early Sunday morning, took the champagne bottles out of the ice water, glued the labels back on, set them aside, and went to work. We got started on the failure analysis, on Sunday.

So you did what you had to. It was such a fascinating job, everything was new, and you were kind of making it up as you went along; there wasn't a road map to follow. Unfortunately now there's all kinds of rules and regulations, which is typical of anything once it gets going, but in the early days it couldn't have been more fascinating. We figured out what was the right thing to do and went and did it. Sometimes we were wrong and had to do something different, but there wasn't any path to follow. We just used our own judgment and engineering sense, and did what we had to.

Did some of that continue on into the 1970s, when you were doing the early work on Voyager?

Sure. It was still pretty good. NASA was starting to grow, and we'd learned a lot. We'd had to go from the high-risk, short-schedule, low-cost concept to low risk, higher cost, more reviews, but that was all good. Whether it's gone too far, well, probably not, when you see Mars Observer, but you can never make something 100 percent reliable. You make the judgment as to what's a good tradeoff. There's an infinite amount of engineering that can be done. There are two things that control the amount that you do on any given project: one, the schedule, and the other, the cost.

A good example is after we had the failure of the Galileo antenna (I chair the Galileo review board) we got very involved in the review of all the work that was done on the analysis of that failure. You would be amazed at the detailed analysis and modeling and testing and simulation that went on, to try and figure out what had happened, and eliminating all the possible different causes and settling on what's the most likely one, and then figuring out, okay, if that's the cause, what are our chances of getting it loose? And then the stuff that went on to simulate, model, and decide what ought to be done to try and get it loose. It is just fascinating what the engineering team did.

I use that as an example, but usually you just can't do that amount of analysis and engineering. But in spite of all that, it was only a hypothesis of what had hung up the antenna and may not be right because the steps that were taken did not release it, so it may be something else that caused it. It's an example of the newness of all this, and you have to figure out what makes the right sense in using this activity, and how much engineering and how much design. That's one of the toughest decisions: how much is enough? How much is enough of any given thing?

One of the most outstanding things done at JPL I think is the analysis of in-flight failures; figuring out what went wrong. They're absolutely superb at that.

What lessons do you think the Voyager project might have for future managers, for future designers?

Certainly to be as thorough as practical in understanding the design of the spacecraft and what its capabilities are, and ensure that it has been adequately tested and qualified.

Wherever "schedulely" and fiscally practical have some flexibility in the design of the system, the capability of the spacecraft, to accommodate unforeseen problems or changes. It wasn't unseen that we put the capabilities in there, but we put them in for a different reason than they were used for. We put them in for a spacecraft that was downstream, but didn't expect that all of them would be used in these actual spacecraft that were flown for this project.

Each case is different in deciding whether or not that's the cost effective thing to do.

One lesson is persevere. When Grand Tour got shot down, we didn't say forget the whole thing; we went back and tried to put together something that would sell and would be approved. So one lesson is "don't give up."

Another lesson is get good, experienced, competent people. We were at a high point when we started on Voyager; high point in the sense of the history that had gone on, the previous projects at JPL, the experienced people there who had gone through the trials and tribulations and knew how to design these things. We had a great deal of these people, so we could select and get the best ones. That's a good part of the reason Voyager was successful.

JOHN R. CASANI
Project Manager, Prelaunch Phase
Born September 17, 1932   Philadelphia, Pennsylvania

John Casani graduated from the University of Pennsylvania in 1955 with a bachelor of science in electrical engineering. The following year he joined JPL as an integration engineer, working on the Jupiter radio inertial guidance system. Subsequent positions included accelerometer development engineer for the Sergeant missile, payload engineer for Pioneers 3 and 4, and spacecraft systems engineer for Ranger 1 and 2 and Mariner Mars 1964.

In 1965 he was appointed chief engineer for an earlier Voyager project. From 1966 to 1971 he worked on Mariner Mars 1969, as deputy spacecraft system manager, spacecraft system manager, and project manager. In 1971 he moved to Mariner Venus-Mercury for three years as spacecraft system manager. Then, after a year as manager of Division 34, he joined the current Voyager program in 1975 as project manager for the crucial prelaunch period. From 1977 to 1988 he was project manager for Galileo. Next he was assigned to Flight Projects, first as deputy assistant laboratory director, and in 1989 he became assistant laboratory director.

His commendations include NASA's Outstanding Leadership Medal for his service as Voyager project manager. AIAA honored him with their Space Systems Award in 1979. He is a member of the National Academy of Engineering, a Fellow of AIAA, and a member of Sigma Xi and the International Academy of Astronomy.

Away from work, earlier activities included service on the Pasadena YMCA Board of Directors, playing a lot of handball, and backpacking. More recently his full-time hobby is doing a complete restoration on a 1965 Mustang Fastback that he bought new, which has been sitting in his garage for 12 years.

Don't ever give up.
The better is the enemy of the good.
The formula for failure is to try to please everybody.
There are lots of right ways,
but in the end you have to pick just one.
There are more wrong ways than right ways.
Count everything.
Patience is a virtue, but timing is everything.
—John R. Casani

JOHN CASANI

When I asked Ed Stone whom I should talk to about Voyager, his first recommendation was Schurmeier and you were his second. He mentioned eight or nine people, but he said that you were absolutely essential to interview.

Well, that was nice of him. I came on the project about two years before launch at a key time, but Schurmeier was really the architect of the project. He was the original project manager, and the one who put the whole thing together. When Bruce Murray came to JPL, he had the idea to create another organization, and Bud Schurmeier was the guy he selected to do that. And that created an opening for me, so it was very good for me.

It was not an easy job.

Yes, but things had been set up pretty well. I didn't have too much of a problem. There were a lot of good people in place, and all the plans and implementation arrangements were in place and working well. We had the normal amount of problems, like you do with any project: pulling everything together, getting hardware together, getting it tested and integrated, and resolving problems as they came up, as they always do and probably always will. It was challenging, but like I said, we had the A team. We had the best people in the lab working on the project at that time. There was no question about that.

How were you able to get such good people?

I can't take any credit for that. Schurmeier assembled the team. I don't mean to imply that other projects were populated with duds, but on the other hand there weren't too many other projects going on at that time. We came off the Viking project, which had been launched in the summer of '75, and there were a lot of people in Viking operations. But there weren't any other projects under development that I can recall, so Voyager got the best project people that were available. Other people were doing other things.

JPL operates in a matrix organization. There's a small project office, and then all of the technical support comes from our technical divisions. We have a Telecommunication Division, a Guidance Control Division, Mechanics Division, and so forth. Each division supports all projects, so within each division there is one person, called the flight project rep, who manages the effort within the division for that project. He only worries about one project. He's the key guy in each of those divisions with respect to the contribution from that division for a particular project.

You've worked at JPL for many years. How did you happen to come here in the first place?

I was graduated from the University of Pennsylvania in 1955 with a degree in electrical engineering. The first job I took was with the Rome Air Development center in Rome, New York. It's a small town, west of Utica. I worked there almost a year, developing character recognition systems and automated direction finding equipment.

I had gone to school with a good friend of mine, Louie Yardumian. He was a chemical engineer; I was an electrical engineer. While we were in college, we used to tease each other and kid a lot about going to California; mainly because in the 1950s if you lived on the East coast, California was kind of a magical place, and was about as far away from home as you could imagine.

So I found myself in upstate New York as a young fellow, but my base was still Philadelphia. Lou was working there in the Atlantic Richfield refining company. I still had my network of friends there, so I would travel down to Philadelphia a couple of times a month for the weekend. One particular week up in Rome it never got warmer than 10 degrees below zero for the whole week. Next time I managed to get back down I said, "Lou, why don't we really go to California?"—I was reacting to that cold. We said "Why not" and we took off.

We left Philadelphia in early May. It took us 56 days getting out to California. We stopped at every place imaginable, especially any campus that happened to have a fraternity chapter that we were members of. We finally got out here. I remember we were in San Francisco on the evening of June 30. We went someplace where they were having a "New Year's Eve" party for the end of the federal fiscal year.

When we got down here that summer, we lived in the fraternity house at USC. Lou got a job at Braun Engineering in Alhambra. I was interviewing with four different companies, one of which was JPL. Another was North American in Downey; they had a big project going called the Navaho, and they made me a pretty attractive offer. I was seriously considering going with them, but the difference was that Jack James, the JPL guy I was dealing with, sent me a telegram every day, so even though it paid less I took this job. A few weeks later the Navaho project was canceled and North American laid off 12,000 people, so it was just a dumb luck choice that I wound up at JPL, and I've been here since 1956.

Several other Voyager people had similar experiences being offered higher paying jobs at other places. What were you working on before you came to Voyager?

Before I came to Voyager, I was the division manager for the Guidance and Control Division. It provided the control system for controlling the attitude of the spacecraft and the power subsystem. Those were the principal contributions from the guidance and control division at that time, so I was very much involved as one of the divisions supporting the Voyager project. The guidance control division had made significant advances in the subsystem technology that we were delivering to the project. It was a challenge for our division to come up with the new equipment designs that were required. So I was very, very close to the Voyager project.

Prior to that I had worked on the Mariner Venus-Mercury mission. I was the spacecraft system manager for it. Before that I'd been a spacecraft system manager, a project manager, on a few previous projects. And I was known by Bruce Murray through his association on projects that I had worked on for the previous 10 years. Maybe that's why he selected me to take the Voyager project after he came on board and moved Schurmeier into this new developing organization that he had started.

What were some of the first things that you had to deal with when you became head of the Voyager project?

When I first came on, there was an overlap that had been by design. I came on as Schurmeier's deputy, with the announced intention to me and to every one else that he would be leaving the project by a certain time, and then I would take over. So my first delegated responsibility was to shadow him. He spent a lot of time with me, making sure, to the degree that he could, that I understood who the players were and what was important, and what the challenges and problems were. He involved me very quickly in the management and all the decision making that was ongoing then.

Some of the challenges we worked with early on had to do with the realization, about the time that I came on the project, that the charged particle environment, the radiation environment in the vicinity of Jupiter, was larger than had been anticipated. A very massive effort was undertaken to make sure that the spacecraft components, the electronics that went into the circuitry, could withstand the radiation exposure that the spacecraft was going to get.

We made a major effort to understand what the dose would be as a function of where stuff was located in the spacecraft, because some of the spacecraft's structure itself provided natural shielding. Then we added shielding where it was required and made special provisions to procure and use parts that were resistant or tolerant of the radiation environment. It was a fairly broad scale program to deal with that concern. That was a major activity.

We had problems with a whole raft of just day-to-day issues. I don't even know if I can bring them to mind.

The project had grown up out of the ashes of the Grand Tour, which was the original concept to send a spacecraft to Jupiter, Saturn, and then Uranus and maybe another, a second spacecraft to Jupiter, Saturn, and Neptune, or Pluto. That had been abandoned and a scaled-down project that would do just Jupiter and Saturn is what evolved. But many of us harbored the notion back then that it was still possible to get the spacecraft to Uranus.

In fact, when I came on the project, I contrived to have my telephone number arranged so that in dialing it—my number is 6578—you put your finger on the buttons, or in the dialer and dial MJSU, standing for Mariner-Jupiter-Saturn-Uranus.

I did that to remind myself and all of us on the project, because when everybody asked me what my phone number was, I told them it was MJSU. We wanted people to be thinking at least beyond Saturn, and even though there wasn't any formal or authorized mission requirement beyond Saturn, we wanted to keep in mind that option. I still have the same number; I'm very proud of it.

So we kept asking ourselves, in terms of decisions or actions that were taken, were we doing anything that was going to prevent or exclude the option, the possibility of the spacecraft going on to Uranus? We did make several specific decisions that subsequently turned out to be very important in enabling the spacecraft to operate well beyond the orbit of Saturn, even out to Neptune.

In fact, even when I was in Division 34, the Guidance and Control Division, we had sun sensors which were part of the attitude control system which provided error signals or information to the spacecraft about its orientation, using input from the sun. Of course, the farther you get from the sun, the dimmer and the lower the intensity—and we had a sun sensor that would work beyond Saturn.

I remember discussions internal to the division that this thing will certainly work beyond Saturn but it will certainly not get to Uranus. So we made, on the sly, a decision to put some special features in the sun sensor, including adding some amplifiers to boost the sensitivity of the device so that we would guarantee that the sun sensor would at least work out beyond the orbit of Uranus. Of course, having that turned out to be crucial, because the spacecraft would not have worked to Uranus, much less Neptune, if we had not done that; even though it was not required to get to Saturn.

Also some of the features that subsequently were used by the mission controllers that allowed them to reconfigure the data system of the spacecraft were put in specifically with the idea that they could be used beyond Saturn. They were there primarily for redundancy and for certain mission enhancement reasons in connection with the Saturn mission, but they were also designed so that they could be reconfigured and used in a different way if the spacecraft ever got beyond Saturn. We hadn't figured out before launch exactly how those elements would be used, and hadn't worked out the details, but at least we built into the hardware, through a conscious decision, the capability of reconfiguring.

An example were the devices for data compression and for coding that allowed us to operate at lower data rates and lower communication signal-to-noise ratio than would have otherwise been possible. It was a combination of these things, plus a lot of things that we hadn't thought of before launch, that the guys who operated the spacecraft were able to use to make the Uranus and Neptune missions work.

I've always felt that we made some tough decisions, but that we made the right ones. It was a balance between not doing anything unnecessary that would add risk or unnecessary complication to the basic mission—the Jupiter-Saturn objectives—but would still allow for flexibility for subsequent decisions and the subsequent actions that actually were implemented.

What would be a typical example of those decisions?

I just mentioned that we put in a couple of features having to do with the way the onboard coders could be configured. We put in data compression circuitry that was really there in case we had a transmitter failure and had to operate at a lower power level at Saturn, so that we could recover part of the Saturn mission with some of these data compression and coding techniques. We could take the two data systems, which were designed to be operated in parallel for redundancy, and string them in series to expand the capability of the system for sequencing and so forth. This capability was built into the system before launch so we could go on to Uranus. It didn't need to be done that way, but we did it that way with the idea that it could be used differently after launch.

I'm trying to convey to you the idea that back before launch we tried to be mindful of opportunities beyond Saturn and implement, where we could, ways that would allow the flexibility to reconfigure the system, provided it didn't add complexity or compromise to the primary mission objective.

I see. But sometimes one has to make a tough choice, that adding capability A means that you have to subtract from possibility B. Did you have to make choices of this kind?

Okay. That's the most challenging aspect of flying these missions: allocating the available resources to the various science instruments or science experiments or science data-taking priorities. We've found, with any spacecraft that we've built with a scan platform—where there's several instruments on the platform and they all have to point together because they're all mounted on the same device—one instrument might want to look at some feature on the planet for 10 minutes but another instrument on the platform may want to see something else during part of that same time. So you have to decide how long you're going to point the platform in direction A and when will you move it to direction B.

There's lots of other examples: the use of the tape recorder and what will be put on it. When will the tape recorder be running? You need to record pictures at certain moments, but the more pictures you put on the tape recorder, the less room there'd be for other types of science data. That is the cardinal challenge, in my opinion, of mission operations, of running the spacecraft.

This involves a long process of identifying, through the active participation and involvement of the science people, what their objectives and priorities are, not only in general but for every specific point in time in the mission. Then, once you identify them, you find there's always conflicts between them of the type that we've been just talking about, so you've got to identify the conflicts and resolve them in some mutually satisfactory way and then build the commands to go to the spacecraft to implement whatever the final decision is. But that's more a challenge in the operation or the use or the flying of the spacecraft than it is in the design of the spacecraft.

When you were working on these capabilities for possible extension beyond Saturn, how open were you about this? Were you forthright about it or was it just something that you slipped in quietly?

It isn't that we weren't forthright about it, but we had to keep in mind that our primary objective was to get to Saturn. We couldn't be talking about, or promulgating, the idea that we were building a spacecraft that was going to be used at Uranus. That would have been unwise and unnecessary. So we did not ballyhoo any of the things that I've been talking about that we were doing. They were done quietly, in quiet council and deliberation. We never advertised. We never talked about a mission to Uranus until shortly before launch when we began to lay out the strategy for the Saturn mission.

We had to pick aim points at Saturn that were a choice; again it was the same sort of conflict resolution thing. In order to get a spacecraft to go from Saturn to Uranus, for example, you had to pick a very specific aim point in the vicinity of Saturn to shoot for, because if you didn't, you couldn't get the trajectory bending that would be necessary to go on to Uranus.

Well, the aim point at Saturn that was required to go on to Uranus was a very particular aim point, and it was not compatible with the Saturn science objectives that had to do with Titan, a moon of Saturn. So the strategy that we evolved shortly before launch was recognizing that the two spacecraft were separated in time by four or five months.

We decided that we would pick the aim point for the first spacecraft at Saturn to satisfy the science objectives with respect to Titan with the idea that if the operations were completely successful and we satisfied the Titan science objectives, we would then have the option of choosing a different aim point for the second spacecraft, one that would allow it to go to Uranus. But if the first attempt missed, either because the spacecraft misperformed or because the operations did not perform as expected, then we would aim the second spacecraft back to the same point in order to have another go at the Saturn-Titan objectives.

So that was the strategy that we evolved and that was the strategy we followed. It was a fairly complex strategy; I didn't explain all of it to you. It also had to do with making sure that we had options that would protect us, depending on what might go wrong or might happen with either of the spacecraft before they got there, and picking the right aim point for the second spacecraft that gave us the flexibility to choose either option.

What sort of publicity or attention did this decision, this possibility of going to Uranus, receive outside of JPL? Was there a lot of notice that you might be going to Uranus?

I went off the project shortly after launch and on to something else, Galileo, but my recollection is that there was not much publicity or discussion or promotion of the Uranus possibility until well after Jupiter, in the period of about two years between Jupiter encounter and Saturn encounter. It was during that period that we began to talk about the possibility of it. We wanted to hold off until both spacecraft appeared headed for success at Saturn, because if either spacecraft had failed or had a problem before we got to Saturn that would have ruled out the possibility of the Uranus mission. We didn't want to build up expectation for it any earlier than necessary.

You're always judged by your success. This was one of Fred Felberg's laws. He's an interesting guy. He used to work here; he's retired now. One of Felberg's "laws" was that the success of any undertaking you try is judged on the basis of whether or not you achieved the objectives that are perceived to have been the most difficult. If it's perceived that the most difficult thing you're doing is X, even though it's only 10 percent of the value of the total mission and you fail to do X, then the perception is usually that the mission was a failure, even though it might have been 90 percent successful. We try to avoid Felberg's Second Law syndrome.

One of the most crucial things was the launch of the two spacecraft. What do you recall there?

We took three spacecraft down to Florida with us: the two that turned out to be Voyager 1 and Voyager 2, and the third spacecraft, which was something we called the proof test model. It was a bed for spares, and we were ready to launch that as a spare spacecraft if something happened to either of the other two. We had to make a lot of changes in hardware subsystems—changing radio systems and switching things around—in the last couple of months before launch, which were unexpected and proved to be quite challenging.

We had lost a number of science instrument detectors, surface barrier detectors, due to the fact that in the building where we were testing the spacecraft they were doing spray painting or something around the outside, and some hydrocarbons were sucked into the building through the air handler. These proved to be contaminates to the detectors, so we were scurrying around at the 11th hour trying to change them out. We replaced some 20 or 30 detectors on a couple of the instruments as a result of that. There was a lot of crisis activity during the last couple of months before launch.

When the first spacecraft was launched, we were surprised by the response of the onboard fault protection and failure detection algorithms. We had not anticipated the environment, the forces that would be imposed on the spacecraft during the launch, and for awhile right after launch it looked like we might have lost the spacecraft. We spent some very difficult and anxious hours before realizing that there was nothing wrong with the spacecraft; in fact, it was performing exactly right. It's just that we hadn't anticipated, and therefore hadn't tested it, in the environment that it was actually exposed to.

During the ascent, the launch vehicle goes through a programmed roll maneuver in which it turns about its roll axis and then does a pitch-over maneuver. The launch vehicle was going to roll, and it was going to roll at a rate which would exceed the linear operating range of our onboard gyros. We didn't properly anticipate the roll rate response of the gyros to the roll rate of the launch vehicle.

I don't remember what the numbers were, but just to give an example, let's suppose the gyros could detect maximum turn rates of, say, plus or minus half a degree per second. But the launch vehicle was planned to rotate at a rate greater than that, say one degree per second. The only thing the gyro could do was go up to half a degree per second, and then its output couldn't go up anymore.

That's called "saturating the gyro." It doesn't hurt the gyro, it's no threat to the safe operation of the gyro, but once one axis of the gyro is saturated (these are two axis gyros—they read out rates on two orthogonal axes) the output from both axes becomes not meaningful.

We had three gyros, sensing rates about two orthogonal axes, arranged so that there were two independent rate sensing measurements about each of the three axes. We set it up so that one axis of each gyro was backed up by one of the two remaining gyros, and the second axis was backed up by an axis on the third gyro, so all three of these gyros backed each other up in this way.

But when one axis saturated, that had the affect of appearing, to our onboard detection circuitry, that the gyro had failed in that axis, and because it had saturated it also caused the second axis to appear to be failed, so the onboard detection circuitry tried to switch over to the other two gyros. Well, one of those other two gyros was also sensing a high roll rate, and it too was saturated, so the onboard system thought that two of the gyros were bad.

The launch vehicle was turning at a much faster rate than the spacecraft would ever turn in space, and that's why it exceeded the linear region of the gyros. We had not anticipated the effect of gyro saturation and therefore hadn't taken care of that in the logic of the onboard fault protection, so the poor dumb spacecraft thought that two of its gyros had failed during the ascent phase.

That was not what happened, and once we got separated from the launch vehicle, and we could figure out what went wrong, then everything was fine. The spacecraft recovered itself after it was separated from the launch vehicle, but in the meantime, based on the indication from the telemetry, we thought we had serious problems.

That was the first spacecraft launched?
Yes.

By the time of the second launch, what happened?

I don't remember that we tried to do anything between the first and second launch. We just knew that this was going to happen and that the spacecraft was in no danger of being damaged or anything like that. It was just behaving and reporting to us something that was totally unexpected with the first spacecraft. We didn't need to do anything about it; the system could deal with it.

However, we did have problems in the first launch with some of the hardware. There's a boom on the spacecraft for supporting the RTGs (radio thermoelectric generators, the power sources) and another one for the science platform. One of those two booms did not fully extend to the correct position. It went almost as far it needed to be, but the telemetry would not confirm that, and we did make some hardware changes, added special springs and what have you, before the launch of the second spacecraft.

What lessons might be learned from Voyager that could be applied to other projects? Why did the Voyager project succeed when others have not?

A lot of things contributed to the success. First, we built three spacecraft: a proof test model or a protoflight model, and two flight units. The first one could have been flown. So there was a learning curve effect, and we were able to move resources from one spacecraft to another during the assembly and test, to play musical chairs so to speak with the hardware. So even though we had more work to do, we had a lot more flexibility. If you had a mechanical problem on one spacecraft, you could put all your mechanical resources on that one and move ahead with electrical work on the second one.

Having two of them provided that kind of flexibility, and that's what we had been used to. We had done that with Viking and the Mariners and all the earlier programs. We typically launched two spacecraft. Voyager was the last project where we did that, and we're not likely to do that any more. The new paradigm is you build one, launch one; people don't want to put the resources into building and launching two of them. In other words, it's viewed as an insurance policy that now we can't afford to take.

Part of the thinking about that is we became so successful. After our early problems got fixed, we weren't having failures anymore, so people said, "Okay, maybe we don't need this insurance policy any more; we're insurance poor, let's just build one and launch one." I think we're committed to that kind of approach for the immediate future.

The other thing is we had a very highly integrated and motivated group of people; we had a cultural mind set among project members that although we were trying stuff that was bold and new and nonconservative, we also learned to be very careful with what we did. People say "be conservative," but I don't think that is the answer. You can't be conservative in this business, but you can be careful. Voyager sustained that concept, or reinforced it.

I think those were two useful concepts: flexibility—find a way to keep flexibility and the capacity to work around problems when they occur; and be careful—you have to be careful.

ROBERT J. (BOB) PARKS
Project Manager, Cruise Phase
Born April 1, 1922   Los Angeles, California

Robert Parks earned a bachelor of science in electrical engineering, with honors, at the California Institute of Technology, graduating in 1944. He then served two years in the U.S. Army Signal Corps, including a tour in occupied Europe. While in the Army, he received considerable additional schooling in electronics and radar at Harvard, MIT, and Fort Monmouth.

After being discharged as a first lieutenant, he worked briefly at Hughes Aircraft. He joined JPL in April 1947 to develop guidance systems for ballistic rockets such as the Army's Corporal and Sergeant missiles. He went on to become chief of the Guidance Division, and later was appointed project director for the Sergeant missile.

After JPL was transferred to NASA in 1959, he was given the task of developing and carrying out a program to explore the other planets of the solar system using unmanned spacecraft. When later the lunar program was added to his assignment, he became head of all spacecraft projects at JPL.

These included the Mariner series to Venus, Mars, and Mercury; the Viking orbiter to Mars; and the Voyagers to Jupiter, Saturn, Uranus, and Neptune—he was on special assignment as Voyager project manager between May 1978 and March 1979.

Other missions he administered included the Surveyor lunar soft lander (in 1965 and 1966 he had been on special assignment as Surveyor project manager); the Seasat Earth Orbiter to demonstrate the value of observing the oceans from space, and the Infrared Astronomy satellite to accomplish the first all-sky survey in the infrared.

He was appointed deputy director of JPL in 1984 and served in that position until he retired in June 1987.

He is a Fellow of AIAA and IEEE and a member of the National Academy of Engineering. He was awarded the Exceptional Service and Distinguished Service medals by NASA, the Goddard Astronautics Award by AIAA, and the Distinguished Alumnus award from Caltech.

He holds a commercial pilot's license with instrument and glider ratings, and is a skier, waterskier, wind surfer, hiker, and tennis player.

The power of teamwork can be awesome.
The team must be well constituted and suited to its task,
but, when motivated, such a team can accomplish wonders.
Team members must believe in the team objective.
Could the human race have a long range common objective?
It should be to learn all that's possible about the origin and evolution
of the Universe including all its life forms. I dream of a large,
sponsored, long duration, international team to pursue that goal.
—Robert J. (Bob) Parks

ROBERT (BOB) PARKS

How did you happen to be at JPL?

I started at Caltech in 1940 to take courses in electrical engineering. A lot of things were happening at that time in electronics; it was an exciting field. When World War II started, I signed up in the Electronics Training Group, which was part of the Army Signal Corps. They promised us that they'd let us graduate and then we would get a commission. We went through summers, so I graduated in February '44 and immediately went to basic training and then Officer Candidate School.

Then they sent me up to Harvard and MIT for a seven month period to learn about radar, which was brand new at that time. After that we went to Fort Monmouth, New Jersey, for training in a specific radar, the SCR 584. Then I was sent over to Europe in the Army of Occupation for awhile.

When I was released from that, I came back, looked for a job, and was hired at Hughes Aircraft Company in the radio division. We were put to work on far-out concepts of missiles for push-button warfare. These were all just paper studies. In the meantime a couple of my classmates who had been at JPL called me up and said how about coming over here? So in April 1947 I went to JPL and liked what I saw. They were actually doing hands-on engineering work for a rocket-type missile, and they offered me the job to be the first and only, at that time, guidance and control person for their work. Dr. Pickering was there; he had known me as a student and wanted me on the team, so I went to work for him at JPL.

After awhile, the army decided to convert the rocket test vehicle, known as the Corporal, into a weapons system. It hadn't been originally designed for that at all; it was just something for demonstrating and learning about the techniques to build and guide and control rockets. Here the training I mentioned earlier on the 584 radar came in handy. We were able to design the guidance system for this rocket around the 584 radar. We had a lot of learning experiences on what it takes to make things operate in the environment of rockets and space.

We then went on to build another rocket weapons system called the Sergeant. The ranges of these things were around 100 miles. If anyone had told me at that time that in a few years we would be launching rockets to the moon and planets, I would have thought they were crazy. But it was all that background that enabled us to learn the hard lessons as well as the proper techniques and approaches to be able to design things that would operate properly under those conditions.

Sergeant was the first project I was director of.

I understand, from Bruce Murray and others, that the problems the two Voyagers were having after launch were serious enough that they wanted a top person to get in there and straighten things out, and that's when you came into the picture. What do you remember about that?

When we transitioned from the Army Ordnance Department to NASA, I got assigned the job of initiating and carrying out planetary exploration at JPL. So we started from scratch, more or less. I was in that job through the early Mariners. Then, because of some problems that developed in the lunar program, it was combined with the planetary, so I had both lunar and planetary projects at that time.

The project managers in both programs reported to me, so I was responsible for Voyager for its full lifetime; but for most of the time I wasn't the immediate project manager for it.

We had lots of challenges with both programs. One of the lunar projects I inherited when this transition took place was the Surveyer, which was a soft landing on the moon. It was a very ambitious concept: to have an automatic spacecraft go to the moon and control itself so it would land softly on the moon and survive and take measurements for an extended period.

That project got in trouble, so they asked me to step in and become the project manager of that one.

Fortunately we were able to pull it out and eventually it worked very well. Of the seven surveyors launched, five successfully landed on the moon and took pictures and other data. It was very important for the Apollo program, because up to that time we weren't sure what the surface of the moon was really going to be like. There were some theories that said it might be so flimsy on the surface that you'd sink right into it; it wouldn't support the weight of the spacecraft or men.

So with that background, when problems occurred with the Voyager receiver, a key critical element, Murray asked me to take charge of the project. He also asked Pete Lyman to join the project. Bud Schurmeier had been the project manager up until just before launch, then Bruce asked him to take another major job, and the fellow who was then the project manager had been the spacecraft system manager, Ray Heacock. I recommended against the move.

Against your making the move?

Yes, because I thought they were recovering from the problems, but Bruce wanted to take the step, so I moved in and took charge of the project from that time through its successful Jupiter encounters.

You obviously have great problem-solving, trouble-shooting ability. How did you fix it?

That is a hard question to answer with any certainty. You come in and assess the state of the problems and the approaches toward solutions. Most of the people I knew and had some calibration on, and I knew what they were doing. You assess whether the right people seem to be doing the right jobs. As much as anything, you assure the people that, despite the problems they're having, that the management is behind them, and that everything will be done to make it a success; boost the morale of the people, and give them confidence, and maintain their motivation.

It's just a matter of going in and trying to keep people's spirits up, keep them motivated, keep them pointed in the right direction, and encourage them to give it their all—which was hardly ever a problem at JPL. There was such a rich team spirit that evolved at JPL; it was one of the major factors in the great successes.

Okay, I'm going to push you a little harder. You're so obviously capable and modest. I know that Dr. Murray had to twist your arm to get you to take over Voyager. Now what aspect of that was really the key? Was it your people-handling ability, or your engineering background, or your experience? What was it? For example, I've heard that you were wonderful to work for because you really listen.

That is an important factor. You've gotta listen to people and make them appreciate that you're listening and really want their opinions, and will give them serious consideration. It's a mixture of being able to understand and contribute to the technical considerations, but to do so in a way that the interaction is effective and motivating.

It's a matter of asking them to really go through the nature of the problem, what the attack is, having group meetings where you'd go from brainstorming to detailed analysis of the situation. What extra work ought to be done, if any. What the schedule limitations were, and trying to understand all aspects.

One of the big successes of JPL, one of the things that gave me the greatest satisfaction from being a part of it, was the fact that objectivity was the theme. When you're dealing with engineering facts, it isn't what you *want* to believe that's going to matter, it's what you've got good sound, solid basis for believing. You've got to be completely objective; you've got to get subjectivity completely out of the picture. Lots of times, human beings being what they are, it's awful hard for them to give up some of the subjectivity.

Another aspect of this consideration is openness and honesty. We fostered an atmosphere of complete disclosure of all events and data that could in any way be pertinent. No one would be penalized for disclosing the full story, even if they had to admit that they had inadvertently contributed to causing the problem.

So probably what I offered the most was maintaining "let's keep it completely open and objective; let's really understand what the complete story here is, objectively determined, and what does that lead to in the way of solutions and actions to take."

You had more group meetings than were previously being held?

There were some we added. I think it was that, plus getting this tone or tenor that we tried to interject into the meetings: objectivity, and let's make sure everybody brings up everything that they can think of that might be pertinent. Get all the facts out on the table; let's look at them objectively and try to interpret what this puzzle is trying to tell us.

I understand that it was a big thing when you retired because you were called the soul of JPL. A speaker at your retirement that evening said you were noted for chairing a meeting and listening carefully to people by the hour, with hardly an interruption, before you made a decision; carefully listening and then objectively deciding the issue. Does that accurately sum up your approach?

Yes.

What lessons or advice could your experiences provide to other managers, not only in aerospace but perhaps also dealing with other problems, even social issues?

That's an interesting point. I think the country made a great big mistake when they said if we can put a man on the moon we ought to be able to solve poverty. That's a complete non-sequitur. Putting a man on the moon is nonpolitical, completely technical, in which objectivity, and not letting subjectivity enter the picture at all, is the key factor. Solving poverty, if it can ever be solved, is a completely *subjective* issue. There's no equations that can help you solve that problem.

Poverty is just one example of the "If you can put a man on the moon, why can't we do that" argument. They're usually things that involve social interactions; subjective people in subjective situations.

To answer your question, what I'm leading up to is I don't think it's easy to translate lessons learned in the type of experience I had to the majority of issues that face the human race.

What about focusing within an organization. What could General Motors learn from Voyager's procedures?

There was almost a complete lack of office politics. That's part of what I call teamwork. People were willing and motivated to subject their own immediate interests to the good of the team.

How did that willingness come about?

I think people believed in the program objectives, and the techniques for achieving them, and that their contributions were important, as indeed they were.

From your own personal perspective, what was the biggest satisfaction and the biggest disappointment? Looking back over the Voyager project, what was the real high, and then we can talk about the real low.

The high wasn't a single event. It was the culmination of lots of events, each one of which was a high in itself. Government projects are not noted for holding to cost estimates. Not only did Voyager succeed in achieving all of its objectives, and more than its planned objectives—going on to Uranus and Neptune, and it had to do it on schedule in order to go on the correct trajectory—but it did it within the original cost estimate. In actual dollars it was not the original cost estimate, but if you take inflation into account, which is only fair, it came in slightly under its original cost estimate. That's just one of the things that culminated in the complete high of the project.

As you approach an encounter with a planet, the intensity becomes very great, the action and the returns; it's a very intense period. Then you calm down to a normal pace for awhile until you get to the next encounter. You go through these great big peaks, so each one of those was a great high as long as things were working, but the real high was the culmination of all these things occurring. And then, just the great increase in knowledge of the outer planets that's been brought about by this project for the benefit and knowledge of the human race.

The biggest disappointment to me was when they decided not to go ahead with what the original plans had been, for the Grand Tour.

There was a price that some people paid. A lot of people worked long hours?

Oh, they do put in long hours. My wife says that JPL was wife Number One for me. The hours and the nights spent, you can't imagine. There were tense periods, like the times around encounters, when you were involved night and day. There were other periods when things weren't so hectic. There were untold times when I flew to Washington to talk to NASA; sometimes night flights, so I would be at the last meeting of the day at JPL and the first meeting at NASA Headquarters the next morning.

Outsiders see these wonderful things happen, but they don't really understand what went into them.

There were divorces, but it's hard to tell what was the real cause. You sometime have to see through a lot of agony before you get to the ecstasy.

What do you remember about the first launch, the launch of Voyager 2? Your feelings and thoughts leading up to launch and then when the problems appeared?

The launch was down at the Cape, of course, but the mission operations were all back at JPL. I was down at the Cape for the actual launches of the two, and we got some telemetry directly from the Cape, but it was only for the very early part of the flight. The rest of it, after the spacecraft was separated from the Titan rocket stage, was picked up by the Deep Space Network, and that was all funneled back to JPL. So we first got word of some unexpected events and apparent problems by telephone from JPL.

Of course you get swept up completely in trying to understand what it all meant, and what the possibilities could be for what was causing these problems, and making sure that the best communication was taking place. A lot of the spacecraft experts were at the launch, and others were back at JPL, as was the telemetry data, so we were having to communicate between the Cape and JPL through a limited number of phones.

We then got back to JPL as quickly as we could and spent many a long hour well into the night trying to analyze and review options for what to do about the various problems.

What happened next, after the bad news came in?

We were able to conclude that no immediate action was necessary. That's the big issue when you first have a problem like that. You have to decide whether it's something to take action on quickly, or do I have reason to believe that I can spend time to analyze it further and make sure that I've really explored all the options of what I can do about it. So you concentrate on that in the very beginning. Of course you're never certain about that, either; you have to interpret the data that you have at that time.

We were able to conclude that no immediate concern was there. Some of the phenomena were intermittent and would come and go but didn't seem to be life-threatening, so we were able to head back to the lab and start doing these reviews.

Was that the same day?

It was a couple of days later.

Let's focus on the period when you were directly in charge of Voyager. What decisions came up that you personally had to make; they couldn't be handled by the regular hands-on people?

I instigated a daily routine that at 8:00 every morning a group of us would get together. It included Ed Stone, Ray Heacock, Pete Lyman. We would go over what the issues were, what the latest test results were, and what actions each of us was going to take that day to help move things ahead as quickly and properly as we could.

Did those meetings continue throughout that year or so?

Yes, while I was in charge, all the time.

So you got these three or four people together every morning?

Every morning, right. It was standard routine. We'd spend only as much time as we felt we needed to, so people could get off and do their other things. It would sometimes last half an hour, or two hours, depending on the need, but we tried to keep it fairly short.

Part of it was to assess what decisions needed to be made and where to get the best information on those decisions. Each of us had a special activity. Ed was primarily concerned with the scientific objectives and the science instruments and the views of the scientific experimenters; how they needed to be communicated with and also what their needs were, and how that entered the situation. Ray Heacock, being the spacecraft systems manager, knew the most about the specifics of *how* the spacecraft worked, so I instigated immediately that Pete Lyman would worry about *what* we should do with the spacecraft; Heacock would worry about how we would accomplish that.

I don't think Murray was ever comfortable about that division of responsibility because he kept needling me about "when are you going to put a regular organization in there?"

But for me it made an awful lot of sense. We were using the best of the people there, and they all had to work together anyway; they couldn't go off and do their thing independently. So part of these morning meetings was to make sure that all of us were on the same wavelength and knew what the signals were for the day, and things like that.

What was Murray needling you to do differently?

When he put Pete Lyman and Ray in there with me, he thought I would come up with some different and more usual organizational arrangement for the people.

When you became head of the Voyager project, there were some decisions that were fairly routine: you could say, "Ray, this is yours" or "Pete, this is yours." But were there decisions you had to handle yourself?

There were a lot of that type. One of the critical things was the command loads that were sent up regularly, and the validation of those command loads. The spacecraft has three computers in it, and each computer has memory, and memories can get altered.

So one of the big discussions we had was about "refreshing the memory" on the attitude control system. This was something that had never been done before, and if something went wrong, you could really scramble the memory. By refreshing, you completely reset the memory from the ground. If anything went wrong in doing so, you could get into a state that you could never recover from. On the other hand, if you didn't do it, you could be impacted in minor to major ways by possible alterations in the memory.

So we went through quite a series of discussions, meetings, and reviews of the pros and cons, how you did it, what the dangers were. That was a typical decision. I finally made the decision to go ahead and let's try refreshing the memory. Fortunately it worked fine.

How often were you faced with that choice?

The big issue was the first time. We hadn't ever done it before. We didn't have to do it too often, but on the other hand, after you did it once and you knew the routine, you were less concerned about doing it again.

At what stage did that happen?

When we were getting ready to encounter Jupiter.

Which Voyager?

Both of them. The very first time it was done was on Voyager 1, but we did it on Voyager 2 also. Just because it worked on Voyager 1 didn't guarantee it was going to work on Voyager 2.

So you were the one who finally had to make the decision. What other examples come to mind that you personally had to decide?

I would usually review and give final approval to all the actions taken to correct major problems. We'd have a discussion at one of the meetings we talked about, where we'd understand what the latest data indicated, what the nature of the problem was, and what action was selected, and then schedule for taking that action. So on things that related to nonstandard commanding of the spacecraft, I would usually make the final decision.

What would be an instance of that?

Whenever we were correcting a problem on the spacecraft. Like the attitude control system needed to have software modified, either to correct a problem or to enhance its performance, either for a special period of time or for continuous operation. Those would come in the category of "nonstandard" command. If it wasn't a more or less routine sequence. "Sequence" is those series of commands which would tell the spacecraft what to look at, what instruments were on, and what data to take.

With these nonstandard commands your staff would do the basic thinking, and would come up with a recommendation?

Right. For each problem a select group of people, each expert in a different aspect of the problem, would do the staff work.

So they'd come to a recommendation but you had to do the final signing off on it?

Right.

Was that a tough thing to do emotionally? Were you pretty confident or did you have to stop and think?

Usually I had been involved in the final discussions of the problem. Part of the recommendation was a detailed discussion of the rationale behind the recommendation.

So by the time it got specifically firmed up, you had a pretty good idea.
Those were the kind of decisions you get used to in that line of work.

An article about your retirement reported that you were praised for your orderly intellect, your knack for quickly focusing on the cause of problems and your comprehensive understanding of missions. What else about your management style?

Not getting distracted. Keeping the objective in mind. Keeping your eye on the ball. Know what you're after and don't deviate. Don't get waylaid by all the other considerations. It's very prevalent in human endeavors to either get emotionally involved so that you lose sight of the objective, or to get sidetracked and maybe get mad at somebody so you take actions to get even with him rather than to remember what the objective is and keep going after it.

So what I brought to it was always to keep in mind what the objective was, not get sidetracked, and be objective and open throughout the whole process. Yet doing it in a way that encourages, motivates, and recognizes the value of all the people involved and the importance of their views and their thoughts.

What would have been a really hard decision you were faced with?

Things didn't seem to work that way. A difficult decision is one where you have to make a decision based on less than adequate facts or analysis. There just didn't develop things we had to do in such a rush that we couldn't take time to get the best opinions and/or analysis done to help guide decisions. Even the best isn't complete; if you know all of the facts, it's not a decision, it's a conclusion. There's always some uncertainty involved.

One thing I'm not, and I don't like the style, is a hipshooter: people who think they know everything and so give immediate orders without taking time to at least hear about all the other opinions and arguments. In emergency situations you have to make snap decisions, but I can't recall any real circumstance where that came up in Voyager. Even when I made the decision on refreshing the memory, for example, I didn't *know* it was going to work; I *thought* it was, but I didn't know it was going to work.

Which computer were you refreshing?

It was the most critical one, the attitude control.

In a leadership position, some of the decisions are technical, some are budgetary. Did you have any tough budgetary decisions that really caused you to think?

Well, yes we had to continually keep in mind the costs and to live within our budget. Each year we got so much money based on what we had asked for at some earlier date. You have to make your money plans two or three years in advance to go through the cycle with Congress and appropriations and authorizations and so on. We were never, at least when I was in immediate charge, faced with decisions that required doing things you otherwise thought were critical that you couldn't do because of money. We were able to find ways to minimize the cost and judicially use our reserve to keep within our budget.

What about personnel decisions? Did you ever get involved in having to make unpleasant or hard personnel decisions?

In my position as head of the flight projects I would be instrumental in selecting the project manager. That one went all the way to the director for approval, but I never got turned down. I was also involved in choosing the next level, like the spacecraft system manager, the mission operations managers, and so on. I was not heavily involved in the lower appointments.

There were a number of times in my career where I had to make unpleasant but necessary personnel decisions. However, when I was put immediately in charge of Voyager, the team was all in there and I had no reason to believe that we would be better off by changing anybody. They had the background and were performing well, so we concentrated on keeping the team working as well as possible and working effectively in concert with one another. So I was not involved at that time in any personnel decisions of the nature you're describing.

Another kind of decision some people have faced is when you have two mutually exclusive choice: you can either point the scan platform this way or that way, and either way one person is going to be happy but somebody else will be disappointed.

Oh yes, we had lots of those kinds of issues, primarily with the experimenters. Your example is a good one. We had to point out that the spacecraft could only do so many things at any one time, and if you did some of these, you couldn't do some of those. In the science aspect Ed Stone was the one who usually ended up refereeing those. I kept abreast of what the issues were, but as long as there weren't other implications of any of them, which usually there weren't, I would let him handle maximizing the science return.

How about other than science? Situations in which you had two things you wanted to accomplish, but you didn't have the time or the funds or resources to do both, so it wasn't an easy choice?

I suspect Bud had a lot more of those, during the development phase of the project, because that was the time when they were making decisions on what new technology would be included in the spacecraft, and how that would effect what the spacecraft was capable of doing, but those didn't occur during the time that I was in immediate charge.

So one of your big contributions right after the problems following launch was just getting the crew that was already there to try to focus better, to work together more efficiently.

And to make sure communications were complete.

Communications within the group, or with other groups?

One of the things outsiders may not appreciate is how much interaction there is. Any decision you make is going to affect lots of different areas. You need to communicate first to understand their interactions, but second to ensure that the consequences of a decision are known and expected by all the people around. There are many times when things would happen where lack of communication could have, and in some cases did have, serious deleterious effects.

What would be an example?

If you do something in one piece of hardware that might affect the nature of the signal coming out of it and going into another piece of hardware, and the other person didn't know about it, they wouldn't work together. And when you've got computers talking to each other as well as the normal kind of electronic intercommunications, there's a lot of potential for that. It's not just the equipment; it's the schedules, the interaction of the scheduled activities. If a schedule is changed somewhere it may require accommodation or changes in other people's schedules, and so on.

The project had detailed schedules, which I used to make everything mesh, to come together. Like the creation of the command sequences. That started off by the scientists all saying what they would like to do during a certain defined period. First they would separate the ones that were incompatible, and then they'd make the decision we talked about earlier, about which would be selected over the others.

Then the detailed time lines were sent out and the sequences of commands were put together that should cause these time lines to be executed. We had a simulator to run these commands through, to make sure they did what we expected them to, and we often were surprised. You'd think you had all the commands right but the simulator wouldn't respond the way you expected it to.

You quickly learn that you don't trust things without testing them first. To get things to work right, you've got to have the theory down right, as best you can, and go through all the analysis you can, but finally you've got to test it in some way before you do it in earnest because the testing is a critical and crucial part of making things work right.

To improve the reliability of parts used on earlier, shorter missions, were these critical parts still made by some external contractor or were they made within JPL?

We didn't make any parts, but we did an awful lot of testing of parts and an awful lot of selection of parts. It was complicated in the Voyager case because of the requirement for them to operate in the high radiation environment we were going to encounter at Jupiter.

Fortunately for us the military was interested at the same time we were in improving reliability, so programs were established to cause the parts manufacturers to set up special lines with more inspection points and better techniques for constructing them, to turn out first-grade, quality parts. We were recipient of some of this from the military, and we were also working with parts manufacturers ourselves. When you do all these special things, it causes the cost of the parts to go up, obviously, but the parts' costs themselves were not a high percentage of the cost of the total spacecraft, so they could go up by several factors and not have a major effect on the total cost.

We had a special group set up to work continuously on the parts problems: how to test parts, how to screen parts, how to set up the lines the best way to produce these parts. When we started on Mariner 1 and 2, none of that existed, so we just had to take parts built for TVs and radios. We did a lot of screening ourselves, and unusual testing, because the manufacturers didn't do that kind of testing. We also derated the parts; in other words, we wouldn't push them to their extreme values. In a transistor you wouldn't put the maximum voltage on it; you'd derate the current you put through it. So all these things were contributing to the success.

Now the other thing was that integrated circuits came along. The process of making them required a lot of painstaking detail and uniformity in their manufacture. So today's ICs can have thousands and thousands of transistors in a little integrated circuit, and it's amazing the reliability they get out of them now. We both contributed to, and benefitted from, the great advances in the components and the electric circuitry.

Did you personally, in your Voyager duties, have contact with the military? Did you trade information back and forth?

Oh yes. Mostly in this parts area we had a lot of interaction. We compared notes, and the AIAA and IEEE, the engineering associations, set up meetings where they'd bring all the interested parties together, but we had direct contacts also. The military, too, was interested in parts that would survive radiation, so we did a lot of note comparing and cooperation in that area, too.

Was there something beside the unique planetary mission that explains JPL's success?

My son is an engineer at JPL. As a student he worked for Rockwell International. Even as a 20-year-old student he recognized, at least in the part where he was, that it was very inefficient. The left hand didn't know what the right hand was working on. When he came to JPL, he said it's miles above—a smooth wheel operating.

Why? What makes the difference?

I think there was a set of unique circumstances. There's a lot of good engineers and engineering companies out in the commercial sector: the TRWs and the Hughes. Many of the aerospace companies did a lot of good work, but the competitive system, which in general I believe in completely and fundamentally, doesn't fit the kind of work that JPL did.

Maybe the fact that the objective was purely scientific and was government-sponsored makes the difference. On the other hand, commercial space programs are undoubtedly best done by commercial companies. Communication satellites are a good example. Competition can, and has, played a natural and beneficial role, and the more competitive companies have reaped financial rewards.

You get a team worked up to where they're very good in a commercial company, but then maybe they don't win the next contract, so that team is broken up, dispersed, and goes off to something else. Whereas we at JPL were able to maintain the continuity of activities over these years, taking bite-sized steps along the way, and were able to maintain and grow the teams. They evolved and improved, and they didn't go through death throes. For a unique activity like exploring the planets, that's very important. It's not a case where competition would help any. There's no point in having two teams trying to send spacecraft to Mars.

What else should people know about you and Voyager?

The opportunity to have been involved in and associated with an activity such as Voyager has really been a tremendous reward to me. I was extremely lucky to be in the right place at the right time, to have been able to play a part and observe the results and see all the fantastic efforts that went into creating something like the Voyagers, and getting the results we got from them. I think that everyone who has ever touched it, was ever involved in any way, felt the same way about it. It was the kind of reward that not everybody gets in life.

PETER LYMAN
Deputy Project Manager, Jupiter 1
Born February 9, 1930   Berkeley, California

After seven years in the merchant marine, Peter Lyman attended the University of California, Berkeley, earning a bachelor of science in mechanical engineering, an master of engineering in naval architecture, and in 1963 a doctorate in mechanical engineering.

He then came to JPL, where he has managed the Spacecraft Design and Integration Section, the Applied Mechanics Division, the Information System Division, and the Deep Space Network. Before his appointment as Voyager's deputy project manager in 1978, he worked for 15 years on Mars missions, including Mariner 1964 and 1969, and was Viking's director of Spacecraft Performance and Flight Path Analysis. He was appointed laboratory director in 1987 and retired five years later.

He has twice been awarded NASA's Outstanding Leadership Medal, as well as other medals from NASA and the American Astronautical Society.

He is a Fellow of the AIAA and is also a member of the American Society of Mechanical Engineers, the Institute of Electrical and Electronics Engineers, the American Astronautical Society, and the International Academy of Astronautics. He has been president of the Pasadena Chamber of Commerce and director of the United Way and Boy Scouts in San Gabriel Valley.

Salvaging and restoring old railroad locomotives is one of his pastimes. He also holds a private pilot's license, instrument rated; FCC amateur and commercial radio licenses; and is a mountain rescue team member and scuba diver.

In order of importance, the three most important factors in selecting team members: 1) integrity, 2) openness, 3) skill.

—*Peter Lyman*

PETER LYMAN

When did you become involved with the Voyager project?

To answer that question, I'm thinking back 20 years or so. It was the same kind of involvement that hundreds and hundreds of JPLers had. We were here working on various projects in the late 1960s at about the time that a mission called "The Grand Tour" was proposed. The alignment of the planets occurs about every 170-odd years that allows a normal propulsion launch from the Earth to go to all the major outer planets: Jupiter, Saturn, Uranus, Neptune, and Pluto.

That mission, which I was not working on at the time, had been proposed, but did not make the budget. Subsequently a scaled-down mission called Voyager was approved. Publicly and legally it was designed to be targeted at Saturn, but there was a gleam in some people's eyes that it could probably go on and get maybe one or two more planets, perhaps even Pluto.

During this period I was a casual onlooker in the sense that I was designing the Viking system that went to Mars. I had been initially the system engineer of the Viking orbiter and then, during the operational phase of the mission, I was the director of spacecraft performance and flight path analysis. In other words, the spacecraft, the two orbiters, the two landers, and the navigation activities were my responsibility. I had no actual personal involvement in the Voyager project up until about a year before launch—August and September of '77.

My first involvement with Voyager was to sit on review boards, looking at the readiness of the mission operations system to conduct a Voyager mission. This was in the year before launch. I was asked to do that based primarily on my experience in developing the operational team for Viking. There was also Gentry Lee, who had an equivalent job to mine, except he was responsible for the science activities and the mission planning of the science on Viking. So Gentry and I were on the review board to assess readiness.

We were the two guys on the review board who consistently said they weren't ready. We based that on looking at what we had gone through and the mistakes we made and the ways we had to fix them. To a large extent we saw the same kinds of mistakes being made by a different set of people who hadn't been through those issues. We made our positions fairly clear; in fact at one point we were referred to as "the Black Hat Twins." That was my prelaunch involvement.

The spacecraft subsequently were launched, and during the next nine months there were a number of spacecraft hardware problems and operational problems with the spacecraft team. The mission was in shaky shape. It was not getting ready for the Jupiter encounter at all well, so the director of the laboratory then, Bruce Murray, decided to change the project manager out, and assigned Bob Parks, who was the assistant laboratory director for flight projects, as the new Voyager project manager. Murray appointed me as a deputy project manager, and the existing project manager, Ray Heacock, was moved to a second deputy position. Parks was given the job to get these spacecraft and team ready for the encounter at Jupiter because there were just too many problems.

So we had a three-man team that was the management team, at the project manager, deputy project manager level. That's when I first started. I was called in on a Friday and told that this job started on Monday. That was in May of '78. I, at the time, was division manager of one of the JPL technical divisions. I'd just come off my Viking assignment. We were about a year from encounter, so we had about a year to get a very ragged operation under control.

I don't mean to denigrate individuals, but, as a team, they weren't a team. They were a lot of individuals, and it took some reorganization, some trading of players, to put the right mix of people together. It was like what a football coach would do in organizing a team: do it in a way that you could handle the ongoing work and then build for the future. So our principal role, which lasted about the year and a half that Bob Parks and I were assigned to the project, was just to assist in getting the whole operations team in shape for the encounter.

We went through the encounter and then, three or four months later, Bob Parks and I left to go off and do other things. So that was my direct involvement with Voyager. I have been associated with the project one way or another ever since.

In 1980, about a year and a half after I left Voyager, I took over the management of the Deep Space Network. It's a worldwide tracking network, and of course that was one of the key technological advantages that enabled Voyager to return the data from Uranus and Neptune, to get the quantity of pictures back. I spent seven years doing that, and since then I've been the JPL deputy director and so I have involvement with all the projects.

Impressive.

Day by day, it was fun. Some days were agony, but when you integrate over a period of time, it's an exciting, interesting place, and fascinating people—the engineers, the scientists, the technicians—everybody you work with. It's hard to find a place that could be more fun for an engineer, for a poor old naval architect.

You're a naval architect? Did you come up through the Academy?

No. In high school I was the kind of guy that got As in mathematics and physics and Cs and Ds in history and Spanish. When I graduated from high school, I didn't have an academic record that said UC Berkeley was going to accept me at the university. So my friend said it'd be fun to go to work in the merchant marine during the summer and sail around, so I went to work for Army Transport Service, as a government employee working on troop ships for the summer. At the end of the summer, the ship was in Italy, I was having too much fun, so I said, "Well, for a year I'll just sail the world." And it lasted seven years, until I met a woman I wanted to marry and decided I'd better go to school.

In the meantime the services were unified in 1949, and we all became civilian employees of the Military Sea Transport Service. They took the ships away from the Army and gave them to the Navy. It's a little known fact that the Army actually had more ships in World War II than the Navy. The Navy had the fighting ships, but the Army had all the landing craft, and almost all the transports that ran back and forth across the oceans were actually Army Transport Service.

Anyway, I worked up through the Korean War doing that. Then, when I decided to go to school, the next question was what did I want to do? Since I had a mechanical bent—I was an engineering officer aboard the ships—I thought I could design better ships, better engine rooms.

So I went back to Berkeley, and they put me through an extension program for a semester and said if I got two Bs and two Cs they'd let me in. I got four As, and they let me in. At that time school seemed easy because I knew what I wanted to do: I wanted to design better engine rooms for ships. At Berkeley marine engineering was a minor in the Department of Mechanical Engineering.

Then, after being in school for about two years, I realized that building a ship was a bigger challenge—the whole ship, not just the engine room. And so when I graduated in mechanical engineering, the Society of Naval Architects made me their national scholar to get a master's degree. I was raising a family during the time; I was married and had three children, so I took two years and with the scholarship plus teaching I got my master degree in naval architecture.

But I decided at that point that I really wanted to go into teaching. I could see the handwriting on the wall for the American merchant marine, which was all downhill. It has virtually ceased to exist today. There are no ships being built in this country, and I can almost name from memory in five minutes all the American ships that are flying under American flags.

To get into teaching, I took my doctorate in mechanical engineering at Berkeley, with minors in naval architecture and physics and math. When I finished that, I accepted a position as an assistant professor of mechanical engineering at UC Davis. It was the first year of the College of Engineering, and there weren't any students, but there were a lot of committee assignments.

There weren't any students?

Well, hardly any. I had one class with about 15 students. It was a sophomore course in engineering statics. Lots of committee assignments. And my wife didn't really want to move to Davis; she liked Berkeley. So I talked to the dean, who had been a professor I'd studied under at Berkeley. He suggested that I go get a year's worth of experience someplace. And so I called two friends, one at Shell Development in Houston and one here at JPL and said, "I'd like to come out and talk to you about doing some research for a year."

I went down to Shell Development. It was 100 degrees, with 100 percent humidity, and it was "Go Texas Week." They hustled me from air-conditioned car to air-conditioned house to air-conditioned building and slapped me on the back with their big hats and boots on. And there were no mountains and none of the things I like a lot. Then I went to JPL and talked with a fellow I knew there. He offered me a job for a year doing research, and I accepted it.

So I came here for a year, and the day I showed up they made me an engineer responsible for some of the mechanical hardware on the first spacecraft going to Mars. I spent the year designing that, but at the end of the year we hadn't launched the spacecraft and I was damned if I was going back to school 'til I saw how my hardware worked. So I stayed another year. Then they made me a supervisor. Two years, three years, four years passed and I finally called up and said, "I'm staying. It's too much fun."

I had that peculiar background of engineering and naval architecture. Naval architects are probably doing some of the world's best work in system engineering. They're dealing with a city afloat—engine rooms, power plants, food handling, cargo handling, people. So the transition into spacecraft system engineering, a very complicated thing working with lots of people, seemed fairly natural. Anyway, it was that funny background.

And both naval architecture and spacecraft systems are dealing with isolated, self-contained units.

Yes, they have that in common. They basically have to take care of themselves. It's particularly the deep space missions—the Voyagers, the Vikings, Galileos, CRAFs (Comet Rendezvous Asteriod Flybys), the Cassinis—missions like that. By the time that the spacecraft can tell you that there's something not right about the spacecraft—in some cases it is "my heart is not working right" or "my brain is not working right"—by the time it tells you that, and an hour later or five hours later, when the signal arrives at the Earth, it's all over.

So you need to design spacecraft that can detect their serious faults and correct them automatically. You can't wait five hours because, by the time you find out about the problem, by the time you analyze the data that says "what is that telling me, what do I do about it?" and send the signal back up, maybe 20 hours have gone by. The spacecraft was dead 18, 19 hours ago.

So you design the spacecraft to detect those faults that are life-threatening. With those faults that are not life-threatening, you just have the spacecraft put itself in a safe state, and in many cases you can take care of them from the ground at your leisure. But there are certain things that have to be taken care of immediately, and that happens aboard a ship, too. When the ship has a fire aboard it, you've got to put the fire out. You can't wait 'til you get to port; you aren't going to make port.

There are those aspects that are similar. I think the real issue is that the naval architect designs not only the shape of the ship, but also the structure of the ship and the hydrodynamics and the interior fitting and the navigation and the handling systems. It's very much of a system. You have to deal with lots of different things, lots of different people, different customers. And a spacecraft—I can see a lot of analogy. That isn't to say that all naval architects would like to build spacecraft, or vice versa.

I was also fortunate that the mechanical engineering curriculum at Berkeley was very much systems-oriented, very much broad-based as opposed to teaching lots of independent things. There was a lot of synthesis, where you had to deal across many technologies in the projects you worked on. Not all schools do that, but they did at Berkeley. It worked out well in my case. It prepared me, although I didn't realize it at the time, to handle a lot of the assignments that I was given here.

Focusing now on Voyager, what were some of the things that you remember about your work on it?

My contribution to Voyager was the year and a half that I was the deputy project manager. The job that the director gave me, and gave Bob Parks as the project manager, was to go over and fix up what wasn't working right, to make that spacecraft work and be a success.

How did you approach that?

I'd just come off of my Viking experience with a total of six years of going through the design of the spacecraft, the design of the operation team of the spacecraft, execution of the operations. And in that six year period, we did a lot of things right, but we made some mistakes along the way. When you're doing complicated things, you know you're going to make a certain number of mistakes, but you're prepared to fix those. You try to do most things right, but you can't do everything right.

I had gone through an experience on Viking of having the operations team designed to one set philosophy. I wasn't involved at the time, but a very rigid philosophy had been stated by the project manager, who said, "It's going to be done *this* way." They tried it for about a year, and at the end of the year they realized it wasn't going to work. At that time I and several others were brought in and told, "Hey, we'd like you to take over these responsibilities under a new set of guidelines."

So we were able to profit by what hadn't worked before, and we were able to tune up a very complicated mission. We had 750 people on the Viking flight team, in a mission designed to land two Viking landers on Mars, which we did. It was a very complicated and very precise mission.

It involved lots of entities. It just wasn't one single organization. It was JPL, it was the Langley Research Center, it was the Martin Marietta Company, both the spacecraft side of Martin Marietta and the launch vehicle side, General Dynamics, the Lewis Research Center. Virtually everybody. There were 20 major players, and we pulled an operations team together with some people from all these different organizations and made them work as a team, not as a bunch of individuals. So those of us who'd gone through the Viking experience had freshly come away from it aware of the importance of the team-building aspect of this endeavor.

Some of us were asked on very short notice, like a weekend, to go and take over the Voyager project. The job was fairly clear. We needed to figure out why we don't really have a team, why the group they call the "team" isn't working right, and fix it.

Bob Parks handled the upward and outward part of the project, the kind of activities a project manager does: dealing with the Lab management, dealing with NASA Headquarters, with the press, with the outside world, and seeing that we were doing the right things. I dealt with examining what was wrong with the operational team, and tried to do those things required to put the operational team on a sound basis.

And Ray Heacock, who was the other deputy, concentrated on getting the spacecraft side of the mission in better shape. He dealt with the day-to-day spacecraft sequence design—what we're going to do with the spacecraft and how to do it, and the engineering. He had been the system manager and one of the chief designers of the spacecraft, and knew it inside and out.

They were spacecraft that had a number of failures and they had to be studied. Why did the scan actuator stick? If it's stuck, how can we use it in the future? Are there some ranges we can use it in, some speeds we can still use it at? Running those tests and projecting ahead as to how we can utilize stuff that might not be just 100 percent perfect? So Ray concentrated on the technical aspects of it.

I started looking at why haven't we got a team. Most teams are built up of groups that have various functions: the spacecraft people who analyze the data from the engineering side, the science part of it, the people that design the sequences, the people that test the sequences. Without getting into personalities, there were some people leading some of those groups for whom team management was really not their long suit. Some people are better designers than they are operators. Some people are better operators, just like some people on a football team are better quarterbacks, some people are better ends. And in some cases we had people in key positions who really weren't the right people for that job at that time.

The best analogy I have is an army: you have peacetime generals and you have wartime generals, and they're seldom the same people. To run an army in peacetime, a general must handle building, training, dealing with politics in a regimented organization. The people who excel at that sometimes are not the people that can cope best in a war, which is chaos, not very well organized. The intent in war is to create chaos for the other guy, so you've got to manage chaos. Managing disciplined organization versus managing chaos really involves two different personalities.

Look at what happens when you build up an army over years, then you go to war—you very rapidly change almost all the leadership. Look at World War II in the Army or the Air Force. Lots and lots of the people who were in leadership positions fell out very rapidly. They weren't suited for the new demands, although they were very good at what they had been doing previously. You can draw those analogies.

We had several cases in which some key people were slotted in the wrong position. They were bright, brilliant, but it was the wrong job for them. So we replaced them and brought in people who had done equivalent jobs in the past very successfully. We redefined interfaces between groups and put discipline into the organization.

What was the situation you found at Voyager? Was Voyager going from a disciplined, stable state toward a chaotic one, or was it the other way around?

I think the leadership in a couple of the key organizations that reported to the project management level were more of the type who could manage peacetime armies as opposed to managing chaos. When you're dealing with a spacecraft in a very complicated mission which is not working right, you're into very complicated, unstructured things that want to become even less structured. It requires the same discipline that a football coach has to have. You have to be very rigorous, very precise. You can't take shortcuts.

Spacecraft are very unforgiving when you make mistakes with them. The Soviets made two bad mistakes dealing with their Phobos spacecraft, and they lost at least one of them by just being sloppy and probably the second one, too, although they're not exactly sure; they've never been able to say exactly what happened to the second one.

We needed people with a mental discipline of precise operational organization and a strong leadership that could understand how to make that work. We made those changes and brought people in. We're only talking about a few people here in the leadership positions, who had been through these kinds of operations before, who had a positive track record so that when they said, "This is what we have to do," the people working for them said, "Hey, that guy knows what he's doing; he's done it before."

And so it was a matter of looking at what each group's charter or role was. Each group has functions they have to do, and they have to interface with one another. Very often groups didn't want to work with one another. Looking at what the problems are, what the interfaces are, redefining interfaces if necessary and just getting a good operational discipline into the team—it seems awfully simple but it takes a lot of dedication, and you need to do it in a way that you don't destroy careers. There was nobody involved in this that wasn't very bright.

One of the hardest things is to remove a very bright person whom you may have worked with for a long time. The ones that had to be changed were people I knew; I'd known them for years and they were very competent, very bright and had been very successful doing other things. Now they just were in the wrong job. They probably had wanted to do it, but it was the wrong assignment for them. They were probably much more at home designing very complicated systems, and less at home working on an operational scenario.

You've heard about what goes on in a hospital emergency room: it's hours and hours of boredom punctuated by moments of sheer terror. You sit around waiting for something bad to be brought through the door, and suddenly what comes in is terrible and you've got to work very hard.

Spacecraft operations is something like that: it requires a lot of training, a lot of stuff that has to be done very methodically over and over again, without any mistakes, running checks and balances to make sure you're not making mistakes. And suddenly you have that punctuated by something very bad happening.

What would be an example of that?

Well, there were a number of things that happened during the first nine months of the Voyager mission. One example where something didn't work right was the scan actuator, which is a servo mechanism that points the platform on which instruments are mounted, so you can point the telescopes or the cameras at some part of a planet. The scan actuator was stuck: it wouldn't move. That was very serious. If you can't move your instruments around you're very limited in what you can do with your spacecraft.

The spacecraft's designed so that, even if that happens, you can still repoint the instrument because you can move the whole spacecraft, but you can't point it very fast. You can't do nearly as much science, but there was a degraded performance you can get out of the mission. When the platform stuck, the spacecraft team went to work trying to understand what the problem was and what they were going to do about it.

I've got to back up a minute, to tell you a little bit about the spacecraft. Remember, there were two spacecraft—Voyager 2 was launched first, then Voyager 1 was launched second. The platform stuck on Voyager 1, and people started focusing on that problem. There was a feature in the spacecraft designed to assist in recovering from failures in the communication system. The communication system on the spacecraft has two receivers, so we can send commands to the spacecraft so it can hear us from the Earth. It has the exciters, which are the devices that generate the signal to be sent to the ground, and the transmitters, the power amplifiers, which amplify it, which results in 20 watts output to the antenna.

This is put together in a subsystem that's called the radio frequency subsystem. This whole thing's called a transponder and a power amplifier. If an amplifier fails, we just don't hear the signal and we, on the ground, say, "Gee, why don't we hear the signal? Maybe the spacecraft's not pointed at us. Maybe the power amplifier failed. Maybe the exciter failed." You can generally fix those things from the ground, given you have enough time to do it. But if a receiver fails, we have no way then of getting a command to the spacecraft. Whatever's in the spacecraft memory is what the spacecraft will do. Voyager has two receivers, and the idea is that if one fails you can use the second one.

From a design standpoint, you'd like to turn them both on and leave them both on and be able, when you send a signal, to address it to one receiver or the other. If one doesn't work, you just change your address code, change frequency and the other frequency gets it.

But you don't have the luxury of doing that, because the spacecraft doesn't have enough power to keep both receivers powered. You're very power constrained in these missions. So you have a receiver turned on and you're using it; what happens if it fails? Then you let the spacecraft understand that the receiver may have failed.

How is it to do that? Well, the only way it can tell if the receiver has failed is that it hasn't heard any messages from Earth. So the spacecraft has a timer in its computer that asks, "Have I heard from the Earth recently?" And you can set that timer to say, "Have I heard in the last five hours? Have I heard in the last seven days?"

During a non-high-activity time in the mission, you set the timer to a week, which means that you have to send at least one valid command to the spacecraft every week. The command doesn't have to do anything, it may be what we call a "no op", but it has to be a valid command that is properly decoded all the way through the system so the computer says, "Hey! That's a real command." It went through and tested the system. But if that command doesn't go in, the spacecraft says, "I assume they've been trying but I haven't heard anything. I haven't received a command, so there must be a problem on the spacecraft."

So it goes to an algorithm that tries to change the configuration of the spacecraft. First it does the easy things. It doesn't change the receiver at first, because that's the last thing you want to change and fuss with. It looks to see if you're on the right antenna. It maybe changes the antenna configurations. It changes the power transmitter because maybe it's putting out a spur, a wrong frequency that's getting into the receiver. You can get something called a self-lock on the receiver, where the spacecraft listens to itself rather than to the ground. The spacecraft has a 20-watt power amplifier on the the receiver listening to the ground, and from the ground it's only hearing microwatts, or billionths of a watt; very little. So you build a spacecraft very carefully so it can't hear itself.

Well, if that transmitter has a spur, it's putting out a little extra energy in the spectrum that it shouldn't. It could lock the spacecraft. I've seen these things in space before. So you have the software go through a series of steps, over a period of days. It'll make one change, then it will wait to see if it now gets a command from the ground. Then it will take the next step. The last thing it does, and it's by design, it changes the receiver.

What happened was Voyager 1 got a stuck scan platform and the team went to work on the stuck platform problem, and they forgot to send the weekly command. Totally inexcusable. The spacecraft started going through its change-out algorithm, and, after checking everything else, it changed the receiver. Normally that shouldn't be any big deal, but about 30 seconds after shifting to the second receiver, the power supply in that receiver failed. So we'd gone from a good receiver, which we shouldn't have changed from, to a receiver that promptly failed.

And now there was nothing you could do from the ground. You had to wait for this algorithm to go back through its process of changing everything else, going down through a chain of events which takes a number of days, to where it would finally in desperation change back to the original receiver, saying, "Well, that's the next best thing to do." So it did, and the spacecraft came back, and eventually, several weeks later, we were able to get a command to it. But now we only had one receiver for the rest of the mission.

That was an example of where the spacecraft team discipline fell apart. That should never have happened. Of course, that might not have changed the fact that, when it eventually might have gone to the second receiver, it would've failed anyway.

But that was a case where we just did things wrong. And there were other cases. It was a cumulation of events like that that led Lab management to say, "Hey, we're going to have to tighten up the operation." As it turned out, the one working receiver on Voyager 2 did have a subsequent failure. It still received, but not the way it was supposed to.

So we have flown the rest of the mission, since before Jupiter, all these years from 1978 until today, with a very difficult receiver on Voyager 2—and that's the Voyager that went to Uranus and to Neptune. A certain amount of luck has allowed us to carry on. The thing we wanted to do in the operations was to build a team that didn't lose sight of things that you shouldn't lose sight of.

What was one of the toughest decisions you were faced with?

What was the most difficult thing I had to do in that year and a half? That's easy. When I first came to work for the Laboratory, I'd been here maybe a year and a half, and I was made a supervisor to replace the person I reported to, Richard Barlow. He went on to a new job, and in briefing me to take over his role, he made it very clear. He said the toughest job there that a supervisor had to do was fire somebody, or remove somebody from a job, and therefore the second hardest job was hiring good people so you wouldn't have to do the hardest job. So he was cautioning me to go out and find the very best people in the world, and then you never had to face this terrible job of replacing anybody.

And over the years I found the hardest job for me, and probably for any manager, is to take some person out of a job. Obviously there are some people that, for various moral or legal reasons, it's *not* difficult to remove. You're happy to see them go and they deserve to go, but that's rare. It is really rare in an organization like this that moving somebody out of a job is easy. There are people who are highly competent, but for one reason or another they're not suited for their current assignment, so you have to do what's right for the organization, which is replace them and put the right people in.

At the same time, you try to preserve for the person being removed a sense of dignity and worth to the organization. These are people we had large investments in, and had good track records. And very often these people are well-beloved by the organization, so when somebody says, "Why'd they remove so-and-so?" It's hard.

It doesn't happen often, but when it does, you are caught up in doing something tough. I think virtually anybody who works in an organization will say that's the toughest thing they do. Fortunately we didn't have to do very much of that, but within the first few weeks we did pull the trigger and then got on with it. The people involved were all very successful people at the Laboratory, but they weren't doing the kinds of things that they do best.

Perhaps related to that, what were some of the most critical decisions that you made?

That was 12 years ago. You know how the memory drapes and blurs things that once stood out like jagged mountain peaks. There was the stress. We were dealing with it each day; waking up three times in the middle of the night to try to figure out how am I going to deal with that one? Getting in the shower in the morning and realizing you've been there a half hour; you forgot you were in the shower because you were trying to figure something out. Doing all those things that needed to be done.

My role was to just guide a lot of bright people, to put the organization into a disciplined, well-defined operational state, getting the right leadership in the subtier organizations. It wasn't all done instantly, but we went through and found the right people, moved some people around. There were just lots of small decisions and tuning the system and writing procedures. I think the critical one was putting the right leadership in place and then providing the resources and the advice to those people and the authority to do it.

There were hundreds or thousands of daily, weekly decisions on how the spacecraft was to be used, and the science sequence and issues. I have a hard time drawing out at this point anything that stands out as critical in my role on the operational side. But let me go back.

The project manager was Bob Parks. Ray Heacock was the deputy handling the spacecraft and engineering side of it. I handled basically the people and the team-building operation for that year and a half. There were lots of other players, but Ed Stone, the project scientist, was a key player, and Dick Laeser, who was the mission director. The five of us had a daily staff meeting. The day Bob Parks took over we started meeting at 8:15 every morning, and it lasted 15 minutes to 45 minutes; generally out at 9:00. Ed Stone came up from the campus, generally two or three days a week, to sit in the meeting. And it was the five of us that made the decisions about all the things that had to get done.

Our staff meeting was basically to be sure we understood what we were doing for the week, and particularly that day, and what the issues were that we were working on. Some of it was just announcement of the things we were doing, and in some cases it was trying to decide how we were going to make a decision. But I don't recall any particularly crucial thing. I think the mountains have all been rounded over so they all look like hills.

What was the highest point associated with Voyager for you personally?

Seeing the first close-up pictures of Jupiter come back. I thought the team was virtually flawless. They had confidence and competence in what they were doing. I remember the first night of the close-up pictures, I was watching those pictures come in virtually flawlessly. We had some excitement because the spacecraft did some things that we'd hoped it wouldn't do, as a result of electrostatic discharge; the Jovian environment was quite severe. But the spacecraft did just what was wanted, and that meant the team was doing what we wanted them to do, and they had confidence in what they were doing. That was the high point, the first Voyager getting to a planet. The second Voyager got there a few months later, and then three months later I left the project to take another job.

In that last three or four months I worked on a new design of the operations team. What had happened in the year before Jupiter was I didn't design the team the way I really thought the team ought to be; I tried to Band-Aid and fix it in the short time we had to get to Jupiter. But since the time from Jupiter on to Saturn was enough, we said, "Hey, now we have time to make the changes we think are the right changes," which involved changing the assignments of various groups. You know, "we'll take this function, put it here, and we'll put that function over there. And that'll make a better organizational entity. The teams will have better defined jobs."

So in that last three months that I was on the project, I proposed a redo of the flight team, which was about the basis for the way it was up until the last planet. It became a real team. There were hundreds of people, and they all contributed. It's hard to single out who had the most critical role or what the most critical decision was. My role, like any manager's, was to get the resources and get the people, pat them on the back, and steer them in the right direction. They do the work.

What was your lowest point when you were on that operations assignment?

I don't know. As I told you earlier, this place has always been fun. Work has been fun. It's always been easy to get up and come to work, even at 3:00 in the morning or on Sunday or Christmas, because it's just fun. Of course, there are some days that it's not as fun as others. The day you call somebody and say, "I'm going to take you out of a job and we'll have to find a new job for you," is not a really fun day. But the work and the things we've accomplished here just make this thing fun. So I don't associate any lows that I had with Voyager. I think it worked well.

What thoughts, what lessons you've learned, should be preserved for the record?

If I were talking to people who were designing missions in the future, the point that I would want them to remember most is that there's a tendency that occurs when you start a mission. I'm talking about back at the day you say, "Hey, wouldn't it be nice to send a spacecraft somewhere, to Saturn, put a probe into the satellite Titan with the mission called Cassini?"

What typically happens is that you want to sell the mission, so you get optimistic about what it's going to cost. The customer says, "Can't you do it for less?" You try to convince yourself, "Gee, we could do it for a little less here, we'll cut this and that." And as you go through the years of designing and building the spacecraft, you begin to run into financial problems. You begin to shortchange the design of the operational phase of the mission, which comes later: the building of the team, the tools, and the apparatus required to actually fly the mission.

On every project I can think of, in the implementation phase the project manager looks around for where can he get some money. And he says, "I can take it out of the operations phase, I'll solve that problem later." The net effect is that very often you get down to the operational phase and you've shortchanged the readiness of the operation: the operational team and its equipment and software.

The Voyager project recognized this early on and set up a team that said that wasn't going to happen again. But in fact it did happen again. They did get shortchanged, and they weren't able to turn it around; and part of the problem was that after launch they weren't as ready, they didn't have the people, they didn't have all the things they needed. They were pretty thoroughly discouraged by it.

I don't know what you do about it other than recognize you don't want to do it. The project manager charged with getting the spacecraft launched has got to recognize that first. He makes the trades, as the customer makes the trades. But as spacecraft get more complicated, it gets more difficult to avoid. Back in the 1960s we'd build a spacecraft in a year or two. Then the mission might only last 66 hours, going to the moon. The same people who designed it also flew the mission, and they put up with it for only 66 hours. It was fun. Maybe they didn't even have to go to bed during those 66 hours. But now it takes three or four years to build a spacecraft and then it takes you 15, 20 years to fly the mission. So it's a different set of philosophies.

The tendency will always be there to solve today's problems with tomorrow's money. We just don't put on enough people in getting ready for operations. That was a strong part of the Voyager situation. It's happened on virtually every other project I can think of, and it becomes a more serious threat as time goes on, because as missions become more complicated they become more difficult to fly safely.

Now we're trying to make them easier to fly, with much improved software tools, expert systems, artificial intelligence—things that make the work of the operator easier. But I continue to tell people, "Don't shortchange the design of the operations." Engineers like to build a toy, but they don't like to spend much time thinking about what you do with it. Yet these toys are very complicated and require every bit as much attention to the team that's going to fly them as to the team that builds them.

I think that if there's one message, that's it.

What should people know about the Deep Space Network related to Voyager?

The Deep Space Network; that's a whole new world. We could talk eight hours on that with no problem at all.

I was privileged to be asked to lead that network from Pearl Harbor Day of 1980—December 7 was my first day—until June 30, 1987. It was a period of change in the Network, of building new facilities, of moving old facilities or closing them down, and combining two networks. Part of the Goddard network was being phased out because of the Tracking and Data Relay System (TDRS) and we were taking part of it over.

It involved moving antennas to collocate them in more convenient sites. They'd grown up over the years scattered over many kilometers of Spain and Australia, so we needed to pull them together, for economies of operation. It involved increasing the sensitivity of the reception of the network; its capabilities to deal in deep space. It involved a great deal of teamwork and cooperation with other government agencies that had antennas, and international activities with the Australians, the Spanish, and with the Japanese in particular.

So there was a great deal of activity, including a complete upgrading of the design of the tracking stations and the software. During this time period we put in more than a million lines of code in new software. This is like changing the wheels on a locomotive while it's running. You can't shut the network down. The network has to be there to support the customers, so you've got to rebuild the network on the fly. It doesn't mean you don't shut an antenna down or don't shut down certain pieces of equipment, but you just don't shut down the whole thing. And so it was a very challenging assignment.

I was not a communications engineer; I was a naval architect, but I was asked to do that job. The people there were super people. In about '82, just after I took the job, I appointed Bob Stevens as chief engineer of the Network. He has managed various of the communications activities of the Laboratory for many years. He was my advisor since I needed somebody close to me to teach me communications real fast. He was on my staff, and he was my powerhouse. He's about 70 years old and is just finishing bicycling across the United States.

He knows the history of the network, and if you want to get a flavor of the challenge of communicating over billions of miles, Bob can tell you in a meaningful way. For example, suppose you want to know how strong the signal from the spacecraft at Neptune is when it reaches Earth, the signal that brought all these pictures back. The spacecraft radiates 20 watts of power off its big antenna, and it comes across 4 or 5 billion miles, and it's received on the ground antenna network here on Earth.

If you took that energy that's received on that antenna and fed it into a perfect capacitor—a capacitor stores an electrical charge; "perfect" meaning it didn't have any leakage and stored it all—and you started doing that five billion years ago when the Earth was formed, and stored up all that energy that came down from that spacecraft for five billion years, stored it on this perfect capacitor, then hooked that capacitor up to the lightbulb in your refrigerator, which is off most of the time, and then jerked the door open, it wouldn't light long enough for you to find a bottle of milk. We're talking about 10^{-18} watts.

We're working with things that are very difficult to deal with, very small, and it's a great technological challenge. There's a great bunch of engineers that have dedicated their careers to building that network. I was able to preside over it for seven years and help them through some difficult changes. Part of the great history of Voyager is in the network, because the communications is what made it all possible.

There's an inverse square law that deals with communication just like with lightbulbs. You move away from a lightbulb twice as far as you are, and you only get one-fourth the amount of light. Same way with communications: you double your distance, you get one-fourth the information back.

So you design a spacecraft to go to Saturn at 10 AUs, say, and it's optimized to return 15,000 pictures from Saturn. Then you go out to Neptune, which is at 30 AUs, and you'll only get one-ninth of the Saturn return. Now we defeat the inverse square law; it's still true, but we build more and more antennas, bigger antennas, so we capture more of that energy.

The energy sent back by that spacecraft is aimed at the Earth, but most of it misses the Earth, because the Earth is a tiny little point out there in that beam the spacecraft is sending. It's like a flashlight trying to look at a small dust speck, and the dust speck gets lit up, and then on that dust speck, which is Earth, we have these very tiny little antennas that try to capture this signal. We improved the network, partly by increasing the size of the antennas, and partly by arraying them in real time with other organizations' antennas.

How would you compare Voyager with other projects you're familiar with?

Voyager has been, from an exploration sense, the greatest single project that we've had.

I think Viking would have been, if it had discovered life on Mars, but we went to Mars and hit a dry hole in terms of life. It was a great mission but people were hoping that it was going to find life. At the two points where we went in and sampled the ground, the atmosphere and so forth, there wasn't anything that we would define as life.

Now I have to say Viking was the high point of my own career because I was close to the design and the operation of the spacecraft. I was on that project womb to tomb, so to speak. So from a professional engineering standpoint, that's where my greatest pleasure as an engineer was.

But, from an exploration standpoint, in terms of what mankind's learned, Voyager is head and shoulders above anything else we've done.

RAYMOND L. HEACOCK
Project Manager, Jupiter 2 and Saturn 1
Born January 9, 1928   Santa Ana, California

Raymond Heacock earned a bachelor of science and a master of science from Caltech in 1952 and 1953, and then came to JPL as a research engineer. In 1959 he joined the newly formed Space Sciences Division and was a co-investigator on the Ranger Imaging Science Team that brought back the first high resolution photographs of the moon.

In 1970 he was appointed spacecraft systems manager for the Outer Planets Missions Project. After the ambitious four-spacecraft project was cut back to the more modest Voyager mission, Heacock became the Voyager spacecraft systems manager, responsible for spacecraft design, development, test, and launch operations support. He was appointed Voyager deputy project manager in October 1977 and project manager in March 1979.

Although he left the project in 1981 to lead an effort to get a U.S. mission to Halley's Comet, he continued as the chairman of the Voyager Standing Review Board, overseeing preparations for the second Saturn encounter, and the Uranus and Neptune encounters. Other assignments involved the Hubble Space Telescope. In 1987 he was named manager of the Flight Instruments Office, and in 1989 was promoted to deputy assistant laboratory director for the Office of Space Sciences and Instruments.

Among his awards are NASA's medals for Distinguished Service and Outstanding Leadership, the James Watt International Gold Medal, and the Space Club's Astronautical Engineers Award. His memberships include the AIAA, and he has been president of the Caltech Alumni Association.

His main hobby is being guardian to his grandson who has lived with him since birth. Otherwise Heacock's work has been his avocation. He reads a lot of science fiction and likes to camp and fish in the Sierras.

Since my retirement on January 31, 1990, I have had plenty of time to reflect on my overall career (life). I feel that luck was involved in selecting and guiding me through my technical training in the navy and at Caltech and in hiring into JPL. My early JPL years provided excellent and very broadbased engineering and engineering management training that served me well. However, my post retirement interactions with my grandsons has made it clear that I neglected my own children. The time lost with the family (wife and children) over the years is my only real regret. Would I do it differently if I had it to do over again? Hopefully, but I am sure I would strive for the same career end results. It is unfortunate that one cannot start a career with an end-of-career perspective.

—*Raymond L. Heacock*

RAYMOND HEACOCK

I was born and grew up in Southern California. After high school I joined the Navy and received electronic training. That motivated me to go to Caltech, from which I received my B.S. and M.S. in electrical engineering.

I hired into JPL directly out of Caltech. I had friends who worked part time at JPL while they were going to school. I was given offers by Proctor and Gamble, North American, Lockheed, and JPL. I told JPL that if they'd match the North American offer I'd come to work at JPL, which is what happened.

My main contribution to Voyager was as the spacecraft systems manager. I actually started as the spacecraft systems manager on the Outer Planets Missions (OPM) project in November of 1970. Bud Schurmeier had been selected as the project manager. He began in October or September 1970 to put together his project team, and I was quite pleased when he selected me to be the spacecraft systems manager, although that was not my background at JPL.

I had worked on analog computers for guidance and control of missiles; tracking, Doppler and time for the first deep space tracking station at Goldstone, California; and then went into the Space Sciences Division when it was formed in 1959. I was responsible for science instruments and science payloads for the early Ranger and Mariner missions, and progressed up to assistant manager of the Division.

Early in 1970 the science data systems portion of the Space Sciences Division was moved into a new Astrionics Division, which had flight computers, data systems, and data processing. I became the deputy manager of the new division. It was from there that I moved when Bud Schurmeier offered me the OPM spacecraft systems manager position.

The concepts behind the Voyager project really began in 1968 when NASA approved an advanced technology development program to look at what was required to take advantage of the gravity assist opportunities in the late 1970s for exploration of the outer planets. That was called the TOPS, the Thermoelectric Outer Planets Spacecraft program. It was a very successful effort, which looked at the spacecraft systems requirements for such missions and defined and supported the kinds of technology developments required to carry them out.

One example is that previous missions had all been close enough to the sun to use batteries and solar panels. Going to the outer planets involved distances ranging from 5 to over 30 AUs. The then existing solar panel technology could not reasonably provide the required power. The TOPS program initiated interactions with the Atomic Energy Commission (now the Department of Energy), which lead to the multi-hundred-watt radioisotope thermoelectric generators (RTGs), which flew on Voyager. The tremendous communication distances also received special attention, calling for larger antennas and higher frequency transmitters to achieve the desired data rates.

The TOPS program looked at several concepts to enhance spacecraft reliability for the Grand Tour mission lifetime of over 12 years. It developed a Self Test And Repair (STAR) computer concept, which would be the heart of the spacecraft. This was accomplished with three online computers with self-test routines and voting logic, requiring three out of three to give the correct answer. A disagreeing computer was assumed to have a failure and would be changed out with one of several backup units.

TOPS also looked at a more sophisticated telemetry system, a Computer Accessed Telemetry System (CATS), which permitted the onboard computer to monitor the telemetry and, using error limits on the data, decide whether or not a subsystem was degrading and/or had failed. So the concepts of fault routines and self test and repair autonomy within the spacecraft grew considerably as a result of the TOPS program.

The TOPS had been going on for two years and was continuing when the OPM project was started. The TOPS program provided an excellent background for the OPM project. We had come up with a plan for the OPM which would send two spacecraft to Jupiter, Saturn, and Pluto and two spacecraft to Jupiter, Uranus, and Neptune to carry out the exploration of the outer solar system. Unfortunately the OPM priced out in excess of $750 million in 1971 dollars.

At the same time the Viking project was also getting started to search for life on Mars. Due to the tightness of the NASA budget, the OPM project was canceled in favor of Viking at the end of 1971. We received notice of the cancellation in early January 1972, and the project scurried around for a fallback position which would be acceptable to NASA and the Congress. That's when the Mariner Jupiter-Saturn 1977 (MJS 77) concept was put forward and eventually sold.

Our MJS 77 proposal involved three options with low, medium, and high budget levels. The lowest budget level was a bare bones mission to go to Jupiter and Saturn to carry out the preliminary exploration of those two systems. The intermediate level added some sophistication to the mission, primarily enhancing the science. The top budget level added more redundancy for reliability, and further enhanced the science. NASA picked the medium level option and then added funds specifically targeted for science enhancements. We were basically given a green light in May of 1972 with the formal project approval in June. Things moved very quickly in early 1972 as we retrenched, restructured, reproposed, and sold the MJS 77 project.

The normal duration from start to launch for a Mariner class program was approximately three years. With the project approval in 1972, we had over five years to launch. We took advantage of the extra time to "re-systems engineer" the proposed concepts for the mission and the spacecraft and to modify the implementation mode. The OPM project had been proposed as a systems contracted effort where JPL would select a prime contractor in industry to carry out the design, development, fabrication, and launch of the spacecraft. JPL would manage the effort for NASA and be responsible for the in-flight operation after launch.

MJS 77 was initially also assumed to be, and costed as, a systems contracted effort. However, upon reconsideration with the Mariner and Viking inheritance and the reduced scope of the effort, it was clear that it would be more cost effective to use JPL's traditional subsystem contracting mode. JPL would act as the prime contractor and let contracts to industry for the spacecraft subsystems, the spacecraft hardware and science instruments, test the system, support the launch operations, and carry out the in-flight operations.

The re-systems engineering, change in implementation mode, the Mariner and Viking inheritance, and the added budget provided by NASA for science enhancements, permitted a large number of significant changes and improvements to the assumed MJS 77 baseline spacecraft and mission.

It was interesting that a large number of the improvements in the capability of the spacecraft were driven by the scientists as science enhancements. All of the changes and improvements were approved by NASA through a series of formal reviews. The significance of these changes is reflected directly in the spectacular results of the Voyager missions.

The earlier three-axis controlled Mariners and Vikings used pressurized nitrogen gas thrusters for attitude control and a separate rocket system for the velocity increments for trajectory correction maneuvers. For the MJS 77 we decided to change to small (0.2 lbs) hydrazine thrusters (rocket engines) for both the three-axis attitude control and trajectory correction maneuvers. (I'll call MJS 77 "Voyager" even though the name wasn't changed until 1977.)

The Titan-Centaur launch vehicle required a final injection stage to achieve the Voyager-required injection velocity. A major spacecraft design change was to integrate the final injection stage into the spacecraft. While this resulted in lower cost and saved significant weight, the most significant benefit was that it permitted the fuel and associated fuel reserves for injection thrust vector control, attitude control and trajectory correction maneuvers to all be consolidated into a single fuel tank. This provided more than adequate fuel margins for the prime mission and enabled the extended missions to Uranus and Neptune and beyond.

While previous Mariners and Vikings used hard-wired analog attitude control systems, it was decided for Voyager to go to a digital computer-based attitude control system. The use of an in-flight digital computer allowed reconfiguring the system as a function of the flight phase and tailoring the performance for special characteristics such as image motion compensation.

Another change from previous Mariners and Vikings was a digital computer-based Flight Data Subsystem. This allowed much more flexibility in terms of data rates and data formats. We reprogrammed the onboard Flight Data Subsystem for Jupiter, Saturn, Uranus, and Neptune, scaling the data rates and the data formats to the exact requirements for each of the encounters. The Command Computer Subsystem was not new, in that it was a direct inheritance from the Viking project. The same computer, with minor modifications, was used for attitude control.

The baseline communication systems consisted of a redundant S-band system and a single X-band system. The higher data rate provided by X-band made it a high priority for science enhancement. As a result a redundant X-band system was added. The S-band system was split into a TWTA (traveling wave tube amplifier) and a solid state amplifier because of concern about the long life capability of the TWTA. (It was the solid state amplifier that failed early in the mission due to thermal dissipation design problems.)

I wrote a paper on the spacecraft in 1980 that lists all the changes, which were made to the original baseline. The changes built on the experiences from previous projects and used the background and many of the people from the TOPS program. But it would not have been possible without the extra time at the front end. Because of the slow start and the change in implementation mode from a systems contracted effort to a subsystem contracted effort, we were able with the approved funding profile to build in a substantial budget reserve at the beginning of the project. This front-end reserve proved to be extremely valuable in expeditiously handling early implementation problems.

In recent years almost every project has had starved front-end budget profiles that really hurt the early phases and make it almost impossible for a project to get a good running start. Voyager did not have those kinds of funding problems. It's one of the unique things about Voyager: its funding support was excellent through the entire life cycle of the spacecraft effort.

How did that fortuitous situation come about?

It came about because we had retrenched from a program that was targeted for about $750 million in FY '71 dollars to a program that was on the order of $250 million in FY '72 dollars, sweetened a little bit by NASA to enhance the science return. The funding profile was set up assuming a systems contracting mode, which would have meant that we had to hire a systems contractor and bring him onboard immediately. This would have required significantly more money than our re-systems engineering effort and, despite the number of changes to the baseline, would have cost more than the subsystem contracting mode for the hardware implementation phases.

As a result we got to where we would like to have been, with the kind of budget profile every project ought to have, but it was not the result of a deliberate process. It was through a set of circumstances that worked out and gave us a unique opportunity. I have already told you about adding the injection stage (propulsion module) to the spacecraft; it's another example of things working out right. It allowed a significant spacecraft payload weight increase for science instruments. We wound up with a larger science payload because we had the ability to accommodate it as a result of the change.

Larger than you had originally hoped or thought?

It was not larger than we had planned for the outer planets mission spacecraft because that was a much larger and more capable spacecraft. It was larger than we had anticipated originally when we put forward the MJS 77 concept. It was not larger than had been hoped for with the approved MJS 77 project with the NASA added funds for science enhancements. We wound up with a science payload that was very close to that desired for the OPM, but it was only possible because we were able to make the additional weight available and the budget supported it. Without the added payload capability we would have had to drop one or two instruments from the payload.

The TOPS program had addressed the charged particle radiation environments due to solar and Jupiter-trapped radiation. Based upon results from radiation workshops, it had been assumed that the Jupiter proton radiation was the principal hazard for the spacecraft. But when Pioneer 10 flew by Jupiter in December 1973, the in situ radiation measurements indicated that the high energy electron environment was three orders of magnitude higher than what had been anticipated. This new data required that the whole radiation issue for electronic parts and instrument sensors be readdressed.

We carried out an intensive program, working all facets of the problem. The program cost the project an additional 13 million dollars. We carried out an extensive testing and screening program on electronic parts and sensors to establish their capabilities and shielding requirements and weed out unacceptable part types. The spacecraft shielding was modified by changing wall thicknesses for electronic housings and chassis and providing localized shielding for particularly susceptible components. A radiation design margin (RDM) of two was established, which required that all parts be able to handle twice their expected radiation at Jupiter. The RDM requirement was met either by the inherent capability of the part or by adding shielding to make the part acceptable.

In the case of CMOS (Composite Metal Oxide Semiconductor) parts, which was a new integrated circuit in that time frame, we restricted their use to just the science instruments and the Flight Data Subsystem. During the manufacture of the CMOS integrated circuits, we screened the wafers for radiation susceptibility and only packaged the parts from the wafers that were hard enough to meet the requirement. That was an elaborate, time-consuming, and expensive process, but it was necessary to achieve the required radiation hardness. If we had flown standard CMOS parts, we would have had serious degradation and multiple failures during the Jupiter encounters. Without the CMOS parts the data system and science instruments would have been far less capable.

As it was, at Jupiter we experienced a timing offset between the Flight Data Subsystem clock and the Command Computer Subsystem clock, from interrupts that occurred due to radiation effects on the Flight Data Subsystem. While there were radiation effects at Jupiter, there were no catastrophic failures. The most serious problem was two failures in the photo polarimeter instrument that restricted its operation. The December 1973 change in the assumed radiation environment was certainly the biggest transient during the development phase of the spacecraft.

The spacecraft came together basically on schedule. However we had not adequately anticipated the fact that this spacecraft had flight computers driven by software: Command Computer Subsystem (CCS), Attitude and Articulation Control Subsystem (AACS), and Flight Data Subsystem (FDS). We had not given enough attention to the early integration and testing of the flight software in the spacecraft. The AACS flight version software wasn't available until after the spacecraft was at the Cape being prepared for launch. Glenn Cunningham and I had to fly back from the Cape on Memorial Day in order to press for the needed attention by the right people at JPL and to get the people and software down to the Cape so it could be integrated and tested in time.

We had a variety of problems at the Cape in preparing to launch the two spacecraft. The most threatening problem was a failure in a 54L integrated circuit. The failure involved a short between internal conductors due to electric field-induced metal migration in the presence of a few percent of moisture in the packaged device. There had been a processing problem, which put at risk the delivered parts during their packaging. Elaborate steps had been taken to eliminate all of the suspect parts, but the failure at the Cape made it clear that we had not done a perfect job.

Since the problem was induced by an electric field, we operated the hardware with 54Ls almost 24 hours a day as a further screen for the problem. We had no further failures at the Cape or, fortunately, after launch. The one that failed was apparently the only one our earlier effort had missed. It was just another thing we had in the back of our minds as a potential threat to the Voyager missions.

And of course on August 20, 1977, we had the problem with the launch of Voyager 2. Voyager 2 was launched first, on a slow trajectory which could support flying the Grand Tour. After injection on the flight trajectory, the normal sequence was for the radioisotope thermoelectric generators and science platform booms to be deployed and locked in place, and then the expended propulsion module solid rocket engine would be ejected from the spacecraft. The separation is accomplished with explosive bolts and springs. The pyrotechnic devices are fired, separating the bolts, which then allows the springs to push away the spent motor.

The Voyager 2 problem was that the science platform boom was not fully deployed and locked in place when the spent solid rocket motor was ejected. The transfer of momentum between the two systems resulted in a turning moment being imparted to the spacecraft due to movement of the science platform. A complicating factor in the problem was that the attitude control software went through an isolation valve process for the thruster system before allowing its use. This was not smart because it delayed the required corrective action.

The onboard fault protection routines detected the induced error and took corrective action. At the first corrective action, the system assumed that the thrusters had failed, and switched to the redundant set. The fault persisted and the system assumed that the thruster drivers were the problem and switched back to the prime thrusters and to the backup thruster drivers.

This process continued through the four available options with the prime computer and then requested the CCS to switch in the backup AACS computer. Switching to the backup computer reinitialized the system, which led to entering the sun search routine and eventually recovery of the spacecraft. The process had one or two steps left when it finally brought the spacecraft under control. If it had exhausted the opportunities, we could have lost the spacecraft, so we came very close to doing that.

We were scheduled to launch Voyager 1 on August 30, 1977, only 10 days after Voyager 2. The problem with Voyager 2 initiated an intensive period of investigation and corrective actions for Voyager 1. The first correction was to delete the isolation valve reset command. Investigation into the deployment of the science platform boom indicated that the science instrument cable at the boom hinge could bump the high gain antenna and slow the deployment.

The deployment force was increased by adding a second negator spring and dash-pot damper assembly to the opposite end of the hinge. The negator spring is a wound-up spring providing a constant force through the turning angle. The dash-pot damper provides a restraining force to control the rate of deployment. The cable clearance over the boom hinge to the high gain antenna was also reworked with some improvement in clearance. All of the investigations and corrective actions were carried out in less than two weeks, permitting the launch of Voyager 1 on September 5, 1977. It all worked beautifully!

However, we soon began to experience a variety of in-flight problems. We had a learning phase to go through to fully understand and operate the spacecraft. The primary spacecraft team was at the Cape to support the preparations and launch of the two spacecraft. The team at JPL for support of the flight operations after launch had less hands-on experience with the spacecraft. Basically, the early flight operations were set up assuming that everything was going to operate right, but everything didn't go exactly right.

In the early phase of the flight operations, we had a sufficient number of problems to force us to revise our plans for cruise to Jupiter. The assumption of long, low activity sequences during cruise had to be abandoned. All of the Jupiter encounter sequences were to be developed while in flight to Jupiter. That was a reasonable and practical thing to do with the one and a half year flight time to Jupiter. But the early problems required the addition of personnel to the flight teams, particularly to the sequence team, in order to respond to the problems and still develop the encounter sequences.

Upon my return to JPL from the Cape after the launch of the Voyagers, I became the deputy project manager under John Casani. At the same time, John was becoming involved as the project manager of Galileo. As the number of in-flight problems grew, it became obvious that added top-level horsepower was needed. John was released to be the full-time Galileo project manager. Bob Parks (assistant laboratory director for flight projects) became the Voyager acting project manager. I became the deputy project manager for the spacecraft and Dr. Peter Lyman was appointed the deputy project manager for operations.

A variety of spacecraft problems was experienced prior to the Jupiter encounters, of which the most serious were the hang-up of the Voyager 1 science scan platform and the problems with the two Voyager 2 receivers. We were able to cope with the problems and successfully implement the required sequences for the Jupiter encounter. By the time Voyager 1 reached Jupiter, the flight team had matured and was fully capable of operating the spacecraft.

After the Voyager 1 successful Jupiter encounter I became the Voyager project manager and selected Esker Davis as my deputy. I was project manager for the second Jupiter encounter and the first Saturn encounter, and left the project in January 1981 to support the attempt to get the Halley Intercept Mission approved. However I kept my association with Voyager, as chairman of the Voyager Standing Review Board, up through the Neptune encounter.

Reviews were held twice a year to look at the status and progress of the project. Prior to each of the Uranus and Neptune encounters, two additional reviews were held to look at the detailed preparations. We looked at the actions being taken to enhance the mission return, the response to existing problems or limitations, and the planning and preparation of the specific science observation sequences. Examples of areas of concern were the hang-up of the Voyager 2 scan platform following the Saturn encounter, incorporation of the FDS Reed-Solomon encoder and Dual Processor mode, and the enhancement of the Deep Space Network capabilities to increase the downlink data rates.

I was pleased to join Schurmeier on the OPM project back in November 1970 because everything seemed so right to me. Here was the gravity assist opportunity, which recurs every 171 years, at just the right time; just when we had the technology and the capability to carry it out. It was so fortuitous to have everything be just right, that it seemed, my god, if I've got the opportunity to do this, and do that spacecraft, there's no way in hell I can turn it down! It was perfect, and everything went on from there to get better and better.

I was concerned about the length of the job. It would be seven years to launch, and I'd never worked on any job for that long. I had worked with Bud Schurmeier on the Mariner Mars 1969 project and knew he would be excellent. I personally believe that he is the best project manager JPL ever had. He is very people- and team-oriented. He spent a seemingly inordinate amount of time putting together his project team. He picked individuals one by one, on the basis of forming a real team and team concept. His management style also fostered and strengthened the team.

We had our problems and had to work many long, long hours, but it was always done with a group of compatible people, all very strongly motivated, and all working towards the same end objective. So it really was a pleasure all the way through, despite the problems and difficulties. When I look back, I don't find any of it unsatisfying: the working relationships, the accomplishments day by day as you progress through the process of the design and development, the finding and solving problems, and always moving forward.

The Voyager relationship with NASA Headquarters was very good. A close working relationship and support existed without excessive involvement in details. When we came up with the radiation environment problem from the Pioneer encounter data, NASA covered the increased cost out of their Allocation for Program Adjustment (APA). The radiation problem was not something we could cover within the framework of our budget plan. NASA also provided added support postlaunch when we had to strengthen the flight teams.

The only thing that I felt unhappy about in our relationship with NASA was the handling of inflation. Because of the long duration of the project, the original agreement in 1972 was that each year to completion (runout) would be adjusted on the basis of actual inflation. However, in 1973 NASA provided us a fixed runout budget assuming a fixed five percent per year inflation rate. It was in that time frame that inflation got up into the double digit category. This change obviously caused problems in tightening the budgets and forced some corner cutting. Fortunately NASA provided postlaunch support when we were really hurting, in order to prevent serious impacts from the budget crunch.

When you look at the overall budget performance for the prime mission, we overran the runout budget based upon the five percent per year inflation rate. However, if you apply the actual inflation rates per the original agreement, we actually underran our budget plan. So we carried out the project within the original budget plan. This was a direct result of having a good funding profile in the early phase of the project; we were never forced into making dumb compromises just because of budget limitations. We were always able to pay the bill as we went and to make our decisions based more on the technical factors and issues rather than on budget constraints.

I know from personal experience how fortunate it was for Voyager. I can cite several projects which were seriously compromised by early funding constraints. Such projects are often forced into compromising decisions based upon budget constraints, which in the long run increases the cost and/or results in serious mission losses.

It's phenomenal what Voyager accomplished, particularly when we look back on scan platform failures and so forth. Roughly what percentage of the hardware on the two Voyagers did go bad?

The Voyager spacecrafts had a very high level of redundancy, almost two spacecrafts within a single structure. As a result, there was very little impact from the few in-flight problems. Through the prime mission there was essentially no mission loss due to failures. There were some science instrument problems.

The photopolarimeter had some radiation induced problems at Jupiter, which limited its performance. The infrared interferometer spectrometer had a materials degradation problem, which restricted the interferometer range. The Flight Data Subsystem lost some memory elements, which were easily programmed around. The Voyager 1 scan platform hang-up appeared to be soft particulate contamination in the gear train, which we were able to drive through, and the problem did not repeat. The loss of the solid state S-band transmitter did not prove to be a problem.

The most serious problem was the Voyager 2 receivers' failures, which occurred while troubleshooting the Voyager 1 scan platform hang-up in April 1978. The Voyager 2 seven-day command loss timer was erroneously allowed to time out. The spacecraft correctly entered the command loss routine and initiated a sequence of configuration changes with the backup receiver to recover communications with the ground.

Normally this would not create a problem. However, when the backup receiver was substituted, we were unable to command the spacecraft, indicating a problem with the backup receiver. The routine recycled (12 hours) back to the prime receiver, and it operated normally for 30 minutes and then failed. The failure was diagnosed as an internal short in its power supply, which blew the protective fuses. We were then left with the failed backup receiver. Its problem proved to be a shorted tracking loop capacitor, which prevented the receiver from locking on to and tracking the ground transmitter signal. The S-band (2300 MHz) receiver had a fixed 100 Hz bandpass through which we had to get our commands into the spacecraft.

Voyager was lucky in that the Deep Space Network had developed a computer controlled local oscillator to drive their transmitters. The problem was to predict and compensate for the Doppler shifts induced by the relative motions of the spacecraft and the Earth (including the Earth's rotation). Using the computer-controlled local oscillator, we were able to transmit a predicted frequency which, when shifted by the Doppler, would fall within the 100 Hz bandpass of the receiver. This added significant complexity to the operation of Voyager 2 throughout its mission to all four of the large outer planets, but it all worked beautifully.

If the Voyager 2 prime receiver failure was a result of an incipient failure mechanism, we were lucky it happened when it did because it gave us time to solve the problem, develop a work around and become proficient in its use. A failure close to or during an encounter would have lost the encounter.

After the prime mission, Voyager 1 lost a block of memory, but it had no more planetary encounters and has easily supported the cruise requirements. Voyager 2's scan platform hung up during the postencounter observations of Saturn. The scan platform drive actuators were inherited hardware from the Viking project. The actuators used sleeve bearings with tight tolerances and little lubrication. With the Voyager longer life and higher usage, there appeared to be enough heating in the sleeve bearing to eventually cause galling and binding. Ground-based disassembly, inspection, and testing of other residual Viking actuators provided sufficient data to allow development of a recovery plan and pre-use conditioning process. The program involved thermal cycling and slew rate limitations, which proved to be effective in permitting operation of the scan platform during both the Uranus and Neptune encounters.

While the Voyagers were built with the primary objective of carrying out the Jupiter and Saturn explorations, the hope of going to Uranus and Neptune was always present. As a hedge against prime mission communication problems and as a potential enhancement for the Uranus and Neptune encounters, we built certain extra hardware into the spacecrafts. We added a Reed-Solomon data encoder to the FDS but did not test it before launch. After the Saturn encounters the Reed-Solomon encoder was verified to be operational and the project set about to develop the ground software to decommutate Reed-Solomon encoded data instead of the baseline Golay encoded data.

Golay encoding for error detection and correction uses one bit per bit. So half the transmitted bits are just for error detection and correction, not actual information. With the Reed-Solomon encoder only a few bits in every 32 bits are used for error detection and correction, so you get almost a factor of two improvement in data bit transmission rate. With the Deep Space Network improvements in enlarging the 64-m antennas to 70 m, arraying 70-m and 35-m antennas and the use of the Reed Solomon encoding, we had essentially no communication capability loss beyond Saturn despite the tremendous increase in range. We were able to get basically all the information wanted from Uranus and Neptune.

Another thing we built into the Voyagers was a dual processor mode in which the prime and backup FDS processors could be operated simultaneously. This mode of operation was also not tested before launch. The dual processor mode was intended to permit data compression (primarily imaging data) to be carried out in one processor/memory pair while the other pair did the standard data handling and formatting. This capability was also verified after the Saturn encounter and was utilized successfully at both Uranus and Neptune.

The decreased intensity of the sunlight at the distance of Uranus and Neptune presented a challenge to the flight team. As an example, images taken with the same exposure as those used at Saturn would have poor signal to noise. Using longer exposures would result in serious image smear. The programmability in the attitude control system was used to provide image motion compensation. An extensive program of calibration was carried out, which provided parametric data on thruster-on pulse-width control versus spacecraft turn rate.

The project was then able to provide image motion compensation by providing the required target orientation and compensating spacecraft turn rates to stabilize the target in the instrument field of view. As a result superb images and science observations were obtained. Without this compensation the beautiful pictures of Miranda and of Triton, for example, would have just been ugly smears.

How many different kinds of commands were you capable of sending to the spacecraft? In an airplane you'd have a rudder for horizontal direction, the ailerons for torque, and the elevators for up and down: three dimensions of commands the pilot would be operating.

Voyager has three programmable subsystems which must be commanded in order to operate the spacecraft. These are the Command Computer Subsystem, Flight Data Subsystem, and Attitude and Articulation Control Subsystem. A large number of the operating commands can be prestored in the memories of the subsystems, but the operational requirements change with the various phases of the mission. In order to optimize the mission return in each phase, it is necessary to send new commands or instructions to the spacecraft. Each of these subsystems required significant commanding throughout the life of the project. I will only provide a gross overview.

The Command Computer Subsystem is commanded to provide the sequences of activity to be carried out by the spacecraft. This can be as simple as a repetitive cycling of the fields and particle measurements during cruise between planets, to a high-intensity encounter of a planet or satellite involving elaborate spacecraft turns, scan platform pointing and complicated sequencing of the science instruments. The CCS issues commands to the AACS and the FDS in executing the sequences. The memory space for sequence command is very limited so a large number of memory loads are required to execute a planetary encounter. The CCS also stores a large number of fault routines that require changes as a function of the mission phase.

The Flight Data Subsystem is commanded to control the rate at which data is taken and/or transmitted, whether the data is recorded or sent in real time, and accomplish the playback of previously recorded data. It is also commanded to change data formats (the mix of data being taken by the science instruments). Commands are sent to the subsystem prior to each mission phase. By having these programs stored in memory, it is easier and quicker to initiate the required changes during the execution of the encounter activities. Thirty FDS data modes were available for use up through the Saturn encounter.

The Attitude and Articulation Control Subsystem, through stored parameters and CCS commands, controls the orientation of the spacecraft and the articulation of the scan platform. It controls the limit cycles (pointing accuracies) of the spacecraft relative to its references (the sun and selected star or the inertial reference unit). The sun sensor is "biasable" to provide an offset angle to point the high gain antenna at the Earth and must be regularly updated to maintain communications.

Under the CCS control, the AACS can be commanded to carry out trajectory correction maneuvers by orienting to the desired direction and firing the thrusters for the required velocity change. It also controls the pointing of the scan platform for target observation by the remote sensing science instruments. The AACS played a key role through its commanded programmability in performing imaging motion compensation for the science observations at Uranus and Neptune.

Trajectory Correction Maneuvers (TCMs) are critical for the success of planetary missions in achieving the desired planet and satellite encounters. We track the spacecraft with the ground antennas, make range measurements, and use optical navigation (photos of the planet and/or moons against a star background) to obtain accurate position and trajectory information.

When a trajectory correction is required, a TCM sequence command load is created and sent to the CCS and AACS. The sequence is initiated to execute the required turns (pitch, yaw, and roll) for the thrusting direction and then fire the thrusters for the required duration. The sequence then reverses the turns to recover the spacecraft's initial orientation. The inertial reference system in the AACS uses gyros to provide three axis attitude control without the sun and a star reference, and is usually used for TCMs.

In December 1979, after the Jupiter encounters, we performed a TCM to correct the targeting to Saturn. The spacecraft did not recover to the initial orientation at the end of the TCM. That is, it was oriented away from the Earth antenna point to perform its propulsive maneuver, and it did not come back to Earth antenna point. We had, in essence, lost the spacecraft.

The Voyagers have a large number of stored fault protection routines which are intended to recover normal operation in the event of a problem. In this case a glitch in one of the commands from the CCS to the AACS aborted the normal recovery to the initial orientation. The spacecraft entered the appropriate fault routine and eventually, with ground assistance, we recovered normal operation. Some weaknesses in the fault routines were revealed as a result of the problem and were quickly corrected. The problem with this TCM came back to haunt me during the Voyager 1 Saturn encounter.

The moon Titan was a major objective for Voyager 1 at Saturn because it has an atmosphere denser than the Earth's. I had approved a TCM as we approached Titan to improve the occultation of Titan. Ultraviolet and radio occultation data was essential to its study. The TCM was to change the trajectory from a slice across the edge of the moon to a pass behind the main body of the moon. The Director of JPL, Dr. Bruce Murray, expressed concern with the maneuver, based on the December 1979 TCM problem. I was convinced that we had corrected the problem and that the Titan TCM was a simpler and less risky maneuver. I proceeded with the maneuver. It worked and provided excellent data. But his concern made me nervous, and I would be twice as careful if I had to repeat something like that.

Speaking of being nervous, what would be the places in the whole experience with both the Voyagers when you were most uneasy?

It was the launch of Voyager 2, the waiting for verification of injection and not getting it. With the Voyager 1 fixes, we got the verification right on schedule. But Voyager 2, because of the scan platform not locking up, we were really in limbo and thought we had lost the spacecraft. We had no data following the separation from the Centaur. Normal operation would have had the spacecraft in the prescribed roll inertia mode about 47 minutes after propulsion module separation. As it was, the spacecraft was one hour and 37 minutes late in entering the roll inertial mode. You cannot imagine how long and anxious those 97 minutes were to the Voyager team.

Voyager 2 was launched on a trajectory which would enable the Grand Tour trajectory. If we had lost Voyager 2 at launch, we would have still been able to launch Voyager 1 on the same trajectory to keep the option open. Having the Voyager 2 receiver problems in April 1978 was also a very nervous time. If we had lost Voyager 2 at that time, we could not have changed Voyager 1 to a Grand Tour trajectory. The loss of both of the receivers would have reduced us to just the Voyager 1 mission without the Grand Tour extension to Uranus and Neptune. Fortunately the failures in the backup receiver proved not to have been a serious limitation on the mission. But it gave us many anxious moments before we understood it and had demonstrated the work around.

One of the protective actions taken against a further failure of the Voyager 2 receiver was to develop a backup mission load for the CCS, which would provide some encounter science even if we were to lose all communications with the spacecraft. These backup mission loads were periodically updated to assure that the latest trajectory and ephemeris data was incorporated in the planned observations. Such loads were only removed during the planned encounters, to permit all the memory space to be used for the encounter sequences.

The short in the backup receiver tracking loop capacitor was a known generic risk, and the failure raised the concern that the problem could occur on Voyager 1. As a result, immediate attention was given to devising a special conditioning test for Voyager 1 which would test the capacitors in both the prime and backup receivers and stress incipient shorts with sufficient energy in the capacitor to clear (burn out) any short that occurred. This conditioning test has been carried out regularly on Voyager 1 since its introduction in 1978.

I have stressed that projects like Voyager are done best as team efforts, and that the individuals that work on such activities have to understand and work within that framework as real team participants. I had an excellent team of people supporting me in doing the Voyager spacecraft. The spacecraft management team consisted of Bill Shipley, development manager (basically my deputy); Bill Fawcett, science instruments manager; Ron Draper, spacecraft system engineer; Glenn Cunningham, deputy spacecraft systems engineer; Chris Jones, spacecraft software systems engineer; Tom Gavin, reliability assurance; Don Howard, quality assurance; Tom Gindorf, environmental requirements engineer; Chuck Reynolds, test and operations manager; and Joe Shaeffer and Gordon Haddock, launch vehicle interface.

Another key area of spacecraft support was the technical divisions project representatives. JPL operated in a matrix organization structure in which the project was in the office of the assistant laboratory director for Flight Projects. The project office was relatively small, with a dozen or so people.

The bulk of the JPL people that carried out the effort are located in the discipline-oriented technical divisions headed by the assistant laboratory director for Technical Divisions. We had a project representative for each technical division who operated as an assistant division manager for Voyager within his division. So he was extremely important in terms of the way the division functioned, and supported the project in working problems, controlling schedules and budgets, working people assignments, etc. Jim Bryden, John Hunter, Harry Margraf, Mike Sander, Joe Savino, Bill Shatz, Dick Spehalski, Fran Sturms, and Frank Wright provided excellent technical division support.

Voyager was certainly a team effort. That includes Dr. Stone as the project scientist. Ed understood and worked within Schurmeier's framework and brought the scientists into the team. It almost appeared that he knew as much or more about their experiments than they did. It was amazing to see him interact with them; there was no bull put forth by those scientists once they got him calibrated!

The team participation of the scientists on Voyager was the best I had seen on any project with which I had been associated. That applied to the design, development, and implementation phase as well as the flight operations. Voyager was a large, long-duration project; it has had a unique team spirit and attitude that has prevailed throughout the life of the project.

The spacecraft achieved everything that we could have hoped for, and more. When we were doing the design and development of the spacecraft, there were no tools for assuring that we were designing a system that would last 12 or more years. In fact the MJS 77 project objective was the three years to carry out the Jupiter and Saturn missions.

What we did was use conservative practices, reliable components, conservative design margins, adequate testing, and as much redundancy as possible. We practically built two spacecrafts within each of the spacecraft structures. Most programs today are not able to be as conservative as Voyager and, as a result, are taking much more risk.

I have been as awed as anybody by the character of the outer planets and their moons as revealed by the Voyagers. None of us imagined how spectacular those early pictures of Jupiter and its moons would be. It was absolutely marvelous to have participated in bringing about those results. I remember those first, Jupiter encounters: I was operating as one of the three mission directors, working 13 hours on and 11 hours off for a couple of months for each encounter. But the whole process of the day-by-day results, the press conferences, the scientists giving their reports, and so forth—I can't think of anything else I would rather have done. There was enough adrenaline flowing that you almost didn't need sleep, and it was a continuous flowing process, with one outstanding result after another.

The initial program was well conceived and was made better by the extra time at the beginning. The up-front enhancements to the spacecraft were completely supported by the scientists. It gave them the flexibility, the programmability, the reliability, the data rates, and the payload capacity; all of those things were the things they wanted as enhancements for the science mission. While they were spacecraft enhancements, they provided the capability needed to achieve the science. So that was a very important period and activity for the project, because it set in place everything else that followed.

Going back to the very beginning of Voyager, how much of the TOPS project was motivated purely by astronomy, and how much might have been motivated by people interested in nuclear energy research, Department of Defense, etc.? What was the impetus for that basic project?

I feel very certain that the motivation was the unique gravity assist opportunity for exploration of the outer planets, and the recognition that conventional spacecraft would not meet the mission requirements. The nuclear power sources were just one element in a well-balanced program looking at the requirements for such a mission, and the status and needs of the technology to carry it out.

We knew we had to have some other power source than solar panels. It turned out that the Air Force was interested in nuclear power sources for satellites, so a parallel and mutually supportive interest developed between the Air Force and NASA for a few hundred-watt power source.

The Voyager RTGs are remarkable devices. They use plutonium oxide spheres in a cermet form that provide 100 thermal watts each. The ceramic/metal spheres are about 2 in. in diameter and are encapsulated in a thin welded shell of iridium. Iridium is a very tough metal and provides a good containment shell for the fuel. The spheres are then wrapped with a graphite yarn to provide a 1/2-in. protective layer as an impact energy absorber. Then the spheres are stacked in a graphite cylinder; six layers of four balls each. The cylinder then provides a heat source of 2400 thermal watts. The heat source is inserted in a protective cylindrical housing made of pyrolytic graphite and polycrystaline graphite. This housing provides excellent structural strength and good thermal ablative properties should a failure force the atmospheric re-entry of the RTGs.

The RTGs' fuel spheres and housings were put through elaborate tests to demonstrate their safety in the event of a catastrophic launch abort or atmospheric re-entry. The assembled heat source was fired into cement walls, dropped from high altitude on to concrete pads, exposed to plasma arc jets to simulate atmospheric re-entry heating, and exposed to high temperature explosions simulating launch vehicle explosions.

Each generator provided in excess of 150 watts at launch and has performed better than predicted throughout the mission. Three RTGs were flown on each Voyager. I remember that we had about 450 watts of power at Jupiter and about 425 watts at Saturn. I believe we had about 10 watts of power margin for the Neptune encounter.

The RTGs were having serious problems with their operating lifetime projections until a gentleman at RCA came up with the approach of coating the thermocouple legs with silicon nitride, thus preventing the sublimation of material which was redepositing in the thermal insulation blankets and shorting them out, thus causing the loss of performance. As a result of this fix, the performance of the RTGs has been spectacular.

But to get back to your question, the TOPS program really was focused on the whole problem of doing the Grand Tour. The RTGs were important and did enable exploration of the outer solar system, but TOPS wasn't done just to open that door. The Air Force program with the Lincoln Experimental Satellites would have produced the MHW-RTG. In fact. two Lincoln Experimental Satellites (LES 8 and 9) were launched with MHW-RTGs ahead of Voyager, on March 15, 1976.

There are lots of stories which could be told about the Voyager project and a lot of detail has been glossed over, particularly with respect to the many contributions of individual team members. I wish you luck with this book.

Thanks to you it's one step closer to success. It was a marvelous project and I feel privileged to have this opportunity to talk with you about it.

It was fantastic, and even if it had ended at Saturn, it would have been well worth the effort. There was a group of people who wanted to do more Titan science with Voyager 2.

In addition to the Voyager 1 redirection?

Yes, to do a similar thing to what Voyager 1 did, and lose the rest of it. They argued that the chance of survival to Uranus and Neptune with the already existing receiver problem was low and that we should go for the high value science at Saturn. But several of us on the project felt very strongly that we should proceed with the Grand Tour.

We might not have known for a long time what a significant loss it would have been. We now know that there would have been no results from a second Titan flyby to compare with the results from the Uranus and Neptune encounters. The Cassini project is providing the kind of capability needed to expand our knowledge of Titan and the rest of the Saturnian system.

Fortunately we were allowed to proceed with the Grand Tour, and our strong convictions about the prospects of success were vindicated.

ESKER K. DAVIS
Project Manager, Saturn 2
Born September 1, 1935 Weston, West Virginia

Esker Davis earned a bachelor of science in geology from West Virginia University in 1958. He then served in the U.S. Air Force, as a material officer, missile squadron support officer for NATO and, from 1963 through 1967, at Cape Canaveral, working on the Manned Lunar Landing Program, NASA's Earth Orbiting Program, and NASA's lunar and planetary missions: Ranger, Surveyor, Mariner, and Lunar Orbiter.

His JPL career began in 1967 on the Deep Space Network as assistant project engineer for Mariner Mars, and then project engineer for Apollo. He was tracking and data systems manager for Mariner Venus-Mercury from 1970 to 1975 and then for Voyager for the next four years.

In 1979 he joined the Voyager mission as deputy project manager, and was project manager for the Saturn portion of Voyager's Grand Tour in 1981 and 1982.

After the successful encounters he spent two years in private industry doing materials research in low gravity. He returned to JPL in 1984 and, after a year in the Office of Engineering and Review, became manager of the Institutional Computing and Mission Operations Division. In 1994 he transferred to the Planetary and Space Physics Program Office to manage the Space and Earth Science Directorate.

Along the way he earned a master of business administration in research and development management from Florida State University in 1965, and a master of engineering from UCLA in 1977.

His awards include three NASA medals for Outstanding Leadership, Exceptional Service, and Exceptional Achievement He also received three Air Force medals: Commendation, Meritorious Service, and Legion of Merit. He is a major general in the Air Force Reserve, serving as mobilization assistant to the Commander, UCLA School of Engineering and Applied Science. He has served since 1978 on the UCLA Dean's Council Executive Committee for the School of Engineering and Applied Science.

He is a member of the Air Force Association, the Reserve Officers's Association, and a Senior Member of AIAA's Space Operations Committee.

He enjoys doing astronomy with his son, using their 10-in. Schmidt cassegrain. He also likes "traveling around different places, exploring; not just to go somewhere and lie on a beach, although that's fun too, but to travel and experience new things."

Near perfection is marginally adequate.
—*Esker K. Davis*

ESKER DAVIS

I was always interested in science, even as a kid, and especially things in the sky; astronomy, the moon, planets, stars, and so forth. I was interested in them from the viewpoint of exploring, not so much, at the time, as a scientist per se, but "could one ever explore, could one ever go to the moon, could one ever go beyond our little Earth moon orbit?"

I was a senior in college, studying geology, when Sputnik went up and then Explorer 1, and I immediately became interested in the whole business of space exploration. Upon graduation I went into the Air Force and became involved first with some missile programs. Then, in the early 1960s, I requested and was assigned to Patrick Air Force Base, Cape Canaveral, where the space program was being born.

The first job I had was in the manned lunar landing program office which the Air Force then was running, but that was soon turned over to NASA. Then I was put into an office called Lunar and Planetary Programs, and that's how I first met people from JPL. Lunar and planetary programs were pretty much the domain of JPL at that time, though there were other centers involved.

My job at Patrick was program support officer. I made all the arrangements for launch support services; such as range control, down range tracking and support, data acquisition. This was how I first met the people from JPL. They were very articulate, very bright—some of them were so bright it was scary; I wondered how one could even compete with them. How could someone just walk out and say, "I've figured out a trajectory to go to the moon and Mars," and yet they knew what they were doing.

I pretty much fell in love with the whole thing in which JPL was involved and when I decided to get out of the Air Force, I applied for a position at JPL and was fortunate to get it. Of course I came here as a known factor, from that four years with them in that job at the Cape.

At JPL I've had various jobs. I was the Deep Space Network (DSN) project engineer for Mariner 4 and for the Apollo first lunar landings. But what really set me up for Voyager was that I became the tracking and data systems manager for Mariner Venus-Mercury, launched in '73. It was the first gravity-assist mission; we used gravity assist at Venus to get to Mercury, and we came back to Mercury three additional times. We did the first antenna array in '74–75 at the second and third Mercury encounters. We collected the signal on two antennas and combined that signal to improve the quality, as if we had a larger antenna.

In '75, right after the third Mercury encounter, I transferred over to this new project called Voyager and became the tracking and data system manager for Voyager. That was two years before launch. My job was to get the DSN and launch support services all ready for mission operations. I did that, through launch and into '79.

You mention personal risk, the personal part of this thing. In the Deep Space Network we had little or no backup. We had some one-of-a-kind boxes for signal combining, antenna arraying, and a lot was riding on it. I remember my boss saying, "You have a lot of neck rolled out here. If this thing fails, if something breaks, you have worse data."

Explain the risk. What was risky?

Once you decide to do signal combining and raise the data rate to a higher rate, if then combining doesn't work, you're receiving at a higher rate on a smaller antenna, and you get very bad, noisy data until we could again send commands to lower the data rate. So it *had* to work. We tried it on the way to Jupiter. We tried test signals through it, and it worked every time, and we were confident.

I think confidence is a big factor here. People work very hard, they're very dedicated, they put in a lot of hours, and end up being quite confident about taking whatever risk is left after they've checked everything out and worked it through.

It turned out that gravity assist and antenna array served Voyager extremely well over the entire span of its mission, so I look back on that as a stepping stone.

Also there are two technical sides to JPL. There's the flight project side, the people that build flight hardware and do the spacecraft; and there's the ground side, the Deep Space Network, tracking and data acquisition. Well, I was over in the tracking and data system part; I was the TDS manager for Voyager.

The interesting question was who here decided to reach out and tap *me*; select me to be the deputy project manager for Voyager? It was almost unheard of that someone from the DSN side of the house would be brought over into the flight project side. When I was notified that I was the top candidate, I couldn't believe it was real, I couldn't believe how fortunate I was.

Had you applied for it?

Not formally. They asked me would I mind being considered as a candidate? Sure, I'd love to be, but I didn't sit down and fill out an application. I wouldn't have been able to fill in all those squares about: "Have you ever built a spacecraft?" No, I'd always been in the ground systems side. But I was selected, and I look back on this whole Voyager experience as being the greatest events in my life. There's no matching it in terms of what came out of it.

Fortunately I was paired up with Ray Heacock, who was the project manager. Ray helped build that spacecraft as the spacecraft engineer; he knew it inside and out. I learned an awful lot from him about spacecraft and what makes them the way they are, particularly Voyager. It had kind of a life of its own, its idiosyncrasies.

But Ray took the time and was very patient and helped me understand spacecraft and instruments. And he helped me put together this book on the different parts of the spacecraft: how the computers work, how the different instruments work, all the fundamentals, a little flip chart. It's part of my Voyager road show that I use when I go out and give talks.

On the other hand, Ray was not much interested in the business management of the project. A lot of my training—both in eight and a half years active duty in the Air Force, and then in the DSN part of JPL—a lot of that included sizeable amounts of business management: planning and budgets, the financial details, so I brought that strength to the project. Whoever chose me probably knew what they were doing, because that was going to be one of the biggest challenges we were facing. Along with the technical job of getting the spacecraft to Jupiter and Saturn and so forth, we had just as big a challenge on the business side.

By "business" are you including personnel, or are you talking more about finance and budgets?

Both, but more the latter: finances, budgeting, approvals, but personnel at least in the way they're used. We didn't have any shortage of talented people for Voyager. We had a lot of people that wanted to get on Voyager, that wanted to stay on Voyager forever, but they had to be organized properly in teams.

So that was what you were doing.

Yes—more efficient organization and re-engineering the organization structure. I'll say more about that. But Ray and most of the other guys were involved in the mission design, the technical plan. "How are we going to carry out the missions and meet the science objectives, and the trajectories and the navigation, and the spacecraft capabilities?"

The Voyagers at times were kind of shaky in terms of their performance. You've probably heard all about the different things that failed—the receivers, the scan platform, and so forth—that caused us to do extraordinary things with people. These bright, talented people really did some extraordinary things, and saved the mission more than once—not only saved it but enhanced it. I don't want to sell that short at all: the sequences, the software loads, the clever ways in which our spacecraft engineers used the capabilities of Voyager.

But that wasn't my strong suit. I applied myself where I thought I could contribute most, and that was on the business side. Ray Heacock had said he wanted me to focus on the project budget. I came on board as the Voyager deputy project manager right after the second Jupiter encounter. I hadn't been on the job very long when Ray showed me a letter—I think it was September '79—a letter he had sent to headquarters, projecting ahead. Looking at all we had to do just to get through the prime mission—the second Saturn encounter plus a year—he had forecast that we would need an additional $17 million over what was in the Voyager budget. To our shock NASA came back and said, "not another dime."

We were pretty jolted by that because we thought, "Hey, we have Voyager on the way, we've passed Jupiter, we're going on to Saturn, our prime objective," and we certainly would get that modest amount of additional money. But that wasn't to be. To what extent NASA was playing hardball with us, I guess we'll never know.

But I picked up the challenge. Ray said "work it," and I started laying out a plan to attack the $17 million overrun. I took a businesslike approach to it. What would I do if I were in a business and were running a deficit? How would I motivate people to get them to want to reduce costs? Not many people want to do this when they're up to their ears in technical challenges and technical problems and working hard, long hours. "What do you mean, I have to work even harder? What do you mean I have to cut costs?" How do I get them motivated to even look for ways to do that?

You must have a plan. It's like budget deficit reduction at a national level. You have to have a plan that says if you get on this curve, if you follow this reduction profile, you'll eventually reduce the budget. So I laid out a profile that would get us to end of mission with no overrun. It meant saving about $5 million a year.

Define "end of mission."

The formal end of mission was defined as Voyager 2 Saturn encounter 1981 plus one year. That would have been August '82, a little over three years. We took the $17 million and straight lined it across and came up with an average of $5 million a year to save.

That's a more manageable challenge. I got people more used to looking at it in terms of the smaller segments. For the next year's budget, rather than just saying, "Tell me what you need," I gave them targets, I gave them budget guidelines that said, "You have less than you had last year." In each area—flight engineering, mission operations, flight science—each office manager received guidelines from me that would help us meet this objective of the first year.

It worked so well that even the GAO came out here and asked, "How did you do this?" I said there's a profit motive here. We're not a for-profit organization, but in running Voyager like a business, I gave my staff incentives. I told them that if you save more than a certain amount, a part of what you save is yours to buy that computer you've wanted.

There's always a long list of things the engineers wanted—there weren't these fancy computers back then, but they were starting to come out. There was always something they couldn't afford, but this profit motive would enable them to acquire it *if* they first filled my corporate coffer here with a certain amount of money. Then there was a trickle down to them. So that was part of the motivation. There was a lot of grumbling, but we built a budget that put us on that curve and finished the prime mission.

The other thing was control. Things were never really out of control, but we were going along month to month, working as hard as we could, and we did our best and we paid the bills—and that's what the cost was. Going along that way, we were close, but that was no longer good enough. So I set alarm limits. Each office manager had to come in for a monthly management review, and if they were $1000 over budget—or under—I wanted to know why.

They said, "Wait a minute, that's ridiculous. One thousand? That's nothing." But I had 120 accounts; what if each of them was over $1000? Then $120,000 times 12 months and I'm going the wrong way in a hurry.

I also set up controls on time. We were really looking at the work force and looking at the dollars spent, down to the resolution of $1000 or one-tenth of a work month, in terms of being off plan. And we began to find things: "What's that charge there?" There's always errors in the system. Someone gets off on the wrong category. Someone was told to quit charging some number, and they didn't—those kinds of things you can clean up.

Because of this incentive, we began to think of ways to cut costs. For example, one of the guys came in with a wild idea. We had a DSN data team, we had a project data team, and we had a science data team. This one would put data on the tape in some format, they'd hand it to the next team and they'd do a little bit to it, and they would hand it to a third team. The guy says, "We could probably use one data team to do it all; combine them." We looked at it; no one saw any good reason why not. That saved a big hunk—$2 to $3 million.

Another idea came in out of the blue: someone noted that the scientists had always asked for what we call virgin tape. Back then we used the big reels of tape, and they wanted it never to have been used before; brand new to put their precious science data on. You can imagine how many tapes we used; they didn't hold a lot of data back then the way they do now, so our tape bill per year was $750,000; just the tape bill! And you could only use it once, and then give it away to someone else. But we got the scientists to accept recertified tape: degaussed and recertified, which turned out in many cases to be better, because you found any flaws. So that saved another $750,000 per year.

It was an effort by a lot of people. Once we started them thinking about, "We have to save money, we have to get this thing on budget," ideas came up when people tried to do things more efficiently. It snowballed. We recovered funds that were on contract but not costed, from back when they built the spacecraft, because we got the procurement people to close out finally those contracts from the spacecraft being built. In '77, '78, and '79 there were still moneys being held as obligated against a potential cost until final audit, and we accelerated that.

Anyway, with those incentives and those controls, people finally realizing we can make a difference like that, we ended up the prime mission $750,000 under budget—three-quarters of a million under budget. I looked on that as a major achievement. We not only met that NASA challenge (again, what kind of hardball they were playing I'm not sure), but it also established one heck of a lot of credibility for us.

All of a sudden NASA stood up and took notice of the Voyager project as being able to stay on budget, get the job done, cut costs, manage it like a business. JPL's relations with headquarters were never better than with the Voyager project. The program managers in Washington couldn't say enough good things about us, in terms of our relationships with them, the way we kept them coordinated and informed about what we did. Our reports to them were top notch, on time, complete.

The other thing is we had a good plan. It was recognized as one of the best project plans NASA had ever seen. We compared our progress to that plan. It stated our goals and budgets and workforce profiles and science achievements; all the right things were in it. That established us well to propose and get approved and carry out this audacious Voyager Uranus-Neptune interstellar continuation.

Of course we had been thinking about it a lot, some during the Jupiter phase, but particularly when it became clear that we were really going to get to Saturn and accomplish our prime mission, we asked NASA about continuation on to Uranus and Neptune, and NASA said "write a plan." So, having had a good plan for the prime mission, we pretty much projected from that, from what we had been doing, on to the proposal for the continuation. But when we showed them a draft they said, "No, no, no, you misunderstood. You have to do it for *half* the cost."

That was such a shock! We were spending about $24 million a year for the Voyager prime mission; $2 million a month, really small dollars these days. But they said, "Get down to $11 or

$12 million a year. It's a long way to Uranus. You may not get there. We can't invest all that; just go simply, don't do anything complicated."

Well, there was an awful lot to do each year to get ready for Uranus and Neptune. But we said if that's what it takes to get the thing approved, then we want to come up with a plan to do that. There was uncertainty that we should even try to execute a plan for Uranus and Neptune under such conditions. There was a lot of resistance, but we did get the cost down. The cost at that time was mainly all people, in mission operations and science ops and spacecraft engineering. So we would cut back to about half the people and we'd fly on to Uranus.

What eventually evolved, as we got closer to Uranus and it became apparent that we might indeed pull this thing off—that it might survive and go on to Neptune and interstellar space—NASA began to add back a little money; they realized that it just wasn't prudent to stay so low. By then I was gone as project manager, but we began to see the budget creep back up. That was good, because you don't want to squeeze too tight when you have an aging spacecraft and these wonderful objectives still to be met out there.

There was another issue about people. Here we are, coming in for the second Saturn encounter; do we announce which half of the people won't have jobs? Do we even breathe a word about this? We had a big discussion about it. I've always been a very open person; I like to get a problem out on the table and work it down. I know there are times when one needs to be discreet and keep things close until they're finalized, but it didn't seem that we were going to solve the problem that way. So I decided we're going to do this out in the open.

I would bet there's a third of the people who are ready to leave Voyager and do something else. There's probably another third that want to stay on Voyager forever, and there's a third maybe that aren't decided yet, and if they were given an opportunity might go either way. I bet that the problem is almost solved if we would just work it. Once people know that they have a job to go to, or they're staying on Voyager, we sort all that out, we won't have any morale problems, we won't have anybody jumping ship, bailing out early because they better go grab that other job now, while they can. Instead, we'll arrange for each person to make sure they have a follow-on job.

We did that, Dick Laeser and I; he was my deputy when I took over as project manager from Ray Heacock. It was one of the really nice inside, unseen achievements, in terms of looking out for people and looking out for the project at the same time, in terms of morale, in terms of placing these people.

I've used that as an example many times around here when the question arises, "Should we tell people there's going to be a downsizing?" I used that example to show other managers there may be an answer to their dilemma, their problem, if they just work it with their people.

And it worked out well on Voyager, pretty much as I predicted. There were those who wanted to stay on Voyager. I wanted to stay. But in early '82, after the Saturn encounter, the Lab's business base was in a big downturn. They were looking for more business with the Air Force. They thought that I, being an Air Force reserve officer, should help do that, so they pulled me off Voyager. I wasn't ready; I didn't have my fill of Voyager yet. But I helped JPL bring in some Air Force business. Dick Laeser eventually took over as project manager.

But the problem that happened on my watch was at Saturn encounter. You talk about going from the heights to the depths; the scan platform stuck. When Voyager came out from behind the planet, we held a late evening press conference, pressured by the press people. They had deadlines; they asked me to come down at midnight and give them a few words. So I went down and said it looks all right; the early data looks like the spacecraft is fine. We don't see any problems in the engineering data, and so forth. What we had not gotten back yet were the pictures.

So we had the press conference, and we were celebrating a little bit in the conference room and decided we'd better call it a night; we'd been up day and night more than two days. I was the last person out; I was holding the conference room door open and switching off the lights when I heard on the communication net, "Ace, this is Bus; we have a problem." I held the door, flipped back on the lights. That's when we heard something was wrong with the pictures. The pictures were not centered in the field of view. Whatever moon we were imaging at that time was drifting off to the side.

So we went back into the conference room and spent the rest of the night there, until about noon the next day. We obviously had a stuck scan platform; it wasn't moving and tracking the target. By noon one of our key people was actually hallucinating at the conference table. He couldn't get his thoughts together. I was really tired myself. That's when we called a halt and said everybody go home and get at least eight hours sleep, and we'll see you back here tonight; we'll start again.

It became a very sticky issue. A lot went on in those few hours. The imaging people were very upset. The imaging scientists wanted to damn the torpedoes, full speed ahead. One of the imaging scientists burst into my office, almost knocking the door off the hinges. "What do you mean, you refuse to command this platform?"

I said I wasn't going to send any commands that are going to rip out the gears or destroy it. We got 40,000 pictures of Saturn; 20,000 from Voyager 1 and 20,000 from Voyager 2. We probably have enough.

"But I've got public and press out there waiting! Send those commands. Move that platform."

I said no, we're not going to do that.

He retorted, "We'll see about that! Call Headquarters."

What he didn't know—and I didn't ever tell anyone or show anyone—I had a letter in my desk, signed by the NASA officials. They were here for the encounter, and before they left they had said stay the course, be conservative. So that's why I was very bold.

That was before this problem arose?

No. As it arose, and we began to see what it was, they said, "You have our full backing in being conservative. Don't send a command until you understand the problem first. Then fix it if you can, but don't put priority for any more Saturn science over trying to preserve the spacecraft and getting it fixed."

That was very useful, having that letter in my desk drawer.

Did you show it to your agitated friend?

No, I never did. I just let it go. He came back later and apologized and said, "In heat of battle, I got carried away. You were absolutely right."

Another interesting thing was that one of the media people, Roy Neal, for years thereafter believed that I lied to him when I stood up at midnight and said, "Everything's fine". He really believed that somehow we contrived that, that we wanted the reporters to go out with their final deadline story and say that everything was fine, because that would be what the public would remember. But the honest truth was we didn't know until after that press conference that the scan platform was stuck. I could never make Roy believe that.

We had developed a really strong rapport with the media. Our credibility was so high, the last thing in the world we wanted to do was lessen that, lower that, destroy it, because it's very useful to us to have the media on our side. They'd ask congressmen, "Hey, what do you mean, you're not going to fund this?" They were very supportive of the program and became attached to the Voyager spacecraft much the way some of the people working here did, in terms of what it had done, and its capabilities.

It's an experience that comes to some people once in a lifetime. I was really fortunate to have many of them grab the brass ring that had arrived on Voyager. It was really super for me.

A lot of the media, particularly the written ones, the newspapers, would try to find an angle. For awhile around here they focused on stress. They were running around asking all of us how do we cope with the stress? Personally, I don't feel stress, but they insisted, "You gotta be wiped out by stress. This must be the most stressful job I can think of."

I think the answer they were getting, not just from me but also from others, was that yes, this was something that was difficult, the challenges enormous, but we were pretty good at what we do, we work very hard, as hard and as good as we knew how, and if that's day and night, fine. We got tired but we didn't feel stressed, because that's the best we could do. And with that attitude it seemed we were just rolling through these extremely challenging and dynamic events.

We had a saying on Voyager, that near perfection was marginally adequate. Think about it; that's kind of where we were. We were operating Voyager on the slimmest of margins, whether that was performance margins, telecom margins, money margins, the accuracy demanded, the performance demanded of people and machinery.

So I got to like that little saying. It put in perspective how we were operating. We did it, and did it without stressing people out entirely. There were an awful lot of real people that made it happen. Some very talented, really bright, but also a lot of ordinary people were involved in the mission, too. It had never been done before; I don't think anyone knew exactly how to do it, yet we carried out that exploration, which had been my dream of exploring the solar system. Going back to those early questions: could we ever really go to the moon? And here we went to the outer solar system!

I came back for all the encounters and helped out with the visiting dignitaries and thoroughly enjoyed it.

The leadership on Voyager was very important. It seemed like the right people came along at different times, whether that was Ray Heacock when we really needed to understand what was going on with this spacecraft and some of its idiosyncrasies, to my coming along with the business approach, and the long haul to Uranus with Dick Laeser and his knack for keeping morale up and people working together in cohesive teams, although I left him that skinny budget to work with. Bob Parks, who was our boss in the program office at the time, was just wonderful. He didn't micromanage us, he didn't get involved if he wasn't needed; if we needed him, he was there. It was a great internal relationship within JPL.

I used to kid Bruce Murray; he was our most difficult, demanding interface because he would pick at little things. When I would go in for the director's review once a month or two, I'd stand up to tell good news about Voyager, and Bruce would get upset because the type on my viewgraph was too small. He'd scold me for something like that. I'm not knocking him, Bruce and I really got along great. It was just one of those things that comes along.

Great mission. I'm sure glad I was involved in it. Many of us got a lot of personal recognition. The whole world tuned in. Many of us on Voyager had a road show. We couldn't satisfy the demand, showing those beautiful pictures, talking to groups in terms they could understand. I made hundreds and hundreds of presentations. I still give them. "What did you give to my wife's group? Can you come and give it to us guys over here?" They'd hear about this Voyager show, and it's still in demand. I've given it to the third grade class in my hometown and to the Academy of Science in Sydney, Australia—that span of interest.

I never turned down a request for it. I'd go anywhere, anytime, give a talk. I didn't care whether it was 12 people or 1200, small community groups, civic groups, large scientific organizations. There was amazing fallout, it was almost like a gratuity.

We won a lot of awards. Voyager probably got the whole list. All the things the project won: Goddard Memorial Trophy twice; Jackson Space Award. It goes on and on. We received so much recognition in terms of the project, and then personally we received things like NASA medals. I received the Outstanding Leadership Medal for my hitch as project manager; it's one of their highest awards.

Dick Laeser and I spent untold hours deciding who should get awards. We built Voyager in '74, '75, launched in '77, got to Jupiter in '79, Saturn in '80, '81, but couldn't give any recognition until the prime mission was over. That was the rule, because what if it fails before the mission is over? Then we would have prematurely awarded somebody. So we waited.

Who established that rule?

NASA Headquarters. The Lab concurred although it was pushing it a bit to say don't recognize anyone, particularly those who figured out how to get there with this gravity assist, this trajectory design. Their job was already done; it was already working.

We worked very hard to sort out all those people and to really focus in on those who were deserving of different levels of awards, from achievement awards to exceptional service medals to outstanding leadership and science medals as well. We had a glorious ceremony out here after the second Saturn encounter. We did it up in a very nice way; we took our time and recognized all those people for all those years. A lot of awards, a lot of medals, a lot of certificates were given out. That was very satisfying.

It was really challenging, sorting them out. If this person got a medal, why not that one? We were trying to make it equitable. We had to bring in some people who weren't even working on Voyager anymore. We recognized companies that helped build it, that supplied pieces.

One of the challenges that happened during your watch was that you had two spacecraft, and when Voyager 1 experienced something—or did not experience something—then there was the opportunity or necessity for perhaps changing, reprogramming what Voyager 2 is going to do. Do you have any thoughts on that?

Yes. It was a double-edged sword, a mixed blessing, in that, knowing we would learn from Voyager 1 certain things—like there's a volcano on Io or there's something about the Saturn rings—that we would significantly change what Voyager 2 would do, in terms of science sequences and operation. This triggered some of our biggest surprises, in that you plan for and budget one spacecraft to do this and one spacecraft to do the other; it gives you complementary coverage and different looks at it.

But the discoveries of the first one though, in many regards, threw a monkey wrench in, and we had to redo. On the other hand we were so excited about what we could add to the science that it certainly seemed well worth doing.

It was a matter of sorting it all out. That's why, when we began to realize that this was likely to be the case, that we got clever at designing these observational sequences in a way we called "putting it in the can." You design them and have them in a form that you can load into the spacecraft, but they are made up in ways where there's blocks or there is flexibility to make some changes. So we would do this preliminary putting it in the can and then later on we could do the update. We would look at the data coming back from Voyager 1. Have we learned anything from its science data or spacecraft performance that we ought to incorporate into Voyager 2?

That's more costly. That's what worried us a lot as managers. That took more people; it took more time to go through the cycle. The people were almost into a production-line mode. It was almost automatic, in that they would produce these products first in a preliminary mode and then as final products for the mission.

That was all tracked very precisely, down to about a day's resolution as to whether we were on schedule or behind schedule, and whether all those instructions that went to the spacecraft as sequence loads were on time and verified as well done—lots of steps; many, many tens of steps to get to the final product. We became very good at that. Otherwise the return from the mission would have been a lot less. It was the kind of mission that just demanded that.

Was the required reprogramming still within your projected budget, so that you could respond effectively to surprises like the increased radiation at Jupiter, or were there surprises of such magnitude that they really threw your planning out of kilter?

Some were and some weren't. There was always some reserve and contingency funds. While the spacecraft were being built and tested, Pioneer went by Jupiter and learned about the radiation belts, so with Voyager we had to go back and use different parts, add shielding; a costly radiation hardening effort. There were other things like that, that we had to go back and improve, change.

When we launched, all the reserve and contingency funds were gone. Also, back then you could mix operations money and engineering development money to some extent. They had probably borrowed from the operations to complete the development, so that operations money was not only no reserve, it was skinnier than what the original plan had been.

Voyager started out being very difficult to operate—I had a saying about that, too: They were wonderful machines but difficult to operate. They were really wonderful, beautiful machines. They were capable of doing all this science across this broad spectrum, yet they were very touchy; and operations were very difficult in terms of understanding what the spacecraft was doing. We had to build them to be autonomous, to take care of themselves, because of the great distances, but we didn't fully understand all those algorithms that were in there to provide the autonomy.

So we had to add people. That was the big surprise. Right after launch for the first year we added 50, 60, maybe 100 people (I wasn't in the driver's seat then) to mission operations, because these things required more watching at the subsystem component level than anyone ever had imagined. Even though the spacecraft are capable of taking care of themselves and going into some safe state, you had to know what that was; you had to know what it was doing to recover and come out of that state safely.

There was an inordinate amount of analysis required on the ground: some experts in guidance and control, some experts in propulsion. What are those thrusters really doing? We fired these thrusters, and the spacecraft didn't quite react the way we thought it would. We would do something and it didn't quite perform the way it was supposed to. There were little things going on in terms of thruster, plume impigment on spacecraft structures, or there was something that wasn't quite calibrated over here.

We were learning to operate those spacecraft for those first two years. About the time the Jupiter encounters were over in '79, we looked at each other and said, "I think we're beginning to learn to operate these things." It's probably like flying a high performance jet. Pilots talk about flying the SR 71. The performance envelope is so narrow, you get a few knots here and a little bit of wing wiggle there, and if you don't keep it within that very narrow envelope, you can lose it.

That is what we began to realize about Voyager: if we didn't keep it within some rather narrow operating envelope, we could lose it, and that was a big fear. We came close a few times to maybe losing one of them. So we added people to watch. We added them in a big way, and those were the people we had to eventually peel off after the Saturn encounter prime mission.

Was that necessary? Or was it just our response? Could we have done something different? I don't know. A tough judgment call. With today's technology we could have computers watching a lot of that stuff and telling us what's going on. That's what we are doing today, to get operations cost down, but back then we were still on mainframes, we still required an awful lot of expensive software programming development to get those kinds of capabilities.

With that level of technology, we didn't have the tools that would help us carry out a mission like this with a lot less people, maybe an order of magnitude less people. That's what we're planning for Pluto, with a computer onboard the spacecraft so small and so capable that it tells us how it is and what it's doing. We won't have to figure it out on the ground. But we couldn't do that with those extremely small computers Voyager had.

Is there any way to export this Voyager experience, the lessons learned, to other organizations, other projects not even necessarily in aerospace?

I'm convinced that "projectizing", forming a project to carry out some mission, could fix a lot of our problems: welfare, poverty, education. Give me a mission, give me an area—1000 square miles, or perhaps a smaller slice. I would fix the problems in that area. I'd do it with trusted agents, with clever people; maybe volunteers, people from churches, social groups.

I wouldn't spend one dime for anything other than carrying out this mission, in the most efficient way I know how. I'll give you an accounting, but I don't need all those people in between. I will get people working for me who *know* who needs help and know who is cheating and doesn't need help.

There is an imbalance in the amount of money we're putting into welfare—not that some of it isn't needed, and God knows, we need to help people, but you help people get *better*. You don't help people to stay where they are, you help them improve. That's the difference. My mission would be to have this project have an end: this area is now self-sustaining in terms of helping its people.

When I was experiencing Voyager, I began to think that's the way to solve a lot of the ills of this country. Of course they don't lend themselves to scientific principles and the rules of physics in the same way. You have emotions, interest groups, politics, and if you let these get in the way you're going to have a failed project. We'd have the same thing here in space if we let too many of those things get in the way.

But if you apply these same kinds of dedicated mission-oriented project principles to these problems, you would really see improvement.

What was the hardest decision that you personally had to make on the Voyager project?

I had a pretty quiet watch in a way, and then I had the scan platform, and I had to plan how to get to Uranus and Neptune and interstellar space at half the cost.

When time is of the essence, the scan platform slaps you right in the face and you have to respond immediately. You wonder if you're doing the right thing. You question your technical competence. You get all the technical experts around you; sometimes they agree and sometimes they don't, and then what?

I think that was the toughest decision, because they didn't all agree. Was it something stuck in the gear, or was it just degreased? Fortunately we have a spare over here in the Lab. We took the spare off the shelf and started running it. It was almost uncanny: we ran it like we were running it during encounter, and it seized up almost the same way, at almost the same time. It was just a matter of running it in the high rate mode and degreasing a few of those bushings, and it seized up. There wasn't a piece of teflon or nylon stuck in the gears; it wasn't some micrometeorite penetrating the gearbox. But all these thoughts were going through our heads, looking at the data. What was it; what could it be? That was the biggest challenge.

The other one was to find a way to continue the mission for half the money. All right, you can sit down and take your time to do it. You might not like it, but it's not the same as being under the gun in real time, with people such as eminent scientists questioning what you're doing. That was the toughest. That's when you feel the loneliest: when you're off over here, and everyone else is saying "do something", and some people were losing their heads, flying off and wanting to do the wrong thing.

But I had a pretty quiet watch. I didn't have some of those early problems when the receiver failed and what do you do about it. But I was there as the tracking and data system manager. We decided on backup mission loads. We decided on all those things we had to live with the rest of the time: what if that other receiver goes out?

Those things we had to live with as we went along were cumulative; they weren't simply isolated, independent. We had a long list of them. This spacecraft is full of consumables. We gave Charley Kohlhase, our mission planning office manager, the specific job of monitoring them. We began to learn a new definition of consumables. Like tape; is it consumable? You only get so many passes across the head and eventually it wears out. How many passes do we have, Charley? When is it going to wear out?

Charley, in his great analytical way, would figure out that we have this many passes, which means we can allocate so many for Saturn, so many for Uranus, so many for Neptune. The hydrazine propellant. How about the electrical power? Those RTGs (radioisotope thermoelectric generators), which provide the electrical power, degrade with time; instead of having 600 watts of power you eventually get down to where you have 350 watts, out where we are now. So you have to figure out that at some point you can't power everything up at once; you've got to share power, time manage it.

All those kinds of consumables people didn't think about when we started out, so some people like Charley did a wonderful job of making those things well understood, under control. We'd review them each month or each quarter. What do we have left, what is the projection? Do we see any strange trends in the curves?

As that list grew—we have to watch this receiver, we have to condition certain instruments to do certain things—that became the personality of each spacecraft; the cumulation of those things. Someone very carefully and in a very dedicated way watched out for all them. But there was a cumulative effect, and as we got more and more of those, it became a significant load on the project, to struggle along with them.

Getting back to that scan platform, was it you who made the final decision to go slow? Your personnel had given advice, but ultimately was it your decision?

Yes, that's right. Beyond the initial flurry of some of the scientists, no one questioned it at all. Bob Parks, the director, NASA Headquarters—no one questioned it. I quickly made a few calls and kept them informed of what we were doing, and they said that's exactly the right thing. They were very supportive, not because it was me, but because they thought it was the right thing to do. This place would quickly tell you if they thought it was wrong. We are very open, honest, candid about saying, "Wait a minute; I don't like that answer; let's get a group together and discuss it." They could have gone that route, but they didn't, and it turned out that we did exactly the right thing.

Could we have recovered faster? Who knows how long it would take for that thing to cool down and the lubricant migrating back. We did recover; we did get to the point where we were comfortable and looked back at Saturn and got some more data for the scientists before we totally left the system.

We planned that so carefully. We didn't want to hurt anything. Come on guys, be very, very careful. We finally got brave enough to send one command, nudge it just to see if it was stuck or would it move. We sent the command—and commanded it the wrong way! Someone in their tired state got turned upside down.

It had been agreed that you would turn it in one direction, but somehow the direction got reversed?

Yes, in preparing a command, we accidentally prepared one that sent it right into the possible obstruction. It wasn't what we intended to do, but it shows what can happen sometimes with people, tired people particularly.

But it turns out that wasn't a big deal. There wasn't something stuck in there, where we were going to further damage things. There were two schools of thought: put it in low gear mode and grind through it, if it was a piece of nylon or plastic. Or we back up and maybe it will fall out. But then this delube theory came on stronger in those next few days, and became very credible when we ran the tests on the bench over in the lab. Most people then forgot about the particle in the gear business.

It was an interesting time. Get the group's input, get consensus, or near consensus. But it's not a group decision; ultimately one person has to decide.

That was you.

Yeah. But I tell you, working here and being involved in exploring the solar system does not seem like work. It's amazing how much joy I get out of this, and I've talked to a lot of other people who do, too. It's not just a job here, it's being involved in exploration on a scale that boggles your mind.

Somebody asks me, "What do you do?"

"I explore the solar system."

"Oh sure".

Just think about it. When I was a student in the 1950s, I couldn't imagine getting to the moon, because it was 250,000 miles away. How would you get there? It would take an old Chevy a long time. And here we have, in our own lifetime, explored the solar system (except for Pluto, and we're working on that, too) in at least an initial reconnaissance mode. It's like Lewis and Clark, and Darwin. I was always impressed by them. I read their reports and a lot of those kinds of things as a kid. I was really impressed by exploring the unknown, being the very first to see something.

The first, not the second?

That's right, the very first. I would be sitting in here at night sometimes when the Voyager mission was flying, and my wife would call up and ask what I was doing. I said just watching pictures come in, being the first person to see this picture. I remember seeing the ones on Io, and being a geologist I thought it looks like a volcanic caldera, but you know there's no volcanoes on Io so you dismiss it. Of course later on they found out that was exactly what we were looking at. I was there and saw that very first picture. Others were in the building too, I'm sure, watching that, but there's that first time you see things. A great experience.

RICHARD P. (DICK) LAESER
Project Manager, Uranus
Born April 8, 1938   Green Bay, Wisconsin

Richard Laeser received his bachelor of science in electrical engineering from MIT in 1962 and a master of science in electrical engineering from the University of Southern California in 1968. Four years later he graduated from the program for management development at the Harvard Graduate School of Business.

He joined JPL in 1964 after two years as a commissioned officer in the U.S. Army Signal Corps, where he managed development of the Army Intelligence Command-Control Information System. At JPL he served in various capacities in the computing, communications, and tracking and data acquisition support of planetary missions, prior to becoming mission operations manager for the Voyager project in 1974. After serving as Voyager mission director from 1977 to 1981, he was named project manager for Voyager 2's 1986 encounter with Uranus, which also included planning for the 1989 flyby of Neptune.

From 1987 to 1991 Laeser established, managed, and decommissioned JPL's space station support office in Reston, Virginia. He then was appointed special assistant to the director, Management Systems, focusing on change of JPL's culture and its underlying management systems and work processes.

He was awarded three NASA medals: Distinguished Service, Outstanding Leadership, and Exceptional Service.

Voyager provided Laeser the platform to be the lead author of the November 1986 *Scientific American* cover article, to be featured in a segment of the PBS *Infinite Voyage* series, and to present a multimedia motivational lecture on the value of teaming to many corporate audiences.

He was president of the Verdugo Gymnastics Club, a nonprofit women's gymnastic center, and served on the competition staff of the 1984 Olympiad in Los Angeles. He is an avid downhill skier, adventure traveler, Mozart fan, and collector of eclectic art.

I thrive on the challenge of understanding the dynamics of complex systems made up of fellow humans and their artifacts, forming mental models of the systems' behavior, and influencing the transformation of those systems to achieve noble goals while navigating a field of obstacles.
Reflecting on the last 24 years, the 13 working with the Voyager team were extremely satisfying; the five associated with the Space Station were wasted on a contrived goal; and the last six on the transformation of JPL represent the biggest challenge and an imposing goal—both yet to be met. Though sometimes gut-wrenching, I love every minute of it.
—Richard P. (Dick) Laeser

RICHARD (DICK) LAESER

After I graduated from MIT, I had an obligation of two years to fill with the U.S. Army because of the ROTC program. I was nearing the end of my two years; I sent my resume around to a lot of organizations. I was interviewed by six organizations and received job offers from four of them. All the offers except one involved me being on the technical staff of the organization, but only the one, JPL, actually told me what I would be doing, and it sounded interesting. I really didn't want to be involved in the aerospace industry, and JPL was the lowest offer among all of them. But the fact that this offer involved an explicit statement of what I'd be doing attracted me to JPL, and I've been at JPL ever since.

What was your MIT degree in?

Electrical engineering, with specialty in computing, but the MIT education is very broad-based, no matter what branch of engineering you're in, and it really set me up very nicely. In the army I was in a project management role, and that fed into my career at JPL.

Please explain a little more just why the JPL offer was attractive.

It was because it was specific and it sounded interesting. The others were vague, they were—jobs. The vagueness of them turned me off. I was intrigued by doing something very specific, and it sounded like fun. It was working in a computing system engineering section, building and specifying the contract specs for consoles that would be used in spaceflight operations.

How did you get involved with the Voyager project?

It was way back at the beginning of the project. I feel like an old man now, but I was a young manager with experience in the Deep Space Network (DSN)—the tracking stations and all that—and the computing and the control center. I had previously been the DSN manager for several of the earlier Mariner projects, and I was doing the project interface work for the computing organization at the Laboratory.

The project manager at the time tapped me to get involved as the mission operations system manager (MOS manager), just when the project was approved. It involved the two areas I had worked in: the control center and computing and the DSN, as well as the mission operations itself, which I had been involved in tangentially but not directly. So it was a new experience, and it was an opportunity that was set down in front of me by Bud Schurmeier. I couldn't pass it up, so I took the job.

About what year was this?

About '74. Up through launch I was the mission operations system manager, which meant responsibility for design of the ground system, the control center, and the plans and procedures that the project would fly the spacecraft by. At launch that role evolved into being what was called the mission director, which is sort of the lead operations job. I stuck with that job through two Jupiter encounters and two Saturn encounters. And for the second Saturn encounter I was concurrently deputy project manager, and then, after the Saturn encounters, became project manager for the long haul to Uranus and including the Uranus encounter.

So I was basically with the project from, not the brainstorm beginnings, but from its beginnings as a project all the way through the Uranus encounter. Then the Laboratory tapped me to give up Voyager, which was very painful, and go out and open up a little branch of JPL in Reston, Virginia, to support the space station, which I've been doing since.

My Voyager career covered a major chunk of my own personal career, spanning well over a decade: 13 years.

What are your strongest impressions of the Voyager project?

There's so many of them. The one that I guess is dominant is the impression of how the flight team, the group of people that flew the spacecraft, jelled. It started out as a collection of people who did not work well together as a team. It was a collection of people and procedures that had been written down but weren't quite polished, and so there was a lot of stumbling; there were mistakes. Yet over the years it not only became a very smooth, polished team with a lot of team spirit, but it became an extremely enthusiastic and tightly knit team.

I've been able to observe a lot of space flight operations in various modes. I don't think there ever has been a team, a flight team flying a spacecraft, that has been as efficient as the Voyager team was, in terms of when we decided to do something, in the vast majority of cases we ended up doing what we set out to do, without getting off on sidetracks. In other words, the work of all the players tended to be productive and be part of the end product.

That was the nature of the team, and a characteristic that made it a very exciting and fascinating place for everybody on the team to work and to spend a major chunk of their career. It's so different from a lot of the aerospace jobs that many of us are involved in. What we did was aimed toward an end product and invariably ended up there; not work that was tossed away or ignored or turned around by an error, a management tradeoff, or something like that. It was just a wonderfully efficient and enthusiastic team. Now that I've been separated from it for three years, I really miss that kind of team playing and team spirit.

How do you account for this feeling? You say it's different from other projects? What made it so?

Part of it was just natural. We had a lot of time to practice. Voyager was unprecedented compared to prior missions in that it was a long mission. It was originally scheduled to be just a Jupiter-Saturn mission, as indicated by the MJS name that it had in its early days, but even in '77 to '81, the four years that were scheduled to get through Saturn, it was, compared to prior missions, a very long mission.

It took the flight team the time from launch through the two Jupiter encounters to really get to the point where we could start fine tuning its procedures, and started to become really efficient, to sort out from the flight team those whose constitutions were incompatible with doing operations, and to get the right kind of personalities involved in the operation. There was a lot of sorting out that took place early in the game.

Once the sorting out took place, we ended up a very stable team of people who tended to be the cadre, the core people who stuck with the program through most of its encounters. There's a number of people who were there from the beginning to end; a larger number of people who were there for most of the encounters—I'm included in that class. Most of the key people were there for three or four encounters, which encompasses a good chunk of time.

In that sorting out process, was it largely self-selection or did some people have to be forced out?

For the most part I can draw a survival-of-the-fittest analogy. It was survival of the people who really enjoyed it; those who didn't enjoy it tended to leave on their own. The people who were best fit and enjoyed it tended to really get into it and become part of the team and became bonded with the team. There were a few cases where we had to ask people to leave because they kind of wanted to stay but they didn't fit personality wise in the kind of team environment that we were trying to build. Not very many cases like that, and those were early in the game. As things sorted out, the enthusiasts just hung in there; we didn't have to make very many changes.

Each encounter we did with fewer and fewer people, even though the encounters tended to be more difficult. That's because the team became more tuned. It became better; it became more efficient and could do as much or more, in terms of doing the operations for an encounter, with fewer people as time went on. That continued on into the Neptune time period.

You've mentioned, as one aspect of the Voyager's distinctiveness, the longevity of the project. What other features of Voyager might have set it off from a more typical project?

Its longevity, and certainly its exploration of the outer solar system, the many planets that the spacecraft visited. That makes it a very unique mission of exploration. Of course, it's never going to be matched again.

Thinking back a bit, one of the parts of Voyager's history is the difficulty we had in getting our arms around the spacecraft after we launched them. The press recently had a story that was fascinatingly analogous to the Voyager story: the Hubble Space Telescope and the problems they have in getting that spacecraft under control, and the funny image that it leaves with the public and the press.

We had some major problems in the beginning. We had some very clever programmers who put capabilities in the spacecraft, and the guys on the flight team couldn't understand why the spacecraft was doing what it was doing. It would take sometimes days or weeks to figure out why the spacecraft did what it did. When they realized it they said, "Oh, yeah! Maybe we oughtta change it a little bit to smooth out these automatics that the spacecraft performs."

I didn't state that right. There are so many paths that you could follow through the automatics that the spacecraft sometimes appeared smarter than the people controlling it.

Some of those same stories have happened on the Hubble Space Telescope, and I've been watching that with fascination. I hope they have the same luck and good fortune that we had in getting everything sorted out and having it turn out to be as successful a mission as Voyager was.

To back up a step further, I have one other impression from the prelaunch days, when the spacecraft was being designed in parallel with the mission operations. I still have the image of the continuing struggle—push-pull back and forth—between the spacecraft designers and the operations designers, and the feeling of always losing, being on the operations side, to the design decisions made by the spacecraft designers. Back in those days it was just natural to be pushing the state of the art on the spacecraft; it was considered to be the more difficult problem, while the operations guys supposedly could fix up any of their problems later.

As a result of losing some of those battles, we had some problems. We ended up with a spacecraft that was limited in some areas, especially in its computing ability, and had some built-in algorithms and built-in capabilities that were at times befuddling to the operations crew. Those lessons were learned and the problems were fixed in the Galileo mission, which followed us.

So that's another memory I have; kind of a fond one. Normally you don't look at losing those design battles as being fond memories, but now that the whole mission was a success I can look back and say, "Well, okay. It all worked out okay."

You referred to lessons learned. Who learned those lessons?

One of the things Galileo did that I wish we had done on Voyager was to have a common data processing system that was used in the system test of the spacecraft and then was transitioned right into the operations time period. Not only was it a common data processing capability for the telemetry data, but also the same *people* transitioned. The Galileo mission operations people were actually involved in the system test, as opposed to being a separate organization as they were on Voyager. So that's how you progress. You have problems on one mission, and you say, "Look, what can you do to fix those problems?" and you do it better the next time. It's all part of the natural evolution of doing projects better.

And it's really twofold. One path is just the carryover of information from one group of people to another group of people. Sometimes the people who were involved in the earlier project get into management positions and they pass on their ideas to their subordinates who are now working on the newer project.

In many cases the people themselves transition. There's a lot of Voyager people who transitioned on to Galileo, some of them before the end of Voyager, so their early Voyager experience was passed on to early Galileo planning. A lot of other Voyager people transitioned after the Neptune encounter and became part of the operations team of Galileo. It's a continuing process.

Ideally JPL and NASA would be able to phase their missions so they could really take advantage of this, but there was a long period when there weren't any new missions, and Galileo was just sitting on the back burner waiting for a ride into space. When you can't get launched, there's not a whole lot of value in passing things on, but it eventually happened, and I think it's all working out fine.

What were your most difficult moments? Were there particular hard decisions you had to make?

There's two of them that stand out above the others. One was early in the game when I was a mission director. It was basically getting this group of discordant people and personalities and procedures all working together; in other words, to get to the point where it was a team. That took years. I set out a personal objective to make this thing work as a team, and I think we got there. That was number one.

Number two was figuring out what the right actions were to take with the Voyager 2 scan platform after it stuck at the Saturn encounter and figure out what the right strategy was for Uranus, trading off being real conservative and not doing very much but maybe, by being conservative, not having the platform fail on you. Or, being more aggressive in using the platform but running the risk of looking like a real stupid jerk if the thing should fail the day before closest approach.

That was a decision that literally took us years to make—about four years—to work it all the way through and do all of the tests and figure out the best strategy, and giving assurance to everybody up and down the line that what we were planning to do was the right thing. The strategy went all the way up to the NASA administrator, and we got everybody's blessing on it. On a scale of aggressiveness, if very conservative is one and really aggressive is 10, we were in the eight to nine area, and history says it paid off.

What is it like to hold your breath over a long period like that, to find how it's going to come out?

First off, decision makers like to make decisions, but they like to make these decisions based upon facts, and we didn't have very many facts, so we had to generate facts. We had to recreate, rebuild from scratch, new actuator mechanisms and go through destructive testing of these mechanisms and build up some statistics, and stuff like that; it was a very long drawn out process.

It was very frustrating not to have all the information that you would ideally like to have. It never was a clean decision; it was always probabilistically based. It was frustrating to have to stretch it out over years, and then in the end, make a decision that you felt in your gut was right but it was still probabilistic.

Another thing that happened with Voyager, and it's an impression that's very strong, is that at each encounter there was an ever increasing interest on the part of the press and the public. When Voyager was launched, there was a lot of negative press about why are we spending the money. Reporters would come. They weren't interested in what was happening; they just wanted to hear justifications of why money should be spent on flying to other planets as opposed to social programs. That continued on to the time period before the first Jupiter encounter, but when those first neat pictures started coming in, all of a sudden there was a positive interest.

At the second Jupiter encounter there was a whole bunch of people who were interested. Then along came Saturn a few years later, and there was the suspense caused by the Voyager 2 scan platform failure, and the question of whether this platform really would work again, so there was a fair amount of pressure on us innocent little guys who were trying to do this thing right—morning press conferences and things like that, and trying to explain it all. At the same time there's a little bit of a ham in each one of us. I really enjoyed leading the press through the anxiety we were experiencing during this period. Several of them made interesting stories out of the whole thing.

At the Uranus encounter we were really in the spotlight, and it was a national or international spotlight. At the Uranus and Neptune encounters more and more the walls of von Kármán auditorium were stretched to their limit in terms of housing the legitimate press, whereas at the early encounters there were a few people from the daily press and the magazines, but just a handful of them. Instead, you had science fiction writers, groupies, and the like in the press room.

The public information people have numerical data indicating the interest of the press, and how it increased from encounter to encounter. It was really an amazing change.

What was your biggest surprise, either positive or negative, of the whole Voyager project?

Miranda. This has nothing to do with flying it; it was science. As we flew into Uranus—and I kind of look at Uranus as my planet since that's the one planet that I was managing the project for; I was captaining the ship for that particular planet. As we came in closer and closer toward the planet, it really looked like it was going to be dullsville. It was not a very interesting looking planet.

We had some indications, as we approached, that Miranda might have some surface features. But when the tape recorder played back that set of Miranda images that included that big scarp, that big cliff, and all of that kind of stuff, I about jumped out of my skin. I happened to be down in the press room at the time, and some people told me my behavior was a bit bizarre. I had set myself up for managing the project for a dull planet and was emotionally prepared to accept it, but when that came through I was really excited. Operations managers are supposed to be cool all the time, of course. That was a memorable experience.

What part did Voyager have in your life as an individual? Again, this is your personal response.

I'm still 13 or 14 years away from retirement here, but I really doubt if, when I retire or I'm laying on my death bed or whatever, that I'll look back and be able to say that anything other than being with the Voyager project for over a decade and being a major player on that team—I don't think anything will beat that memory. When you're over a decade on a single job, on a single project, it is a big chunk of your career, and you must like it; either that or you don't have any alternatives. In this case, my personal career and my life centered around it, and it's a very fond memory.

How do you visualize the two spacecraft themselves? Some people see them as mechanical, robots. Other people seem to think of them as having sort of a personality or life.

No question; they're robots. But, like any good robot, they are extensions of man's intelligence, and that's just the way I view it. We learned to understand them, and we learned how to maximize their utility.

They are robots, designed by a large number of people working on the pieces. There were a few people who were instrumental in the overall architecture, and then there was this team of people I referred to earlier, who learned to be the brains of this robot, and the brains of this team were then transmitted up in the string of bits, and placed in the artificial brain of this robot. That's the way I look at it.

What advice would you have for a younger person coming along, on the basis of your own Voyager experiences?

[Notable hesitation here.] I have really mixed feelings. You're asking me this question while I'm sitting here in the middle of a very expensive manned program, the space station. This is a painful experience for me, coming out of the very efficient robotic mission, which has been highly successful and did the whole thing for a fraction of $1 billion.

Now I find myself in the middle of a big behemoth of a program that has pork barrel characteristics. It has every major aerospace contractor in the country involved in it in one way or another. Politicians are involved, and NASA is spending over $2 billion a year.

You may wonder what this has to do with your question? I'm not too sure where the space business is going. I am somewhat disturbed by the inefficiency of the big massive NASA programs right now, and the fact that they're running over in the budgetary sense. Some of the smaller missions, in terms of getting bang for the buck, do so much better. Magnitudes better.

And I'm upset with the political environment that dictates what is done and how it's done. I'm upset with the general level of intelligence and talent that's applied to these projects; it's a lot different than the environment I came from.

And yet, let's face it; it's reality, and that's probably going to be happening in the future. It's interesting to speculate what happens if we get into a joint venture with the Soviets, when in fact we're already having a terrible time handling a much smaller problem right now with space station, which is a joint venture with the European Space Agency and the Canadians and the Japanese.

These big behemoth programs are not as much technical challenges as they are management challenges, and right now the management challenge hasn't been met.

So in answer to your question, I have trouble advising a young person who's enthusiastic about space because I have concern that they might be disillusioned when they find out what really goes on; that it is more plodding. It's not all adventure. A lot of fumbling around takes place. But at the same time I have to tell myself that the experience I had on Voyager was unique, not just in the space business; it is very rare where work gets done so efficiently, where people work and play together so well for an extended period of time, and where the rewards are so obvious and exciting.

My thoughts aren't really jelled yet on that general subject. I'm going to have to look back on this space station experience a little bit, in retrospect, to understand it better.

To wrap this up, what did the people around you think about your involvement—your family, friends?

My brother, he's younger than I, is a high school teacher up on the plateau of Los Alamos, New Mexico, a kind of scientific, intellectual community. He was with me all the way; really enjoyed it. On the flip side was my ex-wife. I have two daughters, and I was married during the Voyager time period. I haven't sorted out what the relationship between my divorce and Voyager was. Certainly I spent a lot of time on it, but I think there were a lot of other factors, too.

It's kind of interesting that the ex-wife and the daughters never really got involved in what was going on. It was probably because they were too close to it, and because it was consuming so much of Dad's time, and things like that.

Subsequent to the Uranus encounter, and in fact a little bit before, I've done a fair amount of public speaking: technical environments, popular environments, and so on. The populace out there in general—not just in this country, it's worldwide—is just turned on by the accomplishments of the Voyager team and the Voyager spacecraft. It's been a fun experience for me to go out there and talk about Voyager and how we made it happen. So it's the full spectrum of reaction from people.

NORMAN R. HAYNES
Project Manager, Neptune

Born June 14, 1936   Kalamazoo, Michigan

Norman Haynes received his bachelor of science in aeronautical engineering from Purdue University in 1959 and a master of science in aeronautical engineering from USC in 1961.

He joined JPL in 1959 and served as a mission studies engineer before working as a systems analysis project engineer on the Mariner 4 mission to Mars and the Mariner 5 mission to Venus and as mission analysis and engineering manager for Mariner 9 to Mars. In 1973 he was appointed manager of the Mission Design section. He then served as manager of Space Program Development and as science and mission design manager for the Galileo mission to Jupiter before being appointed manager of JPL's Systems Division in 1980.

He was named Voyager project manager in 1987 as planning continued for Voyager 2's 1989 flyby of Neptune. After the successful encounter he became deputy assistant director for the Office of Flight Projects, and in 1992 became assistant laboratory director of the Telecommunications and Data Acquisition Office. He was appointed director of the Telecommunications and Mission Operations Directorate in 1994.

For relaxation he likes to read. He sings in a chorale and plays squash.

*I had always wanted to work on Voyager,
because I considered it to be the best thing that JPL
had ever done and I wanted to be a part of it.*
—Norman R. Haynes

NORMAN HAYNES

I am glad somebody's writing a book about Voyager, because I consider it to be the greatest robotic space mission so far. It's in a different category than Apollo. Maybe in awhile Hubble will catch up, but Voyager made the first exploration of the half of the solar system outside the Earth.

We used to laugh at a 20-year-old astronomy book I had lying around on the shelf. We would periodically open it to the chapter on the planets. There was a fairly large amount written on the Earth, and a fair amount on Mars and Venus because we'd done a lot of missions there. When we got to Jupiter there was a little bit less, and Saturn was a little bit less, mostly the rings, and you got to Uranus and Neptune and there was about three paragraphs. It would basically say they're this far from the sun, and they're in an orbit that's this long—and that was about it. We didn't know anything about them.

Now all of a sudden we found out these planets had rings we didn't expect and all kinds of new satellites we didn't expect. Early on, in the planning for Voyager, the satellites were not thought of as being equal players with the planets, but it turns out that the satellites are as interesting—more interesting in their character and make up—as the planets.

The outer planets turned out to be real planetary systems, with rings and satellites and atmospheres, and they're all different; it's like a crazy family where they're all spawned out of the same initial seed, but they sure turned out different—sort of like my four kids.

So it was an incredible adventure. And you can only do it once, because there aren't four more planets that we don't know much about anymore. There's just Pluto, so it was a great adventure to be able to do that.

Let's look at your own part in it. How did you end up being here at JPL working on Voyager?

In college in my freshman year I majored in business administration, but it turned out I didn't like business administration very well. I was already taking calculus and chemistry anyway, so when I was a sophomore I switched into engineering. But I was not a science fiction reader; I've read only one or two science fiction books in my life.

I came to work at JPL largely through a fluke. I was graduating from Purdue in 1958 with a degree in aeronautical engineering, and everybody in engineering knew about Caltech. It was very prestigious, particularly in aeronautics, and still is. The head of the department at Purdue was a Caltech alum. One thing that attracted me was that JPL was a part of Caltech.

JPL had just been assigned to NASA. They'd been an Army center for a long time, NASA had just been formed, and JPL had been transferred from the Army over to NASA, and they were interviewing at JPL. So I thought I'll go out there and go to grad school at Caltech, and I'll let JPL pay for it because they had a tuition program.

It was a good year for engineering graduates. I had nine or ten job offers from big aerospace companies (they were called aircraft companies in those days). But they had these great big design rooms full of hundreds and hundreds of engineers, and I couldn't quite imagine myself in that setting. I knew a guy who knew a guy who had graduated a year before us and had gone to work at JPL, and all I heard back from him was "Yeah, it's a pretty neat place, and there are only three or four people in an office." Based on that I took the JPL offer, even though it was my lowest one.

So I came to JPL and at the same time discovered that I was going to be doing lunar and planetary explorations. I had not come to JPL to do that; it was kind of happenstance. I now consider myself very lucky that I made that decision.

I showed up here and began working on other projects. I didn't get directly involved with Voyager until I became the project manager of the Neptune portion of the mission, long after launch. Although I had managed JPL organizations that had a lot of people working on Voyager, I didn't personally work on it until then, devoting 100 percent of my time.

But I was in the same section as Roger Bourke. In fact, he and I were in offices practically next to each other. Mike Minovitch and Gary Flandro were also there in the same section, and they were in the offices next to us. They were "academic part times": going to school and working at JPL part time. So we were around when the whole idea of gravity-assisted trajectory started to arise.

I was working on Mars missions and Venus missions, and Roger was doing some advanced study work, so he was more involved in the precursor days of Voyager than I was. I was around the edges. We all knew about the Grand Tour trajectory because it came out of our section, but I continued to work on other missions in other areas. Subsequently I became the section manager and had a lot of the people who did the mission design and flying for Voyager, but again, I didn't work on it personally.

Finally the opportunity came in late 1986. Dick Laeser had been project manager for several years and he decided to do something different, so he transferred back to Washington to start a JPL space station office back there. Someone was needed to manage the Voyager project. I had been managing one of the divisions, so they offered me the job and I said, "Wonderful".

I had always wanted to work on Voyager. Somehow it hadn't worked out before, but this was the right occasion. The reason I wanted to work on it was because I considered Voyager to be the best thing that JPL had ever done, and I wanted to be a part of it, at least for a portion of the mission. It was a great experience. I wouldn't trade it for anything. I got to Voyager very late in my career, and late in Voyager's life; late in Voyager's planetary life anyway.

I was dealing with the day-to-day situations involving several hundred people and systems, getting it all together. By the time I got to Voyager, it was a pretty well-honed machine. I didn't have to do any major readjustments of it. In fact, I realized right away that the best thing for me to do was to not screw up a good thing. The right thing for me to do was make the decisions I had to make, and clear the obstacles out for the other people on the project so they could get their work done because they had learned how to do it very well over the previous seven or eight years.

There were actually very few frustrations. The Neptune mission was a remarkably smooth-running operation, none of which I attribute to any skill I brought to it. What I did was just make a few of the key decisions that had to be made.

For example, when I arrived on the project they had studied but had not yet selected the particular aim point at Neptune. We wanted to fly very close to the north pole, and very close to the planet; just 3000 or 4000 km above the top of the atmosphere, in order to get the trajectory bent way down so we could get a real close approach to Triton as well. There were some questions about could we navigate accurately enough?

You were at the top of the approval hierarchy?

Yes. It turns out that the project manager really had to deal with issues that mostly dealt with risk. The day-to-day operation of the missions was something we had several hundred people for; they were experts at it, and I wasn't going to be able to contribute anything there. So the few areas where I really had to make some fundamental decisions were almost always related to how much risk we wanted to take versus how much return we were going to get. Ultimately somebody had to make those choices, and this was a good example of one.

The mission return we could get would be substantially higher by flying this trajectory but there were some risks because we had to fly through the ring plane, we had to fly very close to the planet's atmosphere, we weren't exactly sure how big the planet was, and we weren't exactly sure how high the atmosphere was above it. So we had to take all these uncertainties into account and ask what is the probability we can actually hit that little spot; miss the atmosphere and duck through the ring plane at the right spot and just zip over the top and get bent down just exactly right?

I looked at the analyses. They had been done well, presenting the pros and cons. I had a lot of confidence in our people, so I said let's do it. Of course the science teams had already decided that they wanted to do it that way, but then they wouldn't have to stand up and explain what happened if something went wrong. Ultimately it just meant that I had to put my approval on it, and I did.

Those are the primary things I did. There were issues that came up related to this is an old spacecraft. By the time we got to Neptune, it was 12 years since launch, so the spacecraft had been flying for a long time. We had a lot of backup systems on the spacecraft, and one of the issues was what happens if one of the primary computers goes down right as we get to the planetary encounter, and we switch to the backup? We've never had the backup computers turned on—at least, certain ones of the computers had never been turned on—so we didn't know if their software was still good and if that computer was good.

There was one set of people that said let's turn on the backup computers and read out the memories and get 'em all reprogrammed. There was another set of people who said that's all well and good, but there's a certain amount of risk to doing that because, if something goes wrong in the middle of this, you might not ever get back to the primary computer and you might get stuck in a backup computer that doesn't have good programs in it. So I decided not to do that one. We just let the thing fly, and it worked fine. In fact, we've still never switched over to that backup computer, and it's now been another six years since.

Those are the kinds of things that the project manager gets called upon to do, which is decide. There's a huge amount of "what if" work that goes on. There's a certain amount of work in planning the basic mission, but then there's a large amount of work that goes into "What happens if this goes wrong? What happens if that goes wrong? How should we back this up? If this maneuver doesn't go right, how do we recover from it and get back on to the time line before we're past the planet?"

You're confident that none of those things is likely to happen, but any one of a whole bunch of them *may* happen. Since this is a once-in-a-lifetime opportunity—it's going to be past my lifetime before we ever go back to Neptune again—you want to make darn sure that you've done everything you possibly can to make sure it works properly, so we spent a lot of time doing that.

I would say 20 to 30 percent of the thought time on a project like Voyager, getting ready for an encounter, is spent trying to figure out contingencies. You know what the basic plan is, but you're trying to understand how to react and what to do in the event that something doesn't go right.

There's some things that, if they go wrong, the spacecraft has to take care of itself, because by the time you find out about it four hours later at Neptune, and figure out what to do, and then send a command four hours late, 10 or 12 hours may have gone by—and that's too late, so certain amounts of those things you have to put on the spacecraft. But there's only so much memory on the spacecraft. There's only so much you can do before you run out of memory, or all you've got are backup plans but you've got no primary plan, so you have to decide how much of that you're going to do.

There are other problems where you've got enough time so that you can take corrective action on the ground. A lot of time and effort was spent on that; fortunately we didn't have to use any of it. Everything went just right according to plan, which we assumed it would. But we knew we couldn't be in a position that if something did go wrong we weren't ready for it. That would be tragic. So a lot of time and effort was spent on making sure we were ready for almost anything.

A lot of your time?

A lot of my time. And I would guess 20 to 30 percent of the time of all the people that worked on Voyager, at least the engineering people, was spent on that kind of thinking. Not just laying out the basic plan for what we were going to do, but also thinking about all the options and side plans, and what if this happens and what if that happens, and planning the sequences and the flight plans such that all those were taken into account. It may even have been more than 20 to 30 percent.

I'll give you an example of what percentage of the people are involved in *planning* for what was going to happen, versus the number of people who actually have to sit down at a console and fly a mission, the people who sit there and watch it and make sure things are going well. For the Neptune encounter, for a couple of years up to and including the encounter, we had about 240 people on the project, not counting the scientists. We had another 150 to 200 scientists. So we had, round numbers, 400 to 450 people on the project, and that didn't count the DSN people. Today those two Voyager spacecraft are still flying, there's no encounter coming, and we're flying both missions with 24 people, and we've got plans to reduce it down to 12.

That gives you some idea of the percentage of people that were *planning* what to do, as opposed to actually doing it. It doesn't take many people to actually fly the mission, but it takes a helluva lot of people to plan it. These are very complex missions, and it takes a lot of thought about how to make sure that things go well. That was primarily what I did: to make sure that all the "what ifs" got done. If there were decisions that had to be made as to should we do this or that, from a risk standpoint, I usually ended up making those decisions.

I think many people have the impression that flight operations is a lot of people sitting around with head sets on, saying "A Okay" and "Roger" and "over and out," but there's hardly anybody that does that. On Voyager we don't have anybody doing that anymore. If there happens to be a tracking pass during the day when people are around, somebody will sit in and watch it to make sure that everything goes right, but if it happens at night, then nobody watches it.

So operations is very different than people think it is. Most of it is people in their offices and conference rooms, planning and running software and doing analyses, developing commands.

I was the primary interface with people from NASA, who were paying for the mission. I had to communicate with the program office and keep them up to date on what our status was and how we were doing, and make sure that the budget came in properly, and all those fun jobs.

Number two was making the risk assessments. Making those kinds of judgments usually came to the project manager. And the third thing I spent most of my time on was developing the public affairs plan. Assuming we were going to be successful, how were we going to demonstrate all this stuff?

Probably those three things are somewhat different from what most people think. They probably think of the project manager as being a guy down there arguing about all these technical details, but there's lots of other people who knew the technical details far better than I did. My job was to make sure that they had enough money so we could have enough people to do the jobs right, and that we made the proper risk tradeoffs, and then get out of the way—and keep the sponsor happy.

Are there any particular moments or experiences that we should note here?

One of the things I remember most is the first time we got an image of Neptune and saw something that looked like a spot. The Uranus encounter for one big reason didn't get much publicity, because it happened two days before the Challenger accident. So it just got blotted out of most people's minds, unfortunately. But also Uranus was the blandest looking thing you've ever seen. It looks like a pale blue billiard ball; you can barely even see any shading across it.

We were concerned that Neptune, being even further away and even colder, was going to be the same blandness, so the mere fact that we saw some kind of an atmospheric feature was fantastic. We thought that was great. And then, as we got closer, we discovered that it had spots, sort of like Jupiter has a giant red spot; it had two or three of these vortices in the atmosphere that looked like they were permanent. It was wonderful to see that.

Another of the things I remember about the encounter is the incredible attention it got. I don't know why or how it worked that way, but it must have been that there wasn't anything else of great import going on in the world. We got incredible attention from all the media. The images were coming out and the whole world was watching. It was a wonderful experience to share the thrill of discovery with a large number of people and realize that there were people out there who actually enjoyed it.

For the encounter night we set up satellite feeds of the data from Voyager into school auditoriums and planetaria all over the country. Afterwards I got several letters. One guy down in Texas invited the public to come in and spend the night because this happened overnight, and they had standing room only. They had people bring little kids in; they had mothers with babies watching this stuff.

That part of it was really exciting, because you felt like the people who sponsored you and paid for this thing were getting some modicum of return, and they got an opportunity to witness history while it was actually happening. We spent a major amount of money out of the project budget to make sure that real time feeds of images and stuff could go to all of these places. The response of the public, their appreciation for the mission, was overwhelming.

Prior to the encounters we always got asked how much did this mission cost? The mission had been spread out over 17 years; five years up to launch and 12 years after launch. The total cost per person in the United States was about $2.47, spread out over those 17 years, and we got several checks in the mail for $2.47. One guy sent us his check and said, "That was fantastic; do it again!"

And we got a check from a lady for $25 because, she said, she didn't have a whole lot of money, but she didn't see us popping champagne corks; she didn't see a big party going on in so-called mission control, so she was worried about that and sent us $25 so we could have a big party. I lost track of the check and we probably misplaced her letter; I always wanted to send a note back to the lady saying how much we appreciated it, so if she ever picks up this book and reads it, maybe she can get some second-hand feedback that way.

The essential thing was that we made significant contact with the public. It wasn't just a bunch of technocrats sitting in their little operations centers; we shared it with a much bigger audience, a much bigger set of people than I anticipated. They were interested, hungry for information. They really appreciated it, and let us know. That was a great part of the experience.

Those were my biggest remembrances of Voyager. I'm having a hard time coming up with much in the way of frustrations. It sounds incredible, because every other project I worked on in my life, I wouldn't have any trouble at all telling you what the frustrations were—overcoming this and getting around that obstacle—but it all worked remarkably smoothly. Part of it was luck, part of it was we had a very skilled team by then, people who really knew what they were doing, and we had a spacecraft that was first rate; it's still performing like a champ.

Did the subject of the solar system mosaic come up at all?

Yes. That was another one, which we had to decide whether or not to do it. It came up before the Neptune encounter, but I didn't have to make the decision then, and I didn't want to clutter up people's time before encounter, so we put it off until after encounter. But I had pretty much made up my mind that we would do it.

After the final encounter there wasn't anything else to do with the cameras. The cameras are really only good for planetary observations when we are close; they don't have anything like the telescopes that Hubble has, so they're not particularly good for anything else. You can't do deep space astronomy with them. There was some probability that we would damage the cameras severely when we finally took the pictures in close to the sun, because we had to point close to the sun, but try not to let the sun get into the camera's field of view, because the sun would burn it out; it would destroy the vidicon tube.

So, after the encounter we decided to go ahead and do it, George Textor and I. He was the deputy project manager. I don't remember whether Ed Stone was in on that decision.

Looking back, what lessons did Voyager teach us, aside from specifics?

Number one is that a whole lot of stuff came together in the right way, but it took a lot of people with a fair amount of vision to make sure it all happened. There was the fact that somebody found the gravity-assist trajectories, and the fact that somebody else, like Gary, found the trajectory that could go past the four outer planets. These possibilities were known fairly early, like 1965 or so, but at that time we didn't know how to send a spacecraft out that far; we were just barely successful with our first Mars mission. It lasted a year. We didn't know if we could build a spacecraft that would last for many years. We didn't know how to build cameras that would work in the dark.

There were huge numbers of such issues, but a lot of people persisted. They had vision, and the vision was being able to explore the four or five great outer planets in one shot, and being able to do it only once every 175 years, and the opportunity was going to be there, and we had to take advantage of it. It required a lot of people to share that vision: people at JPL, at NASA Headquarters, in Congress.

We first proposed a mission that was going to be much more expensive. It got turned down. We finally were successful just with the Jupiter-Saturn mission, but we didn't design anything into the spacecraft that would prohibit us from going to Uranus and Neptune, because we full well knew that we were going to go that far.

It was a real triumph of vision and long-range planning. It was anything but a "faster, better, cheaper" mission. We couldn't have done it much before 1977 because the technology wasn't there. We didn't have RTGs (radioisotope thermoelectric generators) developed yet, which was the power source, and we hadn't yet quite figured out how to build a spacecraft that would last 12 years. So everything came together right, but it wasn't just luck. It was really the triumph of a long-range vision.

Number two, it was a challenge that was almost a little bit bigger than we thought we could really handle, but we bit into it anyway. It makes it a big struggle, but those are the best kind of struggles. That makes it a lot of fun. Everybody kept their faith and their confidence that you could actually do this mission.

Those were the primary things, the high-level things. Then, relative to more mundane things, it was the first time that we put a significant amount of computer power into a spacecraft. It was significant in those days; it's almost humorously insignificant now. It was the first spacecraft we flew that actually had several thousand words of memory on it, and this was in 1972 that we designed that stuff.

One of the funny things that used to happen at all the press conferences during the Neptune encounter, the press loved to ask, "Tell us again, how many words of memory are there on the spacecraft?" And we'd reply, "Well, there's 16,000 words of memory and 16,000 words of backup memory." It seemed so small they would laugh and say, "Gosh, that's the funniest thing." But when we started, back in 1972, personal computers were a long way from even being invented yet, so putting 4000 words of memory in one computer—and there were three of them onboard the spacecraft, two of them with 4 K memory and one with 8 K memory—that was a big step forward.

That was one of the biggest steps we took. If we had not done that, there was no way we could have done the Uranus and Neptune missions, because we had to essentially redesign the spacecraft, and we did it by redesigning the software in between encounters. The spacecraft's original software could take us through Jupiter and Saturn, but when it got to Uranus and Neptune, the distances were much greater, and as you get further away from the sun it gets darker and darker, requiring longer exposures. Our ability to take pictures was dependent on our ability to change the attitude control system so it could control the camera so we didn't get smear in the images.

That was all done by changing the software in the spacecraft, which even two or three years earlier we wouldn't have been able to do, because there wouldn't have been enough software aboard the spacecraft. We made many of the spacecraft functions into software for the first time, and it's the thing that enabled us to do that.

Flexibility is incredibly important, because you don't know what you're going to run into, and it's nice to have a flexible system so that you can adapt to whatever you find. We didn't *have* to put those computers in, and there was a lot of discussion and debate whether or not we should do it. We decided to go ahead, and I think it was one of the fine decisions we made. In retrospect it was an enabling decision.

Computerizing the spacecraft was really critical, and it was the first time that we had ever modified a spacecraft, or in the terminology we used to use, we "re-engineered" the spacecraft in flight. We've done an even bigger job of that on Galileo because of the antenna problem. But that was the first time we actually used that flexibility and adaptability that was built into the software to make a mission succeed.

The other thing we had to do was to continuously change the Deep Space Network. We had to keep upgrading it, making the antennas bigger, and getting agreements from foreign countries to let us use some of their antennas in conjunction with ours, because by the time you get to Neptune, you're 30 times as far away from the sun as the Earth is. The communication distances are measured in billions of miles, and we were trying to send images back over those billions of miles. In my estimation that may have been the finest technological achievement of Voyager: being able to get images back from 3 billion miles away. It's a huge amount of data, and with only a 10 or 20 watt transmitter in the spacecraft.

The biggest thing that comes out of Voyager is that a lot of people had to buy into the vision, and they did. Anybody along the way could have stopped it: NASA Headquarters, if we hadn't convinced them that it was a good thing to do, or if they weren't convinced themselves; or Congress—anything along the way might have stopped it, but they all bought in, and it was a true triumph of long range planning and vision across multiple administrations, and hundreds or thousands of people had to make their own little decisions.

I consider the Voyager mission to be almost more of a success as an exploration and an adventure than as a scientific and technological accomplishment. Of course they were great technological achievements and brought great scientific returns, but I think just the adventure of going places where nobody had been before and seeing things nobody had seen before was an equal part in this enterprise. That's what excited us all, and probably excited to a certain extent the public: we were seeing things that no person had ever seen before. It was great to be present at the unveiling.

GEORGE TEXTOR
Project Manager, Interstellar Mission
Born December 12, 1932   Wilkinsburg, Pennsylvania

After graduating from Pasadena City College in 1952, George Textor entered the U.S. Naval Academy, earning a bachelor of science in engineering and a second lieutenant's commission in the U.S. Air Force in 1956. He received his pilot wings in 1957 and flew multiengine aircraft during his 11-year military flying career.

Upon leaving the Air Force in 1967, he joined JPL as mission operations planning engineer for Mariner Mars 1969 and was DSN project engineer for Mariner Mars 1972. He moved to the Viking Orbiter in 1972, and in 1978 joined Voyager for a series of assignments culminating in 1989 in his present position, project manager for the Interstellar phase of this very long mission.

He has received the NASA Exceptional Service Medal and was twice awarded NASA's Outstanding Leadership Medal.

Hobbies? Golf!

While most people, when recalling the Voyager Mission, will think of the beautiful pictures, the scientific discoveries, and the marvelous spacecraft, I will think of the dedication, skill, and teamwork of the Voyager Flight Team. Through their efforts came the honors and glory of the Voyager Mission.

—George Textor

GEORGE TEXTOR

I went through the Naval Academy, but upon graduation I went into the Air Force instead of the Naval Air, only because I had one landing on a carrier, not as a pilot but as an observer, and decided that was not a very good way to fly. I like a nice long runway that stays in one place.

In the Air Force I was stationed at Colorado Springs in the First Aerospace Squadron. Our role was to keep track of all the Earth-orbiting satellites and debris up there. This was back in the early 1960s. I got an interest in space at that time. Then in 1967 I got out of the service and was going to take 30 days off. I was just going to relax, but I got bored after two weeks. I was in Pasadena, and I'd never interviewed for a job before, so I thought I'd go up to JPL and get some experience. They offered me a job and I've been here ever since.

I've always been associated with operations rather than the building of spacecraft. I've always been involved in getting ready for the launches from the operations point-of-view, and then getting ready for encounters, training people. My only job that was slightly different from that was my second job: I was the Deep Space Network project engineer on Mariner '71, and for that I was on the tracking side. I was responsible for getting the computers and then our Deep Space Network ready to support Mariner '71.

From there I went to Viking, and then from Viking for just a brief period I did studies of advanced missions, and then came onto Voyager. I've been at JPL for 23 years, and 13 of them have been on Voyager. So half of my experience is with Voyager.

When did you first become aware of Voyager?

I first heard about Voyager when I was working on Viking. The two projects were going along in parallel. The Vikings landed in 1976, and the Voyagers launched in '77. At the time that I finished Viking, I was doing a study on a Mars rover mission, and I'd talk to Dick Laeser about some operations things with regard to the study. I'd worked with Dick before. He was the mission director on Voyager at that time, and he mentioned that he could use some help. Since I didn't really enjoy the study I was doing, I came over and started working for Voyager.

That was around December of '77, about three months after launch. At that time Voyager did not have a good name. They were struggling, and people wondered why I wanted to go to work for them. Voyager had a lot of problems right after launch, and the flight team was still developing some of the software needed for the Jupiter encounter. They were understaffed in certain key areas, particularly the science needed to do the planning; not the principal investigators but our supporting science team. And they were lacking some experience in operations.

So I came on as Dick's assistant and developed a plan to get ready for encounter, because our encounter planning was not really going anywhere. We were too busy with the day-to-day operations, and we weren't organized properly to do the encounter planning. I made up a plan and then was put in charge of it; and I was called encounter preparations manager.

At nearly the same time Bob Parks came in as the project manager, replacing Ray Heacock, who stepped down to deputy. People were added to fill some of the areas that needed to be beefed up. The software for encounter sequences was still a struggle, because as the Sequence Team would try to run through a sequence, they would run into a problem and have to go back and repair the software.

The process was moving very slowly, but as we got into the last few months before Jupiter, things started going pretty well, and the Jupiter encounter went rather smoothly. That success seemed to pick the whole team up, and from then on people wanted to be on Voyager.

We had a large staff at that time, around 350 people. The two encounters at Jupiter were almost on top of each other—Voyager 1 in March and Voyager 2 in July—so it was necessary to have that large a staff because we had to do everything in parallel. When we went to Saturn, because the encounters were spread out, we reduced the staffing a little bit and yet we were still able to have a very good encounter. Right after Saturn it went down to a hundred and some. We had about four years to plan for Uranus.

At Saturn, of course, we had the problem with the scan platform that pointed the instruments, so there was a lot of work involved in trying to figure out how to use the platform. That went on for about three years before the Uranus encounter.

After each encounter we'd have a staff-down, and new people would come on, and usually there were new managers. What made it so interesting was how we could pick right up and put the new people in so we were running as a team again. That continued just about the whole way, without any loss in capability. In fact, because we brought in new people we'd get new ideas, and we actually improved. But we also had enough experienced people so that after each encounter they would get a little wiser on how to use the spacecraft. They got very inventive and were able to get more out of the sequence than we did at the previous encounter. So each encounter got better.

Of course, we were lucky. When we had problems, they occurred at a point where they really didn't do that much damage. For instance, as we were coming into Uranus, about four days before closest approach to the planet, we noticed some lines in the pictures were missing. The Deep Space Network had gone through a big upgrade at that time, and the initial thought was that it was a ground problem, because we had had a lot of problems while they were redoing their system.

But it was soon proved that it was on the spacecraft. Within 72 hours we were able to identify the problem and get it repaired so that when we really got close in and did the encounter, we were able to work around it and had no more problem. Earlier, at Saturn, the scan platform stuck *after* we went by the planet. Had it got stuck coming in, we would have lost a lot. So good fortune was with us, out of some bad fortune that occurred.

The most amazing thing was how well the Neptune encounter went, even though it was more complicated than the previous ones. We were going closer and the timing of certain events was more critical, and yet everything went just as smooth as could be.

While I didn't start at the beginning of the Voyager project—it started in 1972—I've been on it 13 years, and a lot of things happened along the way. Several individuals actually died while participating in the project. We lost four to cancer, and some of them were pretty key people. It was a tough experience watching them wanting to stay with the project and yet going downhill as they did.

Do you want to mention any people particularly?

Yes. Charles Stembridge, Jim Long, and Don Acord were three that we lost. There were a couple others that I didn't know, who were involved in building the spacecraft, that were lost along the way, too.

As some people have said, in the 12 years their children grew up and went to college, and some even got through college working on Voyager.

My role hasn't changed a whole lot until recently when they made me project manager, but I was primarily either the deputy mission director or mission director. Directing a mission is one of the most fun jobs on the project, in that you're really involved in the day-to-day operations, and in just about every aspect of getting ready for the encounter. What I enjoyed most, because we were having new people come in and some experienced ones leave, was the fun of going through the group dynamics again with people, getting them all involved, and seeing people who had little experience in the operations take over and do really well.

We were sort of a training ground for JPL in the operations area. We were the only project flying, because of the problems, particularly the shuttle disaster, that kept the other spacecraft grounded that would have been flying during the same time. So we had a lot of people come on the project to gain experience to apply to those other projects, and a lot of our people went off to new projects.

Voyager people were proud that we were able to provide the Laboratory with some experience in the operations area. And we also had a lot of pride in how well we were doing. We were within budget every year, which is always nice, sometimes unusual. We were amazed at how much more the people could do than we ever planned for them to do. You can only do so much thinking about the year ahead and what work you've got to get done; there was always the unexpected that would come along, and you'd have to allow for that, too.

A lot of things were outside our control, such as ground data system changes, that were fairly significant. Our software was so old that, when any new operating system on the ground computers would come in, it was a real challenge to make sure our software was working right. We had to go through some very significant testing, and usually that was not planned for because we didn't know about it when we'd made out the budget.

We kept the people pretty busy for 12 solid years, although they weren't always the same people. There never seemed to be that quiet period someone had predicted, where from Saturn to Uranus we would have a period of quiet cruise, but that cruise was never very quiet. Quiet cruise became a name I wish I'd never heard of because some of the troops believed that maybe it *would* be easier. It never got any easier.

A lot of people don't understand. I'd see some of my friends after an encounter and they'd say, "Well, now you don't have anything to do for four years and the next encounter." People didn't realize the amount of work that goes into getting ready for an encounter.

Give me some idea of what that involves.

First of all, the planning that you go through to just understand what sequences you want to fly, what experiments are going to be made. What makes it so complicated is that we have 11 experiments, and each experimenter has his own parochial view of what an encounter ought to be. For instance, the infrared people want to look at the planet, but some of the imaging people want to look at the planet in a different way, and other people want to look at moons, and some at the rings.

So you've got to time-share the spacecraft, because where the scan platform points, all pointable instruments point; you don't have separate pointing for each instrument. Planning to make sure that we get prime science during this time-sharing, so that each experimenter gets his fair share of what he's after, is a very lengthy process of working all those details out.

The spacecraft itself is complicated anyway and so is working out the engineering aspects of doing those things. Planning the sequence and building the sequence is very time-consuming; it's at least a two-year process. We do it in what we call an advance planning stage where we lay it out as best we can without having detailed information because we haven't seen things close to the planet yet.

In some cases still more information was needed, particularly for Neptune. We were trying to determine exactly where Neptune was, but the ephemeris was not quite as accurate as we needed. So after we'd do the advance planning, then, as we approached the planet, we would redo those sequences and change the pointing, based on our better knowledge of where the planet really was, where the targets were, and where the spacecraft was. Maybe some new information would come up, so we'd modify the sequence to get some other information.

Other things had to be done during cruise. We modified the data system onboard the spacecraft, on the way to Uranus and also to Neptune. We put in a data compression algorithm on the spacecraft so that it would actually do image data compression, which allowed us to send back less bits of information yet have the same information content as before. That way we were able to increase our data, even though we were going farther and farther away from Earth, and should therefore be losing capabilities.

During the cruise period there would usually be large software modifications to some of our key programs. Again, going to Uranus, we had to do extensive work on the scan platform because of it being stuck. And we had to plan the training of the people during this period, then carry out the training program. Each of these encounters had that type of work going on.

Also in cruise itself there were things that we wanted to accomplish. Anything that we were going to do different on the spacecraft, we would check out in cruise first, before we'd do it at encounter. And once Voyager 1 went by Saturn and no longer had an encounter mission for Uranus and Neptune, we used it as a test bed. We would try it first on Voyager 1, and if it worked there then we'd do it on Voyager 2.

There's a lot of engineering tests with the spacecraft that we'd do during cruise, that required a lot of planning and sequencing in parallel with all the rest of the operations, in order to get ready for encounter. There was never a lack of work during this time, but it was the kind of work that made encounters look so smooth.

What are your most vivid impressions of Voyager?

I was always impressed by how well we were able to do each encounter, even with a new group of people to pull together. And how focused people were on it. I've been on other projects and the focus was there, but it wasn't always as harmonious. In most cases people hated to leave Voyager, which was a nice tribute to Voyager. I think it was a good experience for everybody, and that's not always the case in other missions.

Certainly the accomplishments of Voyager is something that stands out. The opportunity to be first to go to Uranus and to Neptune. And even though others had been to Jupiter and Saturn, we were really the first to bring back startling images of the rings and views of the planets themselves, compared to the imaging they had prior to us going there. So the data itself that we were able to obtain was, in a sense, a first, and quite an experience. And the survivability of the spacecraft and how well they've lasted these years is a tribute to the people who built them. The spacecraft are quite amazing.

What were some of the toughest decisions that you had to make with respect to Voyager?

Selecting certain individuals for some of the jobs. We had a lot of good candidates, and some of those selections were among the toughest decisions that I had to make, because it affects people's lives. The nice thing was we did have a lot of good candidates so I could hardly go wrong, but it was just a hard choice.

Another kind of hard decision I had to make involved something that probably would be good to do but we didn't have the resources. For instance, we're trying to get into a new mode of operation in the cruise phase, and we have a lot of work that we've got to get done, and we've reduced our staff significantly, and the Division wants us to update a redundant processor on the attitude control computer.

To do this would require a lot of work with a piece of hardware that hasn't been used very much. While it appears that it would be safe to do, and it's something we would really like to do, we can't do it and still stay on schedule and within the money that we've got right now. But it affects the spacecraft's motions, so should we do it anyhow and let the other work slide and try and figure out how to handle it later? We really couldn't afford to do it, so we elected not to, but it was a hard choice.

There also were hard decisions during the encounters: having to refuse to do some science that we would've liked to do. But, again, our resources were limited, so we would run risks. For instance, if the sequence process was at a point where it would be risky to try and take this work on and still get the sequence ready in time, it was a hard decision, because you hate to turn off science—that's why we're here. But you also don't want to risk more science by not getting the sequence done on time, or make an error because you're rushing. Those tradeoffs occurred quite often through the mission.

Another example occurred after the final encounter, when we took the last series of pictures, looking back at the solar system. Mars had a very bad phase angle for us to capture and, real late in the process, the scientists realized that they had the wrong filter. They wanted to have a clear image where they'd have a good opportunity to see Mars, and instead they were using three colored filters that didn't pick it up as well.

Now it sounds pretty simple to change a filter, but it really isn't because of the software we have. It would require us to run through all the software again to get this change in, and we just had to tell them, "No, we can't do it because it's too late."

As a result, Mars is not in the solar system portrait. We knew we were running the risk of not getting it, but we also knew that if we tried to get it, we might not have got *any* of the images for that portrait, because we may not have got the sequence done in time. So not allowing the filter change was a hard decision to make.

It was that type of decision that I was talking about. It was usually something that was overlooked or came in late and needed to be changed, and we just couldn't. But a lot of times the flight team would volunteer to do it, even though they knew that they were going to have to work way overtime to get it done. I had to protect them from themselves, so there was the decision not to do it, and they were disappointed because they would've liked to have the challenge.

I was always walking that fine line, of either trying to keep them from getting too much on their plate, but making sure that we didn't turn it off too soon if it was something we could accomplish. Those were always difficult choices to make.

It's not surprising that we keep changing sequence team chiefs. That's a key position that changed at just about every encounter.

Was that done deliberately?

No, they wanted to go on to another job. They'd been on that job for three or four years and were looking for something else to do. It's a very pressure-oriented job, and I don't blame anyone for wanting to get off of it.

Voyager personnel have been described as very dedicated. Is this more so than on some other projects?

I don't think it was more than on another project because on every project people usually really get into it, but it's the longevity of Voyager: 12 years. A lot of people have been on it for that full time, and some even longer. For instance, Dr. Stone has been on it since its inception, going on 18 years.

Many people on the sequence team have been on for almost the whole time. That's an area which is very pressure-oriented, because they're the ones that must complete on time, even when people give them input late, and yet they just stay in there doing their job and don't go home 'til they get it done. I think a lot of us get caught up in what we're accomplishing and we want to succeed, because we all feel we're part of history, particularly as we were going through the encounters.

"Part of history"—could you elaborate on that?

Well, when I give talks I sometimes mention that when I was in school, we heard about Uranus, Jupiter, and Saturn, and to show us what they look like we'd be given those fuzzy, blurred pictures from some observatory, and that was supposed to be a planet. Now we're bringing home to people exactly what they look like, and going in a sense where no man had gone before, by courtesy of the spacecraft. That's why I feel part of history.

What lies ahead for Voyager as far you're concerned?

Voyager still has a very interesting mission, although it may take a long time before it can accomplish it because we're looking for where the sun's influence ends and true interstellar space begins. That area is called the heliopause. Scientists don't know where it is; it could be anywhere from 50 to 100 AU. We feel the spacecraft can operate for another 25 years; and the challenge for the next few years is to prepare the spacecraft to do this job and to change our processes so that we can do it with a whole lot less people.

We plan to operate with a sequencing team of about five people, where right now we have 28, and a spacecraft team of anywhere from 10 to 12 people, where at the high point we might've had 75. The total staffing will be around 40-some people, and that's down from an encounter level, even at Neptune, of about 230. Of course, we will be operating both spacecraft, and they're still complicated spacecraft compared to some others that are out there flying around.

Voyager now has a more routine kind of a mission to fly and will not have that excitement that the encounters build. It'll be a real challenge to keep the people interested in Voyager during this time period. Of course, it'll be thrilling if we find this heliopause and are able to get out into true interstellar space. I would like to be around when that happens, although I'm not so sure I will; at least not on Voyager anyway.

But in the next two years there's a lot of work to be done. We're going through a transition phase, and we're also changing our telemetry system and our mode of operation to a more updated way of doing business. That will require an awful lot of work, particularly with the reduced staffing. We've already reduced down to about 60 now. We'll drop to about 50 in October and finally into the 40s in the following year.

Is this reduction in staffing imposed by outside budget restraints or is it from inside?

It is certainly imposed by budget constraints, but the plan always was that after the Neptune encounter our mission really would be mostly complete—no more planetary encounters. The fields and particles experiments don't require a whole lot of hands-on work, and we can make fairly routine the housekeeping activities, the engineering activities we'll do with the spacecraft.

So it was always planned that we would have a reduction. We have had some differences of opinion between NASA and ourselves of exactly how far down we can go. There's a push to get us lower than we think we can operate on, just to reduce the cost, but we're pretty much in agreement now. What remains is just to complete the work we've got started to get into this reduced mode of operation.

Voyager seems to mean various things to various people. What does Voyager mean to you?

I get a very satisfied feeling when I think of Voyager. I think of Voyager as a job well done, from both a team point of view and a personal point of view. And I'm very pleased with the accomplishments and very happy to see Voyager being recognized by others as a truly great accomplishment. I have two more awards to pick up for Voyager this summer; that makes me happy to see that Voyager is getting awards for the effort.

We got one from the National Aviation Hall of Fame, and then the Hawthorne Air Show in California gave Voyager an award. We received an *Aviation Week* award. The Air and Space Museum gave us an award, and we've won two Goddard trophies, and a Pete Jackson award from the National Space Club. Germany sent us a plate that was in the form of an award to the Voyager Team for accomplishments, and we received a scroll from the British people. So, we get letters, lots of letters from various people. Students in other countries have written us asking for information or sometimes for autographs; it's hard to understand why some people actually collect such autographs.

Do they ask for autographs of specific people?

Some have asked for project manager autographs. A lot of them ask for any pictures we could send them, or information on Voyager. I received a letter from a student in China who wanted to know more about Voyager, and we sent him some information. And I receive requests from Italy and Belgium and Germany, England, Brazil; all around. Letters either just congratulatory or wanting information or collecting signatures—things like that.

When you think of the Voyager spacecraft themselves, how do you think of them?

I am truly impressed and amazed at the really good engineering in them, particularly when you consider they were built with very early 1970s technology; how well they've operated way beyond their design life, and how well they've been able to operate out there in space.

How do you perceive Voyager—as something alive, with a personality?

No, I don't. I look upon it as machinery, but certainly a marvelous machine. Even though I think the odds are pretty small, it would be great if there were other humans or other beings somewhere that would come across these things someday—I don't know how many billion years from now—they'll probably still be in one piece. It would be interesting whether they could figure out where they came from and what they were. But that's about the only really far-out thought I have on them.

One thing that impresses me as a layman is the control that you here on Earth have over something that's several billion miles away. How can you do such amazing things with it?

Well, the way we control the spacecraft is by sequences. We put in sequences that run in length from a few days when we're at encounter to as much as six months. For instance, when Voyager 2 was going through Neptune encounter we put a six-month sequence on Voyager 1 so we didn't have to deal with it very much. Of course, we do not deal in a real-time sense with the spacecraft since, like right now, round-trip radio time is over 10 hours to Voyager 1. So you're never going to do any real-time control of it; it's always laid out for what's going to occur at some later time.

The spacecrafts do have capabilities of their own. In case of some problem on the spacecraft, we have what we call fault protection algorithms on it that actually save the spacecraft. They put it in a safe condition until we interact with it from the ground, to determine what the cause of the problem might have been that caused it to go into the fault protection, and then we take corrective action. But again, our corrective action would be through sequences.

If we're building a very quick sequence, like to fix something, it can take maybe a couple days at the minimum. When we build sequences for normal operations, it usually takes around six weeks of work to get it ready. So just about everything we deal with has to be well-planned in advance and takes quite a bit of lead time to get it up to the spacecraft. That's basically the control we have with the spacecraft.

We are always amazed how well the spacecraft does exactly what we tell it, even when we tell it something we didn't really want to tell it. We've at times been confused at why the spacecraft did something, and then we took another look at what we told it to do, and found out that we were the ones that told it to do this strange thing. Luckily that hasn't occurred very often, and usually it's only very minor things.

What things have you learned from your experience that would be good for us to remember?

In any endeavor the right thing to do is to carefully select the people you're going to have to work with, and select the best people you can. We worked hard at doing that. When we were making decisions, we would look very hard for compatible people and good people, and we've been very successful in that. It certainly helps when you get that kind of an environment; we did try to keep a good environment to work in.

The other thing is that if you're going to work on a project, get a spacecraft that's well-built so that you can enjoy the fruits of your labor.

Laboratory Directors

JPL has been the home of Voyager since its inception. The Lab conceived it, constructed it, guided it past four giant planets and gathered in the millions of data bits sent back from the distant craft.

The Laboratory provided the basic necessities for the people who worked on Voyager—office space, lights, parking, cafeteria, telephones .... Yet Voyager, despite its significance, was only one of many responsibilities of the laboratory director. He also had to oversee thousands of employees who were engaged in a variety of tasks, ranging from other science projects to mundane housekeeping functions common to all large organizations.

WILLIAM H. PICKERING
Laboratory Director 1954–1976

Born December 24, 1910   Wellington, New Zealand

William Pickering grew up in New Zealand and came to the United States to attend Caltech, from which he received his bachelor of science in electrical engineering in 1932. After earning a master of science in 1933 and doctorate in 1936, both in physics, he joined the Caltech faculty, becoming a full professor of electrical engineering in 1946.

In the late 1930s he had studied the absorption properties of primary cosmic rays. He was invited to join JPL in 1944 as a result of his experience in the design and use of telemetering devices, which he subsequently developed for the Laboratory's high-altitude research vehicles. Beginning in 1949 he headed the Corporal and Sergeant missile programs, and in 1954 he was appointed director of JPL.

In 1958, a few months after the Soviet Sputnik, JPL successfully launched Explorer 1. That same year JPL, which had been under the direction of the Army, was transferred to NASA with the responsibility for unmanned exploration of the moon and planets. Under Pickering's direction, the Laboratory launched the Ranger missions returning the first closeup, high-resolution pictures of the lunar surface; the Surveyor soft-landers on the moon; the Mariner missions to Mars and Venus; and the first gravity-assist mission to Mercury via Venus. JPL also designed the Viking orbiter to Mars and designed and built the Voyagers for the outer planets mission.

Following his retirement from JPL in 1976, he directed the Research Institutes of Saudi Arabia's University of Petroleum and Minerals. He returned to California in 1978 and established the Pickering Research Corporation for space related projects, ranging from a report on nuclear safety to an image processing system in China. In 1983, responding to the energy crisis, he formed Lignetics, Inc., to manufacture wood pellets from wood waste.

Pickering has been awarded many national and international awards and prizes, including the National Medal of Science, Honorary Knight Commander of the British Empire, Spirit of St. Louis Medal, Edison Medal, NASA Distinguished Service Medal, Columbus Gold Medal, the first Francois-Xavier Bagnoud International Aerospace Prize, and the Japan Prize, among others. He is a member of the National Academy of Sciences and a founding member of the National Academy of Engineering.

He collects New Zealand stamps and is an avid gardener.

I was very fortunate to be director of JPL. The Laboratory represented the best in scientific research in the mid-20th century. We were solving difficult engineering and scientific problems of great interest to the nation and the whole of humanity. I was leading a team of superb individuals who worked together to meet the challenges of our assignments.
—William H. Pickering

WILLIAM PICKERING

Ed Stone commented that you had a unique position from which to observe two key events in the origin of Voyager. One, the initiation of the Grand Tour; and two, the decisions leading to the transition from the Grand Tour project to the MJS 77 project.

He believes that you can contribute something that we won't get from anyone else, because you, as the director at that time, would have been right in the middle of all the issues involved in the Grand Tour proposal and that very important transition.

Yes, I was director of the Laboratory when the proposal for the Grand Tour was made. It seemed like a very exciting opportunity, which was not going to be repeated for 175 years, and we felt that it very definitely should be taken advantage of. However Congress did not support it. We were upset about this, but we did accept an assignment to build a small mission, which originally was just to Jupiter but very quickly became Mariner Jupiter-Saturn, MJS. We got that approved and supported.

So Voyager was initially being built only as MJS. In the construction of the spacecraft, the design modifications necessary to go all the way out to Neptune were relatively minor. These were included in the spacecraft with the hope that, if the Jupiter-Saturn mission was successful, we could go on to the other planets. That is indeed what happened. The control and guidance allowed the trajectory to be selected so that we could go on to these other planets.

Actually we did the Grand Tour, even though NASA did not approve the mission. NASA did provide the necessary funds to carry on to these other planets after we were well on our way. So that's how we did the Grand Tour, even though we were not assigned the mission and even though it was not called the Grand Tour.

When did you first become aware of the idea of a Grand Tour?

I don't recall the exact year, but it was several years before the launching. As soon as we understood the possibilities of these trajectories, which use the gravity field of one planet to help navigate to the next planet, then the Grand Tour was worked up very quickly, since the outer planets were going to be located in the right relative positions.

I've forgotten who did the initial analysis, but it came out of the trajectory group. And once it was pointed out that a Grand Tour was a possibility, we got very enthusiastic at the Laboratory. Everybody decided it would be an excellent mission to do.

Did it come up to your level; did you have to make a decision on it?

Oh yes. We had to sell the notion, the proposal, to NASA. At that time the laboratory assignments were theoretically being given to us by NASA, but in practice were being worked out jointly between NASA and ourselves. So we presented the Grand Tour to NASA, hoping that we would get it approved as a project. I was involved in those presentations.

They were at pretty high levels in NASA; probably at the space science level. In those days the Office of Space Science and Applications (OSSA) was *the* organization in NASA. NASA has changed organizations quite a bit since then.

The argument against the Grand Tour was primarily a financial one: that it was a risky commitment and an expensive one, and NASA couldn't support it. But they subsequently did support it, once we demonstrated that we had a good spacecraft which had the capability to go to the outer planets, and that the incremental costs were not that big. Of course, they were not small because it meant keeping the team together and doing the navigation and the tracking as the spacecraft went out to the outer planets. But at least it came on one planet at a time instead of a commitment for the whole program.

As to who precisely in NASA was involved, it probably was Ed Cortright. He was the man in charge of the space science activities at that time; then I think John Naugle took over.

How did this seem from your perspective as director?

The mission was obviously a very attractive one, and we felt it was a very important opportunity and one which the Laboratory was a natural to do. We wanted to do it very badly, but we were concerned that we couldn't sell it to NASA.

Essentially the spacecraft was already designed, and a certain amount of analysis was done from the very beginning, so we knew how we could go from one planet to the next. When we got near one planet successfully, we had to obtain approval to go on to the next.

We had to make some decisions ahead of time. For example, the flyby of Saturn had to be very precise in order to go on to Uranus. Voyager 2 was programmed to go past Uranus, and Voyager 1 just to go to the outer solar system. That meant that the collection of data in the vicinity of Saturn was modified a little bit, but was not a big problem. We were able to satisfy the scientists because we had two spacecraft. One of them got certain data at Saturn, the other one got other kinds of data.

What sort of decisions came up to your level?

The decisions as to what missions would be carried out and what science would be carried on them. The final decision had to come from Washington, because Washington was the source of funds, and they in turn had to be satisfied that they could get the money approved by Congress.

The selection of science experiments was also done in Washington. Then it was up to us to work with the scientists to make sure that their equipment fit into the spacecraft and would operate properly. In the very early days there was some conflict between the scientists and engineers; the scientists wanted to design the spacecraft around their experiments, and the engineers wanted to design the spacecraft to carry out their mission. Sometimes this led to a certain amount of conflict, but as time went on scientists and engineers learned how to work together. By the time we got to the Voyager program this was not a problem anymore.

Back again at the Grand Tour stage, did the people within your lab come up to your level to settle any of these differences?

No, it usually got settled by the project people. It wouldn't usually come up to my office because it was arguing about details of the design of the experiment, the actual construction of the experiment, the kind of testing that should be done, how much testing should be done at the scientists' own laboratory and how much of it should be done at JPL—that sort of problem came up.

Because the spacecraft was only going to be in the vicinity of a planet for a few hours, the questions of the details of the flyby trajectory had to be considered. Should the spacecraft fly by on the sunlit side or the night side of the planet, north pole or south pole or the equator? Different science experiments have different requirements, and these had to be compromised to arrive at a satisfactory solution; the best solution for everybody concerned. But that was done in the project office.

In missions which have to go from one planet to another, the requirement to fly to the second planet puts constraints on the trajectory past the first planet. Therefore scientists have to accept whatever data they can get from a particular trajectory, because there isn't much choice if the spacecraft is going to another planet. The gravity field of the first planet must be used to put the spacecraft on the right trajectory, because otherwise the amount of rocket fuel required would be prohibitive.

So that trajectory, determined by the requirement of going to the second planet, limits the kinds of observations you can make at the first planet. On the other hand, if the spacecraft is not going on to another planet, then there is a choice how to fly past the planet.

The night side versus the sunlit side question has come up in some missions. On the sunlit side pictures can be taken. On the night side some temperatures and a few other odds and ends can be acquired, so those things have to be discussed and agreed upon. But this is the kind of thing that gets worked out by the project office and the science teams.

Was there anything that came up to your level that was not worked out at those lower levels with respect to the Grand Tour?

Not many of these things came up to my level. In fact, since the science experiments were selected by NASA, there was a tendency for the scientists to complain directly to NASA in case they felt they were not getting what they wanted out of JPL. That's understandable, because NASA appointed them. Sometimes NASA, JPL, and the scientists became involved in discussions but things usually worked out.

Did you make decisions about what resources at JPL could be devoted to developing the Grand Tour?

Yes, that was up to us. NASA would give us an assignment, and we would then put together a plan, estimate the cost, and tell NASA what the bill would be. Then if NASA agreed, it was fine. But sometimes NASA would come back and say there's not that much money; what can you do for X dollars instead of Y dollars? That sort of thing happened occasionally. It was the basic argument against the Grand Tour: the cost. So NASA did not support it in the beginning.

What were your feelings when the Grand Tour was not approved?

We were upset but NASA said, "Okay, go to Jupiter." So we agreed to go to Jupiter.

Was it a surprise when they turned down the Grand Tour?

Yes, it was, because the Grand Tour looked so logical, and the cost increment wasn't that much more. It looked like such a good thing to do that we were surprised that NASA did not support it. At this point I don't know whether it was really a NASA decision or a congressional decision. NASA obviously discussed the program when the NASA budget was being prepared and their programs were evolving. Perhaps NASA got some feedback from Congress that said the Grand Tour wouldn't fly. I just don't know.

What percent of your time as director of the Lab was going into the Grand Tour project? Was it small?

It was small. There were a lot of other things going on at the Lab: the management of the total Lab structure and assuring that we had funding for the people at the Lab and for new projects. In the total picture of the Lab, the Grand Tour was only a part. Of the flight projects, of course it was a major flight project. On the other hand, the Viking mission to Mars was getting started at that time and there were other projects, so that the fraction of the total lab time was maybe a quarter, or something of that order.

Did that percentage stay about the same in the transition to MJS?

Yes, because the actual number of people involved at the Lab didn't change very much when it went to MJS.

The number of people stayed roughly the same through that period: Grand Tour, transition, MJS?

Yes. That transition was made before very much work had been done on the project, so there were not many people involved. Before getting the initial approval just a handful of people were involved. Once we got approval and started cutting metal, then, of course, the number of people involved increased fairly rapidly, but the Grand Tour MJS decision was made before we had cut any metal.

Did you personally have to go back to Washington in the Grand Tour transition to MJS to make presentations there?

I'm sure I did. I don't remember any specific time, but I was back and forth to NASA Headquarters very frequently, for a number of reasons.

Just as a generalized statement, NASA would take the responsibility as far as contacts with Congress were concerned, so that our contacts were primarily with NASA. It doesn't mean that we never talked to Congress; we did. But it was usually with NASA's consent and agreement.

Did you personally talk to Congress?

Oh yes.

What do you remember about that?

Well, I can remember appearing before various committees from time to time, testifying. Basically I would be in a supportive role. NASA would be arguing for their program, and I'd be brought in to talk about some part of that program.

Was this for MJS programs, or generally?

Just generally. MJS would be part of it. I don't remember any specific things at that time. Actually our roughest contacts with Congress were in the very early days when we had six successive failures with the Ranger program. Congress was wanting to investigate us, and NASA was investigating us, and everybody was jumping on us at that time. I had to appeal to Congress to present our side of the picture. But after that, as NASA really got momentum behind it, our appearances before Congress were less. NASA did most of it themselves.

The other side of that coin was that congressmen would turn up at the Laboratory; they'd come out to look at things themselves. We had quite a number of visitors from Congress.

When they visited the Lab, did they meet with you personally?

Yes. I would always be involved with visitors like that. In some cases they would have very specific questions about some project, and in other cases it would be a more informational type of thing: questions of who we were, what we were doing. There were both kinds of visits. As far as specific things about the Grand Tour, I don't remember anything of that nature. I suspect that there were some visits that were Grand Tour related, but offhand I don't remember any.

From your own perspective as director, what was your biggest satisfaction, and what was your biggest headache?

Satisfaction of being the director? I would go right back to the very beginnings, when we put up the Explorer, and when we transferred into NASA. That was obviously very significant, because the Laboratory until that time had been working for the Army, developing missiles, and the transfer over to the space program was very exciting and very important.

The most stressful time at the Laboratory was early on, when we first started our first lunar program, under NASA sponsorship. It was the Ranger program, and the first half dozen Rangers were a real problem. The most stressful time was right after the Ranger 5 failure; we took a year off to try to solve our problems. We had NASA looking over our shoulder with a review committee looking into the project. Then we fired Ranger 6, and it also failed, and that time we had the Congress as well as NASA on our necks. But in a period of a few months we turned that around with a completely successful Ranger 7, and from then on the Laboratory had a very good record of success.

So the low period in my watch as director was the early Ranger series. The high points were the Explorer 1, Ranger 7, Surveyor, the first soft landing on the moon; Mariner 2, the first one to fly by another planet; Mariner 4, the first pictures from Mars. It's hard to say what was the most interesting one after that.

By the late 1960s, when the Grand Tour was starting to be considered, were things going fairly smoothly at the Lab?

Yes, the Lab was doing very well. Mariner 2 flew by Venus in December '62, Ranger 7 was launched in 1964, and by the time we got to the late 1960s we'd had a series of successes. There was a glitch in NASA programs when the end of the Apollo program came, because so much money had gone into Apollo that when Apollo was over, there was the question of what should NASA do.

It was a pretty serious question. Apollo was designed just as a demonstration to prove that a spacecraft could get to the moon and back; it was not part of an ongoing program. So at the end of Apollo, there was the question of what do you do with all of the Apollo people and what do you do with the whole NASA program. We were caught up in that for awhile.

It was a question of what sort of funding we would have. We thought that there might be serious cutback in funding in the early 1970s; while there was some it wasn't serious. In fact, the Voyager and Viking programs were in that period, and the ongoing Mariners—Mariner 10 to Mercury was in '73. The Laboratory was doing very well. That was one of the reasons why we were concerned that we didn't automatically get the Grand Tour, because it looked so good and because of our record. I guess somebody decided that they didn't have the money, but in fact we did the Grand Tour anyway.

Although the scaled down Grand Tour was funded for just Jupiter and Saturn, your colleagues were designing into the Mariner Voyager spacecraft the abilities to go beyond. How much general awareness of that ability to go beyond was there? Was it something you kept fairly quiet?

It was not that we kept it quiet or that we talked about it. It was that the difference in actual design wasn't that great, so we just did it and didn't talk about it really.

So you don't remember anyone challenging you, "Hey, what are you doing this extra work for?"

No, no. The first thing we had to do was build a little unnecessary life into the equipment and that could be justified to get the reliability in deep space, so we automatically had the long life. The nuclear power supply, which had to be used for this type of mission, automatically gave the necessary power supply for a long life. So that was that. As far as the communications were concerned, the ability to have video from Neptune was really an upgrading of the ground stations as much as anything else.

The navigation from one planet to the next, of course, had to be taken care of. After flying by Saturn, in order to go on to Uranus the spacecraft had to pass Saturn in just the right direction, or rather the right location. That had to be taken care of, but that's something taken care of in flight; it doesn't involve anything special, any particular requirements. Maybe put in a little more propulsion in the spacecraft than if it were not to go beyond Saturn, because in order to navigate past Uranus and Neptune, some path corrections would have to be made, which require a little more propulsion. There are a few things like that, which have to be done but they're relatively minor.

During your watch at JPL did you get involved in funding for the possibility of continuing after Saturn, or was that done after you left?

I left in '76, and Voyager wasn't launched until '77, so that happened later.

What lessons might be learned from the Voyager project? Why did the Voyagers work better than some other projects?

They probably were all designed the same way. Maybe Voyager had a little more care in the testing, but I doubt it. You might almost say the others had bad luck. The Galileo antenna problem was the unfortunate consequence of Galileo having to be postponed because of the Challenger disaster. The umbrella-like antenna, folded to fit in the shuttle payload bay, did not unfold as planned. Galileo was going to be launched from a shuttle after Challenger on a direct orbit to Jupiter, but after Challenger everything was put on hold.

The result was that Galileo was eventually launched on a strange orbit which went into the planet Venus first, then came back across Earth a couple of times before it had enough energy to go out to Jupiter. That meant that the spacecraft was sitting around on the ground for a long time, waiting for a launch rocket. Furthermore it was flown on a trajectory toward the sun instead of away from the sun. Therefore the temperature environment was quite different, and various things had to be done to the spacecraft to patch it up for this particular orbit. The problem with the antenna was almost certainly due to this unfortunate chain of events, which changed the Galileo trajectory from the original one to the one actually flown.

The failure of the transmitter on Mars Observer—I don't know what to say about that one. I don't think anyone will have an opinion yet; it's too soon. One thing you learn in this business is when something catastrophic happens, don't rush in with the first guess about what happened; you'd better analyze things pretty carefully because your first guess is usually wrong.

On the Voyager program, Voyager 2, which recently went past Neptune, did have a failure with one receiver only a few months after liftoff, and the whole flight really depended on the number two receiver; it was a piece of equipment which was deliberately put in number two position because it was not considered to be quite as good as number one. Well, the number one failed; number two has been carrying the load for 12 years and is doing fine.

On the Mars Observer one would have expected the backup equipment to take care of things. Unfortunately it didn't. But did JPL learn something from the Voyagers? I hope they did. Specifically they learned that the spacecraft are complex systems, with many interactions among the various parts.

Furthermore they are systems which can't be touched physically after the launch, and therefore any modifications in the system, or any changes or any corrections of anomalous events, can only be made before it leaves the launchpad. After that time, the only modifications possible are those which can be made by radio command.

Therefore the designers have to have an understanding of the system and of what it can do and what it can't do, and know it very thoroughly before launch. Hopefully, if they know it well enough, they can take care of the various anomalous events which may occur, such as the Voyager 2 where the receiver failed.

In the case of Voyager, not only was there the anomaly of the failed receiver to take care of—but Voyager 2, which made the long journey out to Neptune, also had some trouble with its scan platform. The gears in that were binding, and it could only be moved in a limited fashion. Because of this, as the spacecraft was approaching Uranus, there was a decision made to reprogram the computer which was controlling the spacecraft, to allow for some different maneuvers in the vicinity of Uranus. So that computer—which had been flying in space for 9 or 10 years and which had been designed with about 1972 technology (it was at least 15 years old as far as design was concerned)—that computer was reprogrammed from a billion miles away.

That shows several things. First, the validity of the design. Second, the understanding of the design, to where you could do something which had never been planned before the flight. But you could make up your mind during the flight that you would do such and such, and go ahead and do it, with assurance that it was going to work, and it did work.

Voyager, then, is a good example of design of complex automated spacecraft which is well understood by the design team and which can be controlled, modified, almost redesigned in flight because of the flexibility and understanding of the design.

So those are good things to learn from Voyager. Another good thing is that the pictures from Neptune were certainly a communications record as far as distance is concerned, and the ability to operate with very low power—a 20-watt transmitter—over that distance with that band width, is a remarkable achievement, partly due to the programming on the spacecraft and partly due to the design of the equipment on the ground.

The performance actually achieved from Neptune was greater than the performance the team would have expected to get at the time they launched the spacecraft. The ground stations were improved considerably, and the spacecraft was able to respond. So Voyager was a good example of the complexities and the capabilities of an automated system that is well designed.

What added expense did the Lab incur in doing that far-sighted design? Was that a major increment?

It's a fair question, but I don't know how to answer it. If one looks at the history of the Lab, it is clear that every time the Laboratory has carried out a mission it has taken the design of the previous spacecraft and used that experience to design the next one. There has been a continuing evolution of spacecraft.

The cost has gone up, not only in actual dollars but in constant dollars, as the spacecraft have gotten more complicated, but they've gotten more complicated because they are asked to do more complicated things.

Now the present trend at the Laboratory is to try to reverse this, with relatively inexpensive standardized spacecraft which can be used for simpler missions. That's obviously a trend that's going to happen in the future. But I don't know how to put a dollar number or a percentage number on this because the laboratory designs have been consistent with the mission requirements. I don't think the Laboratory has been gold plating things just for the sake of gold plating.

In fact I remember one mission, the mission to Mercury. The Lab asked for something like $120 million. NASA came back and said they didn't have that much money, but they had $95 million; could we do the mission for that? We said yes we could, and NASA made it very clear that it was a fixed price; we couldn't come back for more. We did that mission for that fixed price, and in fact we turned back about half a million dollars. So the Laboratory is conscious of costs and tries to control them, but also tries to be realistic about the costs associated with the mission requirements.

The transition from Grand Tour to MJS and Voyager recalls a notable transition in your own life: from tending horses in a remote village to sending spacecraft to distant planets. Please trace your "trajectory" from childhood in rural New Zealand to director of the world's foremost center for space exploration.

I grew up in a small town on the South Island. My mother had died while I was quite young and my father was often away, so I was raised by my grandparents. Grandfather had a horsedrawn delivery service, and I helped with the horses. After primary school I went to a boarding school in Wellington, where my interest in science developed. A friend and I built the school's first shortwave radio, and we often talked with operators in America.

A relative living in Los Angeles persuaded me to go to Caltech, where I received my bachelor degree in physics in 1932. Job prospects back in New Zealand were discouraging, so I continued at Caltech, getting a physics doctorate in 1936. Afterwards I stayed on to work with my graduate advisor, Robert Millikan. We studied cosmic rays, sending Geiger counters in balloons up to 100,000 feet; I developed electronic telemetering to transmit the data back down from the balloons. I traveled all over with Millikan's team, measuring radiation at different latitudes.

After returning from India in 1940, I was appointed assistant professor at Caltech, teaching electrical engineering. In 1944 I went to JPL as section chief to develop radio telemetry systems for missiles, and five years later I was put in charge of the Army's first long-range, liquid fuel missile, the Corporal. In 1950 I stopped teaching and moved full time to JPL, and soon after became leader of the solid fuel Sergeant missile project. In 1954 I was appointed JPL's director, and I did that for 22 years.

I understand that your leadership capabilities were manifested early: as an undergraduate at Caltech, you were elected class president. How did you view your leadership responsibilities as JPL director?

I considered that my primary role was as a manager organizing groups of engineers. I always thought of the Laboratory's work as basically engineering rather than science: building devices and having them go where they were supposed to go, and communicating with Earth the way they were supposed to communicate. So I thought of myself as organizing engineering teams to accomplish these engineering objectives.

Are there any final thoughts we should have about the origins of Voyager?

I remember the generalities; certainly the general evolution of theoretical appreciation of the Grand Tour, the push to try to sell that to NASA, the disappointment when NASA came with just the MJS, and then the realization that the performance of the spacecraft was such that we could indeed propose going on out to the other planets. Fortunately we were able to sell that to NASA, though that happened after I left.

The assignment to explore the solar system was obviously a difficult engineering task. We had to attract the very best people with the imagination and courage to tackle completely new problems. They had to work together in a team, working to a schedule set by the motion of the planets. They only had one chance to demonstrate the success of their project, and the whole world was watching.

I believe we did build a JPL culture which understood the challenge we faced and how we could solve our problems. We were the best. We knew it, but we also knew we could only maintain our position by remaining true to our JPL heritage.

BRUCE MURRAY
Laboratory Director 1976–1982
Born November 30, 1931   New York, New York

Bruce Murray attended MIT where he received a bachelor of science in 1953, a master of science in 1954, and doctorate in 1955, all in geology.

He joined Caltech in 1960 after working as an oil exploration geologist in Louisiana and an Air Force geophysicist in Massachusetts. A professor of planetary science since 1968, he is the author or co-author of seven books and more than 60 scientific articles.

Before his appointment to head JPL, he was associated with the first space explorations of Mars, Venus, and Mercury. From 1961 to 1973 he was co-investigator on the television science experiments on the Mariner 4, 6, 7, and 9 Mars missions. He was team leader of the 1974–75 Venus and Mercury television science investigations by Mariner 10.

He was director of JPL from April 1976 to July 1982. During that period JPL conducted the operations of the Viking Mars orbiters and landers, and completed development of the two Voyager spacecraft, launched them, and navigated them through their four encounters with Jupiter and Saturn. The Laboratory also managed the Earth satellite projects, Seasat and Solar Mesosphere Explorer; the Infra Red Astronomy Satellite; and developed the Galileo Jupiter orbiter and probe. At the same time, much of Murray's energy was engaged in struggles against cutbacks in funding for JPL.

He has received NASA medals for Exceptional Scientific Achievement, Distinguished Service, and Distinguished Public Service, and awards from Gugenheim, the New York Academy of Sciences, the American Astronautical Society, and the American Institute of Physics. He serves on advisory boards of numerous organizations involved in science and is cofounder, with Carl Sagan, of the Planetary Society.

His favorite recreation is mountain biking: on pavement, gravel, dirt, fenced-off roads—anywhere. He also enjoys flying gliders with his soaring buddy, Bud Schurmeier.

Voyager will live in the history books long after most of the things in our morning newspapers have been forgotten. Buried in this arcane data flowing back from Voyager are clues to Earth's own destiny.
—Bruce Murray

BRUCE MURRAY

I'm glad you're doing this. There was no chronology on Voyager, much less a history. There's not even a document that lists the milestones. The dates of the encounters are listed, and launches and so forth, but all the milestones of developing it—when they made decisions, what the issues were—there's nothing of that.

The reason was the mission was always successful. History is what you write when a mission is over, but this mission keeps going on. I think that's part of it. I think also that Voyager transcended the period of emphasis on robotic exploration, into the period of emphasis on the shuttle, so there was less support at headquarters for getting the Voyager story out, because it wasn't related to selling another mission.

John Naugle was a key person at NASA Headquarters; he was the head of Office of Space Science and Applications, OSSA, when Voyager was developed. A remarkable fellow. There's been nobody like him at NASA for years, who really cared about science and who was worried about building institutional structure. So the consequence is that there's no constituency at NASA for building infrastructure.

At JPL the project managers kept rolling over. It was such a long mission, very successful, and there was nobody who said, "Hey! We need a history of all this." The result is that there is no chronology. There are only bits and pieces. I found that very difficult, and it becomes particularly difficult when you try to trace key decisions because then you discover that the memories of some of the key people differ.

Schurmeier, who was crucial to this, apparently lost his notebook when he moved from Altadena down to Fallbrook. He's a very important guy. If you had to pick one person, the one project manager, the one leader to hang the Voyager trophy on, it would be Schurmeier. Casani would be second, but in terms of the concept of it, the decisions which were made in implementing it, that was all Schurmeier. And bringing together the team, that was all Schurmeier. So he's the guy who deserves the lion's share of the credit for Voyager.

Casani, who was also very good, took over Voyager when I came, because I pushed Schurmeier up; he was too important to let him do something like run Voyager. I needed him for broader responsibilities, and then Casani came in behind. He got all the challenges of the shaky launch and the problems in getting to Jupiter. Then we had to transition him out and that led to this trilogy of Parks, Lyman, and Heacock.

Heacock is a key person. He's a real detail guy. He's good, very able. He reviewed the manuscript of my book, which discusses him personally. He never complained much about the way I treated him; what bothered him the most was that I humanized the robot. I had used terms about the robot getting vertigo, and that bothered him. Really. He wrote long comments on that. It offended him that I was associating him for literary reasons, because it is analogous. So everybody sees Voyager through a different prism.

That's why it's important to talk to them and to you. One of the questions I'm asking Voyager people is how they see the spacecraft. The responses I'm getting are emphatic that it is not alive at all; it's just a mechanical extension. How do you see it?

I certainly think it has a personality. That's perceptual, in the mind of the beholder, not in the vehicle, but I do think it had a very nasty personality for awhile, then lost it. All of us are the product of our experiences. My experience, besides being involved in making it happen originally, being one of the people around who started it—actually its predecessor, the Grand Tour—and then having the responsibilities as director of JPL during the time it was finished up and launched and got to Jupiter and Saturn, my experience included communicating this to the public.

So I developed a way of describing it that I thought made sense to the populace as a whole. The reason that some people like me tend to animate the robot, to attribute human-like attributes to it, is certainly the result of trying to communicate to the public, the larger community, about it.

I also wanted to highlight that it is an extraordinarily advanced machine, in that it takes on so much responsibility to itself. All of this was designed into it. There are not many such machines around yet: autonomous machines that you can cut loose, and that will take on that much responsibility. The Phobos failures were a beautiful example. Phobos 1 and 2 were lost because they didn't have autonomy. They died for simple causes, and the whole mission was lost. This was the Soviet mission, which I was involved with. The contrast between that and Voyager is pretty impressive. So I wanted all along to highlight just how capable these robots are.

A second difference is that I'm a scientist and the other people are engineers, and engineers tend to feel uneasy about human-related things. That gets into other territory, and they like to keep it inanimate. If you talk to Ed Stone, you might ask him; he's a scientist, and the one who knows Voyager the best, since he was involved from the quite early phases to the present, not just as spokesman, which he does extremely well, but also as the person who helped bring people together to build scientific consensus, and to keep scientific consensus.

That is a crucial and very difficult problem. When you have a number of PIs, principal investigators with their own instruments, they're very competitive with each other for use of the communications, for use of the scan platform, instrument mass, power, communications, and so forth. Those are all resources that have to be allocated to competing uses. Somebody wants to know the magnetic field, somebody else wants to take a picture, somebody else wants to take a spectrum. How in the world do you adjudicate that? That takes skill in developing consensus, and Ed has been the paramount person in any mission in doing that.

Of course, the long duration helped. The team worked together for a long time, in fact several generations, so it had a chance to mature, but fortunately it didn't get senile. It just matured and got better as time went on. The robot aged and suffered some losses of memory and other things, but grew more intelligent because more and more sophisticated programming was put in. The personnel grew old and were replaced by younger people, and so there was an organic situation down here on the ground. The wisdom and knowledge of the team grew, accumulated, even though the members turned over because of age and other factors. So you have growth in intelligence at both ends, if you define intelligence generally. Organic change on this side, and smarter software in an aging and somewhat shaky robot as time went on.

Again, these are the parallels that make the engineers uncomfortable that I think are interesting because that's really the way I look at it. I think 100 years from now, when we have *really* sophisticated robots that will make stuff today look like the Boolean machines of the 19th century, that people will very commonly use and associate humanoid characteristics, because they will indeed be appropriate ways to describe it. They won't be self-replicating in the sense that they won't have DNA in them, but they will be very sophisticated devices a century from now, and Voyager will be looked upon as an important precursor step for devices all over the world, not just in space.

I've always felt very proud of Voyager and admired it and therefore the people that have done it, because it was a major step in autonomous systems, and still is.

It also worked for a long time, and that's important. Pioneer 10 and 11 worked for a long time, but they were very unsophisticated devices. When Pioneer 10 or 11 flew by Jupiter or Pioneer 11 by Saturn, we had to send up to 40,000 real-time commands. There was no data storage. There was no memory with all of the commands loaded in. Everything depended on the DSN firing the commands up at just the right time for it to do the thing it had to do. There are many ways in which Voyager was a very substantial increase. In fact it was very complicated and sophisticated, and still lasted. Normally the way you get long life is through simplicity. Witness the shuttle; it got pulled off the pad today.

You can use Voyager as a metaphor; you can use it as an example; you can use it as a precursor. It can be looked at in many different ways. Back in 1980 we had the panel at Caltech called "Saturn and the Mind of Man." We had originally started with "Mars and the Mind of Man," way back in '71. Bradbury and Clarke and Sagan and myself. This became a tradition to have a free-ranging panel. So we got to Saturn in 1980. Fantastic. Just an unbelievable encounter. And at the same time, I was director of JPL; we were losing all the battles. There was this terrible contrast. The Viking landers were here right after I took over, so a hallmark of my tour was to be stuck with these wonderful achievements at the time the society, the political process, was not reinvesting in the future.

One of the reasons I wrote that book was to explain the total ensemble of events and relationships involved. I was pretty depressed, pretty low at that point. In fact it was only a year later that I decided to leave, so I was already getting pretty discouraged. But I had started earlier that year, with the Voyagers going out on their Grand Tour. What's next? Where does that lead?

Space exploration began in my lifetime. I was 28 years old when Sputnik happened. I was involved with the first Mars probe, and I'm involved with the Voyagers going out beyond the solar system. It's incredible. The sudden eruption of exploratory energy and technology and success and results and discovery. It's wonderful. And since I'm a scientist and a teacher, "What's it mean?" So I'm asking myself, and somebody said it must mean going on out to the stars.

I convened a workshop in 1980 on interstellar flight. I had Freeman Dyson, Bob Forward who's a science fiction writer and a "mega thinker" at Hughes research labs, a fellow named Boussard who is the inventor of the Boussard ramjet, Willy Fowler, Nobel prize winner at Caltech. I think Leighton was there, and a couple of people from JPL. We crowded into a room on campus.

This was therapy, in contrast to the difficulties, and I was really expecting to get an answer to how we could go on. What would be the next step? Send automated probes to nearby stars? I think human flight to the stars is problematical at best because the time scales are so long, because you have to travel at relativistic velocities to get anyplace in a reasonable time and come back. So human exploration seems unlikely, but automated exploration seemed to be interesting.

Forward and Freeman Dyson came up with this idea of laser sails, where you'd have lasers on the moon with thousand-kilometer solar sails powered by the laser, and you'd go out to the star and turn around and get a deceleration force and you can slow down enough to see the star. We got into antimatter machines—fusion wouldn't work. They wrote that up, by the way, and Forward has written a couple of science fiction novels on the same idea.

But to me the answer was it ain't going to happen, or only at a very distant time. So I was depressed again, feeling almost claustrophobic, that what I've lived through in my own life, in which Voyager is the epitome, is a very brief anomaly, accelerated as a movie running 100 times the right speed. It has to do with Apollo probably being four generations too early, too. An accidental set of occurrences generated by the Cold War.

In "Saturn and the Mind of Man," I gave a rather somber view, pointing out that going to another star, even with robots, is very difficult. There's certainly not even physics now that works. I don't believe the solar sail is practical. They don't mention in this little proposal that the power used by those lasers on the moon is equal to the total annual electrical power used on the Earth. Somehow I just don't see that happening. Antimatter machines seem very improbable to me. We haven't talked about shielding, about what would happen if you're going along and hit a piece of dust. You're charging along at 0.2, 0.3 of the velocity of light and you hit a dust particle. *Boom*—there goes your ship.

So it's claustrophobic. All of a sudden here I've spent part of my adult life in this wonderful exploratory thing, and to discover that not only *I* wasn't going to do it but maybe *nobody* was going do it, that it kind of ended. Then I said in the same soliloquy in this period, "Well that's okay, we're getting a space telescope, we're going to be able to *see* and so we won't go but we will be able to study it." But then I began to realize that we're near the red limit.

And so I used the simile that we were in a soap bubble inside a red balloon—stuck there, isolated—and that in these several generations post World War II, that the combination of politics and economics and technology permitted us to rush through that at an extraordinary rate; both the physical transfer, represented by Voyager, and the reaching out to the red limit through telescopes, and we'll actually get to the place where we can't observe any further. We can observe within that, but not beyond. That's the red balloon simile.

Please explain "red limit".

There's a book by that title by Timothy Ferris. As the expansion of the universe goes out, as things get farther and farther away, they get redder and redder, until finally the things you want to see are out in an unobservable place. That's the red limit.

Just receding faster and faster?

Right. So what happens then? What happens to society? Here we are, part of Western society, which has always been expanding. Carl Sagan, Phil Morrison, and, I think, Ed Stone were on the panel. Sagan and Morrison are both very good friends of mine, but they were very upset with what I had to say; it violated their intuitive beliefs. I raised the possibility that indeed this was the end. I used the simile that maybe this is the end of adolescence for humanity, that we've been expanding outward and that we're going into another phase where we'll have to learn to live within the space we have, physically and intellectually.

That really got them upset, because they are so deeply imbued with the idea of discovery, as I am, but I was saying, "What happens after that?" That is still a theme. In fact I've got a book concept I've been working on with Chris McKay at Ames Research Center, and I picked up those themes: the book's on exploration. What are the limits of exploration and what are the consequences?

One of the consequences you worry about, and this is true whether it's a permanent limit or whether it's just a hiatus for a few hundred years or a few thousand years, is that western society—our ethics, our approach to things—is based on expanding, growth, and the incorporation of information. If we really reach limits to that, I think we may well be reaching limits on manned exploration. In unmanned solar system exploration we've already reached the limits of what we're going to be doing. We're not going to the nearest star. If we get to a point that we're not expanding outward and feeding on what we learn from discovery and that we still have to feed on what's here, we might get much more into sort of an oriental society: an inward-looking society rather than an outward-looking society.

Voyager as a symbol can conjure up quite a larger perspective, because it's such an extraordinary historical event. It begins to raise questions whether rates are unsupportable. We can't keep exploring like that, which is why I was depressed. I was depressed intellectually and I was depressed personally because I'd been trying to keep it up and it was not even going to stay the same. It was going to fall, which it did, and continues to, despite our attempt.

So there is a lot in Voyager. It's a very meaningful human experience. There's a lot in it, well beyond what we've learned about the satellites and planets and the rings and the interplanetary medium—much more than that—it's about us. What we can do as a people, not just America but as a species, and what some of the challenges of the future are. I was going to put some of that into *Journey into Space*, but it didn't fit. The philosophical stuff I had in but cut out. I'm going to try to put it in this other book on exploration, if I ever get around to getting that one done. But anyway there is a lot there, the broader significance of Voyager.

The other broad significance to me, and I did mention this in *Journey into Space*, is the contrast between the times in America: the circumstances that were involved in Voyager's birth and design, and when it really gets to Neptune. America is completely changed. The world is changed. Voyager is really a product of the 1960s. Although it wasn't commissioned until '71 and didn't get flown until '77, the political and scientific and cultural support for it, the technological basis and the idea of the Grand Tour, of which Voyager is a smaller part, was a late 1960s enterprise. Voyager was the "last hurrah". It was the last of the robotic firecrackers and skyrockets that were set off in the 1960s. We could never do it now. No way.

So by the time it gets to Neptune in August '89, here it is, still reflecting its 1960s origins and 1970s technology. Most of its technology is pretty old. We don't use state-of-the-art technology in a spacecraft; we use proven technology. We don't fly the latest things. It's really flying, in a computer sense, 1970s. Most of those computers aboard are very small, with only about 8000 bytes of memory, much less than the cheapest PC that exists now. One of the reasons that the programming is such a problem is there's so little memory, and some of that memory has gone bad because of radiation, and it's got to be super redundant. So it involves some pretty fancy programming.

I look on the 1960s as being generated by Kennedy and Apollo. That is what made our space program. That's what created the energy and the political support for something like Voyager; it would never have happened without that. If we'd had Eisenhower's space program, we wouldn't have Voyager. From May 1961—which is really the impetus which led to Voyager happening in the late 1960s and 1970s, being launched in '77—to August '89 is 28 years. That's the time scale.

And the contrast between the 1960s turbulence and optimism and aggressiveness in the United States—as manifested not just in Apollo but in the unmanned missions we were doing, state-of-the-art, really pushing things—to where we were when it got to this incredible Voyager encounter .... It was never built to go beyond Saturn, and here it's at Neptune, returning pictures. Beautiful. Incredible. That contrast to me was very moving.

It's a mirror of ourselves. Talk about the red shift and time dilation. Voyager is 28 years, receding year by year. That's us back then, with all that was going on. We had the Watts riots; we had Kennedy assassinated; we had Martin Luther King assassinated; we had Vietnam. It was not a particularly happy time. There were some awful things going on. But in that turmoil there also were some wonderful things going on. To me Voyager was one of the most wonderful.

Are there any other things you would like people to think about?

The unusually autonomous robot aspect. I think that's part of the future. That's going to show up in history books. The significance about ourselves, certainly in a sheer Voyager discovery, which we've just talked about, has just been incredible. The number of worlds observed, the kind of information obtained—there's no precedent in human history. Even the oceanic voyages didn't take place this fast, and the information wasn't disseminated. We've never had an experience where whole, unknown things have been revealed, disseminated, and to some extent internalized by the educated populations. That's unprecedented; that's great.

And then there's this larger, philosophical significance: what does it tell us about exploration? Maybe we are reaching limits. Of course you get the debate between people who don't believe in limits or are offended by them, and people like me who don't like it, and are worried about it.

Please comment on some of the problems you encountered.

I think Voyager really was extraordinary in the sense that the achievements were primarily technical rather than political. The problems were overcome, and the rewards scientifically were very great, and culturally were even larger. That requires a number of things, and Voyager was special in that regard. You had to have very good people, engineers who conceived it and figured out a practical way to take risks; just enough risk but not too much. That's tricky, very tricky, and JPL was at its best for this.

Voyager's problems weren't political; with a lot of other programs we have had political problems. Cassini is a good example of that, and certainly Galileo was political with the shuttle. But with Voyager, they were basically technical: trying to *do* something we'd never done before.

In the unmanned programs, Voyager was the last mission in which the technical challenge was dominant. Since then it's been politics. The shuttle budget, competing objectives, too many scientists' objectives with too little money to satisfy the scientists or the politics, or whatever. But Voyager was driven by the fact that to audaciously explore the unknown in a very wide band way was the objective. It was blessed in that way and permitted to do it that way, and it *did* it. That's why its achievements were technically so far in advance.

It's not going to happen again for a long time, certainly not by any missions now being considered. You have to have a society and a Congress and a president who will support it. It won't work unless people are willing to pay for it. Remember how many Congresses have been involved, how many presidents have been involved. The timing was right. There were people in NASA at that point who believed in this sort of stuff, in the longer view, and they were competent. All was going well, and NASA was riding high in the early part of this—and then it went down.

We had gifted scientists involved, very unusual people. Schurmeier and Stone. I can't overestimate how much of a role Schurmeier had in making this thing work. That doesn't show through very much formally, but to get the right people who could work together and somehow resolve both the engineering and science conflicts (the largest conflicts have been among the scientists themselves)—that's not trivial. The Russians have never done this; their program is a mess because of it. And we did it without a big Voyager bureaucracy. It seemed bureaucratic by our standards, but compared to the Space Telescope Institute and all that goes with it—my God!

And then finally we were lucky. The book tries to describe some of the close calls, like the launch of Voyager 2. If they had swapped the Titan booster rockets, there would have been no Voyager to Neptune. Flip of the coin. Not many people know that, or they have forgotten it.

Three point four seconds of fuel left?

That's right. That was it; it was about that close. We were lucky, and that's important. The Russians have been unlucky in some things. They've done some things right. They've also done things poorly. We've been lucky in a number of areas, and again the book tries to show that there are real laws of chance that operate here, but a good group of people can respond quickly and overcome many challenges.

Finally we've been lucky because of the richness of the discoveries. You go back to those little black and white plates with the band of Jupiter and the rings of Saturn and some dots around them. There was no way to infuse those black and white dots and images with what might be there, and so we've been lucky in that way.

I was involved heavily with the flight to Mercury. Mercury turned out, for the public, to be rather boring, like the moon. It could have been very different. Mercury could have had bizarre land forms on it. It could have had a small atmosphere. At the time we went we didn't know. Scientifically it is very interesting, but publicly it is not. Well, these outer planet moons might have been like that; instead of Io and Callisto they might have been pretty dull. So it's a concatenation of really good people, good ideas, support by society, luck in the engineering sense and luck in the scientific sense.

How did you happen to be at JPL in the first place?

I was not much interested in science in high school. I wasn't a particularly good student, but I was very interested in the out-of-doors and geology. I started collecting minerals, but I didn't see myself as a scientist at that point. I decided early on that I really wanted to go into geology, but not space—nothing space related. I entered MIT in '49, got my bachelor degree in '53, and finished my graduate work early and left in October of '55.

Being in the ROTC, I had an Air Force tour of duty to serve. I had been deferred during the Korean War to finish up my schooling so that's why I went through in a hurry. I was afraid I was going to get called in, and I wanted to finish my graduate work. Then I took what I thought would be a short job assignment with an oil company in Louisiana because I was working with sedimentary rocks. A lot of what we know about sedimentary rocks we've learned from oil wells being drilled. It turned out that I was there for three years.

Although the Korean war ended, the Air Force still wanted me, but not right then. That meant I couldn't find a job somewhere else because I never knew when I was going to be called up, so I worked for the oil company for three years before I finally did go into the Air Force. Then I was sent to the Air Force Cambridge Research Labs, at Hanscom Field outside of Boston. It had two parts: the electronics research directorate, which worked with Lincoln Laboratories on air defense—radar and so forth—and there was the geophysical directorate, which worked on gravity and geodesy and atmosphere. Since I was a geologist and geophysicist, they naturally assigned me to the electronics research directorate.

Fortunately, the personnel officer at the base said, "No, I think you better go over here," so he sent me to an organization called the Gravity and Geodesy branch, which was beginning to worry about how to tie grids from different continents together accurately because the advent of ICBMs was coming up. I got to Hanscomb Field in August of 1958. Sputnik had been launched October 4, 1957. Explorer 1 was January 30, 1958.

My work was kind of boring; I was serving my tour of duty, which was a responsibility I accepted. In the next office was an organization called Space Track, because the Air Force had the job of tracking all the satellites that were up—at that time mainly Russian ones. So I was exposed to those people, simply by chance, and became very interested in that. If I'd gone to the electronic research directorate, I probably would have not. I realized something exciting was happening, so I got more involved in some space-related activities, including the beginnings of what became geodetic positioning satellites.

When my two years were up, which was all you had to serve then, I wrote to Caltech and said, "Hey, I'd like to come work. You've got Palomar, JPL." Fortunately there was an opportunity for me as a postdoctoral fellow for Harrison Brown. I made a career change; I decided I wasn't going to go back to looking for oil, and instead I took a flyer on this new thing called space.

It was a good decision. But I was not obsessed with science as a young person. I was interested in many things: history, politics, but particularly enjoying what you might call the science of rocks and nature which we call geology, which I still do.

What was your role in making Voyager happen?

The precursor of Voyager was the Grand Tour concept. There was a science working group, chaired by James van Allen. I was a member of that, specifically advocating, against considerable disinterest, the satellites, because they were viewed as single points on a photographic plate. I was one of the few people there who was advocating for that part of the mission. That was the specific connection I had with what became Voyager.

You were the lone geologist among 15 members?

Yes. The view of the outer planets from the Earth, before there was space data available, was that Jupiter was the largest radio source in the sky. It had interesting atmospheric phenomena, it was interesting dynamically with some satellites. Saturn had the rings, which made it very exciting, and then there were these funny green things further out, Uranus and Neptune, which were similar in size. But the thinking about the outer solar system then was dominated by the electromagnetic aspects, the plasma, the atmospheres, and things of this type. The view from the Earth doesn't tell you anything about the satellites.

And so, when the committee was formed, it reflected primarily the views of physicists and physical meteorologists and astronomers, but not that of geologists because there was no geology of the outer planets. This was a problem throughout the whole exploration of space, because it was dominated from the beginning by physicists. They were the major advisors to the government as a result of World War II, the H bomb, and things like that.

The committee was under representative of people who had expectations that the surfaces of the satellites might be very exciting. We knew some weird things, but not enough to make a strong case. Therefore my role was to advocate study of these smaller bodies, Io and Europa and Ganymede and Callisto, which turned out to be fascinating objects, to say nothing of little old Triton way out there at Neptune. So that was one of my roles.

The second role I played was that, as head of the Imaging Team for Mariner 10 to Mercury and Venus, we had to develop a new camera system. We had to increase the focal length greatly, by a factor of three, from 500 mm to 1500 mm, in the same physical size. We had to increase the magnification of the secondary mirror by a factor of three. This was using vidicon tubes, which had been used for all missions including Voyager. We did that successfully; it was very difficult, but that gave us this high-resolution imaging system, which permitted us to study the planet from quite some distance away as the flyby was approaching.

When the time came to do Voyager, I decided not to participate as a scientist on board. I was involved with four missions, and that was enough. So I did not go on, but that camera design was what became the Voyager camera. The difference was that they made the optics out of pure quartz so the radiation wouldn't darken them, and they improved the vidicon system a bit—the electronics of it. But it was our Mariner 10 camera way out there at Neptune in 1989.

The key person with us in that, Ed Danielson, did go with it. He and Mert Davies both went from the Mariner 10 team to the Voyager team, so there was a real carryover. Brad Smith had been part of our Mariner 6 and 7 team and Mariner 9 team, which we had worked on before. Larry Soderblom was my prize student from the late 1960s, so I had a lot of inheritance into the mission, both from the people, the optics, and the advocacy for observing the moons carefully.

Your book mentions a crucial phone call in February 1972 to Russell Drew about MJS getting killed.

Right. At that time I had been on a subcommittee of the President's Science Advisory Committee (PSAC), which was formed during the Johnson administration. This subcommittee was concerned with space science. Nixon abolished the PSAC; he didn't like the advice he was getting. But Russell Drew, who had been the contact with this group—he was a naval officer on detail duty to the White House for this purpose—continued on after that committee was abolished. He continued to play the role of getting advice and policy information to the White House as a White House employee.

When this budget cycle came up, the one in '72 which was the end of Nixon's first term, we were still talking about things. Even though there were no official committees anymore, Russell would still use his network to get information. He called me and hinted that something might be up, and I came back with the justification that a mission to the outer planets was the best possible competition with the Russians in space that we could do, because the Russians simply could not carry out a mission like this. That was the kind of argument that appealed to the Nixon administration. It was also likely to be the greatest exploration of our times, although that was harder to prove at that point.

I made the anti-Soviet argument to play into the politics of the time. I targeted it at the Nixon White House, but it was technically accurate: the Russians did not have the capability to get to Jupiter and certainly did not have the capability of dealing with the radiation environment there. They hadn't been able to make a spacecraft live anywhere near that long.

A third crucial part you played was, on your very first day in office, you appointed Bud Schurmeier.

Right. I had known Bud very well, first from Mariner 6 and 7, where he was the project manager, and then in the Grand Tour days. I realized he was an extraordinary person. He was one of the best project managers at JPL. He had set up the systems engineering division at the very beginning, but he also had more lateral vision; he could deal with a variety of people and institutions and sort them out much better than most.

When I came to JPL in April of 1976, my priorities were to build a second, non-NASA business line for JPL, because it was clear from conversations with Fletcher, and others at the head of NASA, that JPL was low on their priorities. Unlike the present, when JPL is considered a model for civil service, in those days, if there was going to be cuts, it would be the non-civil service center, JPL, that would be sacrificed—and they made that clear.

As director I had the responsibility of doing something to minimize that situation. The energy crisis in the Middle East that began in 1973 had stirred up a lot of concern. That led to the creation of a new agency, the Department of Energy, and new programs. I decided we would make a major assault to build a big civilian business for JPL in energy. I wanted the best person available at JPL to do that, and that was Bud Schurmeier, so I took him off the nicest project, Voyager, and put him on a very unrewarding job. When the Department of Energy program got weak, we then took on the defense job that JPL still has.

But the good news was that John Casani, the fair-haired boy of the generation below Bud, then became project manager of Voyager. He was a good person to take over when Bud left. Unfortunately I then took John off and put him on a new project which we needed to start. That became Galileo, and I had to take the best person we had for that, so I robbed Voyager again, of John Casani.

I felt confident enough about Voyager that I was willing to rob it of its two best people for the other two highest priority objectives JPL had at that time. Then, when we had troubles after launch, we had to go back and add people back in. First I had Bob Parks, who was head of all flight projects, take personal responsibility, and we added Pete Lyman in the operations area as a deputy, and Ray Heacock, who had been the project manager, became deputy for the spacecraft. We ran the mission that way up through the Jupiter encounter.

So we took away from Voyager, but we gave back when it turned out that the mission was in deep trouble. It was such a complicated spacecraft that it could not be tested thoroughly enough on the ground. There were too many different computer modes for it to be in, and we began to have trouble operating it. So we had to put talent back in. But Voyager was clearly the number one priority as far as a flight mission at JPL while I was there.

What was the toughest decision that you personally had to make?

The toughest decision I personally had to make was to go back. We had taken Schurmeier off, and then we took Casani off, and that left Ray Heacock as the project manager. We got in trouble. We then had to make Ray a deputy, temporarily, under Parks who became the project manager, and add another deputy, for operations. To do that I had to overrule the advice being given by Parks himself. But talking to various people, I decided we ought to do a maximum effort, and that was the maximum thing we could do at the time. That was the hardest thing personally I had to do, because it was my personal decision, and it was overruling the most respected person in flight projects at JPL, Bob Parks. Actually it wasn't overruling, it was going counter to his recommendation.

Was there a feeling that Ray was being demoted?

Certainly he must have felt that way. It *was* a demotion. But we had to bring in more talent at the top, and that meant in effect a demotion temporarily. After we got those problems solved, Ray became the manager and earned the honor that he deserved as project manager of Voyager through the successful encounters. Finally it was so routine after Saturn that we turned it over to another person, and then another, and there became a whole succession of project managers.

Was it that Ray was not focusing on the personnel?

No. There was nothing he was doing wrong. Ray is an extremely able guy. He'd been involved with the hardware; but we began to have software and operational problems. The crew forgot to send a key command that had to be sent every seven days, so the spacecraft shut itself into a bad mode. We were having a very hard time getting ready for the Jupiter encounter, and we had to add talent that could help in that area. The idea was to split off the spacecraft portion, under Ray, which he knew very well, away from the operations portion, and give that to Pete Lyman, who had been heavily involved in the operations of Viking just before that.

Viking was a very complex, high powered operation because it had two landers and two orbiters. Pete was a key person in the Viking operations, so he had good flight operations experience, recent. And we had Heacock, who is the world's expert on the hardware and the spacecraft part of Voyager. We put on Bob Parks, who was a very powerful person at JPL and could assure that the project got all the support it needed from the Lab in general. So we gave it priority with Parks, we gave it operations with Lyman, and kept Heacock there on the spacecraft side, so that was the maximum amount of power we could give the project.

And that was the decision I made. It was not a negative decision about Heacock; it was a decision that we had underestimated the difficulty of flying this beast, and we now needed to recover, so we recovered by overkill. We added a lot of extra things.

So the focus of the difficulty was the software?

And operations. People on the ground. It's such a complicated spacecraft to fly because it has all these computer loops in it, that you have to train people in very great detail to run the beast. For example, the whole Voyager encounter with Jupiter was completely rehearsed before it ever got there. They flew all the maneuvers, they did everything. That was one of the milestones; we had to prove we could do that.

Was that after launch?

Yes, after launch. After we solved the immediate crisis then we went into a heavy training program and developed a complicated sequence of commands, and then actually did it with the real spacecraft to prove that it worked. That way you know that there's no "I forgots" in the commands because you'd done it—and still we had surprises.

When we got to Jupiter, the radiation belts were so intense that the interaction of the particles in the radiation belts with the metal caused what's called bremsstrahlung; electrons that physically flow through the spacecraft, so there were excess currents being generated as it flew through the environment of Jupiter, right up close.

That's the last thing you want. Imagine your computer if you've got static electricity and charges, and it did; it actually went to a "power on reset." It reset itself in the middle of all that and still survived. That was due to a combination of very good hardware and detailed knowledge and very good software, and thinking through ahead of time "what if", because it's much too late if something actually happens.

By the time you know that it has happened, it's all over. The spacecraft is dead, because you're so far away the radio travel time was an hour and a half or two at that point. There was no way you could sit there and watch the equipment by telemetry and say, "Oh, we've got a problem; let's do something about it." Everything had to be done ahead of time, including providing for bad things. And there were some bad things.

What were your thoughts at the launch of Voyager 2?

I had gone down there for ceremonial purposes; I wasn't involved in the technical decisions about how to launch it, nor should the director of the center be. They kept me informed, but mainly it was great to see, and it was important to show that it was important.

This was a great moment, but it quickly changed to a real serious problem; because right after launch, we began to get signals from it that we didn't understand, and it wouldn't respond to commands properly. That was very serious—the Russians lost Phobos 1 just this way, and Phobos 2 failed, not quite the same way. It was really bad news. It turned out that the software was designed to operate a billion miles from the Earth, not 100,000 miles from Earth.

So what had started out as a tremendous, exciting celebration turned into a lot of long nights in dark rooms with everyone around a speakerphone talking to people at JPL; and people at the blackboards trying to figure out what was wrong and what to do about it. Finally the hemorrhaging was stopped, but the patient was still in critical care. It took quite awhile to work completely out of that situation, very gingerly.

Then there were other problems; the scan platform on Voyager 1 stuck, apparently due to a piece of fiber. So there were some real problems, and there were some misunderstandings that led to real problems. In the meantime you're moving towards Jupiter. Eighteen months sounds like a long time, but not if you spend the first six months just getting the thing to settle down and fly right.

Plus we had to launch Voyager 1 three weeks after Voyager 2 was launched, so we had to fix whatever was wrong. There was a mechanical problem with the big arm that swung out with the instruments on it. They had to do a software fix on that. It's not good to make those fixes at the last minute; you have as much chance to screw it up as you have to fix it, so there was a terrible time criticality, because you had to launch the second one on time.

Those were exciting times, and if Voyager had failed, as it nearly did, that would certainly have put JPL in a very precarious position to continue as the lead center for spacecraft work at NASA. In fact, a lot of people were questioning whether JPL should exist at all, or exist as part of NASA. JPL's back in that situation now, to some extent. If they have a couple of catastrophic failures with other spacecraft, the same issue could arise. So you do bet the store with a big, highly visible mission like this, and I was aware of it. It was not a good time to have a very visible, spectacular failure.

Engineers

After traveling 4.4 billion miles over 12 years, Voyager 2 arrived at Neptune within 21 miles of the point aimed for and within 1.4 seconds of the schedule set years previously. During their journeys through the outer solar system, the two craft sent back hundreds of thousands of scientific measurements, including 70,000 images.

These remarkable accomplishments were achieved by hundreds of JPL engineers working on many teams. Their functions included determining the precise location of the spacecraft and their target planets and satellites, preparing guidelines to govern how resources would be used, developing sequences for observations, monitoring the health of the spacecraft, sending commands, and receiving and processing data.

WILLIAM S. SHIPLEY
Space Development Manager
Born May 6, 1931   Washington, D.C.

William Shipley attended Woodrow Wilson High School where he received honorable mention in the Westinghouse Science Talent Search. He earned a bachelor of science in physics from George Washington University in 1953 and was awarded a Saunder's Memorial Fellowship.

Joining JPL in July 1955, he participated in the early analysis and development of the payload for the Explorer satellite. He held a number of positions in systems engineering, reliability and quality assurance, and project engineering supervision and management. Then he led the Thermoelectric Outer Planet Spacecraft (TOPS) Advanced System Technology project. This 1969, '70, and '71 project provided understanding of the requirements that the outer planet mission imposed on spacecraft systems and subsystems. TOPS also evolved a system architecture and initiated development work on many elements of the Voyager spacecraft.

After serving as Voyager space development manager from project start through launch, he became the first of the Galileo Orbiter managers. In 1981 he was appointed manager of the JPL's Reliability and Quality Assurance Division. His final position was assistant laboratory director for Engineering and Review.

As an undergraduate, Shipley raced NASCAR stock cars. In his retirement he races and cruises ocean-going sailing vessels.

To make or fix something that will work and stay working is to capture truth, knowledge, and joy. I was fortunate to live when and where opportunities for fulfillment in physical entities abounded. Past ages were limited by knowledge; the future appears limited by society. Working on Voyager with such an outstanding team of individuals was a once-in-a-creation opportunity.
—William S. Shipley

WILLIAM S. SHIPLEY

I got my bachelor of science in 1953. I had garage and construction jobs, was an engineering aide at the White Oak, Maryland, Naval Ordnance Lab, and also worked at Intelligent Machines Research Corporation. My degree was in physics, but I worked in electronics as a part of physics.

When I was a graduate student, I was interviewed by a JPL supervisor who was visiting George Washington University, and I was very impressed with him and his attitude. So when I got an offer from JPL, which was exactly the same figure I got from another place, I chose JPL because I thought they were a more interesting kind of people.

So it was the personal contact that did it?
Yes.

What was your first assignment at JPL?
I arrived at JPL the same week President Eisenhower announced that the country was going to launch a satellite. I just happened to be in the group that was going to work on the satellite. So my first assignment at JPL was to compute the temperature of a satellite in a 200-mile orbit. There was no such thing as "space" before that. Space was there, of course, but not space exploration. I came to work on missiles. Some people would say to kill people, others would say to keep people from killing people.

How did you become involved with Voyager?
Since 1959 I had been supervisor or manager in engineering sections. I first became involved with Voyager about June of 1968, when I was asked to be the manager of an advanced system technology program to study the spacecraft that might be used on the Grand Tour. That was known subsequently as the Thermoelectric Outer Planet Spacecraft project, TOPS. It was jointly managed by the Flight Projects Office and Office of Research and Development.

There were two managers: one from flight projects, which was me, and one from research and development, which was Rob Roy MacDonald. We were appointed by the body of assistant laboratory directors who were operating at that time. We had a joint appointment from the assistant laboratory director for tech divisions and the assistant laboratory director for flight projects and the assistant laboratory director for research and development.

TOPS was a system study to define the mission requirements on the spacecraft, the spacecraft requirements on the subsystems, the details of the subsystem requirements, and to do hardware development work in areas where we expected to advance the state of the art and differ from earlier planetary projects.

We had no block of money specifically for TOPS, but we integrated a large number of research and development tasks that were sponsored by different offices at NASA Headquarters into a project. Each subsystem participated in the project to the extent they had their own money. So it was an interesting challenge in itself because what we were doing was taking efforts that had been defined by our line organizations—research and development efforts—and turning them around to solve another purpose. That got considerable discussion at that time.

We did put together a program. We came up with a design for a very elegant Grand Tour spacecraft. In 1972 when the project was started it was dropped back from a Grand Tour to a Jupiter-Saturn mission. We took almost all of that development out of the baseline project; there was almost nothing left in. Then, at the science selection, we replaced the Viking heritage radio with a radio that had many of the features of the TOPS radio.

Subsequently we added many features of the TOPS spacecraft to Voyager. And with the exception of the radio, we did that in a quote "zero risk" mode. We maintained a baseline that used the heritage from Viking and other previous planetary missions, and if other work occurred that brought development to the point where you could bring it in to Voyager at a negligible risk, then we would adopt it.

But we maintained fallback positions in some cases 'til quite late in the game. One of the areas where it stands out the most in that regard is the memory for the flight data system, where our baseline was the plated wire memory that had been used in Mariner Venus-Mercury '73 and Viking. We forced the system manager to maintain the weight and power required for plated wire up until quite late in the game, while we completed the solid-state memory development. There were a great many performance improvements and a lot of flexibilities added to the spacecraft.

In the final analysis there were only a few things that came out of the TOPS design that were not used in Voyager. One of those was a STAR computer. STAR meant Self Test And Repair, which was a triply redundant computer with a voting and replacement system internal to it. Another item was the unfurlable antenna that was not used on Voyager but was subsequently used on Galileo. Another was the silicon vidicon in the imaging. Another was momentum wheels that we'd studied in the TOPS. Most of the subsystems on Voyager reflected some significant results from TOPS, particularly the radio and the attitude control that used a digital processor.

There were a couple of illuminating memos. One document matrixes out what was in the TOPS design, what was in the Voyager baseline design—the MJS baseline of 1972, and then what finally flew. The other thing is a little table we put together that showed cost growth by subsystem, and cost control turns out to be one of the major challenges.

TOPS was supposed to be a three-year effort. We didn't complete all that we wanted to, but we did get a basic functional requirements book. We had got a very good understanding of the system and how it ought to work. And from that we could make a lot of adaptations. We worked a lot of tradeoffs, and there was an awful lot of study and understanding achieved.

We proposed for Grand Tour starts '72, '71, and probably '70. The original proposals, '70 and '71, were ones that I put together. We went through a full budget exercise, like we do for a going project, and we kept coming up with cost estimates that were three-quarters to a billion dollars. It was the size of those cost estimates that caused the Grand Tour not to be accepted, and to switch to the MJS mission.

We were having trouble getting the project approved, so JPL got one of the most experienced project managers at JPL, Bud Schurmeier, to take over the effort, and he got Ray Heacock as a system manager. After the end of TOPS and after I did an SEB, Source Evaluation Board, for another project, I went to Voyager as the spacecraft development manager in 1972 and had a very straightforward job description to go with that assignment.

From a technical standpoint the two things that were the outstanding challenges of Voyager were the radiation hardening and the long life. And the way we dealt with them had synergistic results. Hardening the spacecraft for radiation actually made it more adaptable for a long life.

That came about in two ways. Radiation hardening involved a very extensive worst-case circuit analysis, and that forced all designs to take into account the effects of degradation of parts with aging and with radiation. So, when we did design from a radiation standpoint, we were also doing a good job of long life design.

The other thing was that it forced us to do a lot more parts testing and parts work, and got us very skilled at the selection and acceptance of electronic parts. So our parts program on Voyager—in my view which is obviously a biased view—was probably the best parts program that's ever been carried out on any project in the United States. In spite of that, we have had a number of parts failures in flight over the years, but the fact that they haven't been fatal is the test of the system design.

One of our system design objectives from the beginning was to get a spacecraft that was flexible and was able to survive failures. It had to be able to operate for two weeks without any uplink from the ground. We didn't know the word "fault protection"; I can't remember when we brought that into our vocabulary, but all these things kept churning in our design discussions over the years. That was where we came up with fault protection in a very effective sense.

Fault protection has probably been the feature of the design that's been most visible and most powerful over the years. You could not have the fault protection without an extremely good hardware design, parts selection, and quality test program, because what the fault protection does is allow you to intelligently use the redundancy that you have. It doesn't protect you from *two* failures. Your fault protection just allows you to effectively utilize that tolerance for the first fault in each redundant element.

In some cases, there is more redundancy than that; it comes out of the way the thing is designed. Memory is the classic example. We've had three or four or five memory cells fail, but you can program around the failed cell, so instead of just having block redundancy, you've got a high degree of multiplicity of redundancy just because you've got extra memory. As long you've got extra memory you can keep handling failures. And if you don't have extra memory you can shrink your program and do something less, and operate with whatever memory you've got.

Our original intent was to get a very flexible design, and we talked and thought about that for years. That was critical to making a Voyager that worked.

We had some situations that we could capitalize on that really helped us, too. The folks at NASA Headquarters who were assigned to the Voyager program office were actually helpful. That's extremely unusual. They're the finest people I've ever dealt with at NASA Headquarters. I've never seen another program office like that. I've told the fellows that were with us during development, and during the early days of flight, that Voyager was blessed, that we seemed to always get the best of people, even in NASA HQ.

In Voyager, when we did our cutback from TOPS we did that fairly carefully. We had many situations where we made mistakes in our budgeting, but we kept a very sound program plan. We put in enough hardware; we had enough units; we had an almost complete set of spares in the PTM (Proof Test Model); we had two flight spacecraft. We put in a complete reliability and quality program. We didn't stint on anything to start out with. Where we had uncertainties, we put in an adequate amount of stuff to cover our uncertainty.

One of the classics was the antenna. Nobody had ever built a graphite epoxy antenna that size before. Existing ones were probably a quarter of that size, or a third. So we put in five antennas in our baseline program, because we didn't know what would happen in handling and we were afraid we might damage them.

When we were making the first one, we had a major disaster. The antenna stuck to the form, and the form broke up; and we had some damage to the antenna. But we learned something; we learned it was really very easy to patch. And once we knew it was easy to patch, we didn't need five of them anymore. So we cut down the quantity to three. Two flew on Voyager and the third one went on Magellan to Venus. Cutting the quantities back allowed us to build three for almost the cost that we had estimated for five.

And that was not atypical. We had cost growth in every procurement. Our average on a procurement was a 90 percent cost growth. Our spacecraft total cost growth was only 17 percent. One of the reasons was we had flexibility, and we put in a big enough program in the beginning. We didn't have nearly enough contingency to cover that 17 percent. We only had an $11 million contingency on $150 million for the spacecraft.

But we had some hidden contingencies. We didn't assume that we would have government inspection support provided free. We thought we might, but we didn't know it so we didn't assume it. And each year when we'd negotiate our arrangements with Defense Contract Administrative Services, DCAS, to get the inspection support and we got it, then we'd have basically a little gift.

There were a number of things that we did that way. We didn't assume in budgeting that all the possible things in the world would go right. We assumed that nothing would be better than you ought to reasonably expect. And that was probably the key to the project not running out of money. All these projects today, you negotiate everything with the most optimistic assumptions, and therefore these projects are in money trouble from the outset. But that's required, because that's what everybody expects you to do; that's the kind of assumptions they want.

On Voyager we always had cost pressure. It was big cost pressure, but it wasn't intolerable, because the original plan was good; it had resilience. In the original agreement with NASA, one of the last sentences in the letter from Dr. Pickering was that if inflation exceeded five percent we'd have to have some adjustments in the figure. Well, there were some years in the mid-1970s where the inflation rate was over 10; most of the time it was eight or nine. However, when John Casani went back to NASA to get the adjustment, they just told him to eat it. We never got that adjustment. But we did get the money as cost growth, and we had a 17 percent cost growth. Now most of that vanishes if you put in the right inflation. But that was looked at as cost growth.

NASA Headquarters did provide some money for an increase in scope in some things, and some for the radiation hardening. Like I said, radiation hardening was a major technical challenge. It was not in the original program plan. The estimates of the radiation environment at Jupiter were grossly in error. That was discovered when Pioneer passed Jupiter and discovered that the electron radiation there was three or four orders of magnitude more than what had been predicted by scientists. That was a major impact on the program.

What we did was to break the radiation hardening problem down into small enough pieces and assign them over to matrix organization, just as we would a thermal problem or a mechanical problem. Other people who had tried to radiation harden spacecraft before had not done it in quite that systematic, routine manner. They had made superorganizational structures and things like that. I think that our system was much better. It worked.

When we first got into the radiation problem, many people suggested we needed a radiation czar and things like that. We said that really wasn't what we want. We want everybody to still have responsibility for the thing that they were responsible for. But radiation was another requirement that they had to satisfy. Then we had to build a structure for how you satisfied it.

After we did that it was written up. Some of our consultants worked for the Defense Nuclear Agency, and they put that in the policy and guidelines on how to do radiation hardening. So that was a general benefit to the aerospace industry.

We had many, many problems. It's hard for me to pick out a set because I had a bookshelf full of three-ring binders that were problems. I'd have a problem working in a book. When I got a new problem, I'd open a new section in that book, or if that book looked like it was going to fill up, I'd open a new book. So I wound up with a whole shelf full of three-ring binders full of problems. I used to tell people that was my $20 million bookshelf, because that represented about $20 million worth of problems.

Some of the biggest and best known were with the radio development. On the traveling wave tube amplifiers we had a cost growth, if you look at the original budget estimates, of about a factor of four. Our estimates in our budget were about $2.2 million; by the time we closed the contract it was about eight and a half. It was a problem from beginning to end.

The other side of it is they're probably the finest traveling wave tube amplifiers in the world. At one time somebody else was trying to order copies, and the supplier couldn't make them. One time we even got a communication from a supplier that indicated that, "Gee, we think there might be something wrong with yours, too." The fellow who was in charge of the traveling wave tube at that time said, "God, those guys still can't do good lab work!"

The amplifiers worked very well. We had one that appeared to be degrading slightly, so before the Neptune encounter they changed it because they didn't want to have to change it in a hurry as they got to Neptune. But all in all, the radios worked quite well. There was a failure—a failure that we were not at all surprised at, because we'd seen two of those failures during the ground test program, and we launched with a risk of that happening—and it happened.

The problem was that there was a short in a loop filter capacitor in the radio. For some reason that we don't understand, as long as you keep the radio operating, you don't develop any of these shorts. But if you turn it off and turn it back on, you may find one of these shorts.

It was very mystifying. The ground crew failed to send a command that would let the spacecraft know that all was well and that it was still receiving information from the Earth. They didn't send a command up for two weeks, so the spacecraft thought its radio receiver had failed, and therefore it switched to the other radio receiver, which indicated this kind of a short right off, so it still couldn't receive anything. It switched back, and had some other major failure in the one it switched back to, which we've never really understood.

Ultimately we determined what was wrong with the second receiver and they fiddled around to get the frequency right, and got a signal into the radio. Once they got in they learned how to do that, and they've been basically flying the spacecraft by changing the transmitted frequency to match what they think the radio will pick up. This involves some programming every day to take out the effect of the Earth's Doppler. So as far as the spacecraft is concerned, the Earth is not rotating because there's no Doppler in the signal from the Earth's rotation.

If one really wants to understand why Voyager works, one has to look back over almost the whole history of JPL. Much of Voyager's success was based on the Sergeant and things that occurred in the early 1960s. In Sergeant there were three phases. The first phase was propulsion rounds, the second stage was developmental rounds, and the third stage was the engineering model.

From the engineering model, the military was supposed to make decisions about what they were going to produce. The engineering model program was managed by Mr. Jack James. It was run by reasonable rules. It was an orderly project. It had practically all the elements of a reliability program that we have today: it had problem/failure reporting, it had past stress analysis, it had environmental tests with detailed environmental test specifications for every box. It was a thoroughly thought-out effort.

Explorer 1 was made here. And it was made by a small group of people. There was no systems engineering and little reliability work, but it didn't demand any because it was simple. It weighed 38 lbs. I can still draw much of the circuitry. If somebody gives me a mechanical drawing of the payload, I could put most of the dimensions on it from memory. So it was something a person could carry physically, and you could also carry it in your head.

That does not demand a big systems integration effort. It doesn't demand change control systems. It doesn't demand a lot of documentation and communication. It doesn't demand good functional descriptions. It doesn't demand interface descriptions. It doesn't demand system testing that's defined with objectives for every test.

When we started doing the space program, the general feeling was that we couldn't afford timewise or moneywise to do anything the way we did the Sergeant. So the idea was to build spacecraft more like we built Explorer: with much more independence for all the people, less centralized control. And after we had four Ranger failures, we started another project, called Mariner R.

That happened to be run by the same man that ran the Sergeant engineering model program. This was a nine-month spacecraft development, and he started by writing a book of rules about how he was going to do it. He wrote down his schedules and what all his requirements were. His book of rules was basically that we were going to do this the same way that we did the Sergeant engineering model, and not too surprisingly his spacecraft worked. It flew past Venus and performed our first successful planetary mission.

Immediately thereafter we adapted a similar rule book for the Rangers, and then Rangers started to work. And we basically have enhanced that set of rules over a 30-year period. We added design reviews for the Ranger project; we didn't have those on the first Mariner, at least not with any rigor. Then we added worst-case circuit analysis to Voyager, because of the radiation environment.

In *concept* there's been very little else added, although there's been a lot of *detail* reflecting the additional complexities of newer technology. Things became much more complicated, and there are a lot of considerations that have been added on. When we got to Voyager, we took those documents—the ones that were used in Viking—and reworked them, got everybody to accept them, got the rules all straight, and operated by those rules.

Those rules were fundamental to the thing working. Those rules in combination with a good program definition: of what the program was going to include, how much hardware, how well we were going to do things, what kind of reliability efforts. That's why Voyager works; it's that total combination.

Out of having a good definition, we had the right set of resources to go in with. I had something funny happen. We were in Florida at the Cape, and I was going out to the control center for the first launch. I was standing on the corner at 4:00 in the morning, waiting for a friend, John Hunter, to pick me up. He always comes late to everything. As I waited I began to think, "What if something goes wrong? What would happen if it doesn't work?"

Then I asked myself, "Well, what could go wrong?" I thought of all the problems that we'd worked through, and I came to the conclusion that if somebody came along and offered me another $2 or $3 million and a couple more months, I wouldn't want it because we had worked everything as far as our understanding of the technology would permit. Each of the problems was worked as well as it could be worked.

Then I wondered, "Will anybody ever be in this position again?" because it was obvious that budgets for all future projects were going to get tighter and tighter. In spite of the fact that Voyager had a lot of cost pressures, we never really got forced into doing anything that we didn't think was the best we could do. We worried a lot about what could go wrong, but when we got done with each problem, we felt we had really done as well as the state of knowledge at that time would allow. I was very contented with the status of the problems. I thought that was a very interesting thought to have right before we launched.

How long were you waiting?

About 20 minutes. By the time John picked me up I had gone through each problem, reevaluated the conclusion, and decided, "Okay, that's all we could do about that." So I felt very good about launching it.

I don't think anybody will ever feel that comfortable again, because these newer projects all run with a lot more limits than that. Over and over again people here say, "We can't afford to do anything the way we did Voyager." But the other side of that is, if you're going to have a program that will succeed, you can't afford *not* to do it the way we did Voyager.

We had a topic that turned out to be much more significant than we initially thought: fault protection. The people in the technical divisions, with each group making a subsystem of the spacecraft, wanted the redundancy to be controlled by their own subsystems.

I was asked whether or not we should have central control of the redundancy, and I was told that everybody wanted to study it. I said I don't want to study it, we will just have it, and we have fault protection as a result of that.

I had no concept of what fault protection was going to be, but when we talked about it, it just seemed like the right thing to do, and all the other arguments were specious. If it had been put to a vote, or done in some democratic process, we would have gotten the wrong answer; we would never have fault protection as we have it today. But at that time, and in that environment, we could do things like that.

We were very fortunate that everybody who worked on Voyager had an extremely good background for what they were doing. Some of us hadn't had the same job before, but we'd all had a job that was close. And all of us had worked in areas that covered everything we were dealing with on Voyager.

We were also a very compatible group and by the nature of the way things fell out, people's different personalities were spaced well. Schurmeier was project manager. He is a reasonably pessimistic and nonbelieving person. Heacock was the spacecraft manager. He's very optimistic. I was Heacock's deputy. I'm reasonably pessimistic. Ron Draper, who was the system engineer, was another optimist. So at least we never had two optimists or two pessimists right against each other.

That's more important than you think, because it kept us from being overly conservative, and it also kept us from being absurdly optimistic and overly adventuresome. We'd listen to each other. Heacock would say something one night and come in the next morning and I'd say, "No, I thought about that; I don't think that was a good idea." And he'd think about it and say, "Yeah, it doesn't sound very good to me this morning, either." We did a lot of things like that. But we all got along well, and it worked out quite well as a result.

We did a lot of decision making based on funding, and it doesn't appear that any of it was really erroneous.

What was the hardest decision you were faced with on Voyager?

Well, there isn't a particular one I can pick because I tend not to make decisions. I will keep something working until we get an answer that I feel good about rather than just pick something. I don't ever just pick something; I just keep on working on until we get enough understanding that it's relatively obvious what you ought to do. Occasionally we sort of run out of time, but by that time they were not my decisions to make anymore; they became project manager's decisions.

There are a lot of decisions that you make based on trusting people. There was a tremendous amount of that in Voyager. Where you don't have a whole lot of paper and a lot of study, you just trust people. Some things you do almost in faith. The best example I've got is the problem we had on the Voyager 1 spacecraft, the second one launched. We'd had problems with the computer in the attitude control. It had been intermittent, and we had trouble finding it. We finally found the part that periodically failed while it was on the ground: sometimes it was okay and then it would be bad for awhile. We sent it back to the lab for failure analysis, and they determined the nature of the problem.

In the last couple days before launch we had people going through the screening data, looking to see if any other parts had similar values in their screening results, because this one had been determined to be an outlier from the regular population, but within spec. So we looked for any other outliers. Then we tried to trace back through serial numbers and find out where they were, to make sure they weren't used. This process was all going on the night before the second launch, and early in the morning I went to the control center without anybody ever telling me what the answer was.

So I just sat there, and nobody ever said anything through the network. I had to conclude that it must be all right or somebody'd tell me. Nobody said anything. After the launch I walked out of the control center and one of the fellows who'd been working on the problem all night was standing right at the door. He said, "I wanted to tell you, we found another one of those parts but it was in the spacecraft we already launched." He said, "I didn't think I could explain that on the network; it would sound like a zoo if I got on the network with that story, so I just kept quiet."

There are loads of cases where you just wind up trusting. It's not just knowing *who* you can trust, but knowing who you can trust for *what*. It's not that people want to be dishonest; it's usually that they don't know, or their enthusiasm's got them carried away, or something like that. You have to really understand how the people feel about things to decide how you're going to trust them.

But there are many other cases where you wind up just going with what somebody wants to do, because you think he understands well enough that it seems okay. That gets more and more difficult as you get into a bigger system and get contractors that are more far-flung, but you wind up still doing some of the same things.

One of the things that was extremely important in Voyager was going to monthly contractor's management meetings on all the contracts that seemed to have any problems. People often asked why I always went. One of the reasons was that if the contractor's got a problem, it would get aired at the meeting. Often they're things about which you can do something that solves his problem.

In one case a contractor had this plan that was going to take 18 weeks to accomplish something. It didn't seem like it ought to take that long. So at the end of the meeting, some of the contractor folks and the JPL contract manager, several of us, sat around. "God, why does it take so long to do that?" They went through why it took so long. And I said, "You know, it seems you ought to be able to do something different."

George said, "Gee, I could take that thing and lay it out tonight and take it down in the street and get some artwork done, have it in the shop tomorrow morning, and have it done in about two weeks." I said, "Well, why can't we do that?" George said, "Well, this would be in violation of this and this and that requirement." And I said, "What's it take to override that?" George said, "You just have to sign a waiver."

The problem was that the thing would have to go through the contractor's drawing cycle and come back to JPL and be approved, then go back to the contractor, and so on. I said, "Gee, you're going to work with these guys while they do it tonight. What's different from you getting the thing back at JPL and approving it?" He says, "Nothing". And he said, "I can take it down to the street tomorrow and get the thing made." I said, "What's the risk?" And these guys sat there a minute and they thought about it and said, "No risk. It's not that hard a thing to do, and we got all the people that would ever have to look at it sitting right here anyway."

So I signed the waiver, and we made it; and we saved close to a million bucks that way. But if I hadn't gone there, nobody would ever have thought of doing that. And we would have sat around and kept thinking about why did it take so long and can't we do it better? Here again, I was trusting somebody. But I was going to trust him anyway. He was going to be the one that was ultimately going to sign it off anyhow, and he was just sitting right there and could look at it while it was being done.

There were a lot of cases like that. Sometimes you wind up seeing a problem that nobody else has really recognized. One time on the Galileo we had one like that, on the spin bearing. We wound up sending two guys into another room for the afternoon, telling them to come up with a design, and they solved a major problem.

One of the things I did on Voyager our laboratory procedures wouldn't let me do. I wanted the spacecraft system engineer to approve all the hardware specifications that went into procurements. But there was a feeling that systems engineers shouldn't do that because they're not in charge of the hardware producing divisions, and there were a whole bunch of other political and egotistical type arguments. So my boss told me I couldn't do that.

I said, "Fine," but I was the one who had to finally sign. That was acceptable to everybody, because I was in the project office. I had to sign the procurement package for all these procurements. So I just didn't sign the procurement package until the system engineer had reviewed the specifications. He didn't have his signature on it, but he had to tell me it was all right, and they never looked at one that didn't have some major problems.

It would've all been cost growth if we'd have negotiated contracts, then changed. And most of the problems were in the power interface—the grounding wasn't right. The subsystem engineers just didn't think about that. They looked at how their thing was going to work independently and didn't worry that much about the interface. It was the systems engineers who worried about the interface.

We did something that I think was a significant accomplishment: we electrostatic-proofed the spacecraft between December of '76 and July and August '77. Voyager goes through the electron radiation belts around Jupiter and builds up a surface charge, and there was a concern that the spacecraft wouldn't be an equipotential surface. There could be voltage differences on the surface, and that would lead to arc discharges—lightning on the spacecraft—which could disrupt the electronics in the computers.

At first we thought that the spacecraft design would inherently protect us from that. We did some tests that showed that the spacecraft design was *not* going to inherently protect us, so we did a spacecraft lightning proofing in that six-month period. We spent nearly $1 million.

I had some consultants, and I had known some of these fellows for quite awhile. They worked with us on lightning proofing the spacecraft. We modified the grounding, we modified the shielding, and we put a lot of filtering in some parts of the spacecraft. When we finished, one of these consultants said, "This is the most fun I've ever had."

He worked for one of the major aerospace corporations. I asked him, "Al, why do you think that's true?" And he said, "Because when you guys figure out what you're going to do, you do it. We go through something and study the problem and at that point we write a proposal. And then we have to negotiate. People have to figure out whether it's a fee-bearing change or not, and a whole bunch of other things." He says, "You guys basically have this bag of money that you're trying to operate within, and when you figure out how you have to fix something, you just fix it at that point, and you don't go through all this other crap."

The fact that we're not a profit-making organization allows us to focus singlemindedly on making something that works. There's nobody who's going to get a bonus that's determined by the profit of this month. Everybody is thinking about one thing: is it going to work? And in this case, "work" wasn't anything in the short term. It was in the long term. But everybody was, I think, unanimously focused on that.

One day I was asked by somebody exploring ethics if we had any concern about quality control people falsifying records of tests, because this has come up in industry. And I said, "I have no concern about that, because the quality people here have a big fear that something's going to fail in flight and they're going to see it in the newspaper. And they will all know who was involved in that item." There's an incredible peer pressure to do it right.

Where you might have some questionable ethics is people trying to sneak by on the budget that they have; there's cost pressures. You'll get people not wanting to do certain tests and so on because they don't have the money. But as far as falsifying records when they're concerned about something being faulty, or not fixing problems, or having the wrong results, it's not likely.

The biggest lesson is to have an overall total process. And our process was one of management by requirements and reviews. Your management is only as effective as your requirements are well-defined, and your reviews are able to determine whether the requirements are being met. That process was very, very important.

Of course, you could have a good process, but if you don't have high-quality people it won't work. If you have high-quality people and a poor process, that won't work either. That's been demonstrated in the past, because we had the same people building Rangers as we had building Mariners, yet we had five of the Rangers go down the tube. There wasn't any difference in the people. It was just the lack of a process. I think we've really demonstrated that.

The other thing is really to have a set of resources that fit the total job and the amount of uncertainty. I don't know that NASA and the space program as a whole can justify the finances to do things that they try to do well. They may have to take an absurd amount of risk in order to get to do anything, just because nobody is willing to cough up money. Now on the other hand, it's really cheaper to go first class, because you get it done, and you don't have to redo it.

My observation, in looking at many projects, is that the better total plan and the more thorough it was, the less cost growth they had. Voyager had lower cost growth than anything in the country that was done at that time. The other technical developments had far more cost growth. And Viking Orbiter and Voyager had lower cost growth compared to other JPL things. Nothing done since has been anywhere nearly as good in cost performance. But that was partly because you had a total and complete plan to start out with.

It reminds me of when I was on the team to define Apollo. I was the chairman of the subcommittee for the unmanned part of the Apollo definition study in 1961. One Friday afternoon NASA administrator Jim Webb came in. He said in his classical fast Southern speech, "Fellows, I don't want you to be extravagant, but if there's something you need, I want you to put it in this program. I want a budget for it." And he said, "We're going over to Congress. Congress wants to put a man on the moon and they've got to pay for it." He said, "We're doing this for national prestige, and there's no national prestige in going out and running out of money."

So it turns out that his budget was awfully tight for Apollo on an item-by-item basis; however, it was adequate to cover everything. The program that we executed was nothing like what we budgeted. It wasn't accuracy, it was adequacy, and there's quite a difference. My part of that budget was for not 10 Rangers like we flew, but about 18. For Surveyor, I had 17 Surveyors, and the money I had in for 17 Surveyors was almost enough money to cover the seven we actually flew—not quite, but almost.

A whole program was proposed in there that was very expensive, close to a billion dollars in those days' kind of money: Prospector. It was to be a lunar roving robotic vehicle, but it never happened. The budget was made on assumption of direct ascent and direct return. We never built that big a vehicle. But we made a budget for doing something big, and NASA did a lot of big things.

The best illustration of that was, as soon as that report got accepted by Congress and decisions were made to go ahead to the moon, guys went out and started spending the money. They did the things you have to do first, and it was brilliant. As I think back, I can't imagine how our system could work so well. The first thing they did was to buy real estate. They bought all the ground for the Mississippi test station. They bought the ground on the west side of the river at Cape Canaveral for the Kennedy Space Center.

What could they do? They didn't know the details of the design, but they didn't need to. All they needed to know was they were going to do something great big. So they went out and did what you do if you're going to do something great big. They managed to do that all the way through. The thing that's so impressive about Apollo is they did things as a function of the knowledge they had at that time. They got things out of the way, and those things were taken care of. I really think it was a knowledge-driven program. I'm always amazed at how well Apollo worked out, but I expected it to work.

Voyager wasn't a great big thing like that, but we had a plan and we could work down the plan; and because it was not a wasteful plan but a plan that was robust, we could stand problems and survive. It's never just one problem that kills you; it's always a combination of many, and you have to be able to deal with that many problems. Like I said, I have that bookshelf of $20 million worth of problems.

I always tried to keep it to the point where I had two, at most three, big problems going at once. I really worked hard to get one out of the way if I saw another one coming. If you get more than about three big problems going at once, people decide you really haven't got control of the situation; then you're replaced. But that was a personal goal. When I got more than three problems, the meetings and everything got so totally consuming that it didn't work very well.

Sometimes there'd be something that was run on a low level of effort, to just sort of hold and stay steady for awhile and then get a big bump to close it out. But when you have a big problem, you have a meeting on it every day. On that electrostatic discharge—lightning proofing the spacecraft—that was a meeting-a-day operation for several months.

On the lightning proofing we did lots of things. The spacecraft system engineer and I were cochairmen. He had control of the design, and I had control of the resources. So anything we wanted to do to the system, we could do. That's what made it so powerful and allowed this consultant to think, "Gee, you guys figure out what you're going to do, then you just do it." We had all the controls of our universe right between the two of us. He had complete control of the change request system and I had control of the money.

A very fortunate situation.

No, it wasn't fortunate, it was intended to be that way. When we went to work on that problem, we had the test that was a disaster. It showed we had a problem. John Casani was project manager at that time, and as we were walking down the hill I said, "I think I can fix that." And he said, "Go do it." He never asked me how much it was going to cost; he just said, "Go do it."

One of the funny things. We did have a problem with electrostatic discharge on one of the Voyagers at Jupiter. After we finished spacecraft assembly sequence, I had told John, "I think we got it." But after we had this one noticeable problem in flight, I walked up to him and said, "Well, I guess we *didn't* get it," and he said, "Well, you never said you were *sure* you fixed it." As it turned out, it didn't cost much; just caused a picture to be bad.

The contrast between Voyager and many other projects, like the space station overruns, the C5 cargo plane overrun, is impressive.

Part of the problem is they squeeze down everything in their estimates. The shuttle had no realistic planning. But the Voyager procedure was different: first, spending a couple years with the advanced system technology task, getting a real thorough understanding of what that machine is, and having a lot of heritage designs that you can then take out and put something else in. When you do that you're automatically building all this base of understanding that you benefit from when you attack that new device. When you stick it in there, you *know* what you're putting in.

How was it that the original Voyager budget estimate was so accurate and workable?

It had an awful lot of thought behind it. It didn't have a lot of pad. There was a specific reserve; the reserve in the original budget was $11 million. I think we spent $5 million of that when we negotiated the contracts for the elements, and then the rest of it would be for problem solving. That $150 million total then got increased when we had the radiation problem.

So how was Voyager planning different from later projects?

What was different was there was enough hardware, enough stuff, and also it was not taking the people out of the project right away. We built three sets of stuff, and that provided a long period of time for the people to work with several sets of hardware. You get to really understand what the problems with the hardware are, and you get to fix 'em.

That was one of the things that was maddening with Galileo: in many instances they only had one set of hardware. When Voyager had a problem one of the questions was, "What did the other unit do? How did it work? Did it have this kind of performances?" So you always had a way to compare; find what was really characteristic of the breed versus what was an individual problem.

There's a real advantage to having several sets of stuff. We had about 2.75 sets of stuff for Voyager; maybe 2.9. We didn't have a full third complement. We had a proof test model that had the redundant units in some things missing, but it was a real advantage. We had that proof test spacecraft reassembled in Florida as a test bed while the two spacecraft that were flying were out on the launch pad.

I think Voyager was better understood than any other spacecraft we ever built around here. It was better understood than Galileo was when we launched it, simply because building three pieces of hardware gives you a lot more understanding than one piece. In closing out problem reports, Tom Gavin always made this point. In Galileo, often he would've liked, before he signed off a problem failure report, to have known what another unit looked like. In Voyager he could always ask, "How did the other two serial numbers look? Did they have the same thing? Is this a characteristic of them all?"

Having the extra hardware was important. And nearly all of the extra hardware from Voyager got used. Much of it's been in the Command Development Laboratory, where they test the command software they generate before they send it up to the spacecraft, to verify how the spacecraft will respond. That takes up several pieces of spare hardware that were left over. The extra propulsion went in Magellan. The extra spacecraft frame was in Magellan. The extra antenna was in Magellan. So much of the leftovers from Voyager actually wound up in Magellan.

Is there another way to summarize how Voyager compared with other projects?

One of the fellows that was the environmental requirements engineer on Voyager often said, "Voyager was Camelot." It was the ultimate project to work on. Those of us who worked on it all felt that it was a wonderful experience. They think the things that we produced were about as close to the optimum as you could get. It utilized available technology in a secure manner. We didn't do things that were violently adventurous, yet we were really at the edge of the state of the art, and we weren't that comfortable with it. We were a little *over* the edge in some cases, but we had ways of dealing with it.

That's why we had this bookshelf of problems that we worked on. They obviously were manageable problems. Other projects did not always get in that situation. We also had a number of people in Voyager who were sufficiently paranoid to worry all the time about where do you go when this goes wrong? John Casani always made the statement, "You've got to be appropriately paranoid in this business. If you're not appropriately paranoid you're doomed."

Voyager was in many ways a product of its environment. JPL had worked on the planetary program for a number of years. NASA had a certain amount of confidence in the Laboratory; confidence in the people working on Voyager.

One of the things I think about, as I look at many of these projects today, is that we often could make decisions that had tremendous impact, without a whole lot of overview. We had a review board that looked at everything we did, all the way through TOPS, but they looked and said, "Okay, does this make sense or is there something we think that ought to be done differently?" They didn't necessarily repeat everything. Today these projects have so many reviews that it is very difficult for people to really think in a straightforward way.

Another one that came up was that the review board, when they gave us launch approval, never looked at what we had done for lightning proofing. I don't know that we could have ever convinced anybody that we had it done; I don't know that the business was that mature. As one fellow said, many years later, "If they had examined that very carefully, we'd be sitting there still."

So we got the help we needed, we seemed to get a very detailed review of many things, but we were also very fortunate that some things we just did by how we felt about them—we didn't get reviewed to death about them. I don't see that as a characteristic of the environment today. So Voyager was in many respects a product of this environment.

One of the things that is really important is an attitude that you could easily lose here. There was an attitude that the things we were doing were more important than personal agendas. I haven't ever seen that in any other place I've dealt with, but that was characteristic of JPL up until now; certainly up through Galileo, and it's probably true of Cassini, too. On smaller projects, I don't think that's true, but in big projects that usually gets to be the case.

Another of the big things that was different was we kept the people through the launch, where in the projects now, as soon as people made their thing and delivered it, they're gone, they're off the project.

Why is that?

To cut down the cost. With Voyager we didn't know that you could do it any differently, so in our planning, although we were very pressed, we carried a fair complement of people all the way through launch. We had like 200 people at Cape Canaveral.

I think those are the biggest differences from today. We had the multiple sets of hardware, and we kept the people longer than we would be permitted to keep them now. Those people were always working; they were always busy.

Sometimes on other projects?

No, on our project. Gaining understanding, working the problems out, really getting an in-depth understanding of the machine.

If we were to tell that to someone who's planning a new project now, to Pluto or wherever, they'd say, "Well, that was fine back then, but nowadays we can't afford that."

That's what they'll say, yeah, but that's one of the reasons the spacecraft is still working.

What do you think of the current "smaller faster cheaper" policy?

I think the "smaller" is an incredible advantage. It's an incredible advantage in keeping the cost down. It's an incredible advantage in being able to test the stuff. The thing that's bad is when you combine that with too much of the "cheaper better faster." I think "faster" is good. "Cheaper" certainly comes with being smaller.

For example, Galileo is not quite twice as big as Voyager, but the stands for Galileo in the space simulator, the platform the people stand on to work on around the spacecraft, cost about $300,000. For Voyager it cost about $80,000. If Voyager were a 600- or 700-lb spacecraft, those stands would be down in the $10,000 or $20,000. So there is a lot of cost that comes with being big. In fact, it probably isn't a whole lot different from boat building. With boats the cost usually goes with the cube of the biggest linear dimension.

The technology permits you to do an awful lot of things a lot smaller. You've got to remember that in Voyager the most complicated parts had a little over 1000 transistors per chip, and most of the electronic parts had under 100 transistors per chip. You get into Galileo, and you're talking on the order of 24,000 transistors per chip in the most complicated ones. And when you get into something like Cassini you got millions of transistors per chip. If you could just cut down your appetite for the number of transistors you want, you could make a very significant spacecraft with a very few parts. These newer spacecraft are much more capable, and they're much more like information systems. They're not as simplistic in the way they operate.

We're comparing this with Voyager?

Yes. Even though Voyager was a very sophisticated machine for its time, the electronics in Voyager are much simpler than an ordinary personal computer now. The way it worked was very straightforward, and it can do simple things extremely effectively. It can't do things as complex as Galileo, and not nearly as complex as Cassini, but the things it can do, it does very, very well.

And reliably?

Yes. It has a lot of redundancy and a lot of resilience.

The growth of transistors on each spacecraft has been exponential. I've told the guys on these newer projects that you're making the machine more complicated than the function you are performing, by making it more sophisticated and how it does it. And that's because of our information systems age. Now they do things in the computer world, whereas Voyager was just laid out to do what it was doing.

Do you mean that spacecraft are becoming unnecessarily complex?

Yes. The push that tends to justify that is to make it easier to fly. However my observation has been that even though these things are much more complex internally, they really did not get easier to fly; it didn't take any less people to fly 'em.

So that was an illusory quest?

To some extent, yes.

And did it undermine reliability?

No, because of the way we've done it. I think the electronics in Galileo and Magellan worked well. They're not flawless; a number of parts have failed over the years, but the systems design is such that they can tolerate it. The redundancy and the fault protection are such that you can tolerate occasional failures. It's just that if you ever had a high rate of failures you'd be in real trouble because you couldn't operate fast enough to correct them.

One time I did a little exercise for the engineering subsystems. We looked at board by board, and concluded that if we used Galileo technology and did a couple other things and managed to shrink the instruments proportionately, we could reduce Voyager's weight from 2000 lb down to about 800 lb.

Would there have been any loss in reliability?

No, there wouldn't have been any loss in reliability. It would have been using more advanced technology, with more transistors per chip and putting more things on a board.

So it would have been an outright plus, positive?

Yes, just because of time. Technology changed so you could have done that. Galileo is much bigger, but Galileo does a whole lot of things in a much more complicated way. It has a data bus and all kinds of things that are part of modern technology.

I think it was really significant that, when we launched Voyager, with all the problems we had, I thought we had beaten all of them down, and they were all under control. That was after spending five years thinking about what could go wrong and then trying to fix it.

And after launch your conclusion was still the same during the flights?

Yes. Yes. Absolutely.

HENRY G. (HANK) COX
Tracking Data Systems Manager
Born October 16, 1926   Jacksonville, Florida

Henry Cox attended the University of Illinois after two years in the U.S. Navy and then worked as a field engineer for RCA, Philco-Ford, and Bendix Corporation before coming to JPL in 1967.

He has worked mainly in the Deep Space Network, on the Lunar Orbiter, Mariner Mars, Pioneer, Viking, Helios, and Voyager missions. He has been assistant project engineer and project engineer, CTA-21 manager, group supervisor in the DSN Operations Section, TDA support manager for Radar and Radio Astronomy, and TDS manager for the Pioneer and Voyager missions.

In 1991 he was awarded the NASA Exceptional Service Medal.

He owns, maintains, and races thoroughbred racehorses at Del Mar and other major California racetracks and is a member of the Los Angeles County Grand Jury.

You can't make it perfect;
you can improve it—then improve it again.
You can't eliminate all risks; you can eliminate some, reduce others,
and accept what's left. You don't know whether it works until you
test it—in the operational environment. Achievement is the
realization that you have done everything possible; success is
Fortune's sometime reward for achievement.
—Henry G. (Hank) Cox

HENRY (HANK) COX

I am TDS manager for Voyager. TDS stands for Tracking and Data Systems. My function is to ensure that Voyager gets the ground support from the Deep Space Network that it needs in order to accomplish the mission. Deep Space Network is a series of complexes around the world with big antennas to get the signal. It's a multimission organization that supports Voyager and Pioneer, and now also supports Galileo and Ulysses and Magellan.

My job is to make sure that Voyager gets its share of these facilities, and the facilities are technically capable of meeting their requirements; and the people are proficient in operating the system, so it gives Voyager good data. So I'm Voyager's man in the DSN, and I'm also the DSN's man in Voyager. I'm sort of a liaison between the two.

The way it starts is that a project like Voyager makes a list of support requirements on the Deep Space Network. They put it in a document and give it to us and say, "This is what we want from you."

We respond to that by telling them which of the things they've asked for we can give them, and if we can give it to them on the schedule they need—on time. If it meets all the technical requirements then, generally, yes we can, and we do.

For instance, for Voyager we had to increase the size of our largest antennas from 64-m diameter to 70-m—three of those: one in Australia, one in Spain, and one in California. We also had to go out and get some other antennas that did not belong to the Deep Space Network. One is a facility called the Very Large Array (VLA), down in Socorro, New Mexico. It's a Y-shaped array of 27 25-m antennas. It's used for radio astronomy, but in order to increase the data rate of Voyager, we contracted with them to track Voyager. We took the signals from the antennas at Goldstone and from the VLA and combined them so we could support a higher data rate than we could with just the DSN antennas alone.

We also got the radio telescope in Parkes, Australia, and combined that signal with the one at the DSN station in Canberra to do the same things—get more signal over Australia. And we contracted with ISAS, the Japanese agency that's analogous to NASA, to use their 64-m antenna at Usuda just on encounter day for radio science.

Along the way we had to improve our radio science system. The ground equipment that receives the signal from the spacecraft radio during an encounter, after the signal passes through the planet's rings and atmosphere, has to receive it very particularly so that the scientists can find out what happened to the signal on the way through.

Those are some of the things we had to do, over and above the facility just sitting here stationary from year to year. We did all that for the Neptune encounter. Of course, support of Voyager goes back before launch and Jupiter, Saturn, Uranus, and Neptune. I wasn't involved with Voyager at that time; I came on board just before we got to Uranus and helped the then TDS manager, Marvin Traxler, and after the Uranus encounter I took over and he went on to something else.

So that's basically what my function has been. Before me, someone else was doing the same job: making sure that Voyager had the facilities they needed to get the job done.

What led to your coming to work at JPL?

There's two factors. One is I saw a news conference that Dr. Pickering had from von Kármán auditorium at JPL when Mariner 4 encountered Mars. I saw these guys on television, and I thought, "Boy, wouldn't it be wonderful to work in that environment with all those scientists, and so forth!" So that made me want to go to work for JPL, to work in space.

How it actually happened was I was in India working for Bendix, which was a contractor to JPL, which I didn't know at that time. My contract was over over there and I got a wire that said, "How would you like to go to work in Pasadena on the Deep Space Network at JPL?"

I said, "Hey, man; that would be pretty good," so I accepted and came here working for Bendix and later I transferred over to JPL. I think many people here started out with Bendix.

How did you happen to join the Voyager project?

I had been TDS manager for IRAS (Infrared Astronomical Satellite), another spacecraft that had just about finished its mission. When we got to Uranus, Marv Traxler needed some help, so my boss said, "Why don't you help Marv during the encounter?" I did, and that gave me some insight into Voyager peculiarities, so I took over because Marv wanted to rest for awhile. It is pretty stressful getting ready for an encounter and going through it, and after encounter is finished people just want to kick back and say, "Boy, that sure was great; now let's rest a little bit."

There's a lot of pressure on everybody at JPL, whether you're on the project or not, to make it work, make it successful. A lot of money and time and effort has gone into getting the spacecraft to where it's going, so if anyone lets up and you don't get all the data back that you want, you feel like you blew a unique opportunity.

I used to joke with project managers and say, "Well, don't worry about it. If we miss getting the data at Neptune we'll get it next time." But you know there is no next time, and everyone feels that. It's a one shot thing. We're not going back to Neptune in our lifetime, so what we can learn this time will be very valuable to us, and we won't get a chance to improve our knowledge until we go out there again.

What are the stresses in your work?

In my job it's a little bit different from the stresses inside the project, but they're related in a way. I'm looking towards achieving a state of readiness. That involves procuring new hardware and developing new software and developing procedures to help people get the job done. It's the sort of thing that doesn't make Voyager just one more project.

Voyager is just one of your many projects?

Right. And I'm a programmatic manager, in that I'm not a line manager. People don't report to me, and I don't have the opportunity to say, "Okay, do this and I'll give you a raise, and if you don't do it I'm not going to like ya." I have to persuade people to do things. It's a challenging job to try to get people to be as enthusiastic for this project as you are, so there's some cheerleading involved in getting ready for a project.

One of the things we like to do is to take some of the project people—like Ellis Miner, the deputy project scientist—overseas to talk to the stations and tell them what's going on. Before an encounter we in DSN go over there and tell them what we expect to find at the planet: the new discoveries, the new science that we're going to get out of the encounter.

Then, after it's over, we go back and thank them for their support and also update them. We tell them, "Well, we didn't find out exactly what we thought we would, but we found out *some* of the things and, as always, we get surprises; we didn't expect to see some things we did." So that's another way of motivating people to pay attention to *this* particular project.

Then again, there's always money constraints. There's only so much money in NASA, and you want to get the funding to do the things that *your* project needs, so there's competition between projects for that. Within the Deep Space Network there are conflicts, but what happens is you start out and say, "We've got plenty of time. Here's the encounter coming. We're going to do all these good things." And you write documents and say, "This is the way it's going to work."

As time goes by, some things are done rather easily and well, and there is an end to them. Other things are late. Software is always a problem. Software just doesn't seem to get done on time, and it doesn't work as well as you thought it would. You try to get procedures to work around it if you can, and if you can't you go back and get the software fixed. But then there's a delay because you have to test it and find out how people react to the new software.

Then it's a race to get yourself into the posture you need to be in by the time the spacecraft gets there, because you can't change that. The spacecraft's going to get there according to the laws of physics, not your schedule. There's a great deal of pressure in trying to meet your goals and to live up to your part of the contract, which is to have the ground operations ready to meet the spacecraft when it gets to the planet.

What particular aspects come to mind?

Well, there were two really difficult things we had to do. One was arraying the telemetry data and the other was meeting the radio science specifications.

On the telemetry standpoint we've been tracking Voyager for 12 years, and we know how to point the antennas, tune the receivers, process the data, and deliver it. There's not a problem. But to get the higher data rates they needed for the encounter, we had to array antennas, which means we had to point two antennas correctly and tune two receivers and then combine the data. When we started out, we didn't do that very often; consequently we didn't do it very well.

You would expect that if you had two antennas of the same size, and combined the signal, you'd get twice the signal, if everything worked properly—but it didn't usually work properly. The equipment was new and rather balky, difficult to operate, difficult to maintain, and the procedures were not well understood, so as we started arraying our antennas, we didn't have very good success. Therefore the data we delivered to the project was not as good as it should have been.

Usually it was good enough; our coding techniques overcame the lack of signal level so that they still got all the data, but there was very little margin, which means that if something went wrong—like the antenna went off point, or if the sky was overcast, system noise temperature raised and we got more noise for the same signal—then we would be out of business.

It was a constant worry to me and to the Voyager project, too, whether we, the DSN, would be ready on time; whether the increased proficiency—because of testing and training and just actually doing it—would enable us to deliver the data consistently in encounter. Voyager had a requirement that we deliver 90 percent of the data during cruise periods between planets, but at the actual encounter they wanted 99 percent of the data, which means we had to be very efficient and proficient; we couldn't lose much.

It was difficult getting to that point. We finally made it by attention to detail, a lot of training, a lot of testing, and just focusing our attention. We do a lot of different things for a lot of different projects, but right now it's important that we give to Voyager all that they require to do their mission. By focusing on that, we get more resources than we would normally get from supporting people, and consequently we were able to upgrade the procedures, to fix the software, to do a better job of maintaining the equipment, and to deliver the data. That was one of the biggest challenges.

The other difficult challenge, the radio science, is again something we don't do very often. Voyager requested seven tests of radio science before the Neptune encounter, but we only actually got data in heaps on the day the spacecraft went behind the planet. It lasted a couple of hours and that was the end of it. So here again, the things we don't do very often we don't do well. We need repetition, we need confidence in the equipment, we need expertise in certain areas, and we need to clean up potentially interfering radiomagnetic sources in order to get the radio science data.

As we ran these tests before we got to the planet, we found here and there we had some 60-cycle noise in our system, so we had to find out where it was coming from and either shield it or eliminate it. We also found some spurious radiation that was in the equipment; the way the equipment was designed, interconnected, whatever, caused signals whose frequencies in some instances can multiply and divide and get right in the pass band of where you want to see the signal from the spacecraft. When that happens, you degrade the quality of the experiment, so we had to find those and get rid of them.

Also we had a particular problem in Australia. The geometry of the situation was that the spacecraft went behind the planet when it was in view of Australia. They planned the mission that way because the low declination of the spacecraft meant that you had a longer view period in Australia and the antenna went higher in the sky so you get better data.

They have a long time before encounter to do a small burn and delay the arrival by hours, if you do it three years before you get there. There's an optimal time: if you do it too early, it's not good, or if you do it too late. But they can time it, and they timed it to get the best coverage on the ground, and that was over Australia because we had Parkes and the Usuda station in Japan.

Anyway, we had some noise problems in Australia. Perhaps they didn't have a very good ground, or they had some sources of spurious radiation, or perhaps the equipment, which is theoretically identical to the stuff we have at Goldstone or in Spain, was actually not quite as good, so our radio science equipment there kept failing. That gave us a lot of concern because we were afraid that during that critical time period, when we either got the data or we didn't, the equipment might be broken.

Eventually we ended up by sending a complete new system down there and had two radio science systems sitting side by side; both of them running and recording data so that if one was bad the other one would get it.

So curing those radio science problems and being able to successfully combine the telemetry signals were my biggest concerns and the DSN's biggest challenges for the encounter.

Are there further details you'd like to add? For example, at Uranus there was another spacecraft the Europeans were running, and they were having difficulty, so right during the Uranus encounter they came to us and begged us to give them some assistance.

Yes, I remember that occasion. It was very early in the morning, and I happened to be here at the time. We called up Dick Laeser at home and said they would like to use Fourteen, our 70-m station at Goldstone; can you give it up? and he said no. Well, what about giving up Twelve? Twelve was the 34-m station we were arraying with the 70-m station. Finally he agreed that, after a certain critical time period, we could give it up to ICE (International Comet Explorer), the one that went for the Giacobini–Zinner comet.

That was an instance where we didn't have enough antennas to go around, and that happens all the time. That is not something I normally worry about because we do not decide in the Deep Space Network which spacecraft to track. We say to these projects, "We'll be glad to track these spacecraft you've got out there; *you* decide which one is more important and that's the one we'll track."

It's not our function to discriminate against one project or the other. In this case, when the other project came to DSN and asked, "Can you track us," we said, "Wait a minute; we have to go talk to the other project that we have a responsibility to, and they're the only ones that can release the resources that are committed to them at this time."

Competition for resources and problems; they all seem to run together because they happen so many times. Each project says, "What we're doing is special, and *we* have to have the antenna," and the other project says "No, no. What *we're* doing is special and *we* have to have the antenna." The DSN can't arbitrate in those cases. We prefer to step back and say, "Look, *you* figure it out and tell us which one, and we'll be glad to track it."

Within your range of responsibilities, what are the toughest choices you've had to make? About anything within your job, to give the reader some idea that these marvelous pictures didn't come down automatically but required a lot of hard thought.

One of the hardest times I had was when there were two ways of doing something. It's the responsibility of the DSN to take the physical constants the project gives us, and adjust our receivers to get the signal. The Voyager people tell us how big the planet is, and what the atmosphere is made of, and where the spacecraft will be at a certain time. Based on that, we have to calculate the proper tuning for our receivers to pick up the signal.

They tell us, "You have a pass band that's so big, and because of the gravity of the planet the spacecraft will be accelerating and the frequencies will be changing because of the Doppler shift. We want you to predict what the frequency will be, and tune your receivers to stay within that band pass so the signal will be received and you can put it on tape."

Well, in order to do that we have to take the physical constants and model them into what their effect on the signal will be. It's called "modeling the atmosphere." We have always done that, but there was some concern on the part of the radio science team whether we could do it properly, and whether we could get the "predicts," the frequency predictions, out to the station in time to use them to tune the receiver properly for this last encounter.

The radio science team was concerned because this had to be done at the last minute—getting information on where the spacecraft was and how fast it would be going and all these things, so we would get the data and use that to plan the encounter. And we had to develop software to model the atmosphere. Then we had to use the software to get late information from them on what trajectory the spacecraft was on, to put out predicts to do the tuning.

How much time did you have to do all that?

We had 18 hours from the time that we tracked the spacecraft for the last time and gave that tracking data to the project to the time that we had to get those predicts out to the site. We gave the project the latest information on where we pointed to find the spacecraft and how far away it was from our ranging system, and then the project navigation team used that to update the trajectory. They fed the updated trajectory back to us, and we put it into our software, which told us what the frequency would be at various times.

The question came up, could we turn that around in time? The project had a technique for modeling the atmosphere, which was peculiar to the radio science team. It was not all-purpose, the way the DSN does things. The DSN system was more cumbersome than the specialized project system, and the project thought, "Maybe we'd better use the simple system, because we don't have enough time to use the more complicated system," but we in DSN felt that it was *our* business to send predicts to the station. In other words, the station should operate in accordance with data that *we* generate rather than what the project generates.

At that point you get into political turf: a struggle between what is historically done by project, and what is historically done by the DSN. So we discussed what was going to happen, and we agreed that the project, the radio science team, would generate their predicts, we would generate our predicts, and the project would compare the two together to see if there were any discrepancies. If there were discrepancies, and we found out that they were in the DSN predicts, then we would use the project predicts. But if we agreed that our predicts were right, we would use DSN predicts.

Well, it got to be a matter of politics. We wanted to use our predicts. We had developed the software to do it and we wanted to do it, but I came to an agreement with Len Tyler, who was the radio science principal investigator, that if we could get the predicts together by a certain time—within a minute of the clock, not a day or hours, but a given time—that he would agree that we would use our predicts, but if we couldn't get them on time, then I agreed that we would use the radio science team's predicts.

We got down right to the wire, and we missed it by 15 minutes. We made a run, and we gave them to the radio science team to check, and they said, "Something's wrong with these things; go back and take a look." We went and looked and sure enough, we had made an error. We tried to fix it, but the run didn't go right. And they said, "Hey, the time's up." I said, "Okay, let's go ahead and use your predicts."

It really hurt, because we worked hard to try to develop software and techniques and procedures so that we could provide predicts for this situation, but we didn't make our deadline, so I had to agree to go ahead and use the other ones. It was very painful to me, but I had made a deal so I had to live with it.

Most of the time we do a lot of long-range planning, and because we've been doing this for a quarter of a century we developed knowledge about where things could possibly go wrong. Consequently we have periodic reviews, readiness reviews—Are you ready for this? Are you ready for that?—and progress reviews, and people are experienced enough to ask all the embarrassing questions, so if you haven't thought of things it becomes apparent that some planning hasn't been done.

Peer review is the best thing there is. People who are experienced in this area listen to a guy tell his story, and the guy says, "I did this and this," and somebody says, "Yeah, but did you do *that*?" and he says, "No". "Well, you better take a look at that because we think that could be the trouble."

We've been doing this for so long that there are very few things that fall through the cracks, and we don't get up to the point where this can or cannot work and you gotta make a decision on it, like with the predict problem. We were on course to do it right and everything *looked* good, but still at the last minute it wasn't good.

We don't have too many of those things because, like sending the extra radio science system to Australia, we didn't really think we'd need it, because we'd had problems and the problems appeared to be fixed, but caution said to send another one out there. It cost a little more money, and it caused a lot of work for the guys at the station to install it and test it; but it gave us that bit of redundancy that made everybody breathe a little easier on encounter day, that if the system there did fail, we had a backup for it.

There was a turn of events that seemed catastrophic at the time, at the last big test, the near encounter test, before encounter at Neptune. It was over Australia, and it was the same time of day, and everything was supposed to be the same as encounter day, and we had one of the most disastrous passes in the history of Australia.

What happened was that there was a small gas leak, some cooling nitrogen, up near the antenna where the electronics is, and the leak produced a cloud. The fire detector thought that it was smoke, so it sounded the alarm and turned off the power to the electronics equipment up in the antenna. The people investigating it said, "No, there's no fire," so they reset the circuit breaker that was turned off. They reset one of them but they were not aware that another circuit breaker had been turned off also.

The other circuit breaker was to some air conditioning equipment up in the antenna. As a result of the air conditioning not being on, the electronics started to overheat, and when electronics gets too hot, it does some weird things. Because of that the subreflector controller started moving off and they couldn't control it, and the receiver controller went out also.

They fixed these things one at a time and, absolutely independent of all that, they also had a communications failure. Trying to recover from all of these, they didn't do a couple of things they should have done procedurally, and the radio science equipment failed that day also.

It looked like the situation was horrible. The station couldn't operate; they couldn't point the antenna; they couldn't tune the receivers; the radio science equipment didn't work. They were in terrible shape.

Was this all on the ground, or was some of it on the spacecraft?

All in the ground station. But we discovered some things. We didn't have a very good monitoring system that really told us what was going on up in the antenna. There are rooms up there with electronics, but we did not have a monitoring system which told us exactly what was happening, and therefore we made decisions based on lack of information.

As a result of this particular day, we decided to do some other things that would enable us to know what our status was and be able to take proper steps. One was to have a person in that room on encounter day, so if anything untoward happened, he would know it, and we wouldn't have to worry about some relay clicking in, or, if the alarm light is burned out, you're not going to know anything's wrong. Things like that.

This happened about three days before our big readiness review for the DSN. "DSN, are you ready to support Voyager for the encounter?" Leading up to that we had problem one; we fixed it. We had problem two; we fixed that. Everything had been going great until we had this disaster of a test, and it looked like we were in awful shape. That was my most embarrassing moment in the four years: getting up in front of the reviewers and saying, "It's not as bad as it looks. We had this problem, and we fixed it, and we fixed that problem, too." It seemed like I was just making excuses, and that we were really in terrible shape, but I knew we weren't. And we demonstrated that by having a great encounter.

I told you the project wanted 99 percent of the data; we actually provided 99.6 percent of the data during the whole 119 days of the Neptune encounter. Things looked bad, but they were caused by just a few specific things, and once we found out what they were and they were fixed, then things became good again.

What surprises did you have on the project?

There were a lot of surprises. None of the planetary systems was exactly what I expected. The rings of Saturn were the most spectacular things I ever saw in my whole life. The moons of Jupiter were extremely interesting.

The thing that I feel about Voyager is that you can work long hours and get no particular rewards for doing all this—working weekends and middle of the night and those sorts of things—but then when you see what happens, when the pictures come back, and you look at that and you say, "Isn't that marvelous; isn't that just wonderful," and you say, "I had a piece of that action. I contributed to it. I was integral. They could have done it with someone else besides me, but the things I did someone had to do or we wouldn't be getting this back." The feeling of pride and satisfaction is overwhelming.

A few days after encounter everyone is just walking on air. They paid for their elation by hard work and planning and doing all the things you have to do to get ready. The satisfaction you get out of it is amazing; it really is wonderful.

How do you compare Voyager with other projects you've been on?

I was on the Viking project, and I worked on the Pioneers and some of the previous Mariner missions back all the way to the Lunar Orbiter, in fact. Viking was exciting; you had a feeling this had never been done before and there might be life. And in my opinion finding life outside the Earth would be a great historical moment, if we had found it. Of course we didn't find it, but the anticipation and the thrill of that landing in Viking was really a high point.

But Voyager has had a sustained series of accomplishments. They got there first, and once you're first nobody else can do the same thing. Voyager has contributed so much to scientific knowledge and has done it in a very professional way, a sustained, competent way. I think it is probably the best mission that JPL has ever had, and I've worked on a lot of missions here. I think Voyager is the best because they have sustained excellence for 12 years.

What ingredients went into that outstanding performance?

There were a lot of individuals who contributed, some of them brilliantly, to Voyager. I'm not aware of all the contributions, but from my outlook Voyager was a culmination of how to run a project that JPL had been developing over the years with the moon probes and the Mariners and Viking. Viking had mission operations that were much more complicated than anything we'd ever done before, because we were trying to manage four spacecraft at the same time. They made a lot of mistakes in doing it.

Voyager got that expertise because we did it here at JPL and some of the people from Viking went over to Voyager. Therefore the mistakes had already been made and Voyager was able to take advantage of them: in techniques and operations and control mechanism counterbalances they put into the organization, in establishing mission guidelines and constraints, and having checks and meetings and reviews. They developed a great system during the early years of the project, based on previous JPL experience, that enabled them to do everything so professionally.

Are there lessons in the Voyager project for us and your successors?

I would say perseverance and acceptance of constraints. It is an attitude involved, of saying we want to do this and we want to do it right, but you can't have everything you need or think you need, and you have to live with the constraints the system puts on you, that money and politics and organizations put on you, and just continue to persevere and get the job done. If you accept the limitations that other people put on you, and proceed to do the best you can within those limitations, you'll be successful. Maybe that's the lesson.

One of the things that impresses me is the ability to exert such control over an object that is several billion miles away. How is this done? I understand that the signal coming back from Voyager is only 20 watts.

It's 20 watts when it starts out from the spacecraft. It reaches us at about 1×10^{-18} watts—not much.

So that is far less than what we have in an ordinary refrigerator bulb?

Yeah. A 20-watt bulb is not very big, but 20 watts is what you've got on the spacecraft, and it's billions of miles away. What actually gets here has about the energy of a snowflake hitting the ground.

It's a miniscule amount of energy, but if the signal is more powerful than the noise it has to overcome, then it's usable. By having a very large antenna, you get a big area for the signal to land in; by cooling the receivers down you reduce the noise. If you could cool the receivers down to zero Kelvin you wouldn't have any noise, so any signal would be strong enough to be detectable.

We can't get down that far, but we get down pretty cool, and then, by using coding techniques, you overcome the random noise, because noise doesn't have any pattern to it. It's just strictly random, but the intelligence you're trying to detect does have a pattern to it, so if you can combine bits of the signal so that noise can't hurt it—and that's what the cooling technique does—you can pull the intelligence out with practically no signal at all, and that's what they do. They use Reed-Solomon coding and convolutional coding; they use both, one on top of the other, which enables them to process a very, very weak signal—and it is a very weak signal, even with our big antennas.

Going the other way, what is the strength of the signal you send up to the spacecraft, and what actually reaches there?

Starting out, on the ground we've got 100 kilowatts, compared to 20 watts up there. The antenna gain is the same; the distance is the same. We probably arrive there with about −125 dbm, which is about a million times more than what we get here. The uplink signal when it reaches the spacecraft is much stronger than the signal we receive on the ground, but the spacecraft receiver is not cooled; it's whatever the temperature happens to be out there. It's inherently more noisy than the systems we have on the ground, so you need more signal going up.

Communications-wise, we'll be able to command the spacecraft a lot longer than we'll be able to receive good signals from it. If we really had to, we could build a bigger transmitter on the ground, but we can't change the spacecraft. In fact, as time goes by, we'll have to use more and more power. By the end of the mission we'll probably be radiating 400 kilowatts.

I understand that some people paid a personal price for the success of Voyager. Are there instances that you are aware of?

I *did* make a lot of sacrifices. I worked seven days a week for about four years. I didn't get paid for it, but I did it willingly because I wanted the mission to be a success, and it was a success, and I look back over it, and I have absolutely no regrets.

Were your family and friends understanding?

Yes. The people I worked with closely, I don't know of any sickness or divorces that happened to them.

Any closing thoughts?

Discovery is an ongoing process. It's a continuum. We tend to think of ourselves as being the end results of something: evolution started with this and ends with us. But it doesn't end with us; it keeps going. People will be a lot different a million years from now. They'll probably be more different than we are different from what animal life was like a million years ago, because change usually speeds up.

Things Voyager has done are part of a continuing process, and there are no answers; there's just more questions, and each time that you get a little bit farther, you get a little better grasp of something, you find out that you don't have a good grasp at all, so you've got to go still further to get the answers.

At noon today I was listening to Ed Stone. He said that the theory of shepherding satellites on the rings is very complicated, but some of the motions would require even more complicated theory. When I heard that, I thought every time that we think things are getting a little more complicated what it really means is our *theory* isn't any good and when we learn more we'll find out that the reality is a lot simpler. We're just operating with a theory that's getting complicated.

Ptolemy had a good system of knowing how the planets and the stars moved, and it worked. He could predict what was going to happen, but it was extremely complicated. Copernicus came along; he didn't have these epicycles and all these weird things going on, and his simpler theory turned out to be a lot better. As you learn more and more, your system gets more and more complicated until something happens and you get simple again, and you say, "Oh, why couldn't I see that?"

We're laboring under a lot of misconceptions today that our grandchildren are going to say, "How could they have been so dumb?" That's how I think of the Voyager project. Everything we do is part of a process that gets from here to there, and we've got to make those steps, and at this time, we are making those steps. The first spaceflights couldn't be done when I was born, but by the time my grandchildren die a lot of things will be possible that we can't do today. It's all part of a process.

From your experience on Voyager, are there lessons which could be applied to other projects, either in space or on the Earth?

I think that most of Voyager's success was due to defining problems, understanding the problem, determining what it was that we wanted to do—"where do we go from here?" type thing—and then careful review and analysis.

They do this on all JPL, NASA programs, but I think Voyager paid a lot of attention to the review process and accepting new ideas from outside the project, to do things better, or to be more careful, to take fewer chances.

Some of the later missions have gotten into problems because something went wrong. Though I don't know the details, I'd guess that probably, in a large proportion of those problems, that more care would have kept them from getting into trouble. "Be careful" is the thing I learned from Voyager.

THOMAS R. GAVIN
Project Assurance Manager

Born December 11, 1939   Upper Darby, Pennsylvania

Thomas Gavin earned a bachelor of science in chemistry from Villanova University in 1961. After a year at Electrical Storage Battery Company, he joined JPL as a quality assurance engineer on four Mariner missions.

In 1968 he was appointed product assurance manager for Voyager and for 10 years was responsible for developing advanced microelectronics, radiation hardening, and overall reliability for the spacecraft. After the successful launches he was assigned in 1978 to do the same for Galileo. In 1989 he was appointed manager of JPL's Hardware Assurance Division, and the following year was assigned to his present position as spacecraft system manager for the Cassini project.

His awards include NASA's Exceptional Service Medal for his work on Voyager and the Outstanding Leadership Medal for his contributions to Galileo.

He enjoys boating, softball, and fishing. He has always been interested in historical events and will naturally gravitate to a history book. He is curious about how things happened and the processes that made them happen.

The principal things that drive my view of the universe can be put quite succinctly: an abiding belief in God and his hand in the universe, the responsibility for raising a family with properly defined values, and the necessity for truthfulness in all endeavors.
—Thomas R. Gavin

THOMAS GAVIN

Ever since I was 10 years old and saw the movie *Destination Moon*, I wanted to explore the planets. I got my wish.

Your resume indicates that upon receiving your bachelor of science in 1961 you worked as an analytical chemist, and then joined JPL.
Right.

How did analytical chemistry lead to working on Voyager?
Actually it was quite simple. I knew something about semiconductor chemistry, silicon chemistry. So JPL hired me to be a parts engineer, to worry about parts quality on transistors and so on, in 1962.

My first job at JPL was quality assurance on electronic parts that went into Mariner 4, which went to Mars in 1964. And it just progressed from that: Mariners 5, 6, and 7—I spent all my time on electronic parts. On Voyager I did a lot of parts work, and that was just from a basic chemistry background, understanding the physics of failure, and parts stress, and things like that. A lot of on-the-job training with a basic science background.

I started on the Voyager project in 1968 during the preproject phase and finished at launch plus 30 seconds. That was it. The last time I saw the Voyager spacecraft was July of '77 when it was in SAEF-2, the assembly and test facility at the Kennedy Space Center. There is a looking glass with a viewing bubble into the clean room. I was the reliability engineer, so I was intently interested in whether it was all going to work. I remember just standing there, staring at it—wondering whether it was going to work. At that time I was 39 years old. I'd spent 10 years on the project.

Voyager, for all of us, was a pretty ambitious project. We knew we had a chance to do the Grand Tour mission. In the preproject phase it was originally intended to be the Grand Tour, but there wasn't enough money. But we knew that if we built the spacecraft well enough and designed it with enough margin, it would last far longer than anyone would have imagined. We put in a lot of design margins, so in retrospect its success is not surprising.

For the whole project there were 3500 problem/failure reports from the time we first started putting components down on circuit boards until we launched. And for each one of those problem/failure reports, I was the last signature. When I signed it, no one else looked. It was done. So I had a special sensitivity as to whether or not we had really fixed the problem.

There were a lot of areas where we got into extreme difficulty—points in the development process when you wonder whether or not you're going to get out of it. The Radio Frequency Subsystem (RFS) is an example. The X-band amplifiers on the radios had a number of inventions, including a brand new tube design, new packaging design, and high voltage applications. At one point we had 700 open problem/failure reports against the Radio Frequency Subsystems. During the development process these things just don't come together; they take a great deal of care and energy.

The Radio Frequency Subsystems used components which were found to have high levels of water which could cause corrosion problems. We found filter capacitors that had generic defects. We couldn't figure out which was the good set of capacitors, so we made a brilliant compromise. We put one set in one redundant string and another set in the other redundant string to increase the chances of surviving a generic problem. We made a lot of these types of decisions in the development phase, trying to get to a spacecraft that would work reliably.

About a third of the way into the design, in 1973, we found there was a severe radiation problem. Pioneer 10 had made it to Jupiter and had detected the very significant radiation belts. This was about the time we were doing preliminary design reviews in the subsystems. The original development guidelines made no significant effort for radiation hardening.

In '74, three years before launch and as a result of Pioneer 10's discovery of very severe, trapped electron belts, we determined that we had to radiation harden the spacecraft if we wanted to succeed. So we embarked on a massive effort to make the electronic components impervious to radiation. This job was divided up among three guys: Jim Briden worried about the components, Tom Gindorf worried about the shielding, and I worried about the circuits.

Some integrated circuit families had to have their processes changed so they would work at all. Additionally, we had to change our circuit designs in both the instruments and the spacecraft so we would be sure they would work.

But you still had some worries?

Because of the radiation problem, we put extensive margins in the electronic design, so we knew the spacecraft could last a long time. Ultimately we felt pretty confident that the craft would go through the Jupiter system in fine shape.

Why did I worry about success? In the summer of 1977, there were three spacecraft in Hangar A. O. at CCAFS (Cape Canaveral Air Force Station). On June 8, 1977, an integrated circuit failed in a flight computer due to a time-dependent, migrating gold resistive short. At the integrated circuit level, we're talking microns. The gold moved from one area of the microcircuit to another area and shorted out the part.

That kind of failure mode was inherent in the process, so if one could fail, they all could fail. Some percentage of the parts had water in them—a very, very low level of water. You would consider it a desert, but, from the part standpoint, it could cause a failure. That subsystem had 1500 hours on it when we had that failure, and in the two flight Voyager spacecraft we probably had 5000 of those parts.

We had to stand up in front of the project review board, including the chief engineer from NASA, and the associate administrator, and explain why we thought it was going to work. Our experience had been that if there were going to be problems of this type, they occur in the first 200–300 hours. So two failures at 1500 hours was very disturbing, but we believed we were dealing with the tail end of the distribution.

So we decided to burn in the spacecraft: operate it around the clock in hope that, if there are any similar parts problems, they would fail prior to launch. We operated the spacecraft for four weeks, 24 hours a day in Florida, and had no other failures of that type prior to launch (or in the 16 years of flight so far). I would not have bet on that outcome.

In July 1977 we actually shuffled the spacecraft in the launch flow. We didn't launch what we intended to launch first, because in Florida the flight data system computer had failed, which required taking that computer out and sending it back to California. They tested the computer and couldn't find anything wrong. So they pulled about seven to ten parts which could be responsible for the problem in the computer, put the spacecraft system back together, tested it, and launched. And it worked fine.

In August 1977, we found a bad capacitor. This one was a killer. There was a design application problem with the phase-lock loop capacitor in the RFS, and we found this design problem at the last moment. We argued for a full day whether we should go in and replace the other capacitors.

We had not launched either spacecraft, but Voyager 2 was already out on the pad. Should we do it or not? Finally the critical question, "Do you have anything that's better?" I said, "Probably not." I couldn't tell them that it was any better and that it was worth the risk of pulling the whole thing down and changing it out. So we launched.

The following April 1978 there was a command error which caused the Radio Frequency Subsystems to switch to its redundant half and, as we were afraid, the capacitor failed; the failure mode we had experienced in Florida. But by that time we had a fair idea of what had to be done, so the flight team was able to recover, thanks to a great effort on the part of the Deep Space Network.

All through the summer of 1977 we'd been fighting a processor failure in one of the AACS (Attitude and Articulation Control Subsystem) computers. It was intermittent, and a few days after the launch of the first Voyager, the processor stopped during a test of the third (test) spacecraft. We determined the source of the problem—a very slight separation of a resistor. I came back to California, where data for those parts was stored; we were able to prove from the screening data that it was an outlier and there would be no threat in either of the craft. So we launched the second one.

If you went through that summer of '77, with all this hands-on interaction and with hardware on and off the spacecraft, you would have wondered whether or not the spacecraft would make it. But we knew that if we could get the residual bugs out of it, that it would last a long time. I was pretty confident of it.

It was a fun job. I thoroughly enjoyed it. We had a great project team; very little infighting. We were pretty much devoted to each other. Ray Heacock was the spacecraft manager; Bill Shipley was the development manager; I was the reliability manager; Ron Draper was the system engineer. Everybody was highly focused on making sure it was a good spacecraft. It was exciting. You don't get to do those things very often. I was fortunate.

I came back from the launch and started on Galileo.

How did you get assigned to Voyager in the first place?

I had done a fair amount of work on parts: electronic parts, processor components, and so on. There was a project called TOPS in the spring of 1968. It was preproject research and development, where people could come up with all kinds of ideas on the way to build a spacecraft. At the time, JPL was working on Mariner '69, Mariner '71, and the early Mars Voyager. All the experienced people were working on those projects. I was pretty young, and I was available.

So I was assigned in the preproject phase to work with Ron Draper and Bill Shipley. My job was advanced microelectronics: trying to figure out how we could make long life microelectronic devices as part of this outer planet development effort. I did that for three years until the project was approved. I was still pretty young for a senior project position, but by that time I'd gathered so much experience and so much knowledge that the division I worked for assigned me to the project. It was just a case of being at the right place, having the right set of interests, at the right time.

It was a nine-year job. I thoroughly enjoyed it. With a job like this, you love coming to work every day.

Amongst ourselves, though, we had a kind of gallows humor. With every decision, we'd have to ask what could go wrong. When we'd think of all the things that could go wrong we'd say, "Well, it seemed like a good idea at the time." We have a lot of expertise around JPL, and you try to use it. If you presume to think you know everything and you don't use that valuable input, you're not going to make it. But with the engineering capability, materials capability, and the people around here who are willing to help, you can succeed.

As I said, we had 3500 problem/failure reports; it did not come together easily. You have to be resolute, be willing to hang in and not give in to fear. You can't be afraid to question. When someone tells you something, don't believe it the first time. Often the first information is not accurate or is based on hearsay. If you accept everything you're told, you can quickly get in trouble.

Can you recall an example of that from Voyager?

Yes. Voyager's propulsion thrusters have a catalyst bed heater, which is a 1-mil wire that heats up before the thruster firing. We had a failure in one of these heater wires. We talked to one of the experts, who looked at the scanning electron microscope pictures and said, "Oh, that's a nonpropagating fatigue crack. Not to worry. Couldn't possibly give you a problem."

You hear that and you say to yourself, "Do I really believe this?" My associate, Jerry Lane, and I discussed it, and he didn't believe it. So Jerry started to peel the onion. People were resisting: "Why are you doing this? This is not a problem."

But he kept peeling and peeling the onion. By that time the system was in thermal vacuum testing, and two of the heaters failed. Fortunately, Jerry had done enough preparation work so that by the time we had the failures, we were ready and were able to implement a good fix. We never had a problem in space.

You have to approach quick answers with a fair degree of skepticism, especially if you're the last signature on the problem report. Your signature tells the project manager, "I looked at it and it's okay." You can't be cavalier. You've really got to know that you have a good technical answer. That was a case where we just hung in there, and it required a design change to a 2-mil wire.

There's others where we weren't successful, and we paid for it in flight.

For example?

This falls in the "it seemed like a good idea at the time" category. For Voyager, we employed a new integrated circuit technology called CMOS (Composite Metal Oxide Semiconductor), which almost all home computers, etc., utilize now. But then it was brand new and no one had attempted to build a spacecraft computer out of it. Again, it boiled down to a question of margin. Before we put these chips in the spacecraft, we'd test at temperature and operate them in some higher current or voltage to stress them. We'd throw away the ones that failed. We'd take the ones that didn't fail, that looked strong—essentially the pick of the litter—and we'd fly them.

The manufacturer was very unhappy with the voltage level we picked to stress these parts. They thought since it was a 10-volt part, that we should stress it at only 10 volts; not some higher value. That was contrary to our usual approach. Since it was a new technology and we didn't have a lot of experience with it, we went along with the manufacturer.

Prior to launch, several failures occurred where the glass insulating oxide broke down under the voltage stress. In other words, there was an inherent defect in the part. If we had stressed it during the screening, it would have broken down and been removed. As a result, we had in-flight failures. They would never have failed if I had sufficiently worked the problem.

On Galileo we used the same part but with an 18-volt screen. That's part of the learning process when you're trying to do one of these things.

The flight data system memories were assembled in a hybrid form where the bare chip goes on a substrate. I think there were 128 chips on a substrate, and these were radiation soft so they were in a steel package. In that situation we couldn't do the individual screening. We couldn't stress like we normally would, but we did the best job we could. Although we had four or five in-flight single-cell failures (and we now have screens that get that stuff out, too), I would say it was pretty successful overall. It's all part of the learning experience.

The launch: we finally got the spacecraft together, but we went through a lot of trials and tribulations during the summer of 1977. We were in Hangar A. O., in a control room where we were looking at the spacecraft. I knew that, despite all our efforts, there was some unreliability in the spacecraft. And I'm thinking that I have a family to support; my oldest child is 14 years old. If this thing doesn't work, I don't want to have to meet a lot of important people who are going to be very angry with me. We launched.

Ten seconds into the launch we got a gyro error alarm, and the spacecraft entered fault protection and began systematically swapping everything in the attitude control system. The whole machine went crazy. As the reliability engineer, I'm thinking that the thing didn't last but a few seconds, and my career would likely do the same.

Actually the spacecraft operated perfectly. However, a rate limit of the gyro was not set correctly. So 10 seconds into the launch of the Titan III, it went into a snap roll which saturated the gyros. With every separation we went through an event like that. For the first hour and a half it looked like this might be one of the shortest projects on record—that we were going to save a lot on operations costs.

We were pretty gloomy. It took two or three hours to discover and confirm that the spacecraft was on target and that everything actually had happened the way it should. Then we spent the next 24 hours trying to figure out what had happened. Fortunately, the spacecraft performed flawlessly.

It was a once-in-a-lifetime opportunity to work on something like that. They're projecting that Voyager will run out of propellant perhaps in 2025. I could still be around. If it lasts and I last, that'll be 57 years from when I began working on it. I'll be 80 years old. It would be fun to tell your grandchildren about it.

I still have my Voyager spacecraft model; the project staff got those at launch. And we all have Voyager awards. But the thrill is to see the spacecraft last so long, do so much, and excite so many people. It was fun.

You say officially your job ended 30 seconds after launch, but your own personal connection continues?

That's right. If someone has a question, you're always available. Everyone who worked on it is intensely interested in the mission; we all follow it closely. We're proud of the operations team. Looking at those people gives an interesting perspective, though. Some are 25 years old. They were perhaps in Little League, Girl Scouts, or Cub Scouts when we were trying to make the radios and the circuit boards work.

In fact, we just launched some of the residual Voyager hardware in Galileo. To show you how confident we feel, the radios on Galileo were 17 years old when we launched. There's a lot of Voyager hardware—older stuff that's working great. Now I think old is good.

I'd recommend to anyone that they work on one of these projects from beginning to end. I've done it twice now, and it's just a lot of fun. Sometimes it's amazing to think a person gets paid to do it.

Looking back on it, what were the most significant events connected with Voyager?

Probably the most significant event for me was the Jupiter encounter. Would the radiation hardening really work? And would it really make it through? That encounter was the culmination that let me know that everything we did worked. None of the little victories—solving a small technical problem—counted as much as knowing that it did what it was supposed to do, unfailingly. That gave me a great deal of personal satisfaction.

I've been pleased every time it made it to another planet. During the Neptune encounter last year, people asked whether I worked on Voyager. I was proud to answer that I was the reliability engineer—that it was my job to make sure it worked. Basically we're old history; we've been off the books for years. But because we built it and now see it performing well, that immense personal satisfaction remains.

In fact, some people came to see me last week about the space station. One big issue of the space station is how to make a part last a long time, and a fellow called to get my input. Again, it's exciting to be involved in these things—and very satisfying.

What was your biggest disappointment or frustration?

In March of 1978 the radio receiver failed. I'm still frustrated by that. On the Galileo project we had just finished the inheritance reviews for utilizing the Voyager radios. The very week Galileo was inheriting the leftover radios, we ran into problems with Voyager's own.

Built into the spacecraft is an algorithm which communicates, "If you don't detect in 168 hours, presume that the receiver has failed and switch to the backup receiver."

A command wasn't sent because of a ground error. The algorithm timed out; it switched to the backup radio, and the capacitor was dead, so they couldn't lock up. We knew that six hours later the algorithm would take them back to the primary receiver which had worked perfectly for six months. The primary receiver came back on, worked for 17 minutes, and blew a fuse.

Personally, I still am not satisfied about what caused that primary receiver to fail. I participated in the reviews, and I accept the board reports. But it continues to be a frustration to me. Why did it happen? Other problems were small developmental ones; we had all kinds of those. Name it, we had it. Those ended up as little victories. Have a problem, fix the problem, and it works. But what the hell happened to that radio in March of '78 bothers me. I'd like to get my hands on it and find out. To this day I wish I knew.

How does Voyager differ from other space projects, or do you think it really was different?

I don't think Voyager is that different from other planetary projects we've done, although I wouldn't presume to speak for what other people do. The Voyager project was characteristic of a JPL project: a highly trained, highly motivated, willing-to-work team of honest people. When something is wrong, they tell you it's wrong. That is characteristic of our missions and of that project. Don't shoot the messenger; we've never shot a messenger around this place. When there's a problem, we examine it in a collegial way.

I've been on two flight projects and in the project office on multiple others, and I've seen very few politics involved. In these projects, if your hardware works, it works. If it doesn't, it doesn't. Politics can't really affect that. So, generally, it is a very, very honest process, with minimal politics.

Do you have any final things that you'd like the readers of this book to think about?

I guess one issue. No one person does this thing. It's not one guy, however brilliant, who directs everything. You see that in the movies, but that's not how it is. Nobody is that smart. A project like this requires the collective effort of many people.

And a little flag waving—for anyone who thinks Americans can't build things that last—Voyager was 100 percent American technology.

Is this situation ending?

I believe American taxpayers are supporting us to develop American technology. The realities of funding, however, often require partnerships in which other countries provide their technology at no expense to the U.S. taxpayer. I think this is an acceptable solution.

GLEN R. SOUTHWORTH
Television Recording Engineer
Born September 18, 1925   Moscow, Idaho

Glen Southworth of Colorado Video, Inc., has been involved in television projects since 1953. These have included the design of a wide range of specialized video instruments for use in research and industrial inspection. He has also pioneered in still-image transmission by means of phone line, FM radio, and satellite subcarrier channels as well as in the vertical blanking interval of television station broadcasts. He is particularly interested in telemedicine and teleteaching.

He attended the University of Idaho 1945–1947, where he founded the campus radio station KUOI. Subsequently he became the chief engineer of Empire Office Equipment. In 1953 he was appointed director of Television Studios at Washington State University and later became chief engineer of Radio and Television Services. He also served two years in the U.S. Army Signal Corps.

In 1960 he joined Ball Brothers Research Corporation in Boulder, Colorado. His first assignment was a feasibility study for JPL on applying Delta Modulation encoding techniques to single-frame television. JPL later accepted his proposals for a unique, state-of-the-art Video Modulation Test System. In 1965 he left Ball to start Colorado Video, Inc., where he worked until retirement 31 years later.

A memorable point in his career was the transfer of 117 photographic slides to an encoded audio format that was subsequently rerecorded on the gold-coated recordings carried on the two Voyager spacecraft.

His professional memberships include the IEEE, and other engineering societies involved with acoustics, motion pictures, and photographic instrumentation.

His honors include the 1990 Engineering Emmy award from the National Academy of Television Arts and Sciences, the H. Rex Lee award of the Public Service Satellite Consortium, and TeleSpan Professional of the Year.

He has been awarded 10 patents and has written more than 100 articles and papers.

He enjoys crossword puzzles.

Benign creativity and invention.
Caring for fellow human beings and other living creatures.
Looking at the Universe in more than one way.
Daydreaming.
Communicating interesting and possible useful ideas.
Expansion of human perceptions.
—Glen R. Southworth

GLEN SOUTHWORTH

I want to include in this book an example of someone who was not a JPL employee. However, you have done contract work with JPL.

That's right. Back in 1960 I did a study for JPL on encoding techniques for slow-scan TV. It was a good learning experience, including an acquaintance with a brilliant JPL individual, Dr. Richard Heyser, who provided some excellent insights.

I also had a JPL contract for a video modulation test system based on a proposal of mine. It was probably the best engineering work I've ever done.

There were miscellaneous other activities including working with Elliot Gold when he was an engineer at JPL; another paper relating to the importance of JPL in encouraging technical innovation in the industrial sector; and demonstration of one consequence of JPL early sponsorship, color slow-scan television, at a special recognition meeting in Pasadena.

Note, however, that the Voyager record did not involve any direct connection with JPL.

Why did you donate your time, energy, and expertise to do the recording?

I've been reading science fiction since the age of 10. One of my earliest ambitions was to be a philanthropist but, lacking any money, I eventually decided to become an inventor and contribute to society in that manner. If money also came as a consequence, I wouldn't object. In the meantime I had a notebook in which I tried to enter one invention, one new idea every day.

When Valentin Boriakoff invited Colorado Video to be part of the team that would convert photographic images into a form that could be put on the Voyager phonograph record I was delighted. It seemed something like being one of the stone masons chiseling hieroglyphics into the Rosetta Stone but with far wider implications!

Valentin was a radio astronomer at Cornell University. He was normally based at the big Arecibo radio telescope operation, but at this time he was at Cornell as part of a group gathered around Carl Sagan. These people hoped to communicate with extraterrestrial civilizations and had arranged with NASA to place a phonograph record containing sounds of Earth on the two forthcoming Voyager spacecraft missions. The audio content could be material such as music, languages, birds, lightning storms, earthquakes, and many other things, even kisses.

Shortly before final preparations had to be made, someone in the group asked if it would be possible to put pictures on the phonograph record in addition to the collection of sounds. Nobody knew, for sure, but Valentin was given the assignment of checking into the feasibility of doing so. As part of his investigations, he looked through a large electronics catalog and found an eight-page insert that described a group of unique television products made by Colorado Video. One of these, the Model 201A video converter, was correctly identified as a device that could take a normal television picture and reduce the bandwidth to the point that it could be recoded on a conventional long playing phonograph record.

As a result, Boriakoff contacted us and outlined in general terms what his requirement was. I told him that it was definitely a feasible project and we would be pleased to demonstrate the process for him. Consequently he arrived in Boulder with a small series of 35-mm slides for test purposes in order to personally verify the workability of the concept.

The system used involved a standard Kodak carousel slide projector which was focused onto a high-quality black and white television camera. The output of the camera was then connected to the Colorado Video Model 201A that converted the original video to a single picture with a resolution of 240 by 512 picture elements, 256 shades of grey, and a bandwidth of approximately 8 kHz.

Once made, the test recordings were reconverted to normal television images for visual viewing and assessment of subjective quality. As there was also a requirement for color in some images, we used a simple three-color separation process using red, blue, and green filters in front of the camera lens. This resulted in three separate still images. As the Colorado Video

Model 275 used for conversion back to conventional TV could only do one frame at a time, a triple photographic time exposure of a monitor screen was made in order to verify satisfactory color reproduction.

Following this initial series of tests, Boriakoff returned to Cornell, reported favorably on the technical aspects, and there was begun a rapid search to locate pictures appropriate for inclusion on the phonograph record that might eventually be found by a spacefaring civilization somewhere in the center of our galaxy.

A few weeks later Valentin returned to Boulder with approximately 110 35-mm slides, a good share of them provided by the National Geographic Society (the intent was possibly for *National Geographic* to do a feature article on this, but this never happened). Wyndham Hannaway, a talented video engineer on our staff, Boriakoff, and myself went into a marathon recording session that lasted until about 11:00 that night. I can't remember if we had any supper or not.

Two tape recorders were used to save the slow-scan TV pictures for later conversion to the final phonograph record. The primary machine was a high-quality instrument, the Model 5600C instrumentation recorder made by Honeywell. The quality of each of the signals was verified during the recording by taking the output of the tape playback head and feeding it to a Colorado Video Model 275 scan converter, which then displayed a reconverted picture on a conventional television monitor. Honeywell had also loaned us one of their machines that could make hard copies from television pictures, with one set being given to Valentin.

We worked until 11:00 at night to translate these. There were four separate tape recordings made. As a backup to the primary tape, we also recorded the same pictures on a Teac four-channel quarter-inch audio tape recorder. The whole process was then repeated, giving four separate copies of the pictures to be placed on the Voyager record.

The reason was that this was to be literally a message to the Universe—and we didn't want to risk the chance of something going wrong. Dr. Boriakoff personally carried two of the tapes back to Cornell, and the other pair were dispatched by air several days later after we were sure that Valentin and the tapes had arrived safely. The tapes were subsequently taken to CBS Laboratories in New York who provided the conversion to the final master phonograph disc. We heaved a collective sigh of relief thinking that our part in the creation of a cosmic Rosetta Stone was over. This was not to be.

President Carter became aware of this message to the Universe and posterity and requested that a number of additional frames of information be added to the phonograph record.

Dr. Boriakoff returned to Boulder with the material from the White House, and there was a second recording session that lasted until nearly midnight. Those involved this time were Valentin, Wyndham Hannaway, and Larry McClelland, another of our staff engineers. I missed this one because I was 1000 miles away attending my mother's funeral.

The Voyager record was a labor of love, perhaps mixed with a sense of duty to the human race, and certainly a sense of excitement and speculation as to what might happen when the record was finally discovered. Colorado Video contributed people, equipment, and enthusiasm to the initial picture translation of slides to tape. Honeywell loaned the instrumentation tape recorder and provided the tape to go with it. Cornell provided the initial inspiration for the project as well as the efforts of Dr. Boriakoff and others. CBS Laboratories provided the final steps in the record making process. Many, if not most of the pictures were provided by *National Geographic* photographers, and I'm sure that others played a special part.

To the best of my knowledge, there were no federal funds invested in the entire project, unless it was for a few hours of time that it might have taken a NASA technician to bolt the record to the side of the Voyager spacecraft. Nevertheless, in later years there were some complaints about this government "boondoggle".

The most fun I had with Voyager was retransmitting the pictures from Uranus and Neptune as soon as they arrived at JPL. This had been suggested by Dr. David Swift. For the

Uranus encounter we installed, at Colorado Video's expense, our slow-scan TV equipment and special phone lines at five locations so people could dial in and receive the images. These locations were the Air and Space Museum in Washington, D.C.; MIT; the Los Angeles Museum of Science and Industry; the University of Hawaii; and Colorado Video in Boulder.

It was exciting to watch the images as they came in. MIT set up in an auditorium so that students could watch the images 24 hours a day. In Hawaii, 1000 people turned out on encounter day, and hundreds more came during the following week: students from preschool to university, senior citizens, blue-collar workers, scientists, mothers with infants, and legislators. Typical responses included "amazing", fantastic", "wow", and awed silence. Ten thousand people far from Pasadena were able to see the Uranus images at the same moment as the scientists at JPL.

For the Neptune encounter we worked with AT&T to provide a more flexible setup, using a 900 number to call JPL. In this case anybody with Colorado Video viewing equipment could dial in and get the Voyager images for the price of an ordinary phone call. This allowed people all over the world to view the pictures the instant they were received at JPL.

We took the feed from JPL's scan converter, which provided a normal TV image. At Boulder we set up a computer to store the images as they came in over the phone line, both for later use and also to fill in, in case there was a problem with the connection at JPL—which actually happened. Someone at the Lab, purportedly an FBI agent, accidently hung up our phone line during the vice president's visit, cutting off transmission to a space meeting at Grenoble, France, where 1000 people were watching. So I filled in with some earlier pictures, including one of my cat. The next day I received a magnificent bouquet of flowers from Grenoble.

Subsequently, we did several programs using these stored images. Thousands of people got to see the Neptune images. Colorado Video picked up the tab for both encounters. We didn't make any money but we gained a lot of friends.

ROBERT JOHN CESARONE
Navigation Team Task Leader
Born October 5, 1952   Chicago, Illinois

Robert Cesarone graduated with honors and distinction from the University of Illinois with a bachelor of science in mathematics in 1975 and a master of science in aeronautical and astronautical engineering in 1977. That same year he joined JPL.

As task leader for the trajectory and maneuver functions of the Voyager Navigation Team, he led these activities throughout the planning and operational implementation stages of the Voyager 2 Uranus and Neptune encounters. He was responsible for the actual Voyager 2 Uranus and Neptune flyby trajectories, as well as maneuver targeting to enable their achievements. Prior to these activities, he was a trajectory engineer for Voyager Jupiter flybys and a maneuver engineer for Voyager Saturn flybys. He has 12 years of experience in JPL mission operations.

Other JPL assignments included trajectory design and optimization for Galileo satellite tour and interplanetary follow-on missions; mission design studies for low-thrust ion drive comet missions; and lead trajectory engineer for the Mars Observer mapping orbits, orbit insertion phase, and interplanetary trajectory analysis and design.

Since 1991 he has been assistant program manager for Deep Space Network Strategy Development, involved in program management, strategy development, and long-range planning.

Cesarone has authored 24 technical and popular articles covering the Voyager mission, trajectory design, gravity assist, and space navigation and telecommunications. He is an Associate Fellow of AIAA and a founding member of the World Space Foundation. NASA awarded him the Exceptional Service Medal.

His hobbies include amateur astronomy, collecting classic editions of science fiction and space exploration books, operating model trains, writing songs, and playing five-string banjo in a local band.

Living one's own dream seems most important.
I know of no other way to a joyful life. All the positive things,
success, good fellowship, sense of accomplishment, will follow.
The many reasons to set aside our dreams are best ignored.
Great beauty exists in the universe and its inhabitants.
It is best perceived by adding one's own contribution to these.
Finally, life has a spiritual aspect that can be experienced,
and should not be neglected.
—Robert John Cesarone

ROBERT CESARONE

My interest in space goes all the way back to when I was a little kid. The first thing I ever read on space was an article in the *Chicago Tribune* (I grew up in Chicago) by a guy named Dr. Dan Q. Posen. He wrote some space stuff in the late 1950s and early 1960s. My mom gave me that article, and I felt it was really neat. That was all I had for a while, so I got interested more in space through science fiction. I read all the Heinlein stories and Andre Norton novels in the library where I grew up. When I finished all those and there was no more science fiction left in the library, I still had this insatiable desire to read about space, so I started reading astronomy books.

I eventually started college in astronomy. That was a real love of mine because it was so intensely related to space, but for various reasons I switched to engineering. Those reasons had to do with the fact that in the early 1970s when I started college the classical discipline of observational astronomy had fallen on hard times. Astronomers had evolved into astrophysicists; they were people who spent their days on computer terminals modeling stellar interiors, and they never looked at the stars. I had nothing against astrophysics but it just was not what I wanted to do.

So I eventually switched into engineering when I saw this curriculum that had things like orbital mechanics and fundamentals of rocket propulsion and all kinds of neat stuff like that. Of course I also grew up watching Mercury and Gemini and Apollo. When I saw that, I knew that was for me.

I was tailoring my education towards a career in the space business. My two top choices would have been Johnson Space Center and JPL. I had opportunities to work at both, and for various reasons I chose JPL. I started on June 13 of '77, so I guess you can say I'm pretty happy, having been here 13 years.

How did I specifically get involved with Voyager? I had a job offer to come to JPL and work in the mission design section. I didn't know what project they were going to put me on. When I arrived they said, "You're going to work on Voyager mission operations."

Actually I was a little disappointed, because what I wanted to do as a career would be to go off and crawl in some office and do theoretical studies on how to transfer from one orbit to another with the least amount of propellant expenditure, and every couple of years come out with some paper announcing my latest results. But they said, "No, you're going to work on Voyager." As it turned out, I stayed with Voyager for 12 years. It worked out fantastically for me. It was wonderful.

The first launch of a Voyager was on August 20 of 1977, so two months after I came to JPL I was launching spacecraft to the outer solar system. It was pure dumb luck, but it was just perfect timing: when I got out of school, when I showed up here, what project I got on. You see, that was before all the problems with the shuttle and funding. A lot of people thought it was going to be more fun to work on the advance design of some future mission, like Galileo or whatever. But Galileo didn't launch until last year [1989], and people who made that decision in '77 waited 12 years to get some hardware into space. In those 12 years I launched two spacecraft, flew both of them by Jupiter and Saturn, and one by Uranus and Neptune. (Of course, I didn't do it all by myself.)

At any rate, I just stepped into the right position for me, and I've loved it. It's been incredibly rewarding personally. In the late 1970s through the 1980s, there were really two major things happening in space. The space shuttle started in '81, and it was on again off again, depending on delays or disasters or whatever, and then there was Voyager. Those were the two big things in the U.S. space program, and I got to do one of them.

The specifics are that I was placed on the navigation team because I'm in the Mission Design Section. What generally happens is the navigation team is staffed up with people from the navigation section and from the mission design section. I was a Voyager navigator from 1977 until late 1989 when, having passed our last planet, I made the switch to the mission I'm now working on, which is Mars Observer.

The functions of a navigation team on a flight project are threefold. The first function is what we would call trajectory, or mission design. That's the kind of thing I do. That answers questions like what kind of performance do we need to get to where we want to go—performance on our launch vehicle. How much mass can we launch and still get there, because the more mass you put on your spacecraft the less velocity you can give it, and vice versa. What kinds of trajectories can you fly? What are the science advantages or penalties of those trajectories? What happens with this trajectory as you go through the Jupiter radiation field? What kinds of satellite flybys can you achieve while you're going through the planetary system?

Questions like that relate to the overall planning and design of the mission. I liken that to, in *Star Trek*, what Sulu and Chekov do. The captain says, "Plot a course for Neptune," and we do it and turn around and say, "Course plotted and laid in, sir." So that's the first function of navigation: to define what is the mission we're going to do. That was one of my duties.

The second function, which is actually the biggest function of the team, is called orbit determination. Let's say you've already launched and your spacecraft is going out there and you attempted to put it on the trajectory that you had defined. Well, there's always errors that creep into the system. Our model of the physical world is not perfect. There are little things that happen on the spacecraft. All these things cause it to *not* be on exactly the trajectory that we wanted it to be on, and so you have to go through a process to find out what trajectory it really is on, and that process is called orbit determination. It is a big task; I was not personally involved in orbit determination.

It has two components. Earth-based orbit determination involves tracking the spacecraft with the antennas of our Deep Space Network. There's two things you could do that way.

You would listen to the frequency of the signal that was coming down from the spacecraft. Since we built the spacecraft, we knew what frequency it was supposed to be and because of the radial velocity of the spacecraft, you'd have a Doppler shift, not only due to the spacecraft but also the Earth's motion in its orbit and the tracking station motion on the rotating Earth. But you can model all these, and the windup is that you can tell something about the radial velocity of the spacecraft from the Doppler shift of the signal.

Also, by timing how long it takes the signal to go up and back, and dividing by the speed of light, you can tell something about how far away the spacecraft is. By making measurements like that over a number of days or weeks or months, you can build up a very accurate picture of where the spacecraft is, relative to the Earth, and that's called Earth-based orbit determination, Earth-based navigation.

However, when you're going to do a flyby mission, like Voyager, you also care a great deal about where the spacecraft is relative to the body it's about to fly by, because we didn't know exactly where Neptune was relative to the Earth, and so we didn't know where it was relative to the spacecraft, even though we knew where the spacecraft was relative to the Earth.

When I say we didn't know where Neptune was, I mean we didn't know where it was some months before encounter, to a few thousand kilometers. We could look and see where it was in the sky, but a few thousand kilometers error was too coarse for us. You have to put that in perspective. The planet has a radius of 24,000 or 25,000 km, so to say we didn't know where it was to a few thousand kilometers is just a fraction of its radius, but we needed to deliver our spacecraft to an aim zone in the 50 to 100 km range, so we had to get the errors in where Neptune was down to that level.

For that we used the technique called optical navigation. That basically is the same phenomenon as when you're driving down a road and there's some telephone poles along the road and there's a farm house and silo way out in the distance, and as you move, you tend to see the nearer object shift in position relative to the background object. Well, the nearer object for a flyby would be the flyby planet—Jupiter, Saturn, Uranus, Neptune, or their moons—the far objects are the stars.

So we look at where the target body is, relative to the relatively fixed stars beyond it, and we take pictures. We process those pictures, and after a number of days or weeks, watching how the foreground object shifts, we can build up a very accurate picture of where the spacecraft is, relative to it. Then we combine those two—the optical navigation taken from the spacecraft, and the Earth-based navigation taken with the tracking stations—to build up a total picture of where the spacecraft is relative to the Earth and the flyby body. And that's orbit determination.

Well, that's the second function of navigation. I went through a long-winded discussion, but it is very important; it's the major task of navigation.

So now we've defined the trajectory, and we've tracked our spacecraft, and we've discovered that, "Oh my God, it's not on the trajectory we intended it to be on." It's not very far off, but it is off, and so from those two factors you can probably guess that the next function of navigation is "what are we going to do to correct it?" That we call maneuver analysis, and I got involved in that on the Voyager navigation team also.

What maneuver analysis is basically doing is saying, "Okay, I've learned that the trajectory is different from what I want it to be; what is the right time for me to make a change, to turn the rocket engines in a certain direction and burn 'em to get a velocity increment to put me back on the trajectory I want to be on?" In maneuver analysis we answer those kinds of questions and we design the maneuvers and assist in executing them.

Those three functions comprise navigation. The only thing I want to add to that is you don't just do them once, because let's say you've defined the mission, you've tracked the spacecraft, and you've corrected its trajectory. Well, lo and behold, some new factor comes in. A new science priority is going to make some change in the trajectory you really want to fly, and new orbit errors come in and so you have to do more maneuvers.

So what happens is we do this three-pronged procedure repetitively; we keep doing it over and over again. To quote from a science fiction movie, *The Day the Earth Stood Still*, when the alien was describing to the professor about space flight, he said, "We find it works well enough to get us from planet to planet," and we found that our navigation procedure worked well enough to get *us* from planet to planet, also. That's essentially what Voyager navigation did—and I'll let the results speak for themselves.

What would be a specific illustration of this corrective process? Let's say you were going to a particular planet and you found that you were a certain number of degrees or miles off course. What did you actually do

Let's take the recent Neptune encounter. We had three approach TCMs, trajectory correction maneuvers, within the last 30 days that we potentially could have done for final fine-tuning of our aim point at Neptune to get the trajectory we wanted. The reason we had three of them is that it takes awhile to build up our knowledge of where the spacecraft is relative to the planet.

So you do one about 30 days out, because you will have built up some errors from your interplanetary cruise so you want to take those out right away. But you know it probably won't be your last correction because you're going to get a lot more and better knowledge of the system and the trajectory in the intervening 30 days. You can't do the last one right on top of the encounter because you've got to have a little time for it to take effect, so our last one was three and a half days before the encounter. And we had one about 10 days before the encounter.

We would take this orbit determination information, this tracking data and optical navigation, and we would find out where our spacecraft is, and we would be able to predict from that what its trajectory through the Neptune system was going to be. We wanted to go over the top of Neptune and then down past Triton and pass through the occultation zone of Triton—the zone behind Triton where, if the spacecraft goes through that zone, the sun and the Earth will be hidden by Triton.

We wanted to go through that zone because, as the Earth and the sun are hidden by Triton, you can do some really good science. You can track the spacecraft's signal from Earth as it's disappearing behind the limb of the satellite, and that will tell you something about the atmosphere; or you can look at the sun with the ultraviolet instrument as it's disappearing behind the atmosphere, and you can tell something about what the constituents of it are.

We would have found, 30 days before the encounter, that our pass through that zone wasn't the one we wanted, so we would have computed a certain direction that we would have to thrust in, and a certain time duration that we would have to thrust, in order to correct that trajectory. We have lots of sophisticated computer programs that do that kind of thing: that propagate the trajectory, that put in velocity increments—maneuvers, essentially—and we do that a number of times until we get it just right, just exactly the way we want it.

The output of that process tells us what we have to do. In order to align the rockets in the right direction, you have to turn your spacecraft, and you turn it in two directions generally: you roll it and then you yaw it. That points the rocket engines in the direction that we would have determined is the right direction to point them in. Then you also have to burn for the right amount of time.

How did that actually get implemented? The navigation team—all we did is we said the engines have to be pointed in such and such a direction and turned on for a certain amount of time, and that means the spacecraft has to roll and yaw a certain number of degrees. We would give that information to people on the spacecraft team.

They are more intimately involved in actually controlling the hardware. So they would do various things in order to contribute to a sequence of events which would cause that to happen onboard the spacecraft—that would cause the spacecraft to actually go through this roll, this yaw, turn the engines on, turn the engines off at the right time, and yaw back and roll back to the original attitude so we would be back in contact with the Earth, because when you yaw off the Earth you lose contact.

The sequence team has the responsibility of actually taking the design of these activities and turning them into computer code which the spacecraft's computer can understand. So they got involved in the process too. When that was all done, this command stream would go to the mission control team; that's the interface between the project at JPL and outwards to the Deep Space Network antennas, either at Goldstone in the California Mojave Desert or at Canberra in Australia or at the Madrid area in Spain.

So the mission control team would send this command stream to the appropriate antenna of the Deep Space Network. From there it would go on up to the spacecraft. And, at the right time, because it's all programmed according to the spacecraft timer, Voyager would execute this sequence, and we would have our maneuver.

There are some other things which contribute to making this happen because our spacecraft are so fantastically far away. Even though we have these huge antennas, you have to make sure that you transmit with the right power at the right data rate so that the spacecraft will receive the intended stream of bits, and not get a lot of errors in it. You have to compute all that in advance. You have to understand the telecommunications link, how much power do you need, at what rate can you transmit in order to ensure that the integrity of this command stream arrives at the spacecraft properly.

Those are a lot of the other contributing factors, and various people on the project work those kinds of problems, understanding the telecommunications. You also have to take account of the fact that, as we were approaching Neptune, it took four hours for a signal to go up there, so you have to allow time for that too, and that's all computed in advance. We really have to take account of all of these factors which apply to dealing with a space mission—all these laws of the physical universe—to make it happen.

We did our 30-day maneuver very successfully. It was successful enough that we were able to cancel the maneuver 10 days before Neptune, but we did do the maneuver three and a half days before Neptune, although we did that a little differently. We didn't use the TCM thrusters, the main thrusters, for that. We were able to just use the thrusters which cause the spacecraft to roll, and that bought us a couple of advantages.

One advantage had to do with the fact that way back in 1978, before we even got to Jupiter, we had lost one of the receivers on Voyager 2 and the other one had a failure in a tracking loop capacitor. That meant that the spacecraft, which should have been able to track a broad range of frequencies, was now only able to track a narrow range of frequencies, and it was even less able to track when there were lots of thermal gradients in the spacecraft.

It turned out that if you turned on the TCM thrusters, you also had to turn on the propellant heater lines and things like that. There would be lots of temperature effects, which would make it very difficult for you to reliably send a command into the spacecraft for about 48 hours after you did that. You can see that, a few days out from Neptune, as we're charging down on Neptune, we really want to be able to get a command in to the spacecraft, so we don't want to do anything that close which is going to preclude us from doing that.

That's why we elected to do this final maneuver just by rolling the spacecraft back and forth. That meant we didn't have to turn on any of these heaters, and we got a little bit of sideways velocity increment, which allowed us to very finely tune the trajectory as it flew through the Triton occultation zone. It gave us exactly the trajectory we wanted.

Amazing.
People like to analogize and say what we did was like hitting a hole-in-one on a golf course from New York to California. That's true; it's an amazing feat, but even a golfer would be able to do that if he could, while the golf ball was in flight, every now and then give it a slight little push in the proper direction, to correct its trajectory to land where it's got to land, and that's what we did.

But even the process of giving the golf ball, the spacecraft, the proper push in the proper direction for the proper duration—that is an impressive achievement.
Well, we think so. We're justifiably proud of it. I don't know what else to say than that.

What was your most tense, difficult moment in your work with Voyager?
There were a number of them. Probably the most tense was on Voyager 1 as we were a number of months out from Saturn. We did one of these maneuvers. You recall I said the spacecraft will lose contact with the Earth as it turns away to do the maneuver, and then it's supposed to turn back and make contact again. Well, it didn't make contact again. We were able to recover it eventually, but it took a few days to analyze the problem and understand. I don't recall exactly what the problem was. We finally did get contact back, but it was tense because the possibility existed that we might not get contact back. It was a big relief when we did, so we were then able to have a very successful Voyager 1 Saturn encounter.

It was very important to get a successful Voyager 1 Saturn encounter. People tend to concentrate on Voyager 2 because it flew the Grand Tour—Jupiter, Saturn, Uranus, and Neptune—but if Voyager 1 had not succeeded with all of its goals at Saturn, Voyager 2 would have been reprogrammed to fly a trajectory at Saturn which would have been incompatible with going on to Uranus and Neptune. So it was very important that we achieved our Voyager 1 goals at Saturn.

Other tense times were when, at Saturn, Voyager 2 had the stuck scan platform. We lost some images. It happened when the spacecraft was behind Saturn, in the Earth occultation zone of Saturn, where we couldn't see it. When it reappeared, the scan platform was stuck. That was scary. We lost some pictures, but it wasn't life-threatening for the spacecraft because we had contact with it, but we worried whether we were going to be able to use our cameras.

There are a lot of other instruments on the spacecraft, but many people, including us, really get a lot of understanding out of the images. Were we going to be able to use them for Uranus and Neptune? Were we going to get as much science out of the cameras as we wanted? Or from the other instruments on the scan platform?

That was a tense time—I wouldn't say "moment", because that problem was analyzed for a couple of years before the right solution to it was found. The right solution *was* found because it never seized up again, and we got all the data we wanted from the scan platform science instruments, but there was a lot of uncertainty about that, a lot of tension.

I would say also that charging down on Uranus and Neptune, with the amount of things that we got the best knowledge of only very late in the game—things about when the trajectory events would occur and how we had to update those so rapidly, within a day or two of each of those encounters, Uranus and Neptune—was very tense because the time schedule in which to get things done was very, very compressed, and the reward for getting it right was so great in terms of science, and the penalty for getting it wrong would have been equally as great. But we got it right in both cases; we were well prepared.

You know when you're coming into a planet that this is it. I guess it's like landing a glider; there's no second chance. You must do it right. There's a three-quarter-of-a-billion-dollar mission riding on the fact that you designed this procedure properly and you're executing it properly, at 3:00 in the morning. There's a feeling of responsibility there. I don't mean that it's a painful feeling; it's tremendous fun. It's just a wonderful feeling, but you're keenly aware of what's riding on your decisions.

How does this keen awareness impact on you? Does it have any side effects on your sleep or family relations or general health?

It has an effect on your sleep because it always seems that these key events need to be done at 3:00 in the morning. We joke about that. But I would say that the main effect it has is a positive one. It gives you that feeling of really being alive, really playing the game. You know you are doing something important, not only for yourself or for your organization but really, in this case, for the human race, and you're playing a key role in it.

It's a good feeling. You are aware of the level of responsibility you carry; you realize the importance of doing a really good job at it, but it's not a burden or anything; it's fun. There's a spirit of play you get out of it; it's like when the 49ers play football in the Superbowl; it's an important game, but you're having fun. It's great, and it's the same kind of thing. You are really, really fully alive and doing something that's important and valuable and fun, and if anybody offered you anything else to do, you would say, "No way; this is what I want to be doing."

What would be an example of a really hard decision you had to make? Was there a time when there were two or three possible choices and it was difficult to decide which one to select?

Yeah, although decisions of that caliber are generally run all the way up to the project manager, but let me tell you about one which hit me by surprise.

A few months before Uranus we were projecting our trajectories past Uranus all the way to Neptune. We wanted to fly a trajectory around Uranus which would send us to Neptune and minimize the propellant expenditure that we would need to make corrections. I was investigating some of these trajectories at that time. Really my attention was more on Uranus, because it was coming up in two months and I was only peripherally running these things out to Neptune. I was going to worry more about Neptune after the Uranus encounter.

I was running these trajectories with some new data that we'd gotten on the Neptune system: a new value of its radius and its mass, a new position in its orbit, and a new position of where Triton, its big satellite, was. A refinement in Neptune's parameters had recently come in and this was the first time we were using it, and I discovered a problem.

The trajectory we had planned on flying over the north pole and then down to Triton was supposed to give us a comfortable clearance of Neptune's atmosphere and of any potential rings it might have. But with these new estimated constants everything shifted in a bad direction. Suddenly it looked like we were coming too close to Neptune, maybe going to go through its equatorial plane where there might be some ring material to worry about.

I discovered this while running trajectories just before Uranus, and I thought, "What does this mean? What are we going to do now? We're about to fly a trajectory by Uranus which is going to take us to this trajectory at Neptune; do we really want to do this? We gotta make decisions right away."

It is unpleasant to find out this kind of thing under these circumstances, so I went and talked to my boss, Don Gray, the Voyager navigation team chief. We decided the right thing to do was to raise a red flag to our project management right away. Surprisingly, nobody else was as worried about this as I was. There was another trajectory we could fly at Neptune, which would have been totally safe but it wouldn't have had as much science value as the one we wanted to fly, close over the north pole and down to Triton. So now the question was, could we safely fly the high-science-value north polar trajectory, or would we have to opt for the other trajectory?

What was eventually decided to do at Uranus was to put us on a trajectory which would allow us at a later time, after we'd had more time to investigate this issue, to go to either one of those Neptune options: either the supersafe one which didn't have as much science value, or the "Let's go for it" one which had all the science value we wanted but had a little more risk to it.

We looked at that problem for quite a while after the Uranus flyby and we concluded—based on a lot of analysis by the navigation team, a lot of analysis from the science teams and scientists around the country contracted out to study this—the project decided that it was safe to fly this high-science-value trajectory over the Neptune north pole.

So we flew it, and it proved perfectly safe, although as we went through the Neptune equatorial plane a lot of us had our fingers crossed. You could hear one of the instruments—I think it was the plasma wave instrument—detect little hits from dust grains and things like that. There were a lot of them, as there had been when we passed through the ring plane, the equatorial plane at Saturn and Uranus.

Even though you've done all this analysis and all this study and everybody has looked at it and the project has approved it, when it's really happening you're thinking, "I hope we make it through this," and we did. We made the right decisions, and we flew the right trajectory, and everything worked out beautifully. We did take a little extra risk, but it was prudent risk; it was not like we did anything foolish, and we got all the science we wanted. But there were moments of high tension, high adventure and there were a lot of tough decisions made, not necessarily by me, although I provided the data to support those decisions.

In the final analysis, the decision as to which trajectory to fly rested with the project manager, Norm Haynes, but the process leading up to his decision was quite adventurous. That's exploration in the truest sense of the word. If we knew all about this system and it was perfectly safe to do, it wouldn't be that much of an adventure, but we didn't know all about it; we had to take some risks and it *was* an adventure.

Now that Neptune has been passed, is there any concern for course determination?

Voyager launched in '77, that's 13 years, and we expect to be able to track Voyager until about 2010, 2015. That's 25 more years, so we're really only a third of the way through the mission. We don't have any more planets to fly by; we're just heading on out of the solar system, in more or less a straight line.

The point I wish to make is that we still have two healthy, operating spacecraft with science instruments onboard which can do all kinds of good science out in that region of the solar system—not imaging science, because you're not near something that's a solid body—but you can sense the interplanetary magnetic fields, particles, things like that. To continue to conduct that science we still need to do navigation on the mission, we still need to track the spacecraft and find out where it is, but we don't need to do course corrections because there's nothing out there solid to target it to.

One trajectory is as good as another, so we'll save the propellant we would have used for course corrections and use it for attitude control, which we need in order to keep the antenna pointed at the Earth, so we can keep in contact long enough to cross the heliopause. That's the boundary that separates the region of space magnetically dominated by our sun from the region of space that contains the ambient magnetic fields of the galaxy at large. We will certainly cross that boundary someday because our spacecraft are on trajectories never to return. The question is, will we cross it while the spacecraft are operating?

We would love to, because then we would know where this boundary is. But we don't know where it is, so all we can do is try to keep the spacecraft in communication with the Earth as long as we can and hope that we will cross this boundary while we're still in communication. That's why we want to save our propellant for attitude control, so that we can go as long as we can.

That would be the last feather in the cap for Voyager: to actually cross into true interstellar space (by that definition—there's a lot of ways you could define that) while it's still operating and able to send back to Earth the fact that it had crossed. That would be a nice last accomplishment. I can't tell you when it might occur, though.

After that we just head on out to the stars. Of course we will be out of communication by the time the spacecraft reach the stars, which will take many, many thousands of years.

Are they both headed in about the same direction?

They're not too far apart in what you could call solar system longitude, but Voyager 1 is going north of the plane of the planets and Voyager 2 is going south. If you drew a line between them, there wouldn't be exactly a north-south line between them, but they're roughly in the same part of the sky in that sense, in an east-west sense.

Was there any concern about directing them toward a particular point after they accomplished their planetary mission?

Not a great concern. There is sort of an interesting little thing there because Voyager 2 is going to pass close to Sirius, which is Alpha Canis Majoris, the brightest star in the winter sky. It's the bright star to the east of Orion. Voyager 2 could have come as close as maybe under a light year to that, with a certain kind of flyby around Neptune.

The trajectory we did actually fly past Neptune was similar to the one that would have minimized the Sirius flyby distance. However, it was different enough, for reasons of science and mission safety, to result in a larger Sirius flyby distance: about 4.25 light years.

Sirius is about 8.5 light years from us now, so we're only cutting it down by maybe a factor of two. It's still sort of a romantic concept to think we're heading somewhere near the brightest star in the sky, but nobody was going to sacrifice Neptune science for a romantic concept of what might happen 300,000 years from now. We were going to optimize the Neptune flyby for Neptune science and take what that gave us in terms of stellar flybys which will happen thousands of years from now.

Is it expected that the two craft will travel at a constant speed now, or will there still be losses or increments in speed?

Essentially a constant speed. When you depart a massive body—in this case the body we'll feel the longest is the sun: it's big—it keeps trying to tug you back, but when you're on an escape trajectory, a hyperbola, what happens is you eventually wind up with your "asymptotic velocity." That's something that you gradually get to, but it takes forever to get to because it's an asymptote, but for all practical purposes we'll be going out at constant velocity. As I recall that's about, on the Voyagers, about 3.5 AUs per year. An astronomical unit is the Earth-sun distance, roughly 93 million miles or 150 million km, so 3.5 of those a year.

One of the Voyagers is headed in the general direction of Sirius?

Let me change that. It's not headed in the direction of Sirius now; it's headed in the direction which Sirius is also headed in, so that at some point 300,000 odd years from now, they'll both be somewhere down there. Actually Voyager 2 is heading in the direction of a south polar constellation called Tucana, but the stars have space velocities also, which can be more than the Voyagers'.

Voyager 1 is heading roughly in the direction of a star called Alpha Ophiuchi, as seen on the current star map. In reality it's not going to come anywhere near that star; that's the direction in space where, if we look out, somewhere in the background that star is.

I think the closest star Voyager 1 will actually pass by is an obscure star named AC+79 3888; AC stands for astrographic catalog, and the 79 refers to declination and the other number is just sort of an index number. It's a dwarf star; nothing remarkable about it. I think the date will be something like 40,000 years in the future. It's not a close flyby; it's certainly greater than a light year.

Coming back to the times when you're still planning the desired trajectories, while you're still headed for one of the planets, what confidence can you have, first, in your own theoretical course selections that you made the right decision? Second, once you or your department has decided that this is the desired course, what confidence can there be that these requested course corrections have actually been transmitted and received by the spacecraft, and that it will then do the desired maneuver? How accurately will your instructions be received?

Okay. First question: how do we know the trajectory we select is the right one? In a sense, I never have to make that decision because the navigators, the mission designers, don't operate in a vacuum—I mean in a personnel vacuum; we certainly operate our spacecraft in the vacuum of space. We basically do trade studies, we look at various kinds of options, and we present those to other elements of the project: to the scientists, to people who worry about other aspects of the mission plan, to project management. It generally is individuals at the management level—science management and engineering management—who will make the final decision as to what trajectory we will fly.

I don't mean to downplay my role either because we have the expertise to prepare those options for examination, so the selection of what's going to be done is only going to come out of the set of things we present them in the first place, so we have a lot of responsibility to understand all of the options and present all the options. We're sort of a first filter on the process because we can see that a lot of the options are not desirable.

We'll generally present those anyway and give our rationale as to why they're not desirable—maybe they take too much propellant or maybe they have a bad sun angle when you want to do a certain observation, or something like that—but we realize that the scientists are going to make the observations; we'll let them have the final say on whether they like this geometry or not. We'll let the project manager have the final say on does he want to spend this much propellant.

So we present the options. We have a lot of power, if you want to call it that, because we control the options that are presented, but we don't make the final decision on which option is selected. We always ran a very open shop with our management on Voyager; we always kept them apprised of what we were doing and we gave them essentially all of the options and all of the data.

I didn't really have to have my own personal confidence that it was the right one, because I'm happy to fly any trajectory. I like flying spacecraft, flying trajectories. I tend to like the more challenging ones because it's more of a challenge, so I was glad when the decision was made to fly the north polar flyby over Neptune.

Was it the right one? In that case it was pretty clear that it was right, but you can find some science investigators who would have preferred a different trajectory, because Voyager carried a number of different science investigations as opposed to, say, just one dedicated instrument. You're always going to get this kind of situation where the trajectory which is good for one instrument is not that good for another, and vice versa, so anything is going to be a compromise.

That's why a body like the Science Steering Group, chaired by Ed Stone, the project scientist, is going to make the final decision. Somebody's going to be unhappy; everybody's going to have some level of satisfaction and dissatisfaction with the choice. That's the function of the manager: to take the heat for that and come up with the right one, relative to that criteria. "Right" can only have meaning relative to a criteria like that. What's right for the group isn't necessarily right for *one* of the investigators, so it has to be defined that way.

In terms of our confidence in being able to send up a command or get the spacecraft to execute a maneuver, when we send up commands that have high priority, we tell the spacecraft, "Once you've gotten it into your memory, send it back to us. Keep it there, but tell us what you've got there." That way we can see that it's got what it's supposed to have, so we can have very high confidence in what's been loaded onto the spacecraft.

Now you can have things happen, like when you fly through a planetary magnetic field you get a lot of charged particles; that's a form of radiation, and if one of those comes through a memory location it can cause a glitch. We had problems like that at Uranus; we had some lines in our images which were caused by things like that. We were able to correct that, and essentially we got all the data we needed.

But you plan for certain things like that which might corrupt the sequence that's onboard the spacecraft. We really worried at Neptune that the onboard timer would lose the proper time, and that the instruments would not shutter at the right time because of what we called "power on resets." That's when some of the electronics take a hit from one of those charged particles, and so various things were done in the sequence to cause the timer to be reset at certain times. Maybe we never would have needed it, but we worried about the fact that we were coming so close to Neptune, we were going to go into its magnetic field, we might get high radiation dosages; let's put these in as added protection. We do studies like that.

In terms of the maneuver accuracy model, you gain confidence as you do the maneuvers. The first one you worry about a lot. As you do them and log them successfully behind you, you get higher and higher confidence. We generally had a model which said the duration of our burn would be in error by two percent, so for 100-second burn that would be two seconds. The pointing of the thrust vector would be good to a very small angle; much, much less than a degree. Generally we found that our spacecraft performed better than our model of the statistical errors. We still used that model because it gave us a conservative way to use the spacecraft.

The performance in executing proved to be very good, and so we got to expect that it would be very good. But despite that, any time you'd turn that spacecraft off Earth-point to do a maneuver, no matter how much confidence in it you have, you are going to worry, and you're only going to start feeling really comfortable when it turns back and you re-establish the signal at exactly the time you thought you were going to, and the Doppler shift from the maneuver looks about right. So the answer is we had high confidence, but we still worried. You always worry when you have so high a value on such a one-and-only kind of affair.

What is the tolerance for error on such encounters?

Generally Voyager tried to have delivery errors no greater than about 100 km. We would do our last maneuver at the point where our errors would be somewhere around that order, 100 km. We didn't need to get them down tighter than that to deliver the spacecraft on a flyby trajectory.

I want to point out, though, that you don't stop doing navigation after your last maneuver. Even though there's nothing more you can do to correct the trajectory, there is actually more that you can learn about it, because even though you can't correct it—it's somewhere within this circle 100 km in radius—you're still going to execute science observations which are going to have slightly different pointing, depending on where you're at in that region.

If you continue to do this orbit determination, you can refine, narrow your knowledge of where you are, because you can track for a few more days, to maybe 50–40 km. So even though you can't change the trajectory, you can tweak the instrument pointing a little better, and we would do that. That's what we would do right up until two days, a day and a half before the encounters. We call that the knowledge error.

There's the delivery error, which is how accurately can I deliver my spacecraft. We generally could tolerate 100 km, but we wanted the knowledge error to be twice as good—maybe 50 km, maybe even a little better: 40 km. That's pretty good, when you're talking about something billions of kilometers away.

Timing errors: well, it depended on how we would deliver our spacecraft in terms of our timing errors. It could have been controlled tightly to a few seconds, or loosely up to 20, 30 seconds, something like that. But even in the cases when we would control it loosely, we would always want to know our timing error very precisely. We would want to know it to on the order of one or two seconds, and we would be able to do that by continuing tracking, even after we had done our last maneuver.

What was the level of difficulty of navigation to Jupiter as compared to going farther out to Neptune? Was the challenge greater as you progressed, less, about the same?

It's the same kind of problem. You use the Earth-based and the optical navigation for both of them, so you're doing the same kinds of things. The things we didn't have at Uranus and Neptune that we had at Jupiter and Saturn, we didn't have as good foreknowledge of the system. Pioneer 10 and 11 had been past Jupiter, Pioneer 11 had been past Saturn, so from those spacecraft we had a good idea of what the planet mass was because, as it bends your trajectory, you get a really good handle on the mass. We didn't have that at Uranus and at Neptune. So you have to rely much, much more on that late data coming in, when the spacecraft gets close enough to sense the planet, when it gets close enough to do optical navigation to correct the ephemeris of the planet and the satellites.

An example at Triton. Triton was the satellite behind Neptune, that we were going to fly by after Neptune, that we wanted to penetrate the occultation zone of. Well, the occultation zone is tied to how big the body is, because it's behind the body. Up until a few months before the Neptune-Triton flyby, our best estimate of the Triton radius was 1800 km, but it could have been as low as 1100 and as high as 2500. So we're trying to hit a zone behind it, tied to the radius of this body, but we don't even know what the radius is to ±700 km.

What that tells you is you must, must, must make those late observations of Triton from your spacecraft, because Earth-based observations weren't going to help us. We'd looked at Triton for years from Earth, and 700 km was the best we could ascertain, plus or minus, as its radius error. We had to observe Triton from the spacecraft, just days out, to really find out what its radius is, then correct our trajectory based on that late estimate.

We didn't have to do that at Jupiter and Saturn because we had a good handle on what the Galilean satellites were. The only satellite occultation we had at Saturn was of Titan, and Titan had a pretty well-known radius, but because Uranus and Neptune are so far out the a priori knowledge of those systems was so much poorer, and we had to rely a great deal more on our spacecraft. That complicated it, but by properly factoring it in we were able to execute the same procedure and get essentially the same kind of quality navigation, but we just had to wait until later in the game to make it happen.

What was it like to work on the Voyager project?

It went on so long it became like a family business. You got to know the people. You knew what they needed. The familiarity of a close working group encouraged the interchange of information and ideas.

We were doing things for the first time ever. We couldn't wait to come to work. Other projects don't have near the romance nor the adventure; they can't hold a team together so well. Voyager was the moral equivalent of war. It was a once-in-a-lifetime opportunity.

Finally, what else would you like readers to think about that we haven't covered, about the first visit to the outer planets, particularly if you imagine that this book may be on a library shelf a couple of hundred years from now?

That gets into a philosophical kind of thing, which is always tough to field. This will probably be a little "cliche-ish", but Voyager was a great scientific endeavor and an engineering accomplishment. I'm an engineer so I saw it from that viewpoint. The project scientists saw it as a scientific achievement. But it represents more than that because it's not just something designed to fill the pages of *Scientific American* with obscure articles on planetary science.

It's really a statement about the kind of society we are or the kind of society we should be. Are we going to continue the human experience, which has always been to explore outward, to go out and conquer new territories? We've got this planet pretty much covered now. We're at that place in our history where we're either going to expand or we're going to contract, and Voyager was a step in the direction of expansion.

President Bush stated that we're going to go to Mars and the moon again. Voyager was an unmanned mission. I firmly believe that there is a role for unmanned missions and that they should be followed by crewed missions. I think the benefits of Voyager go far beyond just the data gathered.

This is a mission that is going to be in the history books 500 years from now, but it won't be in the history books because we learned what the harmonics of the gravitational potential field of Neptune were. Nobody's going to care about that. The reader is not going to care, you don't care, even I don't really care. Why is it going to be in the books? Because it was a great adventure of the human race.

It's a little self-serving of me to say that we need to do more of those, because I derive my income from that—that's my job—but we do need to do more of that, and we need to not stop, we need to follow up Voyager with more unmanned missions of discovery. We need eventually to send unmanned probes out to the stars. We even have a mission at JPL which, while it won't get all the way to the stars, will get out to 1000 AUs. It's called TAU for that reason.

We need to follow that up with technology development which will get us to the stars. We need to colonize the moon. We need to colonize Mars, the solar system, the galaxy, whatever is there for whoever can own it, and we should own it, and Voyager and our other space efforts are just steps in that direction.

HOWARD P. MARDERNESS
Voyager Spacecraft Team Chief
Born April 15, 1931   Mohrsville, Pennsylvania

Upon graduation from the Pennsylvania State University in 1953 with a bachelor of science in electrical engineering, Marderness joined the General Electric Company. After a year of assignments at four different GE locations, he worked on jet engine controls and accessories for seven years in Evendale, Ohio.

He then transferred to the GE Missile and Space Division in Valley Forge, Pennsylvania, working on spacecraft controls and accessories including the Viking attitude control subsystem, which was contracted out of JPL. While with GE he earned a master of science in engineering science from Pennsylvania State University.

He came to JPL in 1973 in support of Viking mission operations. In 1978 he joined the Voyager Operations Team as deputy spacecraft team chief and subsequently became the Voyager spacecraft team chief. He transferred to the Galileo project in 1991 as orbit engineering team chief, his current position.

His many awards include NASA's Outstanding Leadership Medal and Exceptional Engineering Achievement Medal.

He enjoys camping and hiking, and has visited many national parks and monuments.

*The Voyagers were significantly more complex
than the spacecraft I had been working on before,
and it was a real challenge to learn their subsystems
and how they operate.*
—Howard P. Marderness

HOWARD MARDERNESS

The Spacecraft Team is responsible for the operation, health, and calibration of the entire spacecraft, including guidance and control, power, propulsion, telecommunications and information subsystems, temperature control, and scientific instruments.

What were your duties as chief of the Voyager Spacecraft Team?

I was responsible for the technical and administrative management of the team; more specifically, for preparation and monitoring of team schedules, leading the resolution of spacecraft problems, controlling team costs, planning in-flight sequence development, and real-time assessment of telemetry measurement alarms. I was the team's representative to the Mission Planning Office. In general I led and directed Spacecraft Team activity.

How did you happen to be at JPL?

Upon graduating from Penn State, I went to work for General Electric in Ohio in 1954 as a design engineer. I developed a low-voltage ignition system for large jet engines, and did testing, analysis, and design modifications, which eliminated the nozzle limit cycle oscillation problem. Subsequent experience as engineer and manager in navigation, guidance, and control brought me to JPL in 1973 as a General Electric contractor working on attitude control for the Viking program.

I was on the Viking project until 1978, when I came over to Voyager, about six months after they launched. They were doing some staffing changes, and they asked me if I was interested in the deputy spacecraft team chief position. I applied for it and got it, and I've been on Voyager ever since.

The Voyagers were new spacecraft to me, and those were rather trying times. There were residual problems from the launch, and soon after I came over here a couple other problems came up. The azimuth actuator stuck on Voyager 1 and not long after that we had a problem on a Voyager 2 receiver, so there were a number of anomalies being worked on.

But we learned how to cope with those. For instance, with the actuator problem, we figured out that some debris that was inadvertently left in the actuator had gotten into the gear train. We were able to work through that, and gradually the material—it was a soft material—conformed or worked out of the gears. The actuators have worked fine ever since, but back at that time it was a rather trying situation.

One of my concerns when I joined the Voyager project was with the problems they were having, and the real-time commanding that was being done to try to resolve the difficulties. I thought there should be more discipline in the way things were being handled. Gradually I was able to influence the manner in which things were done and the amount of real-time commanding.

The Voyagers were significantly more complex spacecraft than the Viking that I had been working on. It was a real challenge to learn all the Voyager subsystems and how they operate, and the sequencing and real-time operations.

Voyager's first encounter was with Jupiter in 1979, about a year after I came on. We didn't have a lot of time to get ready. There were a couple of enhancements made between launch and the encounter. One enhancement involved momentum cancellation for the digital tape recorder. We were aware that when the tape recorder ran at high speed, it would have a significant perturbation on the spacecraft limit cycle, and that was going to cause excessive picture smear at the encounter. So we developed a patch to attitude control and the CCS (Command Computer Subsystem) to cancel the momentum. That was probably the main enhancement done before we reached Jupiter.

At the encounter itself there were several problems. The main one resulted from radiation effects on Voyager 1, which had the closer flyby of Jupiter. We suffered some permanent damage to the star tracker, but we've been able to live with that. We're still using the same tracker that we launched with. We have a redundant unit which we're saving for when it's needed.

The other major effect from Jupiter's radiation was power-on resets in the Flight Data Subsystem. That caused a number of problems with the manner in which the sequence executed on the spacecraft, and some of that did adversely affect the science data return.

A few additional enhancements were made between the Jupiter and Saturn encounters, but between Saturn and Uranus we made some major enhancements, and still more between Uranus and Neptune. As the spacecraft got farther and farther away, particularly the imaging data quality was of concern because they need to have longer shutter open times, which means more picture smear.

Our goal was not to have any more picture smear at Uranus than at Saturn, and no more smear at Neptune than at Uranus, so we put major effort into that, and also into the data rates. As the spacecraft gets further away, the data rate capability drops, so a number of enhancements were made to improve data return. The science community was very pleased with all the enhancements we were able to make.

The Neptune encounter was particularly demanding because of the close flyby, which increased the requirement for image motion compensation, and also for radio science. In fact, the maneuvers for radio science were more demanding than at any previous encounter. But the pictures came back from Uranus and Neptune with very little smear, and the radio science at Neptune went extremely well. It was probably the best of the radio science maneuvers that was performed.

Also one of the real pleasing things was we figured out how to deal with the azimuth actuator fault we had on Voyager 2 at the Saturn encounter. It stuck shortly after closest approach. We initially thought that it was the same kind of fault that had happened several years earlier on Voyager 1, where I was telling you about that piece of material. So we tried to clear the Voyager 2 fault by slewing the platform and working the material out, but this only aggravated the situation. We finally concluded that it was a totally different type of fault, and was actually a bearing-related problem.

Of course all this was a model; we don't know what actually happened on the spacecraft. But there was a lot of work done to come up with a model for the fault, and also a lot of ground testing. It was concluded that the problem resulted from debris that was generated in the bearing and from the lubricant not getting between the bearing and its shaft. It's not a ball bearing type of bearing, but rather a sleeve bearing. Material getting in there and actually closing up the clearance, which was extremely small to begin with, and also using the actuator at the high slew rate, is believed to have precipitated the problem.

We have four different slew rates we could use. Early in the mission, since there had been no indication from ground testing that there was any particular limited lifetime to the actuator, it was used as scientists wanted to use it to get the maximum science, so there was quite a bit of high-rate slewing done.

From the ground testing and the fault model, it was concluded that part of the problem was that high-rate slewing caused heating in the bearings. We found that when you worked the actuator hard, the fault would recur, whereas if you let it sit for a time or didn't use it hard, then it seemed to work fine, so we didn't slew it for over a year. Shortly after the Saturn encounter there was some slewing done, but then there was a moratorium put on slewing.

None at all?

None at all, for over a year. Then we needed to start thinking about what we were going to do at the Uranus encounter in 1986. To do both the Uranus and the Neptune encounter without the actuator would have been a severe penalty to science, so we recommended that we start using the actuator again, in a conservative manner. We started to do that in early

1983, and we found the actuator worked fine. We only used it at the low rate, particularly to get ultraviolet and imaging science.

Primarily the science for which the actuator was used was science that was needed in preparing for the Uranus encounter. We did some testing with it to see that it would work under the type of higher intensity activity that it would experience under the desired near encounter environment, and it went through those tests fine. It went through the Uranus encounter, and we used it conservatively all through the Uranus-Neptune cruise period and the Neptune encounter, with some medium rate slewing in the near encounter period where time was of the essence.

Was that the first time since the problem at Saturn that you used that higher rate? Did you use medium rate at all at Uranus?

We did use it at Uranus and at Neptune. Before Uranus we also used it to do a few slews as part of a test. We were trying to mimic the level of slewing that was going to occur in the near encounter time frame if we did the type of sequence that science wanted to do, and it worked fine there as well.

After we started to slew again, in 1983, we were looking for a means of evaluating the health of the actuator. One of the real concerns was that the actuator might slew all right during the cruise, where we weren't using it too heavily, but that maybe in using it in cruise we might be degrading it as time went on, and that lo and behold, by the time we got to encounter, we'd get a fault condition again.

Encounter preparations started well before the encounter, in all cases. The preparations for the Uranus encounter started soon after the '81 Saturn encounter. If science was going to plan on using the actuator, they needed to have fairly high confidence that it was going to work all right when they got to the encounter period, because of all the time and effort it would take to replan the sequences for not using the actuator.

The alternative was to use roll turn maneuvers in lieu of azimuth slewing, but these two sequences are totally different. To develop both of them would take a lot of time. So we were desperately looking for some way of evaluating the health of the actuator, using the limited telemetry that we have. We thought of trying to measure the voltage that it took to drive the actuator, but there was no way of varying the voltage, and there was no telemetry for measuring it either.

We finally came up with an approach. The actuator is a stepper motor, and the attitude control pulses the motor in order to cause it to step. The frequency of the pulsing determines the slew rate. We have four different slew rates we could use.

Between the pulses, each time the motor is stepped, the power was being applied for the whole width of the pulse. For instance, if the duration of the time between pulses was 120 milliseconds, you'd have power applied to one of the coils of the stepper motor for 120 milliseconds.

We were able to come up with a software patch to attitude control where we could vary this pulse width. We found that a healthy actuator, a new actuator, would start to slew with only a 3-millisecond pulse width. So with this new patch we could vary the pulse width from below 3 milliseconds up to 10 milliseconds. By doing this we were able to evaluate whether the actuator was getting sticky or sluggish, by the pulse width required for it to slew to the next position.

We found that the actuators on Voyager 1 were pretty close to what you'd expect from a new actuator, and we could in fact see the effect from the anomaly we experienced back in 1978. We could see a little residual resistance in the region of the gear train where the piece of material had gotten into the gears. When we evaluated the actuators on Voyager 2, we found that they also were basically restored to a new condition.

We did these tests periodically up to Uranus encounter, and we saw no degradation in the actuator. We did the testing also going up to the Neptune encounter and the actuators worked fine, so we believe that the whole problem was having driven the actuators too hard; having operated them at the high and medium rate too much. The hypothesis was that when we started using the actuator again, a year or so later, the lubricant had gradually migrated back into the bearing surfaces.

That was a real highlight over the long term: recovering from anomalies of that sort, and being able to come up with ways of dealing with them, and having the hardware restored to normal operation.

Amazing. Stepping back to a more basic level, how can a variety of instructions be given to the spacecraft—to attitude control, the computers, and the 11 science instruments—when you were limited to just one narrow channel of communication?

We have the CCS, which is the computer command subsystem. In it we store what we call CCS loads—in fact we loaded one this morning. The CCS memory, at the specified times, then executes to the other subsystems the commands for performing activities. If we want to do a scan platform slew, the CCS issues at the specified time the command to attitude control to slew the actuator.

This is onboard the spacecraft?

Yes. Generally these CCS loads contain all the commands that the CCS is to issue to the other subsystems for about a three-month period.

The instruments are controlled primarily from the flight data subsystem, but the CCS does send commands to the FDS (Flight Data Subsystem) in order to do calibrations. Or if we want to change the state of an instrument we can do that, but we tell the FDS that, and then the FDS controls the science instruments.

The same for the digital tape recorder, DTR; the CCS controls that. If we want to do a playback, CCS tells the FDS to give us the data mode that's needed for the playback, and CCS tells the DTR to run up to the proper speed and start doing the playback of the data through the FDS down to the ground. So everything goes through this central computer.

The attitude control also has a computer which responds to the commands from the CCS. If it gets a command from the CCS to do a slew, it decodes that command, then causes the scan platform to slew.

How do you relay your requests, which cover a large variety of things, to that computer when you have only a narrow range of frequencies to transmit on?

No, it's not a function of the frequency. There's a fixed frequency that's used to communicate through the receiver on the spacecraft. It's just a stream of 1s and 0s, a bit stream which goes into that receiver, just like into your radio receiver. That stream of 1s and 0s is then entered into the CCS computer in such a fashion and with time tags on it so that when CCS interrogates a certain location in memory it recognizes that as a certain command to be executed, like to attitude control. So it's just a bit stream.

Is the command then entirely dependent on the bit stream of 1s and 0s, or in addition is it related to the particular time at which you send that command?

This morning we uplinked a CCS load which spans about a three-month period, so no other commanding is required in order for that sequence to take place. In preparing that CCS load we go through a sequence development process where we define what we want the spacecraft to do on different days, and all of that is then stored with this one load for the full three-month period. The FDS has the prime clock on the spacecraft. Everything is timed off of that. This stream of 1s and 0s has timing information in it which is stored in the CCS so that it knows at what time to issue the different commands to the subsystems.

This load that you sent up this morning; how long did it take to send the complete package?

About an hour. It varies, depending upon how many memory addresses we're going to load. The maximum load length is a bit over an hour. We have some that are only 10 minutes, and when we get further into the Voyager interstellar mission, we'll have loads that will probably be only a few minutes.

Then you have to wait for it to be cleared out of the memory subsystem before you can send up another load?

Yes, you can't write over any of that memory. After, let's say, the first month out of the three months has been executed, you then can write something else into that part of the memory. We do that sometimes with what we call minisequences.

It now takes about five hours and 30 minutes to reach the spacecraft, and then it takes another five hours and 30 minutes to get back to the ground, so it won't be until this evening that we will verify that the load that was sent to the spacecraft this morning has actually been accepted by the spacecraft.

In the meantime you don't send up anything else, because you don't want to interfere with your first batch?

We could send something else, but not to those memory locations.

Is there any final thought you'd like to relay to readers of this book about your work, your impressions of the project?

I think the main thing is that we've got good hardware on Voyager, and our main goal is to not ask the spacecraft to do anything that would be degrading to it. There have been missions terminated because people do things they shouldn't do.

Like the two Soviet Phobos missions?

Sometimes they load things into the memory which cause an unrecoverable event to take place, so we work hard to be sure that we don't make any errors which would cause the spacecraft to do something it shouldn't.

Voyager has shown that even after you launch a spacecraft, there are many enhancements you can make to improve its performance beyond what people envisioned at launch, by taking advantage of capabilities that exist on the spacecraft.

Of course the only thing you can do is reprogram the software; there's obviously nothing you can do to the hardware. If a hardware fault occurs, and we've had some, there's not much you can do about those; you try to learn to live with them, or you try to do something to alleviate or remedy the fault. We have some faults like on Voyager 1, one of the two FDS memories faulted back in 1981. We've given up on trying to recover that; we've learned to operate with the one memory. We do fine. Whereas with the actuators we were able to get those to work again.

The spacecraft, both of them, are marvelous pieces of hardware. Although we've had some faults with hardware, in general the hardware has performed way beyond what was expected at launch. Voyager was initially designed for the Jupiter and Saturn encounters, and beyond that it was really a hope as to what could happen afterward. The Voyager 2 trajectory was designed so as not to preclude the Uranus and Neptune encounters. We've done those, and now we're planning the Voyager Interstellar Mission, which is to go on to the year 2020, so the hardware is amazing.

I've been really impressed with the people on the team. They're always enthusiastic. If there's any need beyond the norm, everyone is there, willing to work. You don't even have to ask them to work extra hours; it just happens.

CHARLES AVIS
Deputy Chief, Imaging Science Data Team
Born February 10, 1951   Houston, Texas

Charles Avis received a bachelor of science in physics from Rice University in 1973 and a master of science in astronomy from the University of Texas at Austin in 1976.

In his 20 years of JPL image processing, Avis supported such flight projects as Viking, Voyager, and Galileo. Early in Voyager operations he was instrumental in developing time-lapse movies as a science product. In his role as the Multimission Image Processing Subsystem (MIPS) Voyager cognizant engineeer, he was responsible for the planning and operation of image processing support for the Uranus and Neptune encounters. For his Uranus encounter efforts, he received NASA's Exceptional Service Medal in 1986. Avis currently is MIPS cognizant engineer for the Cassini project.

His recreational interests include genealogy and the Boy Scouts, and he is a player and manager of a softball team.

Only you lucky readers of future generations
get to view the events of this time in the proper historical context.
Was this the "Golden Age of Space Exploration"
or mere baby steps quickly surpassed?
Did the evidence of fossil Martian life (presented just recently)
become a scientific and social milestone or a humorous footnote?
Just wondering.
—Charles Avis

CHARLES AVIS

What should the public know about how all these wonderful things happened?

That's a tall order. Our support of Voyager is image processing, from real-time acquisition of the images as they are transmitted from the spacecraft, all the way down to making CD-ROM discs of the archival data. But our most visible product is the custom work we do for press releases. For image processing support like that we work under the Voyager Imaging Team. That's the collection of scientists that control what the camera's doing and what happens to the data that are acquired.

We receive instructions from the scientists as to how to proceed: what pictures to process, and what they want to see out of a product. Then we use our programs and creativity to get to that end result from the raw data. We work closely with a representative of the Imaging Team or with the Imaging Team scientists themselves to get the products they're after.

For instance, we get requests for processing. They write out roughly what they want to see and what images to start with. They may be expecting a set of color pictures of Neptune in the next day, so they'll write, "Use these pictures to make a color picture of Neptune." Once the images come in, we'll process them and usually get to a certain point where we think it's what they want. If they're on-lab at the time, we get them to come look on our display device and we ask, "Is that what you're after?" to see if they want some changes or not.

It's kind of subjective for the press releases as to what the Imaging Team is trying to get across to the public in that picture. It may not be just a natural color of Neptune showing features; it may be some false-color version where the red in the picture you see indicates high methane areas. So it's how they want to manipulate the data to show what they want to show. It's essentially an analyst using all our application programs to get the product the requestor wants to see. That's our custom image processing effort.

We do the real-time processing. The telemetry coming from the spacecraft goes to the ground stations, it's transmitted to JPL, and we pick out the imaging records. Other kinds of science records are made, but we pick out the imaging ones. We build up an image and store it on disc in our computers, and at the same time catalog it as to what image number it is, some of its characteristics, and where we actually put it when we created it so we can find it. We make reports of where things are, and something about its quality—how many missing lines it had, things like that. So it's instantly cataloged and made available to the Imaging Team for them when they peruse their data and start initiating some of their requests.

We also have the job called "systematic processing", which means we are required to process every image that was acquired in a standard way, so that it can be compared easily. We also put out photo products in a standard format. That roll of film there contains systematically processed pictures—it is a roll of 100 or so images all processed the same, laid out the same, all produced here, and archived in a library. Also versions of these images are shipped to the scientists. So we make photographic archival products, and also digital products, to store away on magnetic tape and optical disc, and also to distribute to other sites. Those are the main elements of our support.

We do a little processing for the plasma wave subsystem (PWS) instrument and the planetary radio astronomy instrument, mainly because their format that is transmitted to the ground is very similar to the imaging format. So we extract that data as well and can convert the PWS data into an audio signal so the scientists can interpret what their data contained by listening to it. I've never been involved with it enough to know how they can tell what they can tell by listening. But that's another part of our computer system: to allow them to turn it into sound.

How did you get involved in Voyager?

In my graduate studies in astronomy, I did a little image processing on Uranus pictures from ground-based telescopes, so I kind of learned on my own. Then in '77 I was hired out of graduate school to be an image processing analyst on MJS (Mariner Jupiter-Saturn), which turned into Voyager.

When did you first became aware of Voyager?

Last year I was looking through an old scrap book of mine. It had an article I cut out from the *Los Angeles Times* back in '65 which talked about this mission which was going, in '77 or so, to visit all the outer planets.

I've got a viewgraph of it around here somewhere. So I was aware of the space missions back when I was growing up. Voyager was just about ready to go to the Cape when I started here. I started in May and it was launched in August and September.

What was the first thing you remember about it when you came here? What did they have you doing?

One of the first things, a couple of weeks after I got here, was a trip to Wisconsin to check out their atmospheric motion studies software, because we would be studying the atmosphere of Jupiter. The University of Wisconsin had some software that studied cloud motions in satellite pictures of the Earth. That eventually turned into the software package which we use now to determine cloud velocities on other planets. I didn't develop the software. I wasn't even familiar with what we were doing yet when we went out there to check that out.

We had a lot of work to do to get it ready for the first pictures to come back: things like checking out programs, making sure we could find the little reseau marks on the camera. You have to know those to a high accuracy. These are little calibration marks; there's 202 of them in a square grid pattern on the camera. They appear in every picture, but in slightly different places on every picture, so you have to have a program to find them. From previous ground calibration you know where they should be. This allows you to undistort the picture. So we had a lot of work to do in '77 making sure we could find them, find them fast enough, and locate the blemishes in the camera so we could get rid of them.

We had to check out planet center finding. In a planetary image usually there's a sharp edge of the disc that's illuminated, which is the limb of the planet. We were developing software to find that sharp edge. Since you know the range, what angle you're looking at the planet, and how big it is, you know that the points along that edge should fit a known shape. That's one of the major pieces of software we use; one of our critical operations is to determine where in the picture the planet is. This is extremely important for making measurements in terms of latitude, longitude, map projections, and things like that.

We were using Earth-based pictures of Jupiter to see if the software worked—to determine the parameters that it would need to operate properly. We were testing out all kinds of stuff.

Then, shortly after it was launched, the problem arose of not knowing whether the scan platform was latched or not, so we had a flurry of activity here because they turned the camera to take pictures of the edge of the calibration plaque—a metal square of known brightness attached to the spacecraft. If the edge of the plaque came where they expected it, then everything would be okay, but if it was off, then there could be a problem.

We wound up with a big blurry picture because the calibration plaque was only two meters away, so it was not in focus. They wanted to know where the edge of this unfocused thing was, so some of us were busy running filters on those pictures to see if we could determine where the real edge was. If it was in the right place, it would alleviate everybody's concerns—at least until the next problem happened.

What was that?

It was probably when the receiver went out. Other project people know more about that. We're not really involved in uplink and spacecraft problems, unless they affect the images.

Looking back over these dozen or so years you've been with the project, what stands out most vividly?

Well, I feel like I've been on a tour of the solar system, really: seeing the highlights of the outer solar system, filling in all the gaps in the knowledge. I feel like I've been along on the ride with the spacecraft. We pass by Io and then we get a call saying, "You've got pictures showing volcanoes on Io; we've got to have them processed *now*." I get peaks of adrenaline in every encounter, but we accomplished the Grand Tour and it's been great to be onboard.

In working in this facility, we have opportunities to do other projects as they come along, too. I've done Mercury pictures here for other missions, and Venus and Mars. Voyager's been a step up from the previous missions in how much they tried to get out to the public. There was so much emphasis on press products—new kinds of products for the public, like time-lapse movies.

Time-lapse movies applied to space essentially began with Voyager. Jupiter is so big and could be seen so far away that there was a long approach during which we could take many pictures. We weren't just there and gone; we were approaching it for many months.

We've been involved in all the time-lapse movies from the beginning. Nobody was quite sure what kind of product they would be, even though they were planned.

We put together what they call now the "Blue Movie", the Voyager 1 approach movie of Jupiter, where you take pictures at a constant longitude—every time the Red Spot comes around you take another picture. Then you put it together into a two month approach, seeing the planet small, and zooming up and getting big. The Red Spot stays fixed, and everything starts swirling around it. When we first came up with that, people said it was one of the great science products to come out of the space program: seeing the atmosphere of Jupiter speeded up by a factor of whatever. The fact that we made it on a medium that can go out on your home TV brought it closer to everybody.

What were your hardest decisions to make—such as cases in which somebody would be disappointed no matter which way you decided?

A lot of the decisions that I've had to make fall into two categories. One is getting ready for encounters. You have to deal with the facility people to make sure the facility is ready, and with software people who are bringing in new software. You have to stay on their case about the schedules they've made. You can't delay an encounter; you've got to be ready, and so there are decisions about those types of things.

Then there are decisions *at* encounter. We would get a ton of requests coming in during the day. We have a schedule we've made up of personnel to do the work. Deciding who to give what request to can be a little difficult because we have a wide range of capabilities of our people. Some of them are extremely quick and very talented in doing the custom image processing. Others are better suited to grinding out the standard procedures that we relied heavily on. So it was my deciding what resources we have to work with and trying to match them up with requests that come in.

There's not a lot of leeway with the requests because they're wanted at the next day's press conference, and they have to get done today sometime, somehow. I didn't really have any decisions that were the type you're talking about. It's mainly who's going to get this very difficult request that will keep them up most of the night, and whether they can handle it or not, and whether I can go home, since it's midnight, and leave them with it, or do I have to stay and help them.

Sometimes it worked, and sometimes it didn't. Each encounter was more difficult than the last. I was only in charge of the last two at our facility. Neptune was harder than Uranus, but then we had a new computer system at Uranus, so it was more difficult to get ready for it.

I should back up a little bit. The two Jupiters were pretty close together. Then another year, which seemed like a long time, then there were two more Saturns. We kept the same computers the whole time, because there wasn't enough time to change anything in between. After Saturn we had four and a half years, and our entire facility changed from one computer type to another. That was during the time when they expected the Galileo to be coming along at any time. Then it got delayed. We were upgrading quite a bit, so when Uranus came along it became the first encounter with this new system.

There was the big unknown as to support because of our new computer system. The project planners were not sure we had enough computer power to handle the data they expected. There was a lot of getting ready for that encounter and testing performance. It was hard to draw on past experience because it was a brand new computer system. So that encounter was pretty difficult.

On the other hand, the geometry of the planets made it easier in one respect because we didn't have any data coming in after 10:00 at night, so they couldn't ask for any late efforts.

Between Saturn and Uranus the capabilities of the Imaging Team to do image processing themselves went up greatly. They knew more about what we were doing and how to do things on their own. We gave them some work stations in another building and so they did more work by themselves, which they had never really done before. At Saturn and Jupiter they could only browse through the images and write down a request for us to do. At Uranus they looked through the images and then did some of the work themselves, which was a little different. It involved some of our time to help them when they reached snags in their processing, but they were learning how to use our software.

Between Uranus and Neptune their capabilities went up even more, and we were able to pipe the data over to the computer system they set up, so that they would be off our computer system and onto their own. We just piped the data over to them, so if they wanted to play around, they could. Their capabilities went way up, and they were concerned about overloading the system. They didn't want, and we didn't want, all that effort to still reside on our computer because it would swamp it considerably. It was very close to being swamped at Uranus, and we didn't want that same thing to happen at Neptune.

The Neptune encounter was more difficult for us because the geometry allowed data to come in all night long. Even though we still had a deadline of the 10:00 morning press conference—we had to get products to the photo lab for them to make 100 or so prints to hand out—that didn't stop the Imaging Team from saying, "There's a picture coming down at midnight; be ready for it." There was even a set of pictures of Triton coming in at 5:00 in the morning, and they wanted to have some prints by the time of the press conference.

We could plan for that. Even a year in advance we were hoping that, if we knew the identification numbers of the pictures, we could automate the processing, so when the data came in it would just get done. That pretty much worked, though we needed to have somebody there just to make sure it worked. We wouldn't want to slip up and miss the press conference, so the hours were a lot later at Neptune.

Each encounter is unique, with different highlights for each one. At Uranus we were concerned about the computers being so slow. At Neptune our concern was with how many hours people were having to put in, and when is that data finally going to get here? Our capabilities went up, but so did what was being asked for; the detail and complexity of it.

For instance, back at Jupiter pictures came in regularly as we approached Jupiter, but it wasn't until a couple years later that we put them into full planet mosaics, like a map of Earth, with latitudes, longitudes, and the equator. We did that for each rotation of the planet. It rotates once; you see the whole thing, you put it in a map. When it rotates again, do another map, so every 10 hours we'd get another map. It took us a couple of years to put together all the ones reasonable to do—maybe 250 of them. That gives a time-lapse movie of the whole planet in map form.

But at Neptune, they wanted to make the movie as soon as we got the pictures; do it as they came in, not a couple of years later. So there was a lot of planning, a lot of procedure writing, and watching batch jobs crank them out daily. But it did work. That's an illustration of how different a request became; as we could do more, they'd ask for more. We were always performing up to our limits.

How did this affect people's lives?

People put in long hours. When I worked 12:00 p.m. to 8:00 a.m., I had to sleep during the day, so my wife took our baby back to Texas. One of the experiment reps had a heart attack around the time of the Neptune encounter, and we've lost a couple of science managers.

What have been among the hardest aspects of your work?

Well, there are hard products to make and then there are hard situations to deal with.

What would be an example of a hard product?

Trying to mosaic high-resolution Triton pictures fast, like for tomorrow's press conference. We had the capability to do that, but some things went wrong. Some of the geometry information given to us wasn't correct for a couple of days; for instance, the information telling how far we were from the planet. For a couple of days we'd get people telling us, "It isn't right." So we were thrashing about, trying to find other ways to get this information. And by the time we did, they had corrected it. That was one of the more challenging products.

Another example was some of the time-lapse movies we've done, where we're approaching Saturn, or doing a rotation movie of Saturn and its rings. It requires so many pictures, and then they have to be recorded on film or videotape in exactly the right sequence. We have to make sure everything's in the right spot.

Those are probably the more difficult products we've had to do. The more pictures involved, the harder it is. You're dealing with more bytes of data.

Are there limits on the amount of output that your unit can do? Are the limits technological? Are they budgetary? Are they personnel? What keeps you from expanding capabilities?

Our facility is funded by a flight projects support office. They see what computers are necessary for the next mission, for instance. So, in one sense that's budgetary, because if you convince people that this encounter's not going to get done properly without a second computer or something like that, it'll get supplied. There's no way they want an encounter to go bad.

At Uranus was the only time we had concerns about what we could get through the machines. The limitations deal with how much you can get done in one day. A person can only do a certain amount of work a day, no matter how fast the computer is. People must manipulate data, keep track of things, type things in, make decisions.

It still takes a certain amount of time to do these things. We're budgeted only so many people, but that has absolutely nothing to do with how much they're going to request you to do. They will request based on what the data looks like. If there are 20 neat color pictures to do today, they'll request them all. Whether you can normally do 10 doesn't matter; you'll get the 20 requests. You'll make sure they're prioritized, and you get out the 10 most important, or 15; whatever you can manage.

But there's no problem with the imaging people understanding that either. You come to the end of the day, or you come to 5:00, and you look at what's been requested, what's likely to get done, and then rearrange the priorities, and make sure the imaging people know, the Public Information Office knows, and the photo lab knows, what's going to be coming for the next press conference. That was the purpose of our 5:00 p.m. meeting. Things would get requested after that, but they knew at the meeting what those would be, so they got updated on what would be coming up in the morning.

Is your work impacted at all by the great distance and therefore the time lag with which you're working? You're working several billion miles away.
Well, what it meant was we got a lot fewer pictures than we got at Jupiter, and we got them farther apart—not every 48 seconds, but every five minutes. I'd hate to go through an encounter with the quantity we got at Jupiter and the time we had at Neptune. At Jupiter we could see it so far away, because it is so big, that we had time to generate lots of products. And back then there weren't a lot of products requested. During the approach we made one color picture a week. Then we got close and we did three or four a day. At Neptune we got 15 requests a day.

Now if they had three times the amount of pictures to look through, they'd probably request three times as much. So in one way the distance helped us at our end. They're not going to request as many products if there aren't as many raw data frames coming in, because they come down slower. They're going to pick out the best.

It also affected our time-lapse movies. Because of the great distance and the time it takes for each frame, they couldn't take as many frames. So the time-lapse movies had gaps in them because, instead of taking a picture here, they really had to be over there doing some other experiment at that time. They couldn't pack as many data collection events in as we could at Jupiter. So a time-lapse movie might have been 60 or 70 percent filled at Neptune, whereas at Jupiter it might've been 98 percent filled. As a result some of the products might've suffered.

Were you able to make semi-real-time requests for adjustments on the spacecraft because you weren't getting back the kind of picture data that you needed, or was that all cut and dried months ahead?
We don't make decisions like that. What we do is, as we build the frame in real time, we display it on a monitor. Then it goes to the camera engineers. It's displayed in front of the camera engineers so they can see the picture. They can interpret any anomalies they see, then they deal with them. We don't make any decisions about it. But because we see every picture we can say, "That thing doesn't look like what we normally see; take a look at it." Usually they've already seen it by then and are doing something about it.

What lies ahead for you and Voyager?
I think Voyager has done more than its prime mission, and is just going to settle down and look for the heliopause. We'll be getting a few downlinks a year of PWS data as they look for that, but that will be the extent of our Voyager support after this September. We'll just be producing a few tapes a few times a year. I guess it's appropriate to turn off the cameras; there's nothing else out there that we could look at, and it costs a lot of money to keep the capability here on the ground to do the processing.

As soon as we passed the Neptune encounter, we started dropping off people. Magellan kept sucking them up to get ready for its encounter, except for me and a couple of others, and these others will be gone by the Fall [1990].

You won't be keeping any Voyager personnel?
Not in image processing. There won't be any imaging after October.

What would you pass on to somebody else coming on, based on your experience with Voyager?

I'd say, especially for a long duration mission, be ready for major changes in capabilities and what's going to be required. Don't think that what you're starting out with is what you're going to end with. Requirements are going to change, whether they're written down that way or not. With enough planning you can do some very useful things.

You should think of things that ought to be done—even if they weren't required of you; even if they haven't been asked for—and find a way to get them done. For instance, I knew that it would help tremendously if we could tell ahead of time what pictures we should start processing. For example, what Triton pictures would they want press releases of? If they would tell us a week ahead of time, we could be ready. About a year in advance, we started assuming that we would have that information, and we developed an automated procedure in time for encounter. So that as soon as those pictures hit the ground they could start to be processed, without any human intervention. Then when people came in in the morning, these pictures would be out there waiting for them.

Later the request would come in and, sure enough, they'd want those pictures made into color pictures. But we had predicted that and we had some automation to deal with it. Instead of coming in in the morning, seeing the request, and then starting from scratch, we'd be able to do it ahead of time. That was not a required capability; it was just something I knew would make our life so much easier if we could get that done. You shouldn't limit yourself to what's actually stated in requirements.

What is your impression or image of the spacecraft itself?

I don't have any feeling one way or the other. I didn't have anything to do with the hardware.

How will you feel about leaving Voyager?

I've heard that a lot of people feel sad when they have to go on to something else, but I figure Lewis and Clark weren't sad when they got to the Pacific Ocean. I think Voyager was great, and triumphant in its whole mission, but it's time to close the books and go on to accomplish something else.

Scientists

Only a few of the scientists on the Voyager teams were JPL employees. By the time of the Neptune phase, there were 130 investigators at universities, observatories, and aerospace companies scattered about the United States, Europe, and the Soviet Union. They were supported by more than 200 research assistants and students.

JPL scientists provided an interface with their outside colleagues, coordinated scientific activities on other flight projects, and worked in general to build up JPL's scientific capabilities. The four JPL scientists interviewed here include the Voyager project's two chief scientists, a deputy, and a representative who focused on one specific experiment, imaging.

ARDEN ALBEE
Chief Scientist

Born May 28, 1928   Port Huron, Michigan

Arden Albee has been dean of graduate studies since 1984, and professor of geology and planetary science since 1959 at Caltech. He served as the chief scientist of JPL from 1978 to1984, and was the project scientist for the Mars Observer and Mars Global Surveyor Missions from 1984 to the present.

He received his bachelor of arts (1950), master of science (1951), and doctorate (1957) in geology from Harvard University. His research includes geologic field studies: petrologic, electron microprobe, and isotopic investigations of metamorphic rocks, meteorites and lunar samples, and development of techniques for electron microprobe analysis.

In many years of service to NASA, he has been a member of the Lunar Planning Committee, the Terrestrial Bodies Science Working Group, the Physical Science Committee, the Space Science Advisory Committee, the Solar System Exploration Committee, the USA-USSR Joint Working Group for Solar System Exploration, among others. Albee chaired the Lunar Science Review Panel for five years and has chaired a number of working groups and studies on Martian and cometary missions. He served on the review boards for the Magellan, CRAF (Comet Rendezvous Asteroid Flyby), Cassini, Mars '98, and Stardust missions. He received the NASA Medal for Exceptional Scientific Achievement.

Albee has served on the councils of the Geological Society of America, the Geochemical Society, the American Geological Institute, and the Association of Graduate Schools. He is also on the Graduate Record Exam Board and the TOEFL (Test of English as a Second Language) Exam Policy Council. He has been an associate editor for the *Bulletin of the Geological Society of America*, the *American Mineralogist*, the *Journal of Geophysical Research*, and the *Annual Reviews of Earth and Planetary Science*.

He has been active in the Boy Scouts, Girl Scouts, the Presbyterian church, and various charities. With a one-acre yard he is a gardener and remodeler, and an occasional skier and flyfisher.

Change happens!
You can only embrace it and understand it
in terms of past changes.

—Arden Albee

ARDEN ALBEE

There's an incredible amount of footage that was taken at the Lab during the Voyager encounters; taken in the science rooms as things were coming down. Some of the most spectacular events were indeed recorded live, and the scientists' reactions to them were also recorded live. Jurrie van der Woude could point you to that footage. He is a key person for that kind of information. He has been there throughout the whole mission and is very much involved in being the interface between the press and the scientists.

What are your thoughts about Voyager?

It's hard to say without sounding like you're echoing Ed Stone, or Brad Smith, or Larry Soderblom. All of them have been magnificent spokesmen for the project. One of the key things about the project has been the capability of those spokesmen and the way they have been able to transmit science through the media and the project.

If you compare the encounters, you see the improvement of communication during that period because the first encounter was being run much like previous encounters. TV tape wasn't quite as readily available. TV coverage wasn't as readily available. Color was coming for the first time. Viking had some color, but it wasn't available in virtually real time.

Voyager came at exactly the right moment to utilize the advances in the media, the advances in communication. The scientists involved had previously been on Viking and other missions, so they were experienced. It all came together in a very good way for communicating the science results to the public and to the world. It's remarkable when you travel around the world: Voyager pictures are everywhere, from ads to cigarette wrappings to everything else.

When did you first become aware of Voyager, and how did you get involved with it?

I became chief scientist of JPL about a week before the first encounter. As chief scientist I was in effect the ombudsperson in the director's office for all scientists. Consequently, I went on all of the review boards for all of the ongoing projects.

By that time Voyager was really ongoing. Nevertheless, there were review boards for each new encounter, which faced up to problems which had occurred in the previous encounter. The famous stuck actuators prompted an incredible amount of effort at the Lab, trying to understand how to be ready to work around those stuck or sticky actuators, for the next encounter and for subsequent encounters. Thousands of hours of effort were put into running the actuators—under different load conditions and under different temperatures and different voltages and so on—trying to understand them, and then trying to devise a work around that would permit them to get data even if they gave problems.

The sequences had to be designed very, very far in advance. That meant, after the platform stuck, almost all the sequences had to be double or triple sequences. In other words there had to be different routes, different branches to enable you to get results even if something happened. Those work arounds are a remarkable achievement by the Lab—the work around in the radio system, the work around in the actuators to be able to make them function through all this time.

Another interesting aspect was the things that were planned and put onboard which could only be useful if new developments happened. For example, the capability to put on a coding to increase the data rate was put onboard at a time when it wasn't even sure you could ever use it. Also, the increase in the capability of the Deep Space Network: first increasing the size of the 64-m tracking antennas up to 70 m, basically for the second encounter. Then, for the third encounter, beginning to go to arrays of antennas, and the fourth encounter ending up with arrays that used the Very Large Array in New Mexico to get even greater enhancement. So a whole series of things were done.

The spacecraft as built and launched could not do the things it did later. There was a continual development to take advantage of the capabilities for change that were built into the spacecraft, and then to do work arounds on Earth for things that went wrong with the spacecraft. That was a remarkable achievement.

I think that the performance of Ed Stone had a very important influence on how JPL and NASA worked. Until he was project scientist on Voyager, typically the project scientist was sort of a staff person who had certain jobs as spokesman and so on, but did not have the key role that Ed Stone has made of it. In subsequent projects, when it came to selecting the project scientist for Magellan and for Galileo, we did it quite differently.

For Magellan, Steve Saunders is the project scientist; a working scientist working in radar, working on Venus' surface. For Galileo, Torrence Johnson is the project scientist; again a leader in working on Voyager and working at Jupiter. In each case we have changed the role for the project scientist by trying to get a fairly senior scientist who is active and involved in the mission to be the project scientist. I guess I'm the exception, as the project scientist for the Mars Observer.

What were your first impressions when you came to Voyager, during the first encounter?

That's going way back. What you remember is all the excitement at the time of the encounter, the excitement of new worlds, totally new worlds, new ideas. The idea of volcanoes which are not silicate volcanoes but which have processes working similarly. The ideas of icy planets, satellites, with phase changes acting as if they were silicate planets but having liquid water interactions, much as we think of melt-rock interactions on the Earth; Hawaii made out of an ice flow, for example. There were challenges to the way you think as a geologist, challenges to the way you think about processes.

Yet, when finally you began to understand it, the actual thinking was not that different. There were different materials, different conditions, but the things that were going on were really not that different. The deformation of the icy planets or the surfaces of them—these again are not that different from the deformation of rock under different conditions. It requires a slight torquing of the way you think, but not a revolution.

When you think back to the encounters, it seems like they were always in the middle of the night. I remember, at the second or third encounter, that young Governor Brown was at the encounter. Closest encounter was supposed to be at 4:38 a.m. or something like that, and everybody else was practically asleep watching Al Hibbs go on in the Blue Room, the JPL television studio. All of a sudden the governor jumped up and said, "It's 4:38! Why isn't anything happening?!" He had so much energy he couldn't sit still. The fact that closest encounter was going on and Al Hibbs was ignoring it was just too much for him.

What else? Sitting up on the ninth floor, I felt a little bit jealous of the guys who were down there in the heart of it. I was always acting as host to a room or something for VIPs and therefore I seldom got down to watch the data come in.

Where was the heart of it?

In Building 264, the scientists had their room where the transmissions came down, so as each image came up on the monitor they would be looking at it and trying to understand what this new image was. Having been involved in doing the imaging sequences, there were certain images they were waiting for very excitedly. It would get to the point where they just couldn't leave the place because they were waiting for that next image.

Part of it was trying to milk that excitement, to try to get other things happening with NASA, because this glory period of Voyager was during the depths of NASA funding, of NASA problems, of lack of new starts, of the Challenger problem culminating it. Then the

Challenger problem was transmitting on into even further delays. At the same time, Voyager was having these great successes. They were something NASA had built years and years and years before, but had started nothing since.

So a very distinct challenge was trying to take this tremendous public relations success in communicating to the public, and trying to figure out why it didn't communicate to NASA management and to the congressional people in the same kind of way. During that period many of us were trying to make some changes in what was happening at NASA.

During that period we failed to go to Halley's Comet. Proposed mission after proposed mission did not get funded, and eventually there was a change, a total revolution. The Solar System Exploration Committee met and tried to look at a totally different way to do it: at small spacecraft being launched more frequently, and so on. Mars Observer came out of this atmosphere.

But even at the time when Voyager was having such triumphs—having been a very carefully, unique hand-built spacecraft—at that same time Galileo was having problems and getting postponed, because of the Challenger problems in part, but also because of money problems. Ulysses too. This whole string of things which are now about to happen were in just deep yogurt. You were always going from the triumph of the encounter to fighting for the life of planetary exploration. This was a real contrast.

NASA administrators always said nice things about the encounters, but they never backed it up by approving missions; at least not ours. The same with Congress: encounters were great times to say nice things, but the problems that were plaguing NASA in those days were very, very serious.

What were some of the significant things you were aware of in your role as chief scientist?

Without looking at dates it's difficult to know exactly when, but the Venus radar mapper—VOIR, the original one was called, Venus Orbiting Imaging Radar—had a new start and then was canceled. It was eventually replaced by Magellan, which was a much scaled down spacecraft, which was finally approved only after a long period of time and of work. Ulysses—the U.S. part of it got canceled. It was originally going to be a two-spacecraft mission, one part European, one part U.S., and that was canceled. Galileo continued to slip: Space Telescope, a whole variety of things.

So throughout that whole period, the atmosphere in terms of getting new starts was just very, very poor. There's talk about the golden decade of the 1980s but almost nothing was started in the 1980s. It was dependent upon what had happened in the 1970s.

What were some of the hardest decisions that you as chief scientist had to make? Personnel, for example?

Chief scientists don't have to make those decisions because you're on staff. You're not in a line position.

What sort of things, then, does the chief scientist do?

The problem was how do you overcome the inertia that had come about? How do you get a new start, when there'd been such a long period of time without new starts? There was a period when JPL was in danger of dying for lack of new starts, while at the same time it was having these magnificent Voyager encounters. This contrast between them is what I'm referring to.

If you talk to Bruce Murray, you will certainly get this same thing. He, as director, was trying to get almost any kind of a new start. He was trying all sorts of things to make these things happen in order to be certain that the Lab could continue to function and live. Punctuating the struggle to make something happen were these encounters, which came almost as a shock, in contrast to the general pessimistic outlook.

Bruce worked on it one way, as director. I worked on it as a member of a number of NASA committees composed of U.S. scientists who were trying to formulate missions and trying to get them new starts. Particularly the Solar System Exploration Committee then tried to pick up on a new approach to make that happen and eventually succeeded in getting Magellan and Mars Observer. My direct interaction with Voyager was smaller than that of the scientists who were actually on it.

How does Voyager compare with other projects that you've had some awareness of?

I think Voyager is extraordinary, and probably there's nothing else that can compare with it. It was built at the end of a series, so it was the culmination of the development of the whole string of Mariner spacecraft, on up to the Viking Orbiter and then the two Voyager spacecraft. A long series of developments were perfected in Voyager. From a cost point of view, because it was this long string of continual new spacecraft, one every year or so, it was very, very cost-effective. By any kind of measure it was relatively inexpensive, in terms of the capital cost of putting together those spacecraft and launching them.

I suspect that with inflation and so on, the operating costs later on for such a long, long mission probably ended up eclipsing the capital cost, the initial cost. But each of the encounters was treated as being a new opportunity and was recosted and rejustified in that way. It certainly justified itself. Many, many changes were made in the course of it, and that's a long period of time in space science to have a single basic instrument making new discoveries. It's impressive.

It's impressive that you were able to do things with it that were hardly even thought of when you lost physical control of it, right at launch.

That's one of the remarkable things about Voyager. Its total power was about like the lightbulb in a refrigerator, so it is not a high-powered instrument, yet it will never be topped, in the sense that it made so many discoveries of new worlds, new things. When another spacecraft goes to Jupiter, it may get better pictures, it may get better color, it may get more multispectral data but nevertheless it's only doing it a little bit better than Voyager. It's not going to make the initial discoveries.

So that's what Voyager has going for it: the string of unique discoveries. You can't rediscover a new planet, a new satellite. You can only do it once. It's got so many discoveries to its credit that this cannot be matched by any other mission, and will never be. There aren't that many more worlds to be discovered.

What would be some of the lessons learned from the Voyager project?

The lessons that should be taken from Voyager were not taken or will not be taken because Voyager's success was based upon careful planning, with abundant money to put in reserves, to put in the way you designed it and built it, and so on. That day is past. Now everything gets stretched out in a way that does not provide sufficient resources for the planning stages, or for critical times. The very fact that two spacecraft were built shows the difference: now we only build one because we can't "waste the money."

That lesson was how important it is to put the resources in at the appropriate time, and to have the reserve. By "reserve" I mean building the kind of capability into the spacecraft that you would be able to use later on. Being able to work around things that happened to it. It's not as if nothing failed or nothing broke; lots of things failed and broke, but we were able to work around them.

It's not clear that Mars Observer has that extra reserve built into it, which would enable such work arounds. And that's true of all the missions we're flying now. So in this fiscal environment I think it's impossible to apply the lessons learned from the Voyager project.

Thinking back over your whole experience with Voyager, what would be some of your memories?

Late night encounters. The excitement of everyone getting together for the encounters—Governor Brown, as I mentioned. As chief scientist, I end up giving lots of speeches, so I have talked about Voyager many times. A lot of my memory of it is tied up with some of the talks I gave.

You've seen the display in von Kármán with fade-ins and fade-outs. Well, they attempted to reproduce that in a portable setup, a stacked set of projectors with a computer to fade them in and fade them out. Twice I attempted to take that, once to Arizona and once to Santa Barbara, and I never succeeded in making it work when I got there. I would always start, and then something would get out of synch and it would blow up on me. So my memory of that particular device is not very good.

Better that down here than Voyager up there.

Right, right. So you always ended up saying, "We can make it work up there, but we can't make it work here in your lecture hall."

What were the most significant events that come to mind?.

The so-called volcanoes on Io were the thing which got to me the most, in the sense that they illustrated a totally different kind of planetary dynamics than I had thought about. It was a system of totally different gases and totally different phase changes in order to explain vulcanism.

I thought that that discovery and the way that planet had of getting rid of its excess heat was a remarkable discovery. I would rank that high, plus the entire class of icy planets and the fact that deformation, tectonics if you will, occurs on the surface of those icy planets in a way which is understandable; not totally alien to the way a geologist would think.

As a geologist what sort of puzzles still remain for you that were posed by Voyager data?

These are significant discoveries because they open up new areas for thought; they pose problems which continue to be problems. They're not just answers; they are discoveries which point to processes which you get a hint of, but when you really try to get down to the details you get into difficulties in fully understanding exactly how they work and exactly what they mean. That's why Galileo will go back to look at some of these things in much greater detail.

What would be an example of intriguing new data?

The whole cycle of how this happens on Io. Presumably the energy source is basically tidal with respect to the planet, but exactly how these things cycle, whether the material escapes from Io—is Io getting smaller with time? Is the cycle occurring there really cyclic? Is it always showing this kind of vulcanism? The sulphur pools: did we happen to catch them at just the right time or do they always behave this way? I think this whole set of questions about the dynamics of Io is very intriguing.

The same thing is true on the icy planets. What is the time scale of the things you see on their surfaces? And how rapid is that process that you're seeing there? Do the surfaces of those really span the whole of solar system time or are they sort of a miniature, a subset of it that we may be misinterpreting to have them reach way back in time, because some parts of them are heavily cratered. There is a whole set of those sorts of problems.

What was your biggest disappointment with the Voyager mission?

It's hard to point to the disappointments because in the long run it did everything that it was expected to do, and much more. The biggest frustration was the contrast between the triumph of the encounters and what was going on with NASA at the same time. But maybe that just continues forever; maybe it was just that that was when I began to encounter politics.

If, at the very start, you had attempted to say exactly what you expected from Voyager and tried to justify it in very cold ways, it would have been very difficult to do—to put as part of the justification what it did for world spirit, or the impact of the images of the "family portrait", looking back at the solar system.

Voyager did things for the way people think and the way people understand, which could never have been put in as part of the justification for the expenditures at the start. A few people had the vision that these things might be the return, but perhaps no one fully understood what the return could be, what the impact of it would be. So perhaps there is a lesson there: that sometimes the visionaries are their own justification.

What has Voyager meant for you?

It does change the way people look at things. Let's take the ecology movement. I think the ecology movement has as its roots the fact that, for the first time, we looked back at Earth and saw the relatively thin skin of oceans and greenery and atmosphere. That in itself did not come from Voyager, but putting that in context with the other planets and seeing what they looked like as compared to the Earth—that did come to a great extent from Voyager. So it is the contrast of how we look at our own planet, and what are the things that make our own planet livable as compared to all these others. That is an important aspect of Voyager's contribution.

MOUSTAFA T. CHAHINE
Chief Scientist, Jet Propulsion Laboratory
Born January 1, 1935   Beirut, Lebanon

Moustafa Chahine grew up in Lebanon. He came to the United States as a freshman to attend the University of Washington in Seattle where he obtained his bachelor of science in 1956. Four years later he received a doctorate in fluid physics from the University of California at Berkeley and joined JPL as a research scientist. From 1975 to 1978 he headed the Planetary Atmosphere Section. In 1978 he was responsible for establishing the Division of Earth and Space Sciences at JPL and for overseeing the diverse activities of its 400 researchers. In 1984 he was appointed chief scientist for the entire Laboratory.

Throughout his tenure at JPL, Chahine has been directly involved in research in atmospheric sciences and remote sensing. In 1969 he developed an exact mathematical method (the Relaxation Method) for inverse solution of the full radiative transfer equation. The method is used today for deriving atmospheric temperature and composition profiles from satellite observations of Earth, Venus, Mars, and Jupiter. Subsequently he developed a multispectral method using infrared and microwave observations for remote sensing of clouds. These methods were applied in 1980 to provide the first global distribution of the Earth's surface temperature. In 1988 NASA selected his instrument (the Atmospheric Infrared Sounder, AIRS) as a facility instrument to study the Earth's atmosphere from the NASA Earth Observing System.

He has served on the NASA Earth System Sciences Committee, which developed the U.S. program to study global change. He is currently chairman of a closely related activity, the World Climate Research Programme's Global Energy and Water Cycle Experiment (GEWEX), which is organizing a worldwide effort to study the effects of increased greenhouse gases.

He is a Fellow of the American Physical Society, the American Association for the Advancement of Science, the American Meteorological Society, and the British Meteorological Society, and is a member of the International Academy of Astronautics. He was awarded the NASA Medal for Exceptional Scientific Achievements, the NASA Outstanding Leadership Medal, the William T. Pecora Award from NASA and the U.S. Department of the Interior, the Jule G. Charney Award of the American Meteorological Society, and the Losey Atmospheric Sciences Award of AIAA.

His current research activities are in the study of the Earth's hydrological cycle, atmospheric radiation, and water vapor distribution.

My research work determines where I am going to be and what I am going to be doing during most of my daily life. As such, it constitutes for me a framework for interactions and involvement because it imposes a discipline. Yet, from time to time I still take a moment to reflect, to think big and to think small about many things: how to learn new subjects and contribute more to society, how to venture and how to face disappointments, and how to become the person I want to be.

—Moustafa T. Chahine

MOUSTAFA CHAHINE

While I was a graduate student at Berkeley, I saw the picture of Dr. Pickering and Jim van Allen and von Braun holding the model of the Explorer 1, the first U.S. success in space. It became my goal to come and work for JPL, where this thing happened. So before I finished my degree I asked my professor if he knew someone at JPL. He knew aeronautics Professor Hans Liebman at Caltech, and I got in touch with him. I received several offers, but I was always waiting for an offer from JPL.

When it came, that did it! It was very exciting to get that letter, and when I look back at it now, 33 years ago, I still remember it as a very exciting day. It didn't matter how much money they offered me, which was less than what I got offered from other places—that didn't matter.

So I came here to JPL from Berkeley to work on basically anything related to space exploration. At that time it was the re-entry problem; how to protect the spacecraft.

My first personal experience with Voyager was as a co-investigator with a team of scientists from JPL and the Caltech campus who were proposing an instrument to fly. Our instrument was not selected. Instead, Rudy Hanel's interferometer was.

That was a disappointment. But in retrospect I feel that the interferometer was better suited to Voyager than the experiment we proposed, because our experiment was mainly designed to get the best information from the first two planets Voyager visited, Jupiter and Saturn. The interferometer had a broader spectral range, and although it didn't do as well for the first two planets, it was much better suited for Voyager's total tour. The interferometer has done extremely well. The objectives of Voyager changed, and Rudy Hanel's instrument was more suited to adapt for this change.

I continued to follow the progress of Voyager because I'm an atmospheric scientist interested in looking at Jupiter, Saturn, and the other planets to see how each atmosphere is different. Later, my interest and responsibility in Voyager grew when I became JPL chief scientist so that I am now looking at all applications of the Voyager information we have collected from the outer planets.

Historically, Voyager's first disappointment was in not getting approval for the Grand Tour; the mission was approved for only two planets. But, as it turned out, the resilience of the instrument and the ingenuity of the engineers and scientists here managed to turn this decision around, and the Grand Tour was finally realized.

What have we learned from this mission? We know that there is a scientific order in the formation of the solar system, an order in its chemical and physical structure we were not sure of before. The inner planets all have solid surfaces. The outer planets, except Pluto, are gaseous giants. The moons of the outer planets show enormous diversity in their surface composition, with abundant biochemical materials (dark colors). The shepherding moons in the rings of Saturn were not known to us. We haven't yet come to a definitive conclusion on how the solar system was formed, but we have gathered enough information to begin to characterize it. This is a prerequisite in any field before a general theory is derived.

Voyager has brought us to the edge of our solar system and as explorers we wonder what is next? I believe the next phase will be to go past our solar system, not necessarily physically, but going beyond to collect information about other "solar" systems. We want to compare and contrast our solar system with other systems. In order to understand our own solar system we must look at others; it is not enough to study just one.

We became acquainted with all the capabilities of Voyager as a robot or machine, and we taught it new functions. We managed, for the first time, to change the computer language on a spacecraft in flight and to teach the spacecraft new tricks which it didn't know when it left Earth. These machines have a very primitive intelligence, so to speak, but we learned quickly how to interact with them. We've succeeded in extending the human mind, the human reach, beyond the spacecraft that left Earth.

Voyager was not originally designed to do what it did. During its flight our technology here on Earth evolved beyond the level when Voyager was launched. We managed to transfer some of the advancement in technology electronically to Voyager. This is a new experience in human exploration.

Out of this experience we gained a new understanding of machines as extensions of human activities, not just as mere emissaries who go out to do a job they were designated to do. There is no difference in this case between having a human being out there or having a machine which can extend our reach. We can look for opportunities in flight and make machines function as we want to provide us with data.

What does it mean? It means that in the future when we are exploring the surface of Mars or Venus, we are going to explore it without a clear-cut distinction between human exploration and robotic exploration. If you look at the lunar program in the 1960s, human and robotic exploration were clearly distinct activities. The experience from Voyager will translate into new space exploration initiatives in which human elements and robotics elements work together. When we go to the next phase in exploration we would like to develop a plan and approach in which the distinction between machine and human roles becomes fuzzy. And with time I hope this separation will disappear and we'll be dealing with the two aspects of planetary exploration as extensions of each other.

What other lesson is there to be learned from Voyager? Once you send a spacecraft out into space, you cannot get it back. No matter how many problems we have here on Earth, no matter how many budget cuts, no matter how many layoffs, we cannot call the spacecraft back. At a time when we were examining the impact of the shuttle disaster, when we had two or three spacecraft sitting here not able to be launched, Voyager was basically unaffected.

All in all, the Voyager mission has been the most uplifting experience I've experienced at JPL in my 31 years here, from the moment the space program started. What do I compare it to? I compare it to the first launch of a spacecraft by the United States and by JPL. I compare its impact to the landing on the moon. All these were very important events. Voyager is way at the top when you compare its achievements. Next to the landing on the moon, this remains my number one best experience.

Landing on the moon was number one?

Yes, and Voyager would be number two. It comes very close. Both are great, but as a human being, landing on the moon remains my number one experience. But as a JPLer, my most memorable, most uplifting experience through my 31 years at JPL is Voyager. Voyager helped our morale at a time when we probably would have sunk. With Voyager our spirit didn't sink; it took us to the edge of our solar system.

Are there any anxious moments that come to mind regarding Voyager?

The first anxious moment I had was whether the spacecraft was going to survive the strong magnetic field of Jupiter. We had problems in communications with the spacecraft at that time. It survived.

The second critical point was when we lost contact with Voyager for three days. The spacecraft was programmed to shut down when it didn't receive any message from Earth, reorient its own antenna, get a lock and beam back to Earth. Voyager followed this sequence and regained contact with Earth.

These were probably the two most critical moments for me. We waited anxiously to see if this craft was going to make it through these difficulties—and it did. This is Voyager: many little things that work together made it so successful, almost perfect!

A remarkable project. What are the challenges in your position as chief scientist?

Getting an organization like JPL to work on a mission like Voyager for a long period of time, with the same interest and zeal, requires lots of motivation. External motivations do not really survive for a long time—not the environment, the office, facilities; you name it.

To be able to work on a project like this for 20 years, you need internal motivation. Each one of the individuals who started working on Voyager felt after awhile that he or she had put something of themselves into the spacecraft, be it a transistor, or a code in the machine which made the spacecraft function, or in communications in taking care of it, taking its pulse day after day after day.

We have managed in this case to get a small group of people—150 to 200 people—who felt a part of the family. Voyager was like their baby and each one had something to do with it. A contribution has been made and continues to be made in making this "kid" excel and function perfectly. This is the feeling we've had here with Voyager, and it continues until now. We viewed Voyager as our prodigy at JPL, and it's treated as such.

The other question is from the point of view of chief scientist here at JPL. Our long-term goal at JPL is to acquire new knowledge—from the Rangers, Mariners, Viking, and Voyager. We've done that to a very large extent and the solar system is now better understood. We've put it in focus.

When I was in high school or even at the university, Earth and Mars were crisp in my mind, but the rest of the solar system was not in focus. We didn't have enough information. Neptune was described in two pages in a textbook of 500 pages.

The challenge now is to transform the information from Voyager into knowledge. What is the most critical information we need in order to complete our understanding of the origin and the evolution of the solar system? This is one very broad question with at least three requisites: theoretical modeling, detailed in situ information, and a look at other planetary systems around other stars.

Another very important question right now is what part Earth plays in our exploration of the solar system. What is the relationship between Earth, in its macro and micro structure, and the rest of the solar system? We are a part of it and it is a part of our solar system. Earth is definitely unique, but Venus and Mercury have some kinship to us. What kind? How do we differ from them and how do we resemble them?

This is a new focus for Earth and planetary scientists. I have a sense that it will have a strong effect at JPL and the rest of the outside community. I would like to see to what extent, from a planetary perspective, we can bring in a new view of planet Earth and where we (the Earth and us) are heading.

We are working with people in the university community to see how we can extend our current understanding of the solar system to solar (stellar) systems. We are looking in-depth into the relationship between the evolution of life and evolution of the planet Earth contrasted to that of the nearby planets.

We've always tried to look at Venus and Mars globally to see how they have evolved and then compare them to Earth. Now we are taking a very global look at Earth itself, and we are trying to see how it has evolved, how it is changing, and then compare it to the other two planets. This is the reverse of what we have done for 25 years. Looking at comparative planetology from an Earth perspective gives us a long-term outlook on how planets evolve and how anthropogenic activities on Earth change and modify our own planet.

The challenges of the space program are endless. When you look inward, you look at Earth, you look at the smaller scales, you look at molecular evolution, you look at life. When you look outward at the formation of other solar systems, formation of galaxies, formation of the universe, you look at the beginning. Within this infinite spectrum of explorations, we need to select the set of information most likely to answer our questions.

I'm hoping that these interviews will make readers aware that the wonderful achievements of Voyager did not just happen; they weren't just automatic, but that there were hundreds, perhaps thousands of people putting in millions of hours to bring about these results.

The results we got from Voyager would not have happened without the thousands of hours which scientists, engineers, workers, laymen, and other individual supporters gave. They have thought up new ideas and focused on solutions. It was not taken just as a job by those who worked on it. To very many it was a career. I felt that it was a part of me as a scientist here, and I wanted to do my best to make it excel because, in turn, it will make *me* excel. I saw in it, we all saw in it, extensions of our own capabilities. We saw in it a challenge which permitted us to show what we can do.

When you look at the achievements of Voyager, especially during the Neptune and Uranus encounters, it's very hard to differentiate between excellence in human abilities and excellence of the spacecraft. The spacecraft performed as much through our own innovation as with its own capabilities. Voyager showed what can be done, and we showed that we can do it. In fact, every move of Voyager was an element of our own thinking. Our brains went with it wherever the spacecraft went.

What was your title with respect to Voyager?

I was head of the science division. I was also the chief scientist of JPL.

How would that relate to "project scientist"?

If they are from JPL, they are appointed by and are under the overview of the chief scientist. The chief scientist at JPL makes sure that the JPL people who become project scientists are given enough support to perform their functions as project scientists, and also as research scientists maintaining their standing in the research community.

Ed Stone as project scientist for Voyager was not a JPL employee. Before he was appointed lab director, he functioned as a Caltech professor working as a project scientist. As a Caltech professor he was not under the overview of anyone at JPL, he was an independent entity. I had no formal relationship with him, other than to make sure that his needs from the science division at JPL were satisfied. The only official interaction I had with him was when I wanted him to give a presentation about Voyager to JPL employees or to congressional staffers who came to JPL. I had other concerns beside Voyager.

Ellis Miner decided early in his career to devote his energy to Voyager in an administrative, managerial capacity rather than continuing in research, so he is assistant project scientist, supporting Ed Stone.

At the time that Ed Stone was selected project scientist, JPL did not have a science division. It was established in the late 1970s. Before that, there were scientists at JPL but not a science division, so we did not attract scientists of the desired caliber to allow them to assume the role of project scientist and be accepted as such by the outside university community. We had good scientists but not a separate organization for them, so the scientists were always working in support of engineers.

That was one of my accomplishments, before becoming chief scientist: creating and managing the Earth and Space Sciences Division at JPL. Now all project scientists except Ed Stone and Arden Albee (on Mars Surveyor) are active JPL scientists.

So the creation of the science divisions was a major milestone for JPL.

That is true, very true.

What were some of the difficult decisions facing the project?

Take for example the last encounter, with Neptune. We wanted to get close enough to be able to look at Neptune and to look at its moon, but we didn't know precisely enough where we were. The dilemma at that time was do you play it safe, guarding what you have, or do you play it less safe, but still within reason, and get what you want?

This was one of those cases where you cannot easily make a decision. You would like to get the information, but if you don't come close enough, you will not get it. Yet if you do come close enough, you are going to risk the spacecraft.

So an important decision was made: *not* to play it safe and stay out away from Neptune, but to adopt a dynamic approach whereby you modify your trajectory as you learn more about the planet. We began to adjust our thinking, day after day as the spacecraft got nearer and nearer to Neptune, and we determined to what extent our position was risky and therefore we should move away, or to what extent, as we learned more, we could get closer.

These trajectory decisions were complicated by other issues, such as the limited supply of fuel onboard the spacecraft. Do you take more pictures and use more of the fuel in the spacecraft to keep it fixed on the planet, or do you save this fuel for other maneuvers?

Decisions like these had to be made, but there was no happy combination, so our solution was to modify the criteria as we learned more, as we got closer to the planet.

Was that the kind of decision that would come up to you, or would it usually be decided at a lower level?

Decisions are made at the project scientist and the project manager level only if there is a crisis. Usually the question was how do you optimize your situation or your resources? This must be decided only with very strong cooperation of all those who are involved: the scientists who want the information, those who are tracking the spacecraft, and the engineers who want to keep it safe.

Because of the long time needed to send a message to the spacecraft and get it confirmed, we had to do that almost every time. That was probably a factor in creating the intimate family of Voyager at JPL—the 150 scientists and engineers who had to make decisions like this all the time.

Do I understand correctly that most of the time you yourself did not have to make choices? If Infrared wanted one thing and Imaging wanted something else, you were not the one who had to decide?

Correct. That was done by the two leaders, by the principal investigators and by the project, in sessions that lasted beyond midnight.

Did you sit in on some of these?

In some of them, yes, as an observer, but it was *their* decision.

What were the kinds of things that did come to you?

Because of the crunch on facilities at JPL, there was a debate about taking the spacecraft operations and moving them outside JPL. Should we maintain the project office, the support work, within JPL or take it outside JPL and handle it remotely?

I came on the side of keeping it here, because nearly all of the scientists working with Voyager requested it. They descended on me to see if I would be their advocate to keep it here because the situation was delicate. We didn't want to risk disturbing the continuity; we didn't want to make a mistake while transferring this operation from within JPL and putting it outside. We would have only one opportunity for such a mission in our whole lifetime, and we wanted to keep chances of accidents or mistakes near zero. After more debate there was no question that it would be risky to move the operations outside JPL, so the Lab kept it here.

Were there other decisions you were involved in?

This was the only time I came in, but I was backed up by 100 percent of the scientists involved in Voyager.

I probably am not the first person to inquire about your activities.

Correct. In the early 1980s, when I was manager of the planetary atmospheric science section, I was visited by an inspector from the Government Accounting Office who wanted to find out what I do with my time and to see if I followed proper accounting procedures. He asked me questions about what I do; I told him what I do and explained how much time I spend on it. Then he asked me what do I do when I finish this work? What do I do with my spare time? I told him that if I have time left, I think.

This threw him off completely. He thanked me, stood up and left, and I have never seen him since. I guess thinking was not one of the items on his list, and he didn't know how to react to it.

I've repeated that story many times to people here at JPL because, in the rush of doing things, trying to meet deadlines and trying to comply with government regulations, we forget that we really should take time to think. It is the only way to make progress.

One of my relaxations is to sit down with a blank sheet of paper, select a problem, and try to write down a couple of ideas. It could be about science, or society, or ideas for my family, whatever. But if you look at it as a science problem, how would you handle it? You come up with nice innovative approaches that we normally don't use because we tend to simply act and react, without really thinking.

Speaking of innovative ideas, your proposal for Earth-sun images would give us some insight into the complexities of a project. What were some of the issues involved in it?

After we went by Neptune, I had an idea for a final imaging project. We had taken the look back, the "family portrait", the "picture of the century", the 60-frame mosaic showing the sun and six planets. What I wanted to do was take periodic images of just one frame: the sun and its surroundings including us. If you take one every two weeks, in one year they are going to show the rotation of Earth around the sun, and for the first time in human history you would have proof positive that Earth rotates around the sun, not the other way around.

The family portrait was taken at one moment in time, so there was no motion in it, and it is very difficult to see the Earth. But if you take a series of pictures of the sun and its surroundings, what you would see is the sun, which remains in one place against the unchanging background of the distant stars, and there is this tiny dot making a full circle around the sun.

You would need only one picture a month, or every two or three weeks, to portray this yearly cycle and show that, after all, Copernicus was right.

What was involved in this? Why couldn't you say, "Let's go ahead and do it"?

Money and technical feasibility.

How does money come in?

The cost was basically human involvement, to continue to take daily care of the spacecraft, and to examine the camera to see if it could survive 24 exposures.

You can let the spacecraft be dormant quite awhile and then get in touch with it every now and then, but if you want to keep taking pictures every two weeks or so, you have to continue to worry about day-to-day orientation. You have to check the health of the spacecraft day after day after day to make sure that it is ready to function when you need it.

Also, when you have the camera looking at the sun, every time it looks at the sun it's seeing a very hot source, and that doesn't help the system. So now you need a specialist, someone who, before we do it, will look at the camera and determine whether it is going to survive 24 exposures to the sun, or can it survive 50 exposures, or will it survive only five exposures?

What was your estimate of the cost?

A million dollars, half a million. One should view it not just from the cost but also from the benefit. How many students are going to view it? How many people is it going to attract to mathematics, to physics, to chemistry, to science? For two or three generations downstream they would have this as the only visual depiction of the solar system showing the Earth rotating around the sun.

What about technical feasibility?

That was what finally stopped the proposal, because we found that we would not be able to see those planets. The illumination and the photographs were not going to be of good enough quality to allow you to see the planets orbiting around the sun.

So it was that, rather than budgetary constraints?

The two go together: if it would be worth it, then you could get the budget for it. But the images from Voyager's camera at that distance would not be clear enough to see.

You noted that we have more or less completed the initial reconnaissance of the solar system, and the next stage will be to look out toward other solar systems. When do you think this might occur?

We now have the engineering and technical capability to begin to survey some of the stars like Beta Pictoris. Around the turn of the century, we will have instruments in Earth orbit which will be able to study with more detail the structure of the regions around one, two, or three selected stars like Beta Pictoris, which is 50 light years away from us.

Later, when we set up observatories on the moon, we will have a much better chance to look not only for the details of the regions around a star, but zero in on a Jupiter-sized planet, if there is one around it, and begin to study its composition and its structure. Within the coming 25 years, I am confident that we will have instruments in Earth orbit or on the moon with the main objective of looking at and studying other solar systems. Twenty-five years is not very long.

In closing, what would you like to leave for readers of this book from your experience or observations?

In retrospect, what human ingenuity and dedication can do always comes as a surprise to me. No matter how much you've done, you will always be surprised by how much more we can do. Even now, or 10 years from now, someone with dedication and creativity might find something very new to do with Voyager. These two—ingenuity and dedication—are an unbeatable combination. It's always a winner.

It should be applied not only to Voyager; it applies to every aspect of life around us. The American experience is full of examples like this. The United States is a place where people are free to think and advocate. I feel this because I've experienced it.

ELLIS D. MINER
Assistant Project Scientist

Born April 16, 1937   Los Angeles, California

Ellis Miner received his bachelor of science in physics from Utah State University in 1961, and a doctorate in astrophysics and spectroscopy from Brigham Young University in 1965.

He then joined JPL's Infrared Science Team for a series of investigations carried aboard the Mariner 6 and 7 flybys of Mars, the Mariner 9 orbit of Mars, the Mariner 10 flybys of Venus and Mercury, the Viking 1 and 2 orbiters of Mars, and the Voyager 1 and 2 flybys of Jupiter.

He assumed a broader scientific role in 1978, as assistant project scientist of the Voyager project during planning, execution and analysis of the Voyager 1 and 2 missions to Saturn, Uranus, Neptune, and beyond. In 1990 he was appointed science manager for the Cassini mission to Saturn and Titan.

He was twice awarded the NASA Medal for Exceptional Scientific Achievement for his contributions to the success of the Saturn and Uranus encounters, and he received NASA's Outstanding Leadership Award for his role in the Neptune encounter.

As a member of the JPL Speaker's Bureau, he has given more than 300 talks on astronomy and the space program. He has authored more than 50 articles based on scientific results of the unmanned space program and on his own telescopic observations, and has published two books on Uranus.

His love of astronomy led him to purchase his own small telescope for his recreational use. He also enjoys hiking, bicycling, tennis, genealogy, and singing. He and his wife are active in civic and church affairs and have served in various leadership and teaching capacities in the Church of Jesus Christ of Latter-Day Saints.

> *My personal priorities are family, church, occupation, hobbies, communities. I believe it is my responsibility to develop my talents and capabilities and utilize them in the service of God and my fellow men and women. I also believe it is my duty, in partnership with my wife and eternal companion, to teach my children and grandchildren to do the same. I believe that obtaining genuine happiness is the design and goal of our existence.*
>
> —Ellis D. Miner

ELLIS MINER

I came to JPL in 1965 and have been an ardent supporter of the space program since the beginning.

This was a mission that lasted 12 years, and it's a little hard to put your life on hold for 12 years. So in many cases individuals planned some of the events in their own lives around the intensive encounter periods of the Voyager spacecraft at Saturn, Uranus, and Neptune.

For example, we had two of our women on the project who planned their babies so that they would be conceived and born between the encounters of Saturn and Uranus, and then again between the encounters of Uranus and Neptune. I don't know how many individuals away from Voyager would plan their lives around something of that nature. During that period of time, my wife and I also had the last three of our children, so it was fun being involved in similar events in our personal lives as well.

Did you, too, plan it to fit into the encounter schedule?

I didn't have to do that as much, because I wasn't the one who was pregnant, but we did manage to have our children at times which were not the busiest times for Voyager, so that worked out well. But my responsibilities with my church also fit around that sort of a situation. I ended up having those church responsibilities at times when I could handle them more easily, when the pressures on Voyager were not so great that I would not have been able to fulfill my church calling.

I did, however, have the same church responsibility a second time during the Neptune encounter, when it *was* a very busy time. I was trying to write a book on Uranus as well, and so there was a period there of six to eight months where my wife seldom saw me except on weekends. I was here at the Laboratory preparing for the Neptune encounter during eight hours a day, and then another six hours a day plus Saturdays I would be in here working on my book, and then on Sundays my church responsibilities kept me busy. So we said "Hi" and "Bye" from distances quite often. That turned out to be a rather stressful time, but we came through it okay. It did not harm our relationship at all.

A lot of my kids were getting to the point where they were approaching marriage age. My oldest daughter was married just about the same time as the Neptune encounter, so we had that to plan in, among other things. Several of my kids had started in college or in other types of activities which took them away from home for a period of time. I have seven children altogether, so these family activities were something we also had to fit in among all the other things.

My wife, however, enjoyed very much the opportunity for me to be involved in the notoriety, as it were, of the Voyager program. She was able to accompany me on some of my trips to Europe and to Australia and saw areas of the world that she otherwise might not have been able to see. She also enjoyed picking me up on television on occasion and radio broadcasts, and having her friends say, "I saw your husband's picture in *National Geographic*." So it was fun for her as well.

Your kids—what was their attitude?

Oh, you know, a prophet is never accepted in his own country. My kids were aware I was involved in this thing, and I brought them home pictures and other memorabilia on occasion, but for them it was nothing spectacular; it was just normal everyday life. Interestingly, though, of the four that have graduated from college thus far, two ended up in scientific-type fields; not astronomy, but in computer science, which of course I used heavily on the Voyager project.

I understand that you helped JPL Voyager personnel with personal counseling.

I was never officially in that capacity, but individuals, when they had difficulties, seemed to come to me and want to talk things through with me. And since I had a role where I wasn't directly their supervisor, it provided an opportunity for me to do that without it being threatening to them. So I did it, and got to know some of them a lot better than I might otherwise have been able to.

Let's see, other counseling situations. Early in the Voyager flight period, when we were learning to use the spacecraft in the way that it should be used, on occasion we made errors. In one case the errors caused severe damage to one of the scientific instruments. We had a situation where the principal investigator for that investigation was discharged and somebody else was named in his place.

As a result of that problem?

In part as a result of that, in part as a result of his attempting to utilize the experiment in a way that wore it out before it even got to its first encounter with Jupiter. That negligence was unfortunate, and I'm not sure it was intentional. I think it was mainly a matter of the principal investigator just being so enthusiastic to collect the data that he neglected to calculate the effect it would have on the instrument's lifetime, and a few other things of that nature. We ended up for the Uranus and Neptune encounters flying an instrument that was partly disabled, that could only access some of its filters, could only access some of its analyzer positions, and so we had to use it very carefully.

Working with some of the individuals here at JPL who had been responsible in part for the damage was an interesting experience. We had a couple of our experiment representatives quit because they couldn't take the pressure; one was let go as a result of what was felt to be excessive negligence. Those were circumstances that happened less frequently later on in the mission, when we became more familiar with the operation of the instruments and how to check to make sure we weren't doing anything that would damage them. That was a process that involved a lot of learning.

We eventually put together a *Voyager Science Handbook* that we updated two or three times during the mission. It served as a training guide for our investigators and our experiment representatives. It allowed them to quickly become familiar with things they needed to know about the spacecraft.

Between the Saturn and Uranus encounters there was about a three-and-a-half-year period when we cut way back on our personnel, let off some of the assistants, and tried to run with scaled-down science effort for about two years until we staffed up again to do the intensive preparations for the Uranus encounter. That proved to be a mistake. We learned our lesson and did not repeat that error between the Uranus and Neptune encounters. Instead, we kept the entire staff on, but at a lower level, part-time, and worked the sequence process over the entire time interval between the Uranus and Neptune encounters, instead of trying to cram it all into the last year before the Neptune encounter.

I think the Neptune encounter was sequenced a little more effectively than the Uranus encounter was. And Uranus, by the same token, because of our earlier experience, was done better than the Saturn encounters were. So at each successive encounter we became more familiar with the types of things that ought to be done, pitfalls to avoid, and methods of making the sequence more efficient.

Did pressure upon individuals in the Voyager team also decrease because of your learning process through the entire project?

It did become less and less with time. During the Saturn encounters we had to involve the Project Scientist, Ed Stone, in lots of make-or-break decisions. We couldn't resolve conflicts between instruments, and so we were constantly taking things to our science steering group, which consisted of the leaders of each of the investigations, plus the project scientist, and having them try to resolve the issues there in open forum. That turned out to be a very unsatisfactory way of doing things.

So for the Uranus and Neptune encounters we attempted instead to set guidelines such that we wouldn't get requests for more than we could accommodate. We started by bringing in some scientists that were not directly associated with Voyager, asking them what the highest priority observations were that should be done, and putting together a "straw man" timeline. The working group consisted mainly of individuals who didn't have the pressure of trying to satisfy their own private goals. The timeline they put together was used as a guideline for what the investigation teams could then submit as their portion of requested observations. In that manner we were able to hold things in check far better, and we didn't have all the last minute battles that we had during the Saturn encounter.

The result of this revised procedure was that the investigators themselves were more satisfied, there were fewer conflicts between investigations, and we didn't have the battles to resolve the problems. It took a lot of pressure off the project scientist too, because he wasn't put in a position of having to decide between two alternatives, neither of which was very satisfying. In this manner we helped him to be more successful as well, and he's now the laboratory director.

It was a pleasure working with Dr. Stone in that respect because he had a way of being noncaustic and working things out in a way that the scientists would appreciate. He expressed to me after the Neptune encounter how well he thought the process had worked and his appreciation to me and the many others who had made it work, because it saved him endless hours of headaches and problem solving that he had to do in the earlier encounters.

He was referring to this process of setting guidelines before people even put their requests on the table so they would exercise some restraint?

That's right. When we started out planning for the Jupiter encounter, we basically just asked every team to submit what it was they wanted to do, and we ended up with a set of requests *each* of which would have filled the entire time available; in fact, some overfilled the available time. They even requested observations which conflicted with one another. Trying to put those all together and decide what was a valid request and what didn't need to be done was a very difficult process. It was a lot better to scope the process to begin with, and then have the inputs within that determined scope.

By "scope" you mean limit?

Limit the amount of input. Tell the investigations, at the very beginning, when they are free to put in almost as much as they want during some time periods, and when they need to avoid requesting activities because there are a lot of conflicts in that time period. Often we would suggest the things we'd *like* them to submit for those time periods. That process worked very well. We'll use a similar process on Cassini as we plan the orbital mission around Saturn in the early 21st century.

In your capacity as assistant project scientist, did you have any authority problems stemming from your being a JPL employee, while some of these scientists were employed by their own universities and weren't employees of JPL?

Yes, at first. Until I earned their trust there were some difficulties of that nature. By the time we had completed the second Saturn encounter, however, the investigators for the most part recognized that I was fair and that I did not favor one investigation over another, and hence they were more willing to abide by the recommendations that I made at our various meetings.

When there were conflicts, we would work through the conflicts and come up with some alternatives, and then I would suggest that we should adopt a particular alternative, and almost universally those recommendations were accepted. That made things work a lot faster and more smoothly than they might otherwise have done.

We did not have, in the later portions of the Voyager mission, investigators who had a hard time getting along with each other. They had learned to work together well, and they knew about what level of instrument activity they could expect from that encounter. They planned their observations accordingly, so we didn't have individuals saying, "It's unfair. We need a larger percentage of the time. Somebody else is getting treated preferentially." That sort of thing just went away.

Maybe we were just fortunate in having a group of individuals who enjoyed collecting the science so much that they were willing to give up some of their favorite observations in order to assure a better overall science return. But I choose to believe that they saw the wisdom in trying to develop—on the basis of the recommendations of these working groups that we put together beforehand—a sequence that was as efficient as it could be made, and not trying to push for more than that.

You mentioned that one of the principal investigators at the beginning had to be replaced.

Yes.

Did most principal investigators remain on through all four planets?

We had several changes, especially between the Uranus and Neptune encounters. Individuals back in about 1965 were considerably younger than when we passed Neptune in 1989.

1965 or '75?

Well, they were starting to prepare originally in 1965 because we had proposed the Grand Tour mission about 1965. There wasn't actually a selection of team members until 1975, but we certainly had people who were thinking about missions to Jupiter, Saturn, Uranus, and Neptune long before that time, and expressing their interest and being involved in some of the preliminary studies.

Rudi Hanel of the infrared investigation retired after the Uranus encounter, and Barney Conrath, one of his co-investigators, took over as the principal investigator. Robbie Vogt found that his responsibilities as assistant to the president of Caltech were more than he could handle along with his principal investigator role, so he bowed out of that, and Ed Stone as project scientist took on the dual role as the principal investigator for the cosmic ray investigation.

Herb Bridge at MIT reached retirement age and John Belcher took over as the principal investigator on the plasma experiment. Fred Scarf was the only principal investigator who actually died during the mission, and Don Gurnett at the University of Iowa replaced him as the principal investigator for the plasma wave experiment at the Neptune encounter. We had the same team leader for the imaging investigation the entire time period.

So most of them stepped down voluntarily?

Yes. Only one of them was forced.

And the number of experiments stayed at 11 throughout the whole project?

Stayed at 11 from launch onward, and all of the investigations had at least partially working instruments by the Neptune encounter, so we didn't lose any investigations completely. We had to doctor along the infrared investigation because apparently there was some crystallization taking place in some of the bonding material for the mirrors of the Michelson interferometer in that instrument. The crystallization would cause pressures that would misalign mirrors and reduce the sensitivity of the instrument. So we constantly had to heat the instrument up to break down those crystals, to melt the crystals and get them back to a more amorphous form again.

By using that process over and over again, we kept the Voyager 1 and 2 infrared instruments at a sensitivity that was only slightly degraded from the planned sensitivity. That was probably one of the biggest challenges we had—keeping the instruments operating the way they should, other than the damage done to the photopolarimeter that was mentioned earlier.

Voyager was the most satisfying venture that I've ever been involved in in my career at JPL. I have a hard time imagining any other mission, including the Cassini mission, offering anywhere near the satisfaction that the Voyager program did. Number one, it was the first to get to Uranus and Neptune. It was the first to send back really detailed data about Jupiter and Saturn as well, even though the Pioneers had preceded us to those two planets. It was a mission fraught with goodwill even at the time of the Challenger accident.

NASA's woes associated with the Challenger accident would perhaps have been responsible in some way for the demise of NASA had it not been for the concurrent good news that was coming from Voyager as a result of the Uranus encounter at that same time. I'm not saying that we necessarily were the straw that saved NASA, but I am saying that NASA was certainly anxious to have some good news going out at the same time that we were recovering from the shock of the Challenger accident.

The Challenger accident was a personal tragedy to us as well, coming in the midst of our rejoicing over the return from the Uranus encounter. We in fact canceled our last couple of press conferences in deference to the loss of life that occurred in that disaster. Except for that low point and some other small losses of data, I can think of nothing but good to say about the process, the results, and the teamwork associated with Voyager.

We had, particularly on the Science Team and I think it extended to other aspects of the project as well, individuals who were committed to doing as good a job as they possibly could, without limiting the amount of time or effort they would spend on it. Individuals were salaried obviously, but as we were preparing sequences, many of these individuals spent 60 to 70 hours a week just making sure that the product we put out, the process we put out, the data that was returned was as high in quality as we could make it. They took personal pride in seeing that happen.

The engineers on the project had that same spirit about them. If there was something they could do to improve the capabilities of the spacecraft so that we would get even better science, all of them bent over backwards to make that possible. We had a spacecraft for the Uranus and Neptune encounters that was a much different beast than the one that flew by Jupiter or even the one that flew by Saturn. Their dedication, their ingenuity made it a much better mission as a result of that.

When we ran into problems, there had been a tendency at JPL in prior missions other than Voyager to look at it and try to decide, "How long do we have to keep the instrument turned off while we find and test a solution that we dare send to the spacecraft?" But that didn't seem to be the attitude of Voyager engineers if there was any possibility of loss of data associated with the problem.

For example, as we were approaching Uranus, we had some lines appearing in the images where data was simply missing or saturated or some other such thing. We found that it was a bad bit in the memory that stored the processing algorithms for the camera. That was analyzed very quickly by the engineers here. They found a way to program that one bit so it would be located elsewhere in a good location memory and sent that to the spacecraft. We didn't have any of the waiting or loss of data that would normally be associated with such a problem. In fact, by the time we reached closest approach just a few days after this problem was first discovered, pictures were perfect because they had successfully reprogrammed the image processor.

Even the stuck scan platform in the middle of the Saturn encounter for Voyager 2 was not a major problem. We did let the scan platform sit unused for a couple of days just to make sure that we understood the problem well enough. When we made a request for the spacecraft engineers to attempt to move it to a position that would point the cameras back at Saturn so we could at least get some data on the receding planet, they were willing to do that. They devised a procedure which they thought would have good chance of success. Using that procedure, they commanded the scan platform to the appropriate position and lo and behold, when the first picture came back, there was Saturn, right in the middle of the picture!

They didn't need to do that; they didn't need to spend that extra time; it was above and beyond the call of duty. It was just because of their dedication and their determination to make as big a success of Voyager as they could. They spent their own personal time to ensure that.

When some members of the sequence team were putting together a series of commands to be transmitted to the spacecraft, very often they would run into a problem where something we had intended to do didn't work quite the way it was supposed to, or took more time than it was supposed to. When that happened, some individuals would stay the entire night trying to figure out a way around the problem. Instead of coming to us saying, "Sorry, we can't do that," they'd come to us the next morning and say, "We ran into a problem but this is how we worked around it; is this okay?" and in most cases it was. They did this on their own time; they weren't paid extra for that. It was marvelous that they were willing to do that.

The myriad of new findings from Voyager at each of the planets was the most satisfying of all. The scientists of course were heavily involved in coming up with those conclusions and the theories that went along with them. Although we here weren't directly members of the investigation teams, they shared that excitement with me as we met each day to talk about the new findings.

We felt no obligation to restrict our suppositions and theories to our own area of expertise. There was a free exchange of information, and quite often some of the suggestions made by individuals who weren't directly members of the science team went a long way toward directing further analysis of the data. One that was quite well known is Linda Morabito's discovery of the vulcanism on Io; that led to further observations of the vulcanism and ground-based observations that went further.

The theories of how the Saturn ring system worked have been extended to the asteroid belt, to galactic structure, and helped us to understand a lot better some of the dynamics of large aggregates of particles. Some of the studies of the alterations to the surfaces of icy satellites of Uranus and Neptune have helped us to understand that we still have processes going on on what we at one time thought were inert objects. That changes our theories considerably.

The studies of the atmospheres of Jupiter, Saturn, Uranus, and Neptune have helped us to understand meteorology a lot better and will eventually give us tools that will enable us to do weather predictions on Earth a lot better than we would otherwise have been able to do. The studies of the magnetospheres of those giant planets and how particles behave within them, how some of them seem to be frozen in space to the magnetic field as it rotates and others move freely through it—those were theories that we understood in part beforehand, but we've increased our knowledge to the point where we better understand our own Earth's magnetosphere and the protection it provides us and the changes that can take place in it.

Lessons learned from Voyager?

The lessons we have learned from Voyager will be utilized quite widely in other areas and other missions. The successes of Voyager have led us to use procedures which will scope the types of things that we can possibly do at planets, instead of just trying to get independent input from each investigation and then trying to fit them together afterwards. That process has worked very well.

Another lesson learned is the idea of having a dedicated team of support scientists here at JPL who are not simply tied to their own investigation's needs and desires. Our experiment representatives on Voyager, and their equivalents on the Galileo and Cassini missions, have been responsible in large part for putting together suggestions for improving science data collection and analysis that have been and will be utilized by the science team members at their universities and other research institutions. That's worked very well.

We tried with Mariner 10 to Venus and Mercury to get along without that coordination effort here at JPL, and it was only partly successful. So although it cost extra resources, extra dollars that are hard to come by in today's economy, nevertheless the efficiency of these individuals in working through the problems in advance and saving the investigators untold hours of problem solving in the project science group meetings, is a lesson that is well learned.

Another lesson learned from Voyager that can be applied to a lot of space technology, and perhaps in other areas as well, is the most effective type of spacecraft are those that, number one, can be programmed to detect their own problems and solve their own problems, and number two, are reprogrammable to the point where we can change the way the spacecraft operates on the basis of the experience we have with it, to make it more efficient, to make it less vulnerable to problems. Those are some of the major lessons.

I love astronomy. I purchased a small telescope—somewhat unsatisfactory since I had done my observing through professional telescopes—and I love to talk with groups about astronomy, with or without the telescope. I love giving talks about the solar system, the size of the universe. Recently I have shifted my outreach efforts toward the K through 12 ages rather than the adult public, because I have come to realize that if we are to have an adult population that is space savvy, science savvy, that they need to begin that early on in their education.

So I am involved in Project Astro, sponsored by the Astronomical Society of the Pacific and funded by the National Science Foundation, in which professional or amateur astronomers are assigned for a year to one or two fifth, sixth, or seventh grade teachers, and work with those teachers as a resource to provide them with information, to talk with their classes, to help them work out different types of demonstrations related to astronomy that would help to teach the science curriculum in those areas.

Involvement in that program led me to accept the challenge to write a book on Uranus that could be understood by a high school student without necessarily having a science background. To make sure that happened, I had one of our administrative assistants on the Voyager project read through each chapter after I wrote it. If there was anything she couldn't understand, then we went back and rewrote it so she could understand it. She did not have a scientific background.

I like to trace my roots. The discoverer of Uranus, William Herschel, made his discovery at Bath, England. I went over there after the Uranus encounter and gave a lecture to the Herschel Society and found out that 15 miles from Bath was the place, Chew Magna, where my ancestor Thomas Miner lived before emigrating to Stonington, Connecticut.

CANDICE J. (CANDY) HANSEN
Voyager Imaging Experiment Representative
Born August 9, 1953   Pasadena, California

Candice Hansen attended California public schools and earned a bachelor of science in physics from California State University, Fullerton, in 1976. The next year she came to JPL, joining the Voyager Imaging Team as assistant experiment representative.

The Voyager encounters provided milestones marking many events in her life. From 1981 to 1984, the long cruise period between the Saturn and Uranus encounters, she worked at the German Space Operations Center in Oberpfaffenhofen. She worked on the Ion Release Module, the German portion of the Active Magnetospheric Particle Tracer Explorer, a multinational Earth orbiting mission designed to study the Earth's magnetosphere.

She returned to JPL in 1984 for the Uranus encounter. As Imaging Science Team representative and associate, she designed all imaging observations for the Galilean, Saturnian, Uranian, and Neptunian satellites and helped reduce, analyze, and archive these images.

In 1989, in the midst of preparing for the Neptune encounter, she finished her master of science at UCLA in planetary physics and in 1994 completed her doctorate in Earth and space science, also at UCLA. Her dissertation included research on Triton's nitrogen frost and atmosphere, based on data acquired by Voyager during the Neptune encounter. She met her husband while presenting a Voyager talk.

In 1990 Hansen-Koharcheck moved to the Cassini mission to Saturn as an ultraviolet spectrometer scientist. Her tasks there are to define science requirements on camera hardware and mission design. She is also the deputy principal Investigator for the Mars Volatile and Climate Surveyor payload, a mission to be launched to Mars in January 1999 and land near Mars' south pole that December.

Her publications and the activities leading to them have brought her four NASA awards: for Individual Achievement, Exceptional Service, and two for Group Achievement.

She enjoys spending her free time with her husband and daughter, water skiing, four wheeling in the California desert, and learning tae kwon do.

Our solar system holds worlds of incredible beauty just waiting to be discovered. We would try to imagine ahead of time what we might find, and our ideas were never fantastic enough. Neptune, so tiny in a telescope, so achingly beautiful in Voyager's images. It takes my breath away to try to imagine what our galaxy might hold. Being a member of the Voyager flight team was like being a member of Columbus' crew discovering the new world. We tend to differentiate between manned and unmanned missions. Voyager was not an unmanned spacecraft—Voyager carried the hearts and minds and dreams of all of us.

—Candice J. (Candy) Hansen

CANDICE (CANDY) HANSEN

I first got involved with Voyager in 1976. I went to the University of Arizona at Tucson to be Brad Smith's research assistant. I had applied to go to graduate school there, and they assigned me to him. He started me working on a little problem that had to do with Voyager, planning one particular observation. I was not really happy in Tucson; but I knew that I wanted to go into planetary science, and so I talked Brad into sending me here to JPL for the summer.

I got here in May, and Voyager was due to be launched in August. It was an exciting time, and they were staffing up, and I just felt like, "This is where I want to be; I don't want to be in graduate school! I don't want to miss out on all this excitement, I want to be part of it!" I decided that about the second day that I was here. So after that I just started looking for a position, an opening. They were hiring people to do the sequence design. Because I'd already been doing a bit of work on imaging with Brad, I got hired to start doing the imaging sequence design.

The very first thing I worked on was the picture-taking sequence at Io. There's a lot more to this than anyone ever realizes. I continued to redesign the sequence at Io for the next year and a half until we finally got there, and there were no more opportunities to make changes. That's a large part of the job. You keep thinking about things, and you keep learning new things about your target. You want to continually modify your observations to reflect your most current thinking on what the target's going to be like and what you want to learn.

That was when I first started, and it's still true today. You have to be a flexible person if you want to work on a project like this, where you don't know that much about what you're about to go visit.

My first position was just doing sequence design, which is planning what pictures to take at what time and what filters and what exposures. Since then I've been promoted once or twice, but always stayed with imaging. I've been working with the Imaging Team since shortly before launch. It's really been fun.

Every encounter is different. At every one there are different challenges, and I never felt like I was stuck in a rut; I never felt like I was stagnating just because I was still working on Voyager. It was always an evolving spacecraft and an evolving mission design, a different target to plan for. I did take a break between Saturn and Uranus. It was a four-year cruise from Saturn to Uranus, so during two and a half of those years I was gone doing something else, in Germany. Other than that it's been Voyager. It's interesting now dealing with *not* doing Voyager, with the whole thing coming to an end, and forcing myself to look forward to new missions and new challenges.

How do you feel about that?

Well, it's probably time for me to go on and do something new, but it's not easy to let go of something that you've loved doing for all these years, so it's a little nostalgic, a little bittersweet.

What were your earliest impressions of Voyager?

I was hired when I was 23, and a lot of the entry-level staff were about that age: 22, 24, 25. One of the things that impressed me early on was that although we were the most junior of all and doing the lowliest tasks, there was still very good listening in the project from the management down and up again. There was a lot of very good communication; actually more so in those days than later.

I remember writing a memo about some software. I had noticed that different programs gave different results for the same design, or what was purported to be the same design—and I wrote a memo to point this out.

It turned out to be incendiary. People couldn't believe that you could get different answers—how could that possibly be? It had opened this hornet's nest: that the design wasn't quite the same. This or that person had modified this or that along the way, and so what was coming out at this end was not what had been put in at the start.

So rather than this innocuous memo that I had written being thrown on somebody's "to read someday" pile, it was looked at and action was taken. To a 23-year-old in her first job, that was astounding. It made me feel very good, actually. But the project was very good about that, not just with respect to me, but with respect to everyone as far as the communication.

What else from those days? It was a lot of fun. It's been like a family in many ways, because this same group that I hired in with stuck together throughout these past 12 years, and as a result we've gone through various stages together. When we were 23, we were mostly concerned with who was in town giving a concert. Then we all bought houses, and then people got married and had kids. So we've grown up together along with Voyager, and people planned their lives around it. The most common way that you tell time on Voyager is with respect to encounters, like I bought my house between Jupiter and Saturn encounter. My daughter was born after the Uranus encounter. People tell time that way. That's fun.

Jupiter encounter was really exciting for me because I'd never been through anything like that before. When I look back on it, it was haphazard and chaotic but somehow we got everything done. And I marvel at that now because at Neptune we took only 8000 pictures—at Jupiter we took 35,000 with the two spacecraft. I look at the staffing levels and, okay, we had more people at Jupiter, but still ....

Although there was a factor of almost four difference in the number of pictures, we didn't have a factor of four difference in personnel. I remember those days being much more wildly chaotic, and I marvel now that everything turned out okay—well, almost everything. We did overexpose some wide angle pictures, and we didn't do that at Neptune. So there were mistakes at Jupiter that, by the time we got to Neptune, we had figured out how to avoid. But still, in looking back at the numbers, it boggles my mind that we were able to do as much as we did.

This is another early memory: between launch and the first Jupiter encounter, it seemed like we had an anomaly about once a week, and there were tiger teams being formed right and left. Again, this was the way it looked to the person at the bottom, seeing all this activity going on that I didn't really understand.

I chuckle now when I read about all the troubles that Hubble Space Telescope is having. It seems to me that that's just the nature of the game: you launch these spacecraft and you have models of how you *think* they're going to behave, but how they are in reality is not always the same. Models are just not that good.

On Voyager in particular we got into trouble a lot because the fault protection algorithms onboard would get triggered, and the spacecraft would go off looking for stars or the sun or whatever, and shutting down various subsystems or swapping them. Quite often it turned out that there was nothing really wrong with the spacecraft.

What was wrong was that the thresholds were just too tight for telling it to swap or do whatever it was doing; if you expanded the thresholds so that there was more play, the spacecraft was just fine, and it wouldn't go off looking for its navel or whatever. I see that happening now with Hubble, and I just grin and say, "Well, the people controlling it will figure it out."

One of the things that was a significant part of my job was that what we learned with Voyager 1, we tried to incorporate in our plans for Voyager 2. Between Voyager 1 at Jupiter and Voyager 2 at Jupiter was only a three month gap: 1 was in March and 2 was in July. So there weren't that many things that we could even consider changing, but we did manage to work in all of the ring observations and the Io volcano watch.

Resources are always a problem. You never seem to have enough of either computer memory or people time, so you have to figure out what you can change with the limited resources available. One of the problems that was always very critical on Voyager was memory space in the command sequencing computer.

When we started out, the Imaging Team wanted Voyager 1 to make four observations to look for a ring around Jupiter, but we got whittled down to one—just because of the lack of resources. Every time there was a crunch, Ed Stone would say, "Well, we don't know that there's a ring around Jupiter, so why are we doing this observation?" Then we'd lose another one, until we got down to one observation. At that point we said, "We have to hang on to that because we have to see if there is a ring around Jupiter." So, when we found it with Voyager 1, we had the leverage to go back and put in ring observations on Voyager 2.

The other thing that we changed was the Io picture taking, because of having found those volcanoes with Voyager 1. We wanted to get as many shots as we possibly could with Voyager 2. So we did some very creative things, like taking pictures while maneuvering the spacecraft, which ordinarily we wouldn't want to do.

Every time science teams came in and said, "We need to make some changes," Project would moan and groan and say, "Oh, oh, no; we don't have the personnel to make all these changes"—but still they would. Everyone would work just a little harder. The people on Voyager have always been the kind of a group that, if they understand that there's going to be something good coming out of it, they just put in the extra whatever it takes.

So sometimes my job is convincing people, like the Spacecraft Team or the Sequence Team, that there really is a reason, and trying to get them involved in it, or trying to get them motivated to help us out.

With Saturn we had something like nine months between encounters, which was just enough time to literally change everything. The first flyby found so much stuff in the rings that we didn't expect to find, like the kinks and the fact that some of them are elliptical—one picture where the F-ring looks like it's a little braided—all of that stuff was completely unexpected. With nine months between the encounters, we had time to really touch every observation and diddle with it.

I don't think I've ever worked so hard in my life as that nine-month period. We were really trying to revamp the whole set of plans. It was tricky because on Voyager we were allowed to optimize things down sometimes to the second. So a lot of those kinds of changes were really fine tuning the command load to pull the maximum that you could out of a given chunk of time. That was another place where management would moan and groan and tell us they didn't like the idea, but they'd let us do it anyway. That was fun, too.

And then along came Uranus, and we had a different crew. It was very interesting to come back after this four-year hiatus. There had been a skeleton crew, of course, running the spacecraft all during that time, but a large number of people left the project after Saturn, and not that many old-timers came back. It was more new people. (It seemed like it was in the Flight Science Office where most of the people, the experienced people, came back and wanted to do Voyager again.)

We found ourselves in the funny position of sometimes actually knowing more about someone else's job than they knew themselves, because we had been around longer than they had. I used that knowledge a couple of times to go, "Well, at Saturn, we could do blahdy-blah. What do you mean we can't do it now?" That was an advantage.

Typically we start planning science observations about two years ahead of time because, first of all, you want to get a rough idea of what you're going to try and accomplish. We do that with science working groups. Then we take a crack at it and usually turn up a number of problems that then need to be solved in one way or another.

Then we go to a more detailed level of actually doing the individual observations. It just gets more and more detailed from then on. Some of the kinds of things that turn up in those very early stages are things like the need for image motion compensation, or something we call "late stored updates" which is a very, very late change to the command load to take into account very late information.

And what else? Well, stuff like how much DSN performance you're going to need. Then there are those things that obviously need to be worked in. In fact, with respect to the image motion compensation, a lot of the routines, the algorithms that the spacecraft team developed, would be tested on Voyager 1 before they would try it out on Voyager 2. There was also all of this testing onboard the spacecraft before we would actually say, "Okay, we really can do it. Now here's the command load to accomplish that."

In between Uranus and Neptune we revamped our whole exposure timetable, the values that we could use to take pictures. The values that were okay at Jupiter and Saturn, and even Uranus, were too short at Neptune because it's so much darker out there. And so we made some relatively minor changes to the spacecraft software, but they turned out to be enormous changes to the ground simulation software. We spent two years working on that, testing it. That was the big challenge for Neptune. That, and we developed a new kind of image motion compensation, which worked out very well.

Those are the kind of challenges that we've worked through. It's very much a team effort on the project. In fact, one of the things that was hard for me to get used to in Germany was that I wasn't part of a team. I kept sort of looking for somebody to bounce ideas off of. The way they do projects over there is to have one or two or three people working on them, not 20 or 200. Each person is responsible for their own thing, and there's very little overlap, and certainly nobody cross checking that I'm doing the right thing. Everybody just expects you *are* doing it right.

Whereas on Voyager, in particular with command loads, we'd go through the commands one by one and make sure that what was on the way to the spacecraft was really what we intended. The whole nature of developing a command load is error prone. So in addition to computer checking, there's a lot of hand checking. I don't know how you could get around that. It's big cost to any project, and a lot of projects, currently active and future ones, have looked at ways to try to save some money in that area, but it's tricky. The human brain is very good at catching inconsistencies. It's a little more difficult to figure out the algorithms for a computer to do that for you.

Describe your typical tasks.

As the person working for the Imaging Team, first of all it was my job to ferret out from the scientists what it was they were really after. Typically people on my team would have sort of a vague idea. "We want to get the phase curve of Europa," or "Well, we want to do mapping at the highest possible resolution when the terminator is underneath the spacecraft." It's then my job to figure out, okay, what does that really mean in terms of what pictures we take at what time?

To do that sort of thing we have computer programs which, if I put in the time and what I want to look at—Io, for example—will come back and show me how big Io is relative to my field of view.

So a lot of what I do is just joysticking. "Well, I think I'll put a picture here and one there and one here and one there," always trying to keep in mind what the scientific objective is and how best to meet that objective. So I run this program called Scout, with which I can change the time and just look at things. I then come up with a set of pictures which I take back to the scientists and say, "Okay, here's what I've come up with; what do you think?"

Usually, especially in the beginning, they would say, "We don't like this. Why don't you change that? What about this? You didn't think about that." I'd go back and make all the changes they had suggested and come back to them again. "Is it okay now?"

By the time we got to Neptune, I could almost anticipate what they needed and what they would say. And also, I knew the spacecraft constraints far better than anyone on the Imaging

Team did, so I could say, "Well, I wanted to do it that way, but we have this problem and we can't slew that far that fast," or whatever the problem was. There was a lot of one-to-one interaction with members of the Imaging Team: on designing, figuring out what pictures to take, and when to get the data set that they needed to do their analysis.

After that, my interaction was more with the Sequence Team and the Spacecraft Team, to actually turn my graphics into commands to the scan platform to move—right, left, up, and down—and the timing of the picture taking and what filters and what exposures to use. All of that, what we call the "uplink process", was the predominant part of my job at Jupiter and at Saturn, and then less so at Uranus and Neptune because by then I had been promoted to a point where I worried about everything that affected the Imaging Team, not just their observation designs.

At Uranus and Neptune I also worried about the downlink, what pictures were we getting, how are they going to be processed, how are they going to be displayed, making sure that we got all the calibrations we needed, who needs to get what data when, what should our interaction be with the press and with the Public Information Office, and coordinating things through the photo lab. Also making sure that the team got offices; with the way space is around here, that was a battle.

An experiment representative, which is what I am, just has to be able to juggle a lot of balls simultaneously. You have to be able to switch from one task to another without getting flustered. You can't be the kind of person that has to concentrate on one thing at a time. You have to be the kind of a person that can do three things at a time. Sometimes my office would look like a tornado had hit it, because I'd be working on 20 things that day. But that's part of the fun too, to have that kind of variety.

Anyway, that pretty much sums up what I do. In a nutshell it's my job to make sure that the interests of the Imaging Team get represented on a daily basis here at JPL, because the members are scattered all over the country and their job is teaching classes at a university or whatever. Their main job is not to have to worry about Voyager every day; it's my job to worry for them. And I really like the people on the Imaging Team, so it's been a good association all these years.

What were some of your surprises?

The Jovian ring was one of those great things which gives you the satisfaction of saying, "Yes, we knew we should look for it," and sure enough, it was there. Now we have pictures that we wouldn't have had otherwise. The volcanoes on Io were another great discovery.

I think that Voyager completes the Copernican Revolution, where you really quit being Earth-bound with your ideas. We went to Jupiter with some preconceived notions about the moons; notions that were very Earthly in nature. Voyager has opened our eyes to the fact that you can't just go in and say, "Well, on the Earth, it's like blahdy-blah so therefore out at Jupiter it should be similar." It's not going to be similar. It's going to be different. We've learned that lesson over and over and over again at each planet, and I think it's finally sinking in.

It's amazing how many things, even in planning sequences, are preconceived notions. We didn't think we'd see structure in Saturn's rings because you can't see structure from the Earth, things like the kinks and the braids—all that radial structure.

Then we'd think we were getting smart, so when we were planning the Neptune sequence, we were going, "Okay, we have to rid ourselves of any kind of expectations and cover all our bases." But we still were in this post-Uranus depression that said, "Oh, it's so cold out there, Neptune is probably going to be as boring as Uranus was." So we had this Uranian expectation that affected the way we designed our observations for Neptune, and we were wrong again. We were really trying to be open minded, but we weren't. I don't know how you overcome that.

Any further comments on misleading expectations?

This one about the atmosphere being boring because Uranus is very bland, speaking from an Imaging Team point of view. We only found a couple of clouds to track, and so it was just said, "You're so far from the sun there's so little energy available to stir up storms or whatever," and that colored our thinking at Neptune, even though we knew from Heidi Hammel's data that there is some kind of activity. It still colored our thinking in the sense that, when we got into conflict situations over resources, we didn't fight as hard as we would've had we known what we were going to find. We just kind of shrugged it off and went, "Well, there might not be anything there anyway," which was the wrong answer.

What would be examples of conflict over resources?

All the scan platform instruments are bore-sighted, so where one person or one instrument looks, everybody has to look. But a lot of times the science teams have far differing ideas of who wants to look where and when, so the first thing we have to resolve is the time line: where are we going to point as a function of time? And so the first set of conflicts comes with the science teams on the scan platform jostling to get their observations.

Typically there are compromises that can be worked out. But it always seemed like we would come down to a couple of things that were just really sticky issues where everybody, or each team, had some really good justification for what they wanted to do, but there was no way that you could do the two observations simultaneously. Then Ed Stone would have to make a decision.

Once the time line was laid out, then the issue of computer memory came up. Of course, we could always use more than we had, as far as commanding the spacecraft. Then we'd go through another round of compromising where we would simplify observations. In regard to trajectory selection, at Neptune the Radio Science Team wanted to be 68,000 km from Triton and the Imaging Team wanted to be 28,000 km from Triton. We compromised at 40,000 km, which was where both teams said, "Well, that's it; any further and our performance is just going to be nothing."

Those are the kinds of decisions that have to be made. We would always try to work them out at our level first, and only if we couldn't find a compromise there, would it get handled by Ed. He's very fair, very fair, so everyone would always feel like they had had their day in court, and that the judgment or decision that was made was the right one. No one ever felt that Ed had a favorite, or a bias towards any particular kind of science, or anything like that.

What kinds of decisions did you personally make?

I have done all of the satellite encounter designs. My strategy was always to try to be the first one to look at the geometry. I would try to think ahead, to anticipate the teams' needs, and incorporate those in my imaging designs so that then I could say, "Okay, here's the mosaic we want, and it incorporates this drift across the planet for the ultraviolet spectrometer, or this integration on the subsolar point for the infrared," so that I would get someone else on my side. Just by having some kind of a grasp on what other teams want to do and trying to incorporate that in the very beginning, you can already avoid a conflict.

Or sometimes I would come in with a design and say, "Well, the Imaging Team likes this design." And the photopolarimeter guys would show up and say, "But we want to go this way and not that way." I would know that my team didn't care whether we went this way or that way, and so I would say, "Okay." I'd make sure that the Imaging Team really didn't care, and then I'd redesign it. That would take care of that kind of conflict. So a lot of those kinds of things really weren't conflicts; they were just situations where you come in with two separate inputs that could be merged, and nobody had done it yet.

You were the one that would suggest how to merge these inputs?

Yeah, to optimize the science for everyone.

You were sort of coordinating these various teams?

I didn't really represent the Ultraviolet Team or the Infrared, but I had an idea of what they would want. It was a strategy more than anything, because you don't like to always go crying to your dad to resolve a conflict. And I think it's creative, too, to try to understand all of the science and then try to come up with designs that are pleasing to everyone.

If a compromise was really impossible, then I would know what my team was after and I'd stand my ground and say, "We have to have it this way, and here are the reasons." I'd try to do a really good job of representing them, so that the scientists wouldn't have to come in and do it themselves. If I lost, I'd call somebody up and say, "You've got to come out to JPL; we're losing this time here," or whatever, but if I won, then I saved everybody a trip.

The most negative time for me was when the scan platform stuck. That was really devastating. We lost our best Enceladus and Tethys observations. We lost our highest resolution of the F-ring with image motion compensation. We lost some of our occultation data. These were things that were all close to closest approach, so they were all unique science that could not be replaced by something of lower resolution or a different time period.

The geometry was unique, and we had just spent nine months between Voyager 1 and Voyager 2 redoing all those command loads; we had all really put our hearts into optimizing that Voyager 2 encounter. Then to have the scan platform stick like that was devastating. Obviously the whole encounter was considered a success in spite of that, but when you personally have put so much into it—that was probably the worst time of the project, or the worst surprise.

There have been personality things as well. I've had management that I didn't think very highly of and gotten into situations where I was very unhappy, but these were short-lived and never enough to make me think seriously about changing jobs. There are times, too, when you just have to give people some slack. When you know that everyone's been working 20 hours a day and tempers are getting a little frayed, you can't take things too personally. That's just something else that's part of the nature of the job.

Personal relationships suffer also. I don't know how you get around that because it seems like everybody on the project is so dedicated that their spouse or boyfriend or girlfriend just has to understand, and if they don't, they don't. I'm sure that there are stories like that, and perhaps even divorces that have come about for that reason.

What are the main things you'd like to convey to the public?

That's a tough question. I think what's really been important about Voyager is opening our eyes to the variety in our own solar system and finally letting go of our Earthly preconceived ideas, recognizing that the Earth isn't the only place in the universe. It's sort of like going and living in a foreign country for a while—you discover that there are other equally valid ways of living, and there are other equally valid lifestyles compared to how we live here in America. You don't always understand that until you leave, and I think it's the same thing with the solar system—there are other equally valid expressions of physics that you find when you leave the Earth. So I think that's maybe the most important thing.

Other than that, the way the mission was run with flexibility and good communication. Prior to or at Jupiter and at Saturn, when we would suggest engineering changes, the engineers would always come back and say, "That's not in the specs. You didn't put that in the specs. Nope, you can't have it."

But after we got past Saturn, it was a whole new ball game, because now we were into a mission that no one had really planned for. Obviously we had put the spacecraft on that trajectory, but none of the design goals were for anything beyond Saturn, so there was an entirely different attitude, sometimes from the exact same people who had previously refused to make changes. But now they said, "Well, let's see how we could solve that problem. Let's take a look at it." I thought that was wonderful. It was a very pleasant change.

After their original mission was accomplished, they were open to new possibilities?

Yes. I gave a talk last year to the Attitude Control Group here on-lab, and it was mostly men who had worked on the spacecraft while it was being built. They hadn't seen it since and were now working on other projects. I was just talking about one thing after another: "Well, we changed this, and we changed that, and we changed such-and-such." At the end, one of them remarked that a lot of what we had accomplished was never foreseen at the time they were building the spacecraft.

I would love to see us go back to Neptune. That was a planet that we knew nothing about. You've probably seen the pictures that Heidi has from her ground-based observing. What you glean out of those pictures versus what we actually found out there is just incredible. You can't do anything on Triton from the Earth, and now we know there are those crazy geysers. Are they just something that happens in the spring? What are they? I could think of a lot of things that, if we sent an orbiter back, we could spend some time investigating.

I have heard that there are a series of launch windows that open around the turn of the century. I really do hope we send a spacecraft that gets there around 2015, because I won't be 65 until 2017, and I figure I could work on it for a couple of years and then retire. So, we'll see.

I love to read about Voyager. I pick up practically everything that hits the newsstand or the bookstore that has anything to do with Voyager, because I love to see it through other people's eyes. I've been so close to it all these years that I'm not the least bit objective, and I enjoy seeing it from a different point of view.

I think now, more than I thought back when I was on the project, that Voyager was very special, that there was a feeling of teamwork, and we were all aiming at the same goal. I don't see that now. The projects I'm working on now are in different stages, so maybe it's not really fair to compare flight operations and when you're still building a spacecraft. But the level of teamwork we had on Voyager and the cameraderie were very special, and I appreciate it even more now that I don't have it, than I did at the time.

In fact, we were thinking about powering on one of the Voyager cameras to watch the Shoemaker–Levy comet impacts of Jupiter. It turned out that we couldn't, for technical reasons. But I had a tremendous amount of fun designing what the experiment would have been, before we ran into technical difficulties. I was really excited about working with the Voyager folks again.

Some of them are still around?

Oh yes. The two spacecraft still have a mission to do, and it's a small group that's left, but it's basically the same people. The Voyager experiment reps, the six of us, get together only on rare occasions anymore, but tomorrow's going to be that day: we're going out to lunch.

The teamwork on Voyager was special.

Yes. Having a common goal, everyone knowing what that goal was, and then working together, each with their own specialty, but all heading in the same direction. That's not unique; it's just a matter of having a very clear vision of where you're going. That's really the key. Then the right set of skills and expertise to get you there. It's not magic; it's setting your sights on that star. The top priority was getting to Neptune and being ready when we got there, and everything else had to fall into place. With Voyager it was very, very clear where we were headed.

People have their own personal agendas: they want to advance their career, they want to write this or that spectacular paper—there's always going to be that aspect, but is that the controlling aspect or is that a sideline? On Voyager all of that was there, but it was so overwhelmed by the ultimate objective that it never got in the way.

On other projects I've worked on more recently, those personal agendas and ambitions sometimes lead to not the right thing happening, and issues get clouded. I think that is the problem with a lot of the bigger social issues, too. People gather aid for victims in Africa, but then it sits in a harbor someplace or gets absconded, because there were other agendas and the overriding one did not prevail.

Of course Voyager was a mission where you could come up with technical solutions to problems; you didn't have to come up with political solutions. If there are things that can be solved by inventing just the right widget or understanding the physics well enough to make something happen, then it has a much better chance of succeeding in a Voyager-style environment than things that need political solutions. As Americans we like to think that technology can be applied to solve all sorts of problems, but that's not always true.

Do I understand correctly that you were one of the people who pushed to get the solar system mosaic? A little birdie over in Public Information said you were a "true believer".

Yeah, we started calling ourselves that because it was not easy to get the approval to do that observation.

Why was it not easy?

One of the problems was that it was never really seen as science, and therefore wasn't really justified to use the resources to do that observation. It was first proposed not in the form it ultimately took. Somewhere between Jupiter and Saturn we came up with the idea, floated it, and it quickly got shot down. Then, between Saturn and Uranus we proposed it again, but the project was staffed at such low levels they couldn't consider it. Never giving up hope—that's how we became this band of "true believers"—we proposed it again after Uranus.

We got pretty close to getting it approved then, but it was still complicated. We envisioned doing it by blocking the sun with the high gain antenna, to avoid the danger of exposing the instruments to direct sunlight down their bore sights, which would have been damaging.

We would have to be very careful. It involved maneuvering the high-gain antenna to block the sun and pointing the scan platform to just peek above the top of the high-gain antenna. We worked on it for awhile, but with the Neptune encounter approaching, people just got cold feet about taking the risk of pointing the instruments at the sun and doing this tricky maneuver.

Never giving up hope—actually "nagging" is another term that comes to mind—after the Neptune encounter we proposed it again. Then, because we did not need to protect the instruments for any future observations, we proposed not even using the high gain antenna; not even doing that maneuver, but just going in and doing the mosaic, and that made it simple. That was something the Project could get behind. Then it was just a matter of deciding when the alignment of the planets would be best.

We were really after the picture of the Earth, so we had to look for the point in the Earth's orbit which would have it as far away as possible from the sun. It turned out that we could also get Neptune, Uranus, Saturn, Jupiter, and Venus. Mars was not in a good position—it wasn't too close to the sun, but it was at a high phase angle so we didn't have a very good shot at it, and also we didn't detect it. No matter where Pluto would have been, it would be too small and too dark for the cameras to detect.

At this stage were there any objections to what you were trying to do?

After Neptune, people were sort of resigned to it. Sagan had gone to headquarters and gotten approval, so there was no getting out of it anymore; but there were still some funny things that happened. For instance, we used to have a meeting where all of the observation requests for a given computer load would be presented, and usually all of the science was presented together by one person from the science office; quite often the team chief, or his deputy, or sometimes the cruise coordinator. They would ordinarily stand up at this meeting and say, "We've got these 10 requests: the purpose of these 10 requests is such-and-such. In order to do them the spacecraft will have to do so-and-so."

They refused to put the mosaic observation on that list, so I presented it myself, which was fine. I enjoyed the chance to get on a podium and say, "Here's why we want to do this." Funny things like that kept happening.

People were still dragging their feet?

They were still dragging their feet, and it was all this business about, "Well, it's not really science; it's something for the public."

Okay, it wasn't really science, but it was something that we thought would strike a chord, that might affect people's view of the world in a way that would promote science.

Were there any valid technical reasons for opposing this, like consuming scarce fuel?

No. Once we removed the maneuver, we removed the risk of doing something with the spacecraft that would be nonrecoverable. There was still a problem with resources, with just plain not having enough people to do the sequencing, but even that worked out fine. That is, however, why we didn't do it twice, because the level of staffing down after the Neptune encounter was pretty steep. We did the portrait from Voyager 1. There was also an opportunity to use Voyager 2 four or five months later, but that was out of the question because we wouldn't have enough staff to handle it.

How would that picture have differed from what Voyager 1 obtained?

Neptune would have been big and bold, as opposed to a tiny little dot, which is what it is from Voyager 1. We would have been able to get Mars. Those were the two main differences.

I liked Ed Stone's description of the solar system portrait. He used the term "last light" through the Voyager telescopes. At observatories people always celebrate the first light. We were marking the occasion of the last light, and it really did turn out to be the last light. Although we had considered doing the comet impact, we couldn't, and so the mosaic really was the final use of the Voyager cameras.

Support Services

Jet Propulsion Laboratory with its thousands of employees is like a small city, and as such has many needs not directly related to science. So along with directors, engineers, and scientists, JPL also has truck drivers, telephone operators, secretaries, security guards, students, photographers, mechanics, and public relations personnel. Although far from the limelight, their dedicated work, year after year, contributed to the success of Voyager.

ROBERT POST
Photo Laboratory Supervisor
Born October 29, 1936   Pasadena, California

Robert Post attended Rosemead High School and Pasadena City College, earning an associate of arts in 1957 with a major in lithography and minors in photography and graphic arts.

He then joined the JPL photo laboratory, where he served almost all of his 38 years until retirement, interrupted only in 1960 by a two-year hiatus with the U.S. Army in Germany.

During his years in the photo lab, he participated in every JPL flight project from Ranger 7 through Magellan. During the Voyager encounters with Jupiter, Saturn, Uranus, and Neptune, he was the supervisor of the photo lab operations group and served as the interface between the photo lab, the Imaging Team, and the Public Information Office.

His interests range from coin collecting, traveling, and target shooting to an avid interest in the ancient cultures of the Americas and various other places around the world. He especially enjoys visiting ancient sites he has been reading about for many years.

A formal education free of political influence and personal biases.
The ability to read and listen to clearly understand others.
The ability to speak and write
so that others will clearly understand you.
Living your personal and public life
with patience, compassion, reason, and integrity.
Project Voyager was the result of many people
working together for a long time, putting aside
personal pettiness, jealousy, selfishness, and prejudices.
The accomplishment was phenomenal!
—*Robert Post*

ROBERT POST

JPL's photo lab is an integral part of everything that anybody sees out there because we do all of the imagery that comes down from the spacecraft. We've produced the imagery for at least 35 missions, starting with Ranger, and it all comes out of this photo lab.

When the imaging data arrives from the spacecraft, the image processing lab massages it, transfers it to film, and gives it to us. We do the actual processing and printing. There is a crew of 12 or 13 people in my group.

The photo lab has been here a long time. It was one of the first sections established at JPL, back in 1940–41. I myself have been here 33 years, so I've been through a lot, and Voyager was a biggie, a 12-year mission from the time it was launched. I just wish I had kept notes of what happened through the whole thing.

How did you happen to come to JPL?

After high school I received my associate of arts degree from Pasadena City College. My interest centered on the copy camera which is used to shoot the line negatives and half-tone negatives used in the lithographic process. My instructor at Pasadena City College was a personal friend of George Emmerson, who was the JPL photo lab manager. George called my instructor and asked him to send someone up to JPL to work the copy camera in the photo lab on the night shift. I was sent, I was hired, and I have been here ever since.

The photo lab is involved in everything at JPL, from retirement parties to imagery from all the planetary probes. I have been involved in everything from the Ranger project to project Voyager. Since 1980 I've been the lab operations group supervisor. The Lab has been involved in other things as well, from consulting with outside police departments to doing all the film processing for the Shroud of Turin study.

How did you get involved with the Voyager project?

We got involved because we're the photo lab and they needed pictures. You can only go so far with electronics; eventually those electronics have to be turned into some kind of visuals so people can see exactly what is going on, and that's what we do. Yes, they have their images on the TV screen, but there's quite a difference in the quality of a TV image, as opposed to a nicely finished color print or black and white print in your hand.

We produce all of the imagery that goes out to the press at the press conferences, plus a lot of science stuff: for scientists who may want specific images of a planet or a moon for geological study, and things like that. So we are an integral part of the Voyager project because without the photo lab, nobody sees anything. We work with Jurrie van der Woude closely because he's the Public Information Office (PIO) interface with us, and we work closely with the project.

What are your most vivid memories of Voyager?

I think the whole thing was memorable. A 12-year odyssey like that is just unbelievable. My most outstanding memory was the Voyager 1 encounter with Jupiter, the very first one, because nobody knew exactly what to expect press-wise or image-wise.

We had a basic concept of what it was going to be, and the Public Information Office had a budget for so much money for so much imagery for press releases, but when all these images started coming in—these gorgeous pictures of Jupiter and different moons and all that—everything just went out the window and we started producing things. It was so nerve-wracking. All these fantastic pictures are coming in; you're seeing things that nobody has ever seen before, and it is just blowing your mind away.

The demand for these pictures almost buried us. We didn't expect that tremendous public interest. We were producing 24 hours a day, seven days a week, phenomenal amounts. It almost blew us out of the water, because there were many times when we were working until late at night, and everybody was so exhausted; you're not even sure where you are, but you keep going because you don't want to miss anything. You don't want to go home because you don't want to miss anything, and yet you are so tired you can't see anything anymore anyway. I still to this day don't understand how we did it, but we did it.

So the most memorable thing was that first Jupiter encounter. It was horrible, but it was wonderful horrible. It was too much, way too much; it was chaotic. It almost snapped the system. You get to a point where everything explodes and then nothing happens anymore. We got just about to that point. But every negative was kept; nothing was lost; all the numbering system stayed intact. I plain don't understand how it was done; I guess it was because we had a bunch of good people that saw that it got done.

Being involved with something like Voyager happens only once in a lifetime, where you start the initial exploration of the solar system. And to be involved with it, in the photo lab you're right smack in the middle of it. I can't think of a better place to be. That's why a lot of people who work here never leave; that's why I never left; that's why a lot of people have been here longer than I have. They just don't want to leave.

There's no place else that is as exciting, if you are going to be in this kind of business, than right here. You can do photography anywhere; you can be on the outside doing baby pictures and portraits and people's vacations and all that. But if you are going to do it, there's no place better than this. I feel very fortunate to have been part of it. I have no intentions of leaving because there are too many more things coming.

It would be interesting to review what happens during a typical encounter, if there is such a thing. As the imaging data, the raw stuff, comes down from the Deep Space Network and into your department, what do you do with it then?

We don't get involved with it until after it has gone through the image processing lab (IPL), where they do the computerization of the data coming back. They do their thing; manipulating the imagery and cleaning it up and whatever else they do down there; that's not part of our job. From there the data goes onto a tape, and that tape goes to a film recorder in the building next door. The film recorder transfers the data onto film; color or black and white film. At that point it becomes just a normal product that any photo lab could handle.

Then we just process it like normal film and print it like normal film, except that the constraints we are held to for quality are quite a bit tighter than what any lab outside would have. We have to guarantee to the project that our quality is going to be up to a certain standard. We did it strictly the way Kodak said to do it: their control specs, and that kind of thing. We weren't doing anything exotic; just keeping our controls tight, extremely tight.

Of course the image processing laboratory has to keep their quality up too, because if they put an image on the film that isn't right, color-wise, there's nothing much we can do about it here. It's like a regular camera; if you go outside and you have the wrong light or the wrong exposure, you can't print the film. Or you can print it, but it's going to look funny because the colors are not right. As long as their quality is set up to line, we'll get a negative that will fall within our parameters for printing, and then we print.

We print to the gray scales that come off of the film recorder which receives the data from the spacecraft, because we don't know what that image is supposed to look like. If we get an image of, say, one of the moons of Jupiter or Saturn, we don't know what the colors are supposed to be. So we leave it up to the Imaging Team to set those parameters. That's all done through image processing lab, and then we get the negative. If we print to that gray scale the way we're supposed to print it, the imagery on that film will be exactly what it's supposed to be according to the scientists.

Then they can come over and look at it, and they may tell us it's not quite right. Then we can make the color corrections, with somebody from the Imaging Team looking at it and saying, "Add a little blue, take out a little yellow." Then, when it is exactly the way they say it is supposed to look, that image is locked into place color-wise, and that's how we will print it for press release.

Once it's released as that color, it must never be changed because we're also printing for science people too. If we sent out a picture of Saturn, say, that looks just exactly the way Brad Smith says it's supposed to look, and it's this particular color, and then next time we print that negative, somebody prints it a little greenish and it goes off to some other scientist, and he compares that with the other scientist who got a yellow one, then there is a lot of explanations of why is mine green and his yellow.

So once a color is established for a particular image, it has to stay that way, and we can't vary from it, no matter when it's printed. Even Voyager 1 images at Jupiter, if we print that stuff today, it has got to be the same color it was back at that point when it was first released.

We have a master print that was made back then that we keep under good conditions, and we go back to that every time somebody pulls out that negative to print this picture. They'll also pull that master print out and have it right there on their table beside the enlarger, and print to that. They'll have to match it visually with their eye, so that the new print looks just like this one. But eventually the master print will get to the point where we can't do that anymore.

It's just a problem with color film; it's not going to last. Even in storage the dyes will change and shift. Black and white will pretty well stay forever, but color won't. I imagine they have computer tapes somewhere so this stuff can be regenerated. I don't know what that would cost, but the images we have on file here that we use to reprint this stuff will eventually all disappear.

What time scale are we talking about?

I don't know. We'll just have to wait and see. We're noticing now we're having problems with the originals from Voyager 1. Of course, these are the most popular images. They are in the enlargers all the time, with a lot of heat and light going through them more often than the others. This will degrade the negative quality also. It's an ongoing battle. So far it's not a real problem yet, but I foresee down the line it will be.

We're here just pumping all this stuff out. We don't print it with automatic equipment; this is a completely custom photo lab, and everything is printed to a customer's exact specifications. If they want an image to look like a particular shade or hue, then that's the way we have to make it.

When we do produce large amounts of prints, say like 600 or 700 copies from each negative, to hand out on a press release day, we do have automatic roll easels once the image is set up to what everybody wants. Then we can set up a roll easel and put a big roll of paper in there and just let it automatically print until the roll is done. Then we can add another roll and keep right on going.

For example, during the Voyager first encounter of Jupiter, we would have 12, 13 press releases a night, at 600 to 700 prints per press release, plus 100 35-mm slides, plus duplicate negatives and everything else. The amount of production that is going out of here every night during these encounters is tremendous.

And it all has to be captioned. The captions are manually taped on the back of each print before they can go out. So we generally borrow a lot of people from everywhere; any place there is a spare desk, somebody's sitting there taping on captions. We do that in waves; we get so many ready, like 100 of each, and send them down to the Public Information Office. While they are getting ready to pass those out, we're doing another 100 until it's all over with. It's extremely hectic, it's extremely exasperating, and it's a lot of hard work, it's a lot of running around, but everybody stays right in there. Nobody ever gets sick during this period; they all want to be here.

Anyway, when we get a load together, somebody who isn't involved with doing any particular thing, mainly me, packages this stuff up, and we hand carry them down to the auditorium in boxes, give them to Jurrie, and he takes them from there. He collates them into packages, and that is what gets passed out to the press. We come back up here and just keep going until we get all the stuff done. Then we start it all over again the next night.

This goes on for two weeks or thereabouts. It's a hectic time, but I wouldn't be anywhere else. I've always been interested in what is out there, even as a kid. And I am finding out what's out there, and there's no better place to be than right here. You have access to all the scientists, and if you have a question to ask them, they'll take time to answer it. The majority of them enjoy talking about what they're doing. They'll bend your ear to the point where you don't understand it anymore. Working with them, seeing all this stuff, being the first people ever to see this—it's just unbelievable.

When you are in the thick of doing all these jobs, do you have a chance to take a few seconds out to get the significance of what you are doing?

You sort of lose it, because it's like being in the middle of the forest and you can't see the forest because of all the trees around you. You're just here doing this job. You sort of understand what is happening, but you don't get a chance to really stand back and take an overall view of just what historically is going on here. This is the first of it; this is the beginning. It's like being with Columbus or the Vikings when they were out exploring a new world; that type of thing. I just can't describe it.

One day during the Voyager first Jupiter encounter I had a couple of images from one of Jupiter's moons. We needed an okay; we had done our initial print of it and then we had to wait for one of the Imaging Team members to come up and look at it and give his final judgment on yes, that's the way it is supposed to look, or what corrections to make. So I was waiting up in the front office. It was about 7:00 or 8:00 at night, and we were sort of in the doldrums for a few minutes, waiting for one of these guys to come up and okay it.

One of the ladies was sitting up there too, and we were just talking. I had this print in my hand, and all of a sudden I started getting a funny feeling about looking at this picture. I showed it to her and said, "Do you see what this is?" She says, "Yes, that's so-and-so." I said, "But do you know the significance of what you are looking at here? You're probably the third person in the history of the world to ever see this image."

That's when it started sinking in a bit just exactly where we were and what we were doing. It probably won't really sink in for another few years yet when we look back in retrospect at where we've gone and what we've done.

And then putting together this solar system picture, the last picture from Voyager. That's a project in itself that we're heavily involved with.

Voyager looked back from roughly 4 billion miles out and took a picture of the solar system. The planets were tiny spots, so small that they could not be represented on a small print that could be handed to anybody because, if you get down to an 8 × 10 print, you can't see any planets. Their images, the dots that represent Earth and Venus and all the other planets, are so small that they totally disappear.

So in order to have the Earth visible to somebody, what is the smallest we could make it and have it all together with all of these spots relative in size and on one picture? What's the smallest we can make it and have it so that everybody can see it?

Well, we took the pictures of Venus and Earth, which were the smallest spots, and we started reducing them until they got down to the point where they disappear; you can't see them anymore. Then you start bringing it back until, okay, this is the magnification that's minimum. If it's any smaller than this, you can't see it; at this point you can see it if it's pointed out. It's a tiny spot, but it's there, and if you know where to look, you can see it.

Now if we do that with Earth and Venus, which are the smallest spots, how big will this picture have to be? Well, it turns out that it would be 21 ft long. We're in the process now of putting together a mosaic of this. Each frame will have to be made independently and then mosaicked on the board 21 ft long, which will be in von Kármán auditorium here at JPL.

What will be the photo lab's participation in the outer space, interstellar segment of the Voyager project?

The actual encounters are over with; however we still do a lot of work for follow-up talks, reports, and everything else. As of now, the cameras are shut down, and they will never be turned on again, though other kinds of data will still be coming back for another 20 years or so.

What was the high point, or low point, of your acquaintance with Voyager?

There was no low point; it's all a high, a total high. I can't say any particular thing, other than that first Jupiter encounter. The imagery was so gorgeous and the demand for our products almost buried us, but we survived and we got better from there. Problems cropped up that we didn't expect. Things are always going to come up that you don't expect.

What would be an example?

The main thing was the public demand for these pictures; we didn't expect it to be that high. We were stretched to the absolute limit trying to get the stuff out; there was so much demand. Afterwards we came up with a lot of new ideas on how to streamline the operation. Now that we know how to do it beautifully, we don't have any problems.

How did Voyager differ from other kinds of activities?

The only difference was its duration: 12 years. During that 12 years we had the two spacecraft that encountered Jupiter and Saturn, and they also had an encounter with each moon. Each one of those basically was a separate encounter. We imaged Saturn, then we'd image a moon, then another moon. And of course, the rings. That was a lot of imagery each night of different objects. There was an awful lot out there of interest, close together in one place.

As we went further out, the amount of moons diminished, which meant we did less work. For instance, with Saturn we had the rings and seven moons. Then when we got to Uranus, we had five moons. Uranus was a big dud visually, because there was nothing there to see; it's like looking at a blue bowling ball. The only real fascinating thing at Uranus was the five moons, Miranda specifically, because it turned out to be quite an unusual moon. So Uranus, with only one spacecraft (the other had gone out of the ecliptic plane), was an easy encounter. By the time the last press conference was over with, we were all caught up and ready to go on with daily work, whereas with Jupiter 1 we spent weeks afterwards picking up the pieces.

We learned a lot from that and by the time we got to Neptune, everything went so well it was like an oiled machine; just like we didn't have an encounter, it went so smoothly. Which made it better because you could relax more and enjoy what was going on, without racing around trying to correct all the problems. Everything went beautifully.

What would you like people to know when they look back to the 1980s and your work with Voyager?

I honestly don't care what anybody else thinks about the whole thing; I know what I did and what I was in the middle of. I may not have tangible things laying around that say, "Yes, I did this; yes, I did that," but it's all in my head, and that's all that counts.

There are very, very, very few people in the history of the world that can be in this place at this time and can be involved with something like this, and I feel so fortunate to be one of them. It's just this photo lab; a small number of people who are working here. A lot of them here now were here through the whole thing.

The people outside the photo lab—your family, your friends who are not at JPL—what was their understanding of what you did?

A lot of them are really interested. They're always asking me about this and that. At a party where you are meeting people you don't know and you are being introduced for the first time, during the course of the conversation it comes up, "What do you do for a living?" I just say, "Well, I run a photo lab at JPL." In most cases that's as far as it goes. But sometimes people recognize "JPL", and they ask questions and it gets deeper and deeper. Some people are intensely interested, and when they find out I run the photo lab involved with all the imagery, their interest peaks and I get bombarded and they really want to know what's going on.

And there are other people who don't know a doggone thing. They don't know what JPL is; never heard of it before. They don't know what Voyager is; they don't know anything about what has been going on out there. I feel sorry for them.

Maybe they're doing something important to them that I don't know about; there's a lot of things I don't know about. But if I had my choice of doing my whole life over again, I wouldn't do it any different. I would want to be right here doing the same thing. Even if I could be a member of the Voyager project but had a different job in it, say as an administrator or something like that, I still would rather be here. This is where the action is. This is where it all comes down to what are we going to see, and this is where you see it, right here. This is the final version of that image, and we see it before anyone else in the world does, because it starts here.

Was there any really difficult decision you were faced with?

Not really. We pretty well know what's coming, so we can get things arranged in advance. And as far as which pictures are selected to be released, I'm not involved with that at all. I sit in on the meetings, but I don't have a specific input as to what images are selected. I'm just there because I have to know what's coming, what to look for. Nothing drastic ever came along, luckily, that I had to make an on-the-spot decision about. Everything just fell into place.

Because this is where the prints finally come to fruition as a finished image that somebody can look at, we get a lot of prestigious people coming in here. Not so much personalities but science types. You never know who might be sitting out there at the table, waiting for the pictures to look at. Carl Sagan was in and out periodically, different types of scientists, the lab director—you never know who is going to walk in. We don't much care about that because we are just doing what we are supposed to. But that can be a high point.

I always thought it would be interesting to have a behind-the-scenes look at what went on during an encounter. What happens with all these average people out behind that nobody ever sees, who are actually doing all this work? All these pictures that go out to the public—does anybody really know where they come from, or who does them? Sometimes it bothers me, but then again, it shouldn't. I'm doing what I want to do, and that to me is enough.

But I would like to have the photo lab get some kind of recognition. We're the ones who did it. We are the ones who made those images so that other people could see them. It's like these montages here. This is Neptune, that's Uranus, and here are two of Saturn. I'm the one who put those together. These pictures are actually cut out and pasted down. That's the art work that was photographed. That's exciting for me. Now that I think about it, that's sort of a high point, because I put these things together and I see them published. I see them in magazines and newspapers and book jackets and record covers.

There's no indication whatsoever of who put these together, who made them; nobody knows. But I know: I did. Now how many artists would give their left arm to have their artwork published like this? I've done something that practically everybody in the world has seen.

That is personally satisfying. At first I tried saving them everytime I saw one on a magazine cover and put them away in a box, but I finally gave it up because it got too big, too heavy, too much.

These pictures on the wall. I was watching a movie on TV one night, and some guy walked into a big office. It was all walnut paneling, with a big gorgeous desk, and behind the desk on the wall was that Saturn picture right there. That was satisfying, to see your stuff blazing away on the screen.

I still see them. You open an ad for K-Mart, and there's a television set and on the screen there is that Saturn image. They're always cropping up. That's satisfying. The second Saturn on this wall we did specifically for the cover of *Science* magazine. That's why the black space at the top is so large: so they can get the title in.

It's hard to believe it was me that was involved in it. I never planned on this, never thought about doing this as a living; I didn't know what I was going to do. It was all handed to me on a silver platter. "You want to work here?" I said, "Sure". "Okay, go up and talk to so-and-so." I did, and 33 years later I am still here, doing this. I can't imagine doing anything else.

My wife is a librarian over at Alvarado Junior High School where we live. I would take pictures home to her, and she put them up in the library, and the kids would be intensely interested. One time we had a combination of three pictures. It was one of the Apollo images looking back at Earth, and the United States was in the picture. Then we had a picture taken from a spacecraft looking over the Southern California, Los Angeles area. And there was another photo, taken by the U2 flight over Alvarado, and the picture detail got down to the point where you could make out individual homes.

So I took all three of these pictures and laid them out on the board. I had a line running from the picture of the Earth down to this picture of Southern California. Then down here was the U2 picture of our neighborhood, and I put a caption underneath it, "Can you pick out where you live?" It created a tremendous stir at the school because if you look close enough at the U2 picture, you could see the school, the streets, and individual homes. I was able to find mine.

From 70,000 ft up?

Something like that. It gives me a good feeling when I can pass some of this on to kids who'd never get a chance to see any of this stuff. A picture like that in a newspaper or a magazine is really destroyed. It doesn't come out well because the lithographic reproduction—unless it's an extremely good one, which newspapers don't have—is a very bad reproduction. They're designed to reproduce just black and white. Somebody like the *National Geographic* can reproduce them quite well, but they still don't do it justice; it's not like having an original color print in your hand, made from the original color negative of that image. So when I get a chance, I haul them home, and she takes them to school. That's satisfying.

My wife has been very interested in what is going on and very supportive, and we've had no family problems; everything has gone along fine. Of course, I haven't been putting in the horrible hours like some of these project people did during this whole thing. Some families and marriages have been destroyed; heart attacks, everything else, all the way down the line. The most I put in would be 10 or 12 hours a day, and that was only during an encounter, while the press is here, maybe just for a two-week period. Everybody just grits their teeth and bears it—but they don't grit very hard because it's so fascinating.

Aside from technological issues, any thoughts on administrative or supervisory matters?

One thing we learned as we went along was that communication between everybody involved was of primary importance. From the Imaging Team right on down to the printer who has to actually make the image on the paper, everybody has to know precisely what's going on and what's in the other person's mind. We have to know what it is they're after. We have to know which images they want. Give us all the information that will help us to better produce what you need.

It didn't start out that way. Nobody really knew what was going on as far as this whole loop was concerned; everybody was new to the show. At the first Jupiter encounter, we ended up one night with images coming in, but we couldn't get an input from anybody as to exactly what they wanted. We couldn't just sit there and wait until they decided because we were running out of time; we had a deadline for getting this stuff done, so we made a unilateral decision to just go ahead and print *everything*, in order that they would have what they wanted when the time came.

So when the scientists came in the following morning, they had these thousands and thousands of images to peruse through; they got what they needed, but they also had a lot more than they needed. Because we couldn't let them down, we had spent an awful lot of time and effort producing an awful lot of stuff that eventually just got thrown away.

After that encounter we tried to figure out what had happened and what to do to make it easier the next time. We talked it over with the Public Information Office and the Imaging Team, to see how we could cut down on the confusion, cut down on doing more work than was necessary.

So we decided that during encounters we would have a 5:00 p.m. meeting each afternoon to plan for the next morning's press conference. We sat down with the Imaging Team and the people who would be involved in the press conference, and they'd tell us what images were coming down and what they wanted for releases the next morning. That way we all knew what everybody was doing and what to expect.

The scientists and the Imaging Team designated which images were going to go out for public release, and they worked with the Public Information Office on what type of imagery the public would be interested in. So working together they decided which images were going to be released for the following morning press conference, but the Imaging Team had the final word as to whether that image would go or not.

So we started getting better organized, better put together. Voyager 2 at Jupiter was still hectic, but it went a lot easier. Out of the mistakes we made, when Saturn came along we had it a little easier still. And then a little easier for the second Saturn, and as we went on it got to the point where everything was working rather nicely.

PETER W. WELLS
Head Dispatcher, Transportation
Born January 2, 1940   Los Angeles, California

Peter Wells graduated from Franklin High School in Los Angeles and has taken courses in management, running a division, electricity, small engine repair, auto brakes, computer DOS, computer Windows 3.1, and total quality management.

He worked for Sears Roebuck and Company in Pasadena from March 1966 to October 1970 as a furniture salesman, warehouseman, furniture touch-up specialist, forklift operator, and sewing machine repairman. Then he joined JPL and served six years as a medium-duty truck driver, forklift operator, and VIP chauffeur. For the next 18 years as head dispatcher, he supervised 30 drivers and took calls to set up freight, courier, and VIP runs. Since retiring in 1995, he is an independent travel agent and insurance agent.

He is a member and office holder in Kiwanis and the Juniper Hills Community Association.

He enjoys music, travel, and observes that he is active, still can dance, and can ride a dirt bike as good as any kid in town.

Respect your peers regardless of what he or she looks like.
You never know what wonderful thoughts and ideas
are hidden way down in the recesses of their minds.
Don't ignore the suggestions of new members of your peer group.
They might have had some experiences you aren't aware of
and could come up with a better way of doing things
that you could never have thought of!
Why, in today's society, do most people never
seem to remember The Golden Rule?
—Peter W. Wells

PETER WELLS

For Voyager, or any other project, transportation is usually on the ground floor. After the R&D we start moving parts back and forth, and whether it's ground support or flight or whatever, we move it and handle it the same way. Instruments go back and forth to calibration, flight gear goes back and forth. After things are built, like parts of the spacecraft, we take them from testing facility to testing facility, building to building, with guard escort. We have spacecraft convoys.

People moving back and forth, we move them. The engineers and scientists that are involved in the project, we give them cars to run out and go different places. Different facilities off-lab are usually involved in making things; we loan our QA people cars to go out and check what is being made at those facilities.

What is QA?

Quality assurance. Like if Boeing is making a spacecraft component for us, our QA people go out almost daily to monitor to make sure it's up to our specifications. For anyone involved in the Voyager project, coming and going, we get them a car or transport them. When our lab engineers and supervisors have to go back to Washington and NASA, we take them to the airport, put them on the plane, and a couple of days later we pick them back up, bring them in.

You provide drivers?

Yes, we provide chauffeur service for VIPs coming and going. And practically every bit of the spacecraft, the physical gear, is handled by our drivers. Our padded instrument van and our forklifts have all been fitted with what they call an "air ride," a hydraulic ride, so just about everything we have up in space right now has ridden in our air ride forklifts for an easy air-over-oil cushioned ride.

The logistics of it is horrendous. Say a project manager wants to get all his people together and finds space. We would move the people's furniture and belongings around on weekends into one location, so they can get together and talk and work on the R&D. Sometimes they will get a whole building, and we'll move people from seven or eight other facilities into that one area for a spaceflight project. So we get into even that level of logistics: moving furniture and people.

At times of encounter we take people back and forth from the airport. We take them to hotels and motels and out to be wined and dined. Sometimes they go to Caltech for award ceremonies and so forth. We get involved quite a bit with each project, and that's besides the day-to-day business of JPL.

Tell me more about the day-to-day activities.

If you name it, we move it. These tickets here are from people calling in. We have radio pickup drivers and forklift drivers that read tickets everyday, plus we have a fleet of cars we give out, 25 cars that come and go, for people going out to different vendors on lab business. And we have regular transportation jobs, like delivering all the stationery, delivering all the instruments to people—just regular set-up jobs. The mail has to go back and forth. The interlab mail is horrendous, with the amount of employees we have. We deliver supplies to everybody: paper for the duplicating machines—it's a papermill.

Transportation has around 26 drivers at any one time, supporting a 6000-person lab, and we can't stay caught up; we're always moving back and forth. We have shuttles and buses taking people constantly to and from the different sites we have off-lab. There's Caltech students that we run back and forth daily between here and Caltech.

There's a driver that goes around the lab all day for 10 hours, from each parking lot, taking people around the lab, to their buildings, back to their cars. We have another one that goes to the IPC (Institutional Processing Center) facility. There's four or five buildings over there; people come back and forth all day. We have another that goes to 525, our Sierra Madre Villa Avenue facility, and two that go between the Foothill complex and Caltech.

I can't conceive of any project being accomplished without transportation. If you got to get from A to B, if you got to transport it, we transport it.

We have a warehouse in Bell, California, called Cheli warehouse. There's a lot of flight gear stored there, and some of the equipment they used on a previous flight project would pertain to this one, so we're going back and forth, picking up different gear down there, bringing it in on the tractors and trailers, delivering it to the buildings, and they either use it or copy it.

We have long trips also. We got a guy up at Ames, near San Francisco, picking up pieces for a project being brought in on a plane. He drove up there, waits for the plane, and will bring the stuff back here. So we have overnight trips. We got another guy who went to New Jersey to pick up a van that they bought down there. Another one's in Oregon. We just had two Chevy Blazers come back from a month in Canada. They went up there on a JPL experiment; they came back with equipment and personnel. It just goes on and on and on.

Of course supporting us there has to be a mechanic's shop. There's about 12 or 13 mechanics and service men; they have to service the equipment, fix the equipment, give preventive maintenance. The equipment that's loaned out to different sections has to be maintained. We have a large fleet of cars and trucks loaned out to different sections, so that people can operate themselves, without calling us for every single thing. That all has to be maintained and serviced on a schedule.

Are the servicing people under your jurisdiction?

No. I'll introduce you to Marty Greer, the supervisor of the shop, and he can give you his position on the support. They have to support everything we do, and everybody who orders a car, right down the line, it has to be supported. Or a truck or even a forklift. They all have to be repaired or maintained.

I talked to Larry Williams when I was here before. He's gone now. What was the relation between your office here and his?

Larry Williams was in charge of shipping, receiving, the mail department, and transportation. So we worked like a three-sided triangle, all working together towards the same aim. When things were coming from shipping and receiving, we would deliver it, or we would pick up things from the lab to take to shipping and receiving so they could send it out by vendor. Of course the mail department has a horrendous job, the amount of mail they deliver.

Do you deliver their mail?

We give them the vehicles to deliver it. They have people who work in the mail room, sorting and putting it in bags, and our drivers go down several times a day and deliver the mail to the various buildings.

So the mail department doesn't actually have carriers.

They had for years, but that changed last year. They did have trucks and delivered their own mail, but now they have what they call "just in time" teams that deliver stationery supplies, mail, small packages from shipping and receiving, and Federal Express, priority mail. We have five trucks with five teams of two that do nothing but hustle all day long. So the mail department is left to just receive the mail, sort the mail, make sure everything's okay, then we pick it up and hustle with it.

Are there any incidents particularly with Voyager that we should note: challenges, successes?

We are pretty successful with all our flight projects. We have a glitch here and there, but we're the ones they call on to straighten out the glitches. "Hey, we've got an emergency; this has to be taken from A to B immediately," and we respond.

The word "flight" gets priority over everything else. If we have two or three moves to do, they take a back seat while we do the "flight" first. If we have a flight convoy, it's escorted by guards with blinking lights. There's certain rules for a spacecraft convoy: you can't go over five miles per hour. Nobody can pass. If a rear guard vehicle sees someone try to pass, the guard moves over and cuts 'em off so they cannot pass.

This is out on the highway?

No, on-lab. They don't want even as much as a pebble being kicked up. And if so, they have to record every little knick, so if they have trouble with the spacecraft a year or two down the line, this is all recorded. There was another spacecraft they drove in a convoy from here to Florida. They did not want to fly it there because there is a chance the plane might crash, so they wanted it all on the ground, in a supported convoy. They drove from here to the Cape with it.

Was that a new procedure? Do they usually fly stuff there?

It depends on what it is and who's in charge. We have the system of "this is the way it's always worked and this is the way we'll do it, rather than take a chance on a new way."

Do you have a voice in that decision, or is it somebody upstairs?

It's somebody upstairs. It's up to the engineers and the flight project managers how they want to do it. We have done it, and we know how to do it, but we sit back and keep our mouths shut and let them say, "let's do this." That way, if anything goes wrong, they're the ones that made the decision, not the transportation driver. He might have moved spacecraft 20 times a certain way, but if a manager wants it done another way, okay. If our driver is standing there and the manager says, "put a strap here," the driver will do what the manager says.

What sort of records do you keep? Do you have to write down that it took 17 minutes to move from point A to B?

Yes, we file a ticket on everything we do. For example, here's the person's name and extension, the date, the time, from, to, and what we're moving. And under "special instructions" we would put "flight." Everything is kept and filed and recorded.

How far back does this go?

We're supposed to keep records for one year, but we keep them longer than that, just to be on the safe side. Also, anything we move back and forth, from on- and off-lab, we have manifests. This one says "From Cheli warehouse in Bell, California, to Pasadena." These go to four different places. Property keeps a record, shipping and receiving keeps a record, and the sender and the receiver. It's a papermill, but it has to be because we have calls like, "Four years ago in April I was expecting a package from UCLA; what did you do with it?"

We get those all the time, and we find it: "So-and-so signed for it."

"Oh? She was an academic part-time summer student. Just a minute; let me look in her desk. Here it is!"

So it's not just the paper; you actually find the objects.

The items. Yeah, we have calls like, "Three years ago I was supposed to get three instruments. I don't know where they are. What did you do with them?"

We got one right now, for a computer. A lady just called and wonders where her computer is, so Robert Curiel, the leadman, has to trace it down and satisfy this customer. It turns out that the person who's looking for the computer is going on a trip, and he has to take the computer with him because it's a laptop.

I've been at JPL for 24 years. I started in on the ground floor as a driver and chauffeur and heavy equipment operator. Then I got into the dispatch end of it and found it quite a challenge. You have so many pieces of equipment and so many men and so much workload, and you have to coordinate it and make it work out. It's never the same; it's quite a challenge every day, which has kept me on it for so long. There's always a good feeling of satisfaction when you get the job done and you've made your customer happy.

It's frustrating when it doesn't work out?

It is frustrating. You try to make it work out. The drivers who support you are very understanding when you pull them off of one job and put them on another, even though it's an inconvenience to them and it puts a load on them. Sometimes one driver is supposed to do one thing for eight hours, and he ends up doing 10 different things because of situations that come up.

Sometimes there are pressures on the job, and sometimes people paid a personal price for doing their work.

I have a real good example. The pressure sometimes is horrendous. I got a call one day from the director's office. He wanted to go to TRW in El Segundo. I very politely asked "when?" The secretary replied, "I forgot to call you before, but he's waiting downstairs now and he needs to be there in 20 minutes."

So I yelled at somebody to watch the phone, jumped in the car, picked him up, ran the gate, and made it down there.

We do get these hurry up, last minute calls. We call them "fire drills." They'll call up and say, "Hey, we forgot to call you, but we've got 37 people standing at the main gate and we need to go on a Lab tour." This happens all the time.

So what happens then? Obviously you can't tell them to start walking. Do you send a bus there?

Yeah, right. Many years ago, when I was a driver and a chauffeur, I got a call on the radio from the dispatcher who told me to go to the main gate; there was a VIP down there and he needed to be taken around the Lab. I went down there and there was this gorgeous 18-year-old girl. She turned out to be the princess of Iran, the Shah's daughter. That was when the Shah was in power, and he was supposed to come to JPL to look at different projects. He couldn't make it, so he sent his daughter as a token. Here me, being a lowly driver, had to take this girl around to show her different buildings and things.

At that time she was interested in our clean air facility, where we were developing clean burning auto engines. So I showed her that facility and the man in charge of the building showed her around and answered questions, but she was here more as a token. I went to the main gate to pick up a VIP—and it was a princess!

Another time they sent me to LAX (Los Angeles International Airport) to pick up a VIP. I didn't have his name; I was told he would recognize the JPL uniform and car. So I picked up this funny little man to transport back to JPL, and as we were going down Sunset Boulevard I said to him, "If you'll pay close attention to people getting out of cars and walking down the sidewalk, maybe you'll see a movie star." His reply was, "I had enough of movie stars while I was waiting for my Oscar."

Realizing I'd made a mistake, I said, "Well sir, in what capacity were you related to the industry?" He said, "I'm Arthur C. Clarke; I wrote *2001!*"

I kept my mouth shut for the rest of the run. He added, "I was nominated for an Oscar but never received it, but I had to sit in the audience with those idiots, and I had my fill of those movie people during the Academy Awards night."

I was not told who I was picking up. Our drivers do get into things like that. They've hauled William Shatner back and forth, and different people. It's quite an interesting part of the job. We'll go along for days with very mundane things happening—then all of a sudden, *wham*. We're responsible for bringing in congressional visitors, high NASA officials. Once in awhile high politicians will land by helicopter up on our Mesa antenna range, and we'll pick them up and bring them down to work in different buildings, so you have quite a broad spectrum of things you're responsible for.

Does this sometimes require somebody working overtime on short notice?

Yes, it does. Sometimes we'll go to LAX to pick up someone coming in at midnight. They've taken a late plane, so our driver will go down there and wait.

How do their families feel about this?

They learn to live with it, and I try to spread it around enough where I don't use one guy all the time, so he may go on a run like that every three months. They just know it's part of the job.

Other people I've talked to mentioned heart trouble, cancer, and things like that, not necessarily related to the job, but I guess, with the pressures you guys work under, that could happen, too.

We pretty well have a crew that can handle the stress. They know about it right away so we keep the people who can handle it. The old-timers that are left are the people who have dealt with it for years, and it bounces off one side and goes out the other. But if you are very sensitive and shy and introverted and can't handle stress, you soon find out this is not the job for you, and you either transfer to another job or do something else. That leaves us with the ones who can handle pressure, stress, and "fire drills".

ANITA M. SOHUS
Documentation Representative
Born August 9, 1951   Scottsbluff, Nebraska

Anita Sohus started at JPL as an academic part-time during her first year of college and became a technical writer shortly after graduating from UCLA in 1973 with a degree in English. In 1980 she received professional designation in public relations from UCLA Extension.

In 1977 she became the documentation representative to the Voyager and Galileo projects, thus gaining experience in both development and operations phases of projects. In 1986 she worked at NASA Headquarters in the Solar System Exploration Division and then joined JPL's space station office.

Returning to Pasadena and the Voyager project in 1988, she happily completed the Grand Tour, a total of 12 years and six planetary encounters. From 1990 to 1994, she was an information specialist in the Flight Projects Office and then was selected to support the Office of Space Science and Instruments. She is currently the acting deputy manager of the Outreach and Education Office of JPL's Space and Earth Science Programs Directorate.

She is a member of the American Astronautical Society's Division for Planetary Sciences, the American Geophysical Union, and the National Science Teachers Association, and has received publications awards from the Society of Technical Communication as well as many NASA group achievement Awards. She has also sold a number of freelance articles to various magazines and for several years wrote a weekly public radio program on space.

She is a dedicated horsewoman and also enjoys camping and backpacking, cross-country and downhill skiing, and genealogy. In deference to Earthquakes, fires, and living in a megalopolis, she is also an amateur radio operator.

Go out some night, away from the lights.
Gaze up at the canopy of stars.
Among them you will see
other worlds: worlds of
extreme temperatures
and crushing pressures,
noxious gases and
precipitous cliffs.
Worlds where
several moons rise
and set each day, rings
cut icy swaths across the sky,
superbolts of lightning slice the air,
and auroras light up the horizon for
thousands of miles. Worlds vastly
different from our own, and yet
in many ways, the same.

—Anita M. Sohus

ANITA SOHUS

How did you become involved with Voyager? How did you fit into it?

In 1977 I became the Voyager "doc rep," the representative to the Voyager project from the Lab's documentation section. That meant I was the designated editor for all of the project's documents, but it grew into more than that. I had been assisting the previous Voyager doc rep, Shozo Murakami, since 1976. I got the job when he left the Laboratory.

Historically, the doc rep had always produced the flight status bulletins, which started out with the Mariners, maybe even earlier. The intent is to get something to the employees as soon as possible on the status of the mission. With Voyager, it got to be a 12-year mission instead of a three-day mission like some of the Mariner programs. Over the years the flight status bulletins on different projects have evolved differently, and now they are more of a public information thing than just for employees. Jim Wilson did the Mariner ones; Duke Reiber did the Viking ones. I've done all the *Voyager Bulletins*; the one I just sent to press is number 99.

I have a complete set of them; so you're the person who did them?

Yep. I loved doing the bulletins because I learned so much, but I usually ended up writing them after hours, the only time I had any quiet time to concentrate. My job on Voyager evolved into three main parts: technical documentation, mission status bulletins, and coordinating the press conference graphics.

The documentation part of it sounds boring, but it has to be kept up and it has to be kept up to a standard because they constantly make changes to the documents, and when somebody who is new to the project comes in, they need to know what's the latest revision to the document, what's the latest change, so they can get up to speed quickly. There needed to be some standards set, and some consistency. That's where I came in.

You're talking now about technical documents, not just for public information?

Yes, technical documents. There are hundreds of documents: plans, procedures, policies, requirements, specifications, user's manuals .... For example, the software has to be documented so somebody new coming in can know how the program is written. Someone impressed upon me when I was a really green editor that a spacecraft had to be destroyed at the Cape because somebody had written a minus sign over Snopake on a printout; the Snopake flaked off, revealing a plus sign. And that's what caused the rocket to go off course, and the range safety officer had to destroy a rocket that had a JPL spacecraft on it. True? I don't know, but it sure made an impression on me.

With that as a warning I try to be very careful about proofreading all this stuff and checking it. It sounds kind of boring, but it gave me an overview of the project that a lot of people don't get because virtually every technical document or conference paper written at the Lab for Voyager came across my desk. So I got to meet people from all parts of the project and I learned something about their jobs, which then helped when I needed to write the public information stuff. I knew who to go to, for what, on short notice.

I realized, too, the documentation sometimes didn't really matter when it was finally published. What was important was the process of getting people to sit down together and talk about what they thought things meant, to reach an understanding and to state things clearly. The Space Flight Operations Plan, for example, contains all the agreements between each team: what information each team is expected to supply to the other teams, what information each team needs to receive from another team, what each team needs to do their bit, like running a trajectory file or something. The Space Flight Operations Plan also includes operating plans on how each team is going to perform. This is another case where the process is as important as writing it.

Of course, the final product, the publication, has to be important too, as there was a turnover in personnel. There were five years between Saturn and Uranus, so a lot of new people came on for Uranus, and a lot of new people came on for Neptune. They needed to be able to go someplace and read about how things were done before they actually had to do it, so that was the value of that.

I've been told that much documentation has already disappeared.

That may be true of the informal documentation, like memos. I noticed that everytime there would be a change of secretaries, the new secretary would clean house and maybe would throw out old files or send them to records storage, not recognizing that they might have some historical significance. I would always cringe and try to stop that when I saw it happening, saying at least leave us a record of where you sent the stuff, so we can get it back.

The Lab doesn't have storage space onsite, so a lot of stuff is sent to a federal records center. Supposedly there's a certain length of time stuff will be held, and then they question you whether to keep it or not. I don't know what happens if they can't find the person who sent it to them. Theoretically, the routine for all the editors all over the Lab is that, for anything that's got a document number, two copies are sent to vellum files; one becomes a shelf copy and the other is microfilmed or microfiched. All of the formal documentation is available; it's just a matter of plowing through it.

A third facet of my job was born during the press conferences for the first Voyager Jupiter encounter. Management and public affairs realized that it wasn't working to have scientists stand up there on camera and drone on about graphs. That's when we started trying to explain things better by drawing things, making graphics pieces overnight for the press conferences to help explain the results and why they were important and what the scientists were seeing. I worked behind the scenes during all six Voyager encounter press periods (roughly 10 days each) to coordinate getting all those graphics done overnight for the press conferences.

It's 24 hours around the clock to get them done, because the scientists need time to look at their data and try to figure it out, and they need time to figure out how to explain it to a lay person. Then we need to meet with the art vendor and discuss how to do the piece of art, and then the art vendors work all night and bring the art in early enough the next morning so the scientist can see what he's got. If there's a problem with it, then he knows beforehand; he doesn't get a rude surprise seeing it for the first time on national TV.

TV graphics are different than the technical art that we normally produce at the Lab. The aspect ratio for a TV screen is different than for overheads or slides. The usable image area is smaller. The art has to be extremely simple, and the colors have to show up well in relation to each other. The colors look different on TV than they do with your eye. The line weights have to be heavier, and the type size has to be fairly large. Tain't easy to take technical art and condense it down to TV graphics. It seemed like we had to break in a new bunch of illustrators each time and educate them about doing TV art. Some nights we sent out as many as 50 pieces of new color art to be delivered as slides for the next morning's press conference (and we didn't have computers to do this). The art was also used by the scientists and project managers for their public talks.

It also isn't easy to take a complex concept and make it understandable and exciting to a lay audience. Especially at Jupiter and Saturn, Ed Stone was very involved in helping the scientists do this. Every night he would sit down with them and help them decide what story to tell and how to tell it. At Uranus and Neptune they were more on their own, but by then they'd learned a lot better how to present the story. Ed is wonderful at explaining things to people. Between encounters I would help him with whatever graphics he needed for his public talks and articles.

Those have been the three main parts of my job on Voyager.

What are some of the outstanding memories or impressions that you have about the project?

Well, the people. There are some really outstanding individuals, without whom it just wouldn't have been the same project. I think Voyager learned a lot, and I'm hoping that the projects that are coming up will learn from Voyager's experiences instead of having to relearn everything themselves. But I don't think that's going to happen, except where there's Voyager people who have gone on to those other projects and will be listened to. Sometimes people won't listen to others who speak with the voice of experience.

Ed Stone certainly has been one of the driving forces on this mission. He's very popular, very well liked; he can explain things to people. He's not pretentious at all. He's very willing to give people his time.

I felt like I was his foil; for example, when we were working on a piece of art for a concept that he wanted to present, if he diagrammed it out and looked up and he saw my face was a blank, he'd start again. He'd play with it until I said, "Oh, that's really neat!" Then he knew he had something that communicated to a nontechnical person. And I could make suggestions or ask questions that would help him communicate a particular concept better.

I also saw myself as a translator between scientists and artists, because I'm neither, but I would sit in on their meetings and let them talk to each other. If I saw that they were not communicating, I could say to one or the other, "You've gotta do a better job of talking," or break in and explain one's jargon to the other.

I feel really lucky to have been on the front seat at all these encounters, the front seat for the excitement that was here. For example, when Voyager went through the rings of Saturn, audio recordings of the waveforms were made. Fred Scarf, who was the plasma wave subsystem (PWS) principal investigator, came roaring into Ed's office with his little cassette player, about the size of yours, with a big grin on his face, so excited because he had the sound of the spacecraft being bombarded with particles as it went through the ring plane of Saturn. Ed just lit up like a Christmas tree, and they were all excited. The excitement is so contagious.

Ellis Miner, who is the assistant project scientist, is a very quiet, laid-back person, but he's really been a bonus to the project because he has a very well-organized mind and he has an amazing ability to keep track of details. He was the leader ex officio of the Science Investigation Support Team, which, when it started, was a lot of very young people. Now, it's all people that are in their 40s and have a couple of kids apiece. They all grew up on that team. Ellis was the guiding hand, helping them make the tradeoffs when they had conflicting requirements between different investigations. Of course the final authority on that would be Ed, but on smaller things Ellis was. Ellis could keep aware of all these possible conflicts and raise a flag to the people: "Hey, you should be aware of this." So Ellis is another person without whom the project wouldn't have been the same.

Howard Marderness on the spacecraft team is another key person. He is the spacecraft on Earth. He knows exactly what's going to happen in those computers if you do this or that. He's a very shy, quiet man, but he's very well respected.

Lonne Lane was another one. Initially he was the assistant project scientist for Jupiter. After that he became the principal Investigator on the photopolarimeter instrument. He was another one who was a guiding hand for this group of young people who were doing the science investigation support work. He has so much energy. He works himself to death almost. When the photopolarimeter instrument on Voyager 2 was damaged, nobody thought it could be used, but he salvaged it and the whole experiment and got some really valuable data with it.

Richard Laeser and Ellis Miner were the two main reviewers of my bulletin, so I would interview Dick often. He'd tell me things, I'd write it up and take it back, and he'd help me whip it into shape a little better. He was Mission Operations Director before he was project manager. He was on the project a long time.

That's one thing that I think contributed to Voyager's success: people have tended to stay with the project rather than move on to something else. Many of the people who developed and tested the spacecraft stayed on to run it and talk with it. Without that body of "corporate knowledge", a lot would have been lost.

People could take advantage of periods between encounters—almost five years between Saturn and Uranus, for example—to go off and do something else with their life for awhile, but a lot of people wanted to come back for the next encounters. I'm glad they did.

This was not just the scientists who have a regular job elsewhere?

No, this is the flight team. The scientists, especially the principal investigators, get all the press, but it couldn't happen without the engineers and the other people here; people like Candy Hansen and Andy Collins. There's so many people, each with a specific job that needs to be done, and if they don't do it well, there's trouble.

Charley Kohlhase is another interesting character, the mission design manager. He is a very bright and imaginative guy. Working with him was interesting to see how his mind works, to see the wheels turning.

A lot of team members enjoyed telling the public about Voyager. We had a speakers' bureau, and we'd send people out to talk to schools on request or to different companies or professional organizations. A lot of the flight team would do that, often on their own time. Those who were very articulate and presented well were put on TV during the encounters.

What meaning does Voyager have for you?

I was lucky to fall into it. It's been a constant in my life. It's been nice to work with really intelligent people and also very nice people for a long period of time. Whatever was going on in my personal life, the ups and downs, this was still here, and I liked the stability of it. People like Dick Laeser and Charles Stembridge, among others, guided me along. In the earlier days of Voyager, I reported to the project's administrative assistant, Maynard Hine, and he was a great help to me, too. I was lucky to have a group of people who overlooked the fact that I was young and green, and who helped me along.

If I could remember everything that people have told me or taught me over the years, I would be a genius. I have had the opportunity to essentially be tutored by some of the best scientists and engineers in the world. I sometimes think it should have happened to somebody better than me, who could do more with that.

Philosophically the main thing from Voyager for me is just how fragile Earth is, because all these complex systems have to work together. This is the only place that they've seemed to work together where we can exist. That's why we need to go out and study these other planets because they are laboratories that haven't been contaminated by what man has done. Understanding some of their chemical systems and atmospheres may help us understand a piece of the puzzle on Earth, so we can try to save ourselves.

What are your most vivid impressions about the Voyager mission?

As the status bulletin editor, I got to go to Florida for the launch of Voyager 2. I was really wet behind the ears, really unconscious of a lot of things, and just learning and awakening at the time. I hadn't watched Voyager being built or anything, so it was really an experience for me to go down to the Cape.

The safety officer, Larry Montgomery, saw me standing around and said, "Want to go see the launch pad?" and I said, "Yeah!" and he said, "Grab a hard hat." So I got a hard hat, and we drove out to the pad. Voyager 2 was on top of the rocket, on the pad. It was the day before launch, and Monty was making the final safety check. I got to climb all over the gantry with the rocket in place. That was really exciting.

The launch went off really well. I was out on the causeway with a bunch of other people, taking pictures. Then I went back to a postlaunch press conference. The flight controllers hadn't gotten any indication that the scan platform boom was locked in place. This was pretty awful; it could ruin the whole mission if that boom wasn't in place. Some of the guys were nearly in tears, from the stress. They finally decided that it was locked into place, but as a precaution they delayed the launch of Voyager 1 several days so they could take the shroud off and check its boom mechanisms.

Voyager 1 was in the spacecraft encapsulation facility before it was moved out to the pad to be put on the rocket. I stood outside the clean room and watched through the windows while they lifted the shroud off of Voyager 1 so they could check out the spring on its boom, and I watched as they put the shroud back on. This was the only time I actually saw Voyager, as I came on to the project right before launch and did not watch the spacecraft being built at JPL.

Voyager had some great parties, mostly at work. One of the project secretaries, Layne Whyman, has a great sense of humor and is great at organizing. They used to have picnics up here in the conference room where they'd have bowling in the hallways with soda cans as pins. Pie day, when everybody would bring a different kind of pie, was fun. Pie day got tied in with Halloween, when everybody came with a different kind of hat. Ellis Miner would always come as Major Miner, with a major's hat on. They did fun things like that, which helped draw the team together.

Before the first Jupiter encounter, a bunch of the girls like Sue Linick and Candy Hansen and Jude Diner started what they called the John Travolta Look-Alike Contest. It was a great morale booster, at a time when everybody was under a lot of pressure and things were a little fragmented. It was about the time of the movie *Saturday Night Fever*. John Travolta had that distinctive pose, in the white suit with his thumbs up. So the girls went around with a Polaroid camera and tried to get all the guys on the project to pose in their version of John Travolta. Then they posted all the pictures, and they had a vote as to who was the winner.

I think Norm Ness was the winner. I would have picked Ed Stone: I thought his pose was perfect. Some of them were really funny. One guy was too embarrassed to pose, so they held his arms up for him. Another guy reclined on a desk a la the Burt Reynolds centerfold. (This was slightly out of character for this guy!) It was so much fun. It was a glue to bring everybody together and get everyone talking and laughing and being a unit. I don't think they could ever top that as a cohesive type of thing, bringing people together.

Things like that, playing, relieved some of the pressure. There were tensions, interpersonal conflicts. Since for most of the life of the project I lived in another building, I wasn't here all the time; I didn't see all of that. The past couple of years I've been housed with the project, so I saw more of it. They needed that play as a safety release valve sometimes. Everybody got to know the personalities and lived with it if somebody was a little odd, because they were valued so much.

The science investigation support team used to play a lot of practical jokes on each other. I didn't actually see the one I liked the best: they turned someone's desk drawers into an aquarium. She comes in, she knows they've done something, but she doesn't know what. She opens up her desk drawer, and there's goldfish swimming around in there with a pump and the whole bit. Another time they took up all the floor tiles around someone's desk, leaving two vertical feet of crawl space all around it. This person was on a killer deadline at the time and did not think it was funny!

The press conference graphics thing started at the first Jupiter press conference, where one of the Principal Investigators got up in front of this crowded auditorium with a verifax viewgraph of a chart of his data that had no redeeming features except to himself, and droned on about it. Everybody was groaning, "Oh, this is awful, this is terrible." People like Lonne Lane and a few others were standing in the back of the auditorium where I was. After that press conference, a few of those people took Ed and slipped him into an office for about 15 minutes. When he came out of that office, boy, that was it! We started producing color graphics to explain the concepts. It was unheard of to produce color graphics so quickly; from a hand-scribbled sketch at 5:00 or 6:00 or 7:00 at night, to a color slide by 7:00 the next morning. Our art vendors were from all over Los Angeles, so there was travel time involved, too.

The slides, when projected, still had moisture on them, and the heat of the lamp caused an embarrassing black blob to grow across the projected image—we were very upset about it. Finally they realized that it was the type of film they were using, combined with the glass-mounted slides, that was causing this. It was retaining the moisture inside the glass mount, and every time you projected that slide the moisture would spread across the projected image.

That was a disaster we finally solved, but at first the speakers weren't very forgiving. Some of them would make nasty comments while they were doing their presentations, which would make us not too happy. We were trying the best we could, but we were new at this. The behind-the-scenes activity in the projection booth really is a lot like the scenes in that movie, *Broadcast News*.

People like Ed Stone would just flow with it. If something goes wrong, he doesn't need the slides; he can just keep talking. There's often technical problems. Slides tend to pop out of focus when projected for a long time under a high-intensity bulb. Slides in different sorts of mounts (cardboard or plastic, for example) have a different focus point. The people up in the projection booth would really take exception when a speaker would get nasty about the focus and cry "focus, focus.". So the projectionist would turn the focus knob on the projector massively back and forth until it was obviously way out of focus on either side before they zeroed in on the right setting.

How does Voyager compare with other projects you know of?

They're structured similarly. There's the basic structure: the project manager and the mission planning office and the DSN people and the Flight Operations Team. I was doc rep to Galileo and Voyager at the same time. Galileo was then being built and designed, while Voyager was already in the operations phase, and I learned that there is a big difference between development people and operations people.

I wasn't involved with Voyager in the planning and building stages. I came on just before launch, and so my look at Voyager was all operations.

There's always lots of meetings. The mission director's meeting, scheduling with the DSN, the SCORE meetings—that's science operations report—those are all the weekly or daily meetings. The Science Steering Group, which would be all the principal investigators, would meet on varying schedules, sometimes once a year and sometimes, during an encounter, every day. All of these funneled information in to Ed and the project manager and whoever else needed to know it to make decisions.

What sort of decisions did you have to make?

None of my decisions would have had a major impact on the project. Decisions are rated as to whether they would have a catastrophic impact on the spacecraft; I was never involved in those. My decisions were more mundane, like what should I put in the bulletin, and how should it be said, what pictures should we put in the brochure and how should it be done—fairly small things that everybody does on their own job every day.

I'd heard that at the early encounters security was quite informal, and someone swiped a pack of instruction cards to feed to a spacecraft.

I hadn't heard that story, but the crowds were getting so large that people couldn't get their work done. Everybody, especially the press, wanted to be over here with the imaging team, watching pictures come in and getting instant analysis of the geology or whatever. So they had to institute a badging system and a guard to let only approved people on the third floor where the science teams were generally housed.

In the spaceflight operations facility, where the DSN stuff is done, they had key cards, or before that, just guards. They instituted that on the fifth floor of this building, where the Voyager operations area is. Those would be turned on during peak or critical periods, when they didn't want people roaming in and out, interrupting their work.

When you've got people roaming around the Lab, there's always the possibility of theft. There's a lot of people who are real Voyager space cadets; they are such fans. They want to know all about it. These people sometimes scare me.

Last summer there was some concern about computer security because of viruses and things like that. Hackers break into a computer system just for the challenge of it. We were worried that somebody might try to break into the JPL and Voyager computers just for the challenge and this could seriously disrupt our activities. Dick Rudd, who's now the mission director, was in charge of Voyager from the standpoint of looking into that. During certain periods they cut off access in or out through these networks, so that people couldn't get in. They also made everybody change their passwords.

There was one particular guy who was from back East. He got my bulletin, and he really wanted to know all about Voyager: when would we be sending the sequences, all this other stuff. He made me nervous, because hackers are pretty smart. I am not a computer nut myself, but I know that with not much information they can figure out a lot of things.

I had only heard this guy on the phone, but one day he appears in my office with a visitor's badge on. I wondered, "Who let this guy in?" It said on his card, so I went to that person and asked about him. He didn't seem concerned about the guy. He wasn't a threat, but I was nervous enough. Why would anyone want to disrupt Voyager Neptune? I don't know, but there are strange people around.

There have been demonstrations, not necessarily against Voyager. I remember long ago when the Deep Space Network still had a tracking antenna in South Africa, the Lab was picketed. At a press conference at Saturn or maybe even Uranus, some reporter brought that up. Most of the people on the podium had long forgotten that the DSN ever had anything in South Africa. They were wondering, "What's he talking about?"

Another time the Lab had a bomb threat, and everybody had to stand outside the buildings in the sun for an hour or so until they decided it was a crank call. Why would anybody do that, except to say that they did? It makes me worry about the weird people around. What goes on in their minds? How did they get that way?

Any particular lessons that come to mind?

Voyager taught me more than I realized. I worked at NASA Headquarters for a year (between Uranus and Neptune) in the Solar System Exploration Division, and I became aware real fast that the rest of the world didn't necessarily do business the way JPL did business or the way Voyager did business. The ways of operating that I took for granted didn't exist back there. It's not to say that we do things perfectly here, but to me it was a surprise that they weren't organized the way we were. Systems engineering was one area that I didn't realize was so specific to JPL and the way JPL works.

Dick Laeser could tell you more about that. He and his people were invited by Admiral Truly to give a pitch to top management at NASA Headquarters about systems engineering, a structured approach to solving a problem. I'm not an engineer so I couldn't explain the whole thing to you, but it was real obvious to me when it didn't exist.

When I was back at headquarters, I also put on my other hat and said, "JPL's pretty provincial. They've got their view of the world, and they don't look beyond themselves sometimes."

Headquarters is being bombarded with all kinds of views from the Hill or from the White House or from OMB (Office of Management and Budget) and has to respond to all of those views. Now that I'm back here I may get upset when headquarters calls up and wants this impossible information on a short fuse, but I try to remember where they're coming from. I know what forces are out there and what they are having to contend with.

They get these requests from Congress or something, and it sits on somebody's desk a few days each step of the way, and by the time it gets to the person who can answer the question, they've got one day or one hour to write the testimony for the Hill.

People at that level are sometimes more interested in what the engineers around here would call trivia. It's statistics; it's what Charley calls "gee whiz" things. That's what sometimes they need to know back in Washington to make an impact, to explain to somebody who knows nothing about it. They can say it costs 20 cents a year per person to do a project, and people can relate to that better than they can to "$550 million."

They may want to know what are the lift weights of the Ariane and the Energia compared to the shuttle? It's pretty involved to work up that information and get it from the right source and compile it, but sometimes they want to know that on a short fuse.

So you get a lot of requests that you don't really think of as being JPL's mission?

Well, I think that's probably true of any high-tech organization. I wouldn't say its not our mission. It's part of our job. You always get these questions that require an answer from a perspective you aren't used to taking. Answering such questions is an art, and people like Charley Kohlhase can think that way very well. Others aren't quite as good at taking a technical thing and breaking it down into something a layman can understand; something that's interesting and understandable at the same time.

Ed is very good at it; Dick Laeser is pretty good at it. I wish I were better. The people who work with Charley on his staff are all pretty good at taking a technical concept and making it more understandable to the lay public.

You have to come up with analogies, and it's really difficult. There's got to be some real fact behind that analogy because there's always some engineer who will try to work backwards from the analogy to the fact and catch you up; they just love to do that. I had a guy come up to me last week from the DSN and say, "There's an analogy in there that I don't think is right. I think it's off by an order of magnitude." He's probably right.

How do you view the Voyagers?

I think they're extensions of the minds and souls of the people who operate them. They are very well-designed machines. The designers deserve a lot of credit, and they're forgotten. The people who designed them and planned what they should do, and the people who carry out those plans—they have to be the dreamers and think through the hard problems.

So in that sense the Voyager machinery out there in space is just an extension of the people here on Earth and their minds. Without instructions through the DSN every once in awhile, the spacecraft would be nothing. It would just be an object out there floating through space. It needs instructions from the human mind to tell it what to do, and it needs the human mind to interpret what is sent back.

I'm not a romantic about the spacecraft. I give a lot of credit to the people who designed them and built them so they have lasted so long and been able to do so much, also to the people who knew the spacecraft well enough to work around any problems, like the failed radio receiver, and the failed tracking loop capacitor, and the sticky platform, and the aging instruments. The people who knew how to work around those problems really deserve a lot of credit and recognition.

It has been fun, too, that the whole project has ignited other creative minds, like artists and song writers and movie producers. That says a lot for the creativity of the engineers and scientists. They created something, and it sparks a different kind of creativity in people with other kinds of minds; they've gotten more lyrical and romantic about it. People here approach it just as a problem to be solved, so it takes all kinds of minds and different kinds of creativity.

Do you feel that the Voyager project is over?

I don't think it will be over until its over, until they lose contact with the spacecraft for whatever reason. I hope we can keep Voyager funded and operating with the full complement of cruise science, including the ultraviolet astronomy, for as long as the spacecraft can stay alive and healthy.

And I hope that we can keep the project in the public's eye, as the spacecraft are going out to find the edge of the sun's magnetic influence and perhaps enter interstellar space—cosmic rays and that kind of stuff. It's all pretty esoteric for most of us. A planet's magnetic field encloses it and sweeps particles in space around with it as the planet rotates. The sun has the same kind of magnetic envelope, and we are in that magnetic field. As the sun's solar wind sweeps past Earth's magnetic field, it is distended like a wind sock in the solar wind, created by the planet being there.

Just to start thinking about the immensity of that is really exciting. I'm really hoping that Voyager 1 does reach the heliopause, does cross the termination shock, and does enter interstellar space. It will be really exciting to hear the scientists decide what's the composition of interstellar space—how does it differ from what's inside the heliosphere?

You mentioned Voyager 1 only?

Voyager 1 has probably the best chance of being first. It's the fastest of the four solar-system-escaping spacecraft: Pioneers 10 and 11 and Voyagers 1 and 2. Voyager 1 is going above the ecliptic, Voyager 2 is going below the ecliptic, Pioneer 10 is going toward what would be the bulge at the head of the windsock, and Pioneer 11 is going down the tail. Pioneer 11 probably will never exit the sun's magnetic influence, but the other three probably will if we can still be in contact with them when that happens. They used to estimate that Voyager 1 could reach that boundary by 1991, but they've spread that time line out a bit.

Ed showed a slide that was a poll he took at a meeting of about 100 experts on where is the edge of the sun's magnetic field? When is Voyager going to cross this? It ranged anywhere from 60 to 100 AU. It was all over the map, as far as these experts could agree. The moral of the story is that nobody really knows. There was a big cluster around 60 to 80 AU; that had the most votes.

That's another thing: to think that it took 12 years for Voyager to get to Neptune, traveling at 40,000 miles per hour on the average. That's fast. That's a long way. That's 12 miles per second. It's like from here to Arcadia in a second. Zoom, it's gone. I use that explanation to illustrate to kids how fast and far it's going.

There is this whole thing about the spacecraft being autonomous. It's supposed to be able to take care of itself without human intervention, because the flight time is so long. They have these fault protection algorithms resident in the software on the spacecraft. "Gee, if I notice my cameras are getting too much sunlight, it's going to burn out the optics, so I better turn my cameras off, tell them that I've done it, and wait until they tell me what to do."

They build that kind of logic into the spacecraft. There's seven or eight fault protection algorithms on the spacecraft at all times, and then they do these contingency loads like the backup mission load in addition to that. They have to plan. They can't lose this multimillion-dollar spacecraft with all the eyes of the world on them.

Ever since the primary radio receiver failed in '78 on Voyager 2, they have kept resident in the spacecraft's computers what they call the backup mission load. That has been changed seven or eight times now. It is designed to provide minimum mission science in case we lose the remaining radio receiver on Voyager 2, and we can't contact it anymore to send it any more instructions, but it can send us information. It was first designed to do a minimum Jupiter mission. When that main Jupiter mission was successful, the backup mission load was changed so the Saturn science mission could be done.

The backup mission load is changed after each major milestone has been successfully accomplished so that the next milestone will be done if we happen to lose the spacecraft receiver. This is part of the contingency planning that always goes on: What would happen if we lost part of this computer or a bit in that computer, or what if we lost capability to move the scan platform, or what if two things happened at once? What if we had a power surge and a battery failure at the same time?

What is the probability that any of these things is going to happen? What is the probability that any two or three of these things will happen at the same time? What should our response be? And based on the probability of whether or not it is going to happen, how should we use our resources here on Earth? Which problem should we attack? Which problem should we be most prepared to handle? I think they had some advance, contingency planning already done when the scan platform stuck. They had thought of that as a remote possibility.

They're doing this over a tremendous distance with an infinitesimal amount of power.

Yes. Somebody gave me an image that I've really liked: It takes 512 million bits to capture the information in an imaging frame. A scene is electronically changed into 512 million bits, and when those bits are sent back to Earth those are spread out over 3 or 4 billion miles. As the information comes through space, it spreads out in a cone. To be able to know when those radio signals are going to hit the Earth and where, at what frequency, at what signal level, is really a black art to me. The DSN people can really tell you more about that. Our current DSN representative is Hank Cox.

I'm just one of the very small contributors to Voyager, compared to most of these people, so I hope some of these insights help people understand what a great achievement Voyager is.

JOHN E. BRICKETT
Plant Protection Chief
Born October 25, 1931   Shelton, Connecticut

John Brickett graduated from Shelton High School in June 1949 and enlisted in the U.S. Air Force January 5, 1951. After graduating from Airplane/Engine Mechanics School at Wichita Falls, Texas, he was assigned to an air base in England where he worked on C-47s and C-54s and cross trained in the Air Force security police field. Then he attended the Military Police School at Oberammegau and Air Base Defense School at Grafenwohr, both in Germany.

During his 26 years in the U.S. Air Force he was stationed in England, Germany, Italy, Vietnam, and numerous stateside locations, and achieved the rank of police superintendent. He performed presidential security for President Nixon during his visit to Rome, Italy, in 1970. Brickett was awarded the Meritorious Service Medal with Oak Leaf, the Commendation Medal with Oak Leaf, the Vietnam Cross of Gallantry with Palm Leaf, and the Presidential Unit Citation for Valor.

He joined JPL August 12, 1977, as a security officer and was promoted through the ranks to become plant protection chief in 1986. He was involved in security and planning for space encounters, open houses, and visits by distinguished visitors. He has enjoyed meeting Vice President Quayle, Prime Minister Margaret Thatcher, and other interesting visitors from various countries, including their security personnel.

He has been instrumental in upgrading JPL plant protection uniforms, equipment, communications, and vehicle fleet. He published the *Plant Protection Officers' Procedures Manual*, updated the JPL Emergency Plan, published a Distinguished Visitor Plan, and coordinated security for distinguished visitors with Secret Service, State Department, and White House staff. He introduced a computerized alarm system for security, fire, and mechanical systems, and a Meridian telephone system for recording Emergency telephone calls within JPL.

His main off-work interest being travel, he has visited extensively through North America, the Bahamas, Jamaica, France, and England. He and his wife like weekend excursions, staying at bed and breakfast accommodations, and visiting California wineries and craft shops. He also enjoys gardening.

Grade and high school are the basis towards professional avenues;
college hones these skills.
We owe to the employer dedication to tasks, giving best effort,
enthusiasm, timely completion, can do attitude, improving personal
performance and knowledge, plus loyalty.
Knowing in your own mind that you performed
to the best of your ability is reward enough.
Family support is important in maintaining
job performance and satisfaction.

—John E. Brickett

JOHN BRICKETT

How did you come to be working here at JPL?

I retired from the Air Force in 1977 and went to the Veterans Administration looking for a job, and they directed me to JPL. They said they were hiring security people and that my expertise in the Air Force would help. So a month after I got out of the Air Force I began working here, just prior to Voyager being launched. I started here as a regular security officer and was promoted up through the ranks until finally I became chief.

What were your duties when you first came to JPL?

Just regular security officer duties: checking entry and exit of personnel through the laboratory gates; patrolling the laboratory; walking through the buildings during non-working hours; checking for safety and security violations; looking for fire hazards, safety hazards; making sure the gates were secure; responding to any alarms we might have, either mechanical or security type alarms; working on special projects; parking automobiles, like with the Voyager encounter.

What is the first you remember about the Voyager project?

I remember the launch of both Voyagers, Voyager 1 and Voyager 2. They happened right after I arrived. I remember the launch explicitly because we got into the thing with overtime. I was just a security officer then, doing safety and security inspections in the laboratory, just like the guy standing at the gates who brought you in. We had to make accommodations for parking, and we had to provide for security of the equipment that came in with the press, and such as that. It was the same situation in the original launch that we had during the other Voyager encounters. All the encounters and launches were basically the same as far as we were concerned.

What was your biggest problem during encounters?

The biggest problem we had was to ensure that there was sufficient parking for the visitor-type people—not employees—but press and other outsiders, to make arrangements to bring in their equipment, to make spaces available to park their cars and trailers and such. Of course, parking is a problem here all the time.

Over and above what we normally encounter on a daily basis with our regular employees, we were inundated with a large number of visitors, so we had to give certain parts of the laboratory to the press trailers. We lost some parking area that we would normally use during special occasions, so we had to generate additional parking space to accommodate everybody.

There was some security required for the press room in the auditorium, and the additional trailers that were brought in for the press people. And we had some other activities taking place, other than the actual encounter itself, with some other visitors who came in that we had to support security-wise. It meant additional work hours for a lot of us; additional overtime and personnel to make sure we had a sufficient number of people to accommodate the activity, whatever it may be.

I had to make sure that everything that was required was taken care of. I've got a staff of six people working for me, supervisor-wise, two per shift. My main job was to put out the requests that were directed towards us and just oversee to make sure that everything was taken care of.

I've got a good staff. I don't have many worries with them. I usually put out the word, and I know that it's going to be done in an exemplary manner. They have different types of jobs, but everybody works as a team on things like this. They understand what the importance of it is: the importance to the Laboratory, the importance to the nation. It puts us in the limelight.

They understand that, and they usually put out a little extra effort to make sure that everything is done properly and there's no glitches.

We consider ourselves to be a professional organization. We have people here with different backgrounds in security, from police departments, military, all the contract guard services, and we look for people who will fit in here at JPL. Not because of their experience, but personality-wise. We're a service organization, and we're here to supply a service and to give the technicians a secure environment to work in.

Security is a bitter pill to swallow, and in many cases you have to force feed people, but we want to feed them security in such a way that they come away with a good feeling. We're not trying to operate with a police department hard-nosed kind of concept; we are trying to let the people know we're here to help them, not hinder. So when we recruit our people, that's what we're trying for. We look for people with experience who can fit in that type of program. JPL is a small community, so that's what we try to instill in our people.

We had very little problem with the visitors because they were here on a special task: they knew what they had to do. We tried to assist them in getting their job done versus hindering them. We tried to handle everything in a diplomatic way instead of, "I'm the Big Brother and you are going to do it the way I say" type thing. We knew they had a job to do, and we tried as best we could to make sure they got it done properly.

When we had the visit of the Vice President Quayle, we had to provide security for him during his visit, and that worked very well. There was a lot of planning put into it; preparations for it with other agencies, on-lab and off-lab both. It was a very smooth run operation, very smooth. In fact, you can see pictures around here, and I got a letter from him thanking us for his visit. He's a gentleman, a unique man to talk to.

What were the low points of your work at JPL?

During the encounters, to me there wasn't really any low points. It was all anticipation of the scheduled events taking place. Our job is never the same from one day to another; there's always something different going on when you work in security. We may have certain functions we do on a daily basis, but basically it's a different thing every day.

It's like the Voyager encounter; we planned ahead of time; we had a plan on how we were going to conduct things. But a plan is only as good as it operates. It doesn't go exactly as it is on paper. There's always something else that is going to take place at the last minute. There's always the anticipation of having to accommodate someone that "forgot" something.

What would be an example?

Someone forgot to tell us that they invited 50 people to come here, and they are all going to be driving their own privately owned automobiles and they're all expecting to park in the visitor lot. Someone else calls up at the last minute and says, "We forgot to tell you that a visiting scientist is coming and we need to accommodate him." Or, "We need von Kármán auditorium; we've got 250 people coming in for a press conference."

We react well, and we always have. We have to handle these various things calmly because it happens to us all the time. It's not because people are trying to undermine us or sabotage us; it's just that in their busy day there are certain little things that they may forget. And many times they call us at the last minute about something, and we have to react promptly.

In any agency you work in, you are not authorized to man your section based on things that *may* happen; you man your section based on things that are normal routine daily functions. So when something extra comes up, you have to come up with the extra bodies, the extra personnel to handle the situation. Many times you have to really jump into a quick time motion to get things done to catch up, because you got a late notification. But that's part of the business. Security does operate that way. And from my past experience in the Air Force, it's the same no matter where you go.

Does it require things like phoning guys on their day off and telling them they have to come in?

Oh yes. We've worked time and a half and double time. When we get into the serious portion of an encounter, we go into two 12-hour shifts, which requires we work overtime. In some cases it was a six-day and seven-day stretch. Of course, they get paid for it. And it is important enough that they see it that way, and then there is very little grumbling when you come to a situation like that.

But we programmed ahead of time. Because most of the staff have been here during most of Voyager's encounters, including the launch, we anticipate ahead of time what we require. We had generated a plan for the first encounter, and all we did was use that original plan and then programmed the extras into it, based on what the inputs were. We had a basic plan to start with, and we knew exactly the number of people we would need.

I gave each one of the staff members a different project to oversee. I had a sergeant and lieutenant who were in charge of nothing but parking; that was their job. I had another sergeant and lieutenant who were charged with internal security of the Laboratory only; that was their job. I had another sergeant and lieutenant who were in charge of just the mobile patrol areas of the laboratory; that was their responsibility. Then all I did was oversee.

In fact, the staff members did the plan. I assisted them, but they did the plan. Then my job was just to oversee and make sure everything was done the way it should be done. And I did all the contacts with the different agencies and attended all the meetings. We had a number of meetings to discuss what was going to take place.

Keeping you very busy.

Yes, we were very busy. At the time of the after encounter party, I started out in the morning, and I didn't get home until late that night when it was all over.

When that happens, how does your family react? Do they understand?

My wife, yes, she understands. She knows that I have a second life. I come here in uniform, and when I go home I take it off and I assume my second life. We've been married for 36, going on 37 years, and she put up with the same thing in the Air Force, so she knows what it is. It doesn't happen all the time, but she knows when there's something special and I think I should be here, I will be here. She just adapts to it, and I think that's the case with most of the employees' families.

She's been out here when we had the big open house; in 1978 and 1979 we had that big four-day open house, and she thought it was tremendous. She saw what I did here; I took her on the tour of the laboratory to see what we're doing, and she understands. She think's it's wonderful. Some of the wives probably don't think that way.

This is a tremendous place to work. I've enjoyed all the time I've been here.

What makes it that way?

It's a nice environment to work in. We have great people here on the Laboratory that we are working with. We're into projects that keep it interesting. Security-wise it's probably one of the best. The location works so well for us. We're at the end of the road; we have nothing but mountains behind us.

It's hard to explain this place because it's not an industrial complex and it's not an institutional complex, but to me it's a cross between the two. We have some tremendous people here. Of course, you're going to have your small problems with some people because with a complex this large and that many people you're always going to get a bad apple. But basically, no.

The way it works here is everybody realizes that everyone has a job to do, regardless of how minor it is. They appreciate the effort that people put in, regardless of whether it be a plumber, a carpenter, a security officer, or a technician. Everybody is involved, and everybody knows that what little they contribute fills the overall picture of the project, whether it be Voyager or something else.

I think the Laboratory recognizes that, all the way up to Dr. Allen, the former director. They recognize that everybody who puts what little bit they put in, it's important. And you are treated that way. That has a lot to do with the enthusiasm of people; they know no matter what they do, it's appreciated, and they receive something to indicate that it is appreciated. We all got our plaques after it was all over, like that one on the wall with our names on it, which is nice.

Do people off the job—your neighbors, relatives, friends—ask you about Voyager?

Oh yes. It's surprising; the population is very interested in space exploration. I hand out a lot of information to people I know. I usually keep a stack of the Public Information Office folders with all the little goodies, the pictures and everything. If I get a visitor in, I hand it out to them. A lot of people discuss JPL with me: my son, my daughter, and son-in-law, my grandkids, they're always asking, "What do you do out there, Grandpa?"

What has been the high point of your work here with the Voyager project?

Voyager has been exciting all the way through. The launches were exciting; all the encounters were exciting. I never stopped being amazed at some of the things that take place here, even though I've been here almost 15 years. It's an exciting place to work. The excitement of the laboratory builds as we get close; everybody is interested.

When things like encounters happen, you get a chance to meet different people. We have a lot of visitors come through here that people would just love to meet personally: movie actors, the Vice President Quayle, the news media. We see them all.

The anchormen and women and the Vice President visited us during the last encounter. After the encounter we had a party put on by The Planetary Society and we had Chuck Berry here that night. We had a good time. It was hard work, but most of our guys enjoyed it; they enjoyed the party. After it was all over we said, "That was great."

Of all these, who or what was the most memorable?

The Vice President. I was fortunate. Usually in my job we have an important part, but in some instances we don't get the opportunity to meet the distinguished visitor. You're there, but then again, you're not there. You're kind of in the background, watching what is going on and making sure everything is conducted properly and the individual is in a secure environment.

And although I got close to him, I never expected to meet him personally. But when he was in the building one of the agents came over to me and said, "Come on, I want you to meet the Vice President." I was the first one in line that he stopped to speak to. It was different; it was really different.

I met other famous people who visited here. I met Sidney Poitier when he came out; I met Angie Dickenson, Stephanie Powers, Mariette Hartley; I met all the anchor people from the the TV stations, but the Vice President was the most memorable.

What were some of the things that impressed you about Voyager?

That that thing has been out there all those years, since 1977, and we just had the last encounter with it, and it operated flawlessly. It's almost impossible to visualize how many billions of miles it is out there, and it is still sending signals back on a system that has the same power as what's running your wristwatch, and we get excellent pictures and excellent information from it. And every command that they sent to it worked just right. It's almost unimaginable that a piece of equipment that has been out there floating that many billions of miles in that many years still operates flawlessly like it did. It's mind boggling.

I think the scientists throughout the world are looking at it the same way. I don't know anything about it; my job is security. We just get involved when they're putting these things together, to assist them in the security. But to think about it, it's tremendous. It's a shame that a lot of the normal population of the United States don't realize what goes into these things. All they look at is the bucks that put this thing up in the air, but think how much hard work, how much ingenuity, and how flawlessly they operate over so long a period of time, and what comes back from them.

It's going to be interesting to see if they pick up anything else. I understand the last pictures that came back here may be the last ones it takes. Be interesting if they're not the last ones—what else is out there?

It's tremendous. It really is. And just think about the next one—Galileo. Five years before it arrives, and it is supposed to be right on target. To think that the guy who is on a computer sends a signal out to that thing billions of miles away, and it reacts to it and sends back a message. What is the time now between going out there and coming back? Nine hours: four and a half hours to get the signal off, and four and a half hours for it to come back answering. To realize that this thing can be that far away and that the parachute is going to open just on a split second—almost impossible when you think about it.

Yes, Voyager was great. I grew up with Voyager at the Laboratory. I came here in August '77, and they launched the first one that same month.

BOBBIE FISHMAN
Visitor Control Greeter

Born July 2, 1947   Los Angeles, California

Growing up in Los Angeles, Bobbie Fishman attended Fremont High School and Los Angeles Community College. Her employment career began in the University of San Francisco admissions office in 1967. She subsequently worked at Cedars of Lebanon Hospital and the University of California, Los Angeles, business office. From 1969 to 1972, she was employed in personnel at California State University Los Angeles and at Huntington Hospital from 1972 to 1974.

Then, after five years as a homemaker, she joined JPL in 1979 and the following year moved to the front desk in the Visitor Control Center where she has been ever since. She is responsible for greeting and processing 150 to 200 visitors to JPL every day.

She was awarded the 1994 Medal of Excellence from Women at Work, a nonprofit career and job resource center. In nominating her, the JPL Advisory Council for Women observed that "Bobbie provides an extraordinary first impression of JPL" and noted her "willingness to go out of her way to make everyone feel welcome and informed while promptly resolving problems in an effortless manner."

She is mistress of ceremonies at JPL events such as talent shows and Heritage Week observations. Her recreational interests include acting in JPL-Caltech musicals, singing, and dancing.

*I feel very fortunate that I have a wonderful husband,
Mark Fishman, and a wonderful job. I really care a great deal
about this company and can't imagine working anywhere else.
I've met so many wonderful people here, from world famous visitors
to lab employees. I certainly try to maintain a good positive image
for JPL and provide a role model for my race.*
—Bobbie Fishman

BOBBIE FISHMAN

I was born and raised in Los Angeles. My mother was from Oklahoma, and my father was from Arkansas, and they both met here. I couldn't imagine living anywhere else but California.

When I was a little girl, my cousin gave me encyclopedias. That's when I first became interested in dinosaurs and the stars. Then in school I studied astronomy, I took music lessons. I've always been interested in the arts.

Media coverage prompted me to see what was going on at JPL. One day I came here. I tested and then interviewed for a position in professional development and was hired that same day. I worked at the lab for about ten months, and the position here in visitor control became available and I was hired.

I've been so fortunate. Originally I thought I'd just stay here for three months and then move on. I never had stayed at a job for longer than three and a half years, but I've been here since the beginning of the 1980s, so that does say something about the company. I still live in Los Angeles, but I commute every day to Pasadena.

We are family, and within the family circle you do have problems from time to time, but, honestly, everyone's attitude is so wonderful. It's like a place to escape from home when you come here. In fact, when my first marriage was breaking up, working at JPL saved my life.

So you've been here from the Voyager Saturn encounters.

The people here work around the clock. The media is here. It's an exciting time. Our guests, our visitors that come in, ask, "What's going on?" and we tell them about the Voyager, and what it's doing, and it's amazing. From my religious background I always tell everyone Voyager is kissing Heaven, going to meet God. I just feel that way. To know that God has given man a brain to concentrate, work, and develop an instrument that can go out there without anyone onboard, to be able to do this, is really the most wonderful thing in the world.

I wish we could follow it even longer. I'm sure they are still getting information, but I wish we could continue to have the encounters and know what the information is that's coming back. I'd love to be part of that. I will try my best to follow as the information comes back, so I can pass it on to the people who are interested in the travels of Voyager.

Some of the most brilliant minds in the world have walked through those doors, and I've been fortunate enough to meet them, and they work here also. Sometimes when they talk to me they will break things down into layman's terms since I'm not a scientist or an engineer. I have managed to get so much knowledge from them and maintain it in my head because they've taken the time to explain it to me. Just reading it I don't understand, but when they explain it to us it really helps.

I always want to work for encounters; it's so exciting. I've met people like Stephanie Powers. One time Stephen Spielberg came through, and Angie Dickinson. I have met William Shatner, Gene Roddenberry, Perry Como—I can't tell you, so many celebrities.

One of the high spots was the big party at the end of the encounter last year [1989]. The Planetary Society put on a planetary fest, and Carl Sagan was here, and Chuck Berry came in. Carl Sagan has been very kind, and he manages to come down and say hello to Adeline (Deet) Stoffers and myself, and that's fun, too.

The Saturn encounter was when I met Neil Armstrong, and that was really, really exciting. That was the highlight of my life here at JPL. I will carry that with me always. Not only did I meet him; I had a chance to spend about 25 to 30 minutes actually talking to him. He was wonderful, and I hope and pray that someday I will get to meet him again.

David Low, the astronaut with the shuttle when they did the Galileo, used to work here, and he's coming back for a visit next week, so I'm looking forward to that. He's a very nice young man.

When the Voyager encounters came around, we would put in a lot of overtime. There was so much activity going on. The actual encounter itself would often be at midnight or early morning. And always at the end of it they would include us in their "thank you" for the participation. What we do is so tiny, so small compared to the actual people who are really involved.

Who would do that?

The project managers from the Voyager, and also Dr. Allen's office, would give us thank you notices because when we have the other outside guests and the media and the outside activities, it puts a hardship on the Lab. Because we have more visitors using our facilities, such as our cafeteria and our parking, we have to go out of the way to accommodate all these extra people. And so we get a big thank you from the director's office.

So, you have found this an exciting job?

Deet and I have been on television, I'm told, in Germany and Japan and Mexico. People come in and say, "I've seen you on TV," and I say, "How do I sound when I talk in Japanese?" I recently had a chance to do a little part in a PBS special, *Newton's Apple*, so they sent me a t-shirt; that was really exciting. And interfacing with Caltech; I work with their theater arts group sometimes. There was a man who recently passed away who was a Nobel laureate, Dick Feynman. When I did the Bloody Mary character in South Pacific, one of the highlights in my life was his compliments toward my work in that character. Then we had a chance to talk, and I asked him about his bongo playing. I have pictures of him.

He and men like Al Hibbs who would always narrate the encounter, these are men I actually did plays with. We had parts, and we had to rehearse, and over a six-week period I had a chance to know them. We went to the wrap parties together and things like that. I could not even imagine this before I came here myself—these people were almost like outer space, they were like untouchable, yet in reality they were just so down to earth.

I remember telling Dick Feynman—during that time I was going out dancing at discos—telling him about what was going on in disco clubs, and one time I tried to show him a dance. He got a kick out of that. He was a very nice man. You'd never know how brilliant he was. One night he came in when I happened to be working late, and I had his book in my car, and I said, "Oh my God." And we talked and we thought someday we're going to do a play together again. And Dr. Bruce Murray was just through here yesterday. He has a lovely wife. The wives are so nice. Dr. Lew Allen's introduced herself as Barbara, and she's always nice wherever I see her, and we talk about the arts and crafts fair.

That's another thing we have here. I emcee the arts and crafts fair, the entertainment for that. That's sponsored by our child daycare center. They put on a big function once a year, and for the last seven or eight years they've asked me to emcee the entertainment. That's a couple of hours Friday evening, all day Saturday, all day Sunday. We bring in entertainment from all over, and most of it's volunteer. So these are some of the highlights I've had working here. I've had opportunities I never would have had, had I not worked at the front desk.

Does this desk have to be open 24 hours during encounters?

No, it doesn't. Arrangements are made with the von Kármán auditorium so that the media and the people over there monitoring the Voyager can enter directly from the outside. During the actual encounter itself, when there are going to be a lot of celebrities and special people, VIPs, then they set up a tent over in the director's parking lot, on the right side of our building here, and they make special arrangements, special badges, to bring them in through there. As it turned out, this time it was on a Saturday, and we were offered a chance to come in and sign them in that Saturday.

What does your family think about this?

My son has been working here since he graduated from high school, in 1986. He worked in our gift shop, and and he's worked encounters; he went away, and now he's back again. He works at the mail room, in an area called "distribution." He works at night, so he's just leaving as I come in in the morning. He's been exposed to JPL since he was 11 years old, so it's sort of inevitable that he would end up working here.

My mother has come here a few times on visits. She's seen the film; she sat through it twice. It's really funny because so much of this is just so alien to her, but you mention Voyager and she can tell you Voyager. She's one who never goes anyplace, but she comes to JPL, and she wants to see the film. She loves it here.

I'm going to get married in August [1990], and my fiancé was working here. I have lots of friends here. It's a great place to work. In fact we're going to get married at Caltech.

What is the hardest part of your work?

This may sound funny: I come to work every day prepared in case I have a problem.

First of all, some of the most wonderful people in the world walk through those doors, so that helps make my job a lot easier. We realize that they're anxious to get to an important meeting, and we try and get them through as quick as we can. If our employees inside notify us and let us know that they are expecting these visitors, it does make our workload a lot easier. When we get all the information we need from the employee on the inside and from the visitor on the outside, things flow smoothly.

Sometimes it gets very hectic. Once in a while we have people come in who are very, very concerned about getting in as soon as possible. We have to remain professional at all times, stand firm with them. Usually they're very patient because we try and have a good attitude. If sometimes you have to say "no", it's not so much that you say "no"; it's the way you say it. So we try and maintain our good disposition down here. We're trying to do the job we must do, for the security and safety of the Lab. That comes first.

Everyone is usually happy and satisfied, both the employees from the inside and the visitors.

Can you tell us some more about what it is like for you when several anxious people are at the counter and the phone is ringing? I've seen this a number of times.

I sure can. In fact this morning there was some remodeling going on, and we had some carpenters and electricians here. At the same time we had visitors signing in for a SIRTF (Space Infrared Telescope Facility) meeting, a Cassini meeting, we had foreign nationals; we even had flowers that were being delivered at the same time. We had some students here who were involved with taping astronaut candidate Stephanie Wilson, and a tour was just coming through. We had a lot of visitors in here at one time.

If the line goes out the door, I constantly remind people I'm working as fast as a I can and I'll get with you as soon as I can. I will pull out a lot of visitor badges at one time and put them on the counter, and I'll get several visitors started on completing their badges while I pull out the sign-in log and get other visitors. And I still have to watch that gate: that's top priority—no one gets in until I can verify that each person is expected. I was fortunate that for two of those meetings we had the representatives from JPL here to escort them in, and that helped.

When you've had as many years at the desk as I have, you learn a few shortcuts, and you recognize a lot of people that have been here in the past, and that helps too, to get them in a little quicker than some visitors we are not familiar with. My co-worker has been here five years, and she recognizes a lot of the regular visitors, and that really does help speed up the processing.

But even with the shortcuts, there's certain things you don't let get past you: you must check their credentials and verify that they are expected.

You still have to do the paperwork for everybody, no matter how familiar they are?

That's right. When we have tour groups, we don't always have to badge all of the visitors that are with the tour. Such as if we have 10 visitors or more, and the JPL person has cleared the paperwork through their supervisor and is taking them on a tour through the open area, that's fine;

that's been cleared through security, and with his or her badge they can take them in and escort them without being delayed with signing in everybody. But that's if they only go to the open areas. If they're going to other areas, even if there's 100 people, they all have to be badged.

So we sometimes have to call in secretaries, or even some of the managers will come down and help us with the log-in and signing of each of these visitors. In that paperwork there's five questions they must answer: their name, their citizenship, who they're representing, who they're here to see, and the purpose of that visit. That's what we need for each and every visitor that comes here.

We do have other tours: with schools, clubs, companies, organizations, and we have a lady, Kay Ferrari, from public service, and she will come down and take a large group of 40 or 50 for a tour of an open area. That's okay, and that's a case where we don't have to do badging. But we do get a list of each and every person on that tour.

No matter how hectic it gets, we must maintain our workload, processing the people for the safety and security of the Lab. And behind the desk, you must maintain your smile and continue to be cordial, even though they're thinking "Hurry up, hurry up." No matter what, you keep moving, you never dally, you let them see that you're working as fast as you can. And they're wonderful. In fact, a lot of the visitors will go over to the lab phone and do their own phoning and will tell me, "I have so-and-so on the phone for you already," and that helps us, too.

It seems merciless from the outside, from the visitor's point of view. They're here to visit someone. They're not so much concerned with the fact that you have 100 people; they just know that *they* have to get in. So you have to do your best to get them in. It's not their fault that there's maybe 20 or 30 people ahead of them.

They think "How come there's only one? Shouldn't there be two or three receptionists?" Well, they are right. It is a two-person desk. There are two of us, but one of us must take a break and a lunch, and it just so happens that break and lunch times are around the time of the meetings. We get a lot of people at the last minute for a 9:00 meeting. My co-worker and I are very good; we never ever take a break until the lobby is free. We don't leave each other in the lurch. I won't walk away if she's got 20 people coming in the door. We have the flexibility of our breaks and our lunch.

I've worked with this lady for five years, and I consider her one of my very best friends. She was a manager before she came here, and I think that helped her adjust to the workload. She works very quickly, and we work well together.

This is a very hectic desk, but you never take anything for granted, because the one time you take it for granted, something could happen.

What would be an example of taking something for granted?

There might be someone who comes on a regular basis. As I said earlier, we know the ones who come regularly, and we're pretty sure of them, but we still have to verify. There's some people who have been coming here for years, and we know them but we still say "Would you flash that ID for me again."

We still need to see their ID, and we still need to verify that they are expected. Because maybe something happens. Maybe someone who came in today might have a problem, a blowup. I haven't experienced this, but just for instance. They may come back the next day, and .... You never know, so we just have to do our job to make sure that nothing like that happens, that that person *is* expected.

I have had instances where somebody from inside calls me and says, "I see so-and-so on-lab; how did he get in here? Who is his contact?"

I can cover myself and say, "That person is here to see so-and-so and I called and that person said 'Send him up' or 'I'll be down to get him.'"

In each instance we're covered because we've done our part down here. If the employee asks who is he here to see, I can't say, "I don't know; we just sent him in," or "isn't he here to see you?" I can't do that. I'm held accountable. It's a very big responsibility at this desk.

You see, one of our main functions is to protect our employees inside. Everyone that comes through here we must screen, find out the purpose of their visit, and check with the employees inside to see if they are expecting these visitors. And any undesirables that are not expected or wanted on-lab, we have to make sure that they don't get beyond these doors. Once in awhile it does mean we have to contact our security office, but we try not to go that far. If we can handle it here, fine, but sometimes that help is needed.

We get some undesirables that want to cause problems, or want to get in and see JPL, or they claim they have an invention and they want to go talk to the director. If it is legitimate business, we ask for it in writing, and we'll pass it on to our Public Information Office. But sometimes these people get a little bit irate. They claim it will save the Lab so much money, and they insist on going inside to present their project.

We want to be diplomatic, even with these poor souls. We still want to handle them as courteously as we can. We don't want to cause problems. We don't want to upset the other visitors who are not even aware that this person is here, and we still try to maintain a peaceful calm. I cannot tell you the times we've snuffed a situation by one of us, if we see someone having a problem, we quietly step over and notify security to come.

There was one visitor who was continuously calling the director's office about an invention. We were told that if this happened again to alert security. He did return, and we alerted security very discreetly. They came over, and there was a bit of an altercation in the lobby. The lobby was full of visitors, but there was no getting around it because this person would not leave peacefully, quietly, so they had to escort him out. It was sad that happened, but the visitors were wonderful; they stepped aside. It left me pretty shaky, but sometimes those things happen.

Another instance a person came in, again with an invention, and he just sat down and refused to leave. We tried to channel it, get our Public Information Office involved, we tried, but if there's no getting around it, our plant protection people will come down and talk to the person. We go above and beyond the call of duty to keep things like this from happening. We don't want any scene; we don't want any problems; we don't want to be in the news. We just try to snuff out the situation.

We don't have time to sit down and listen. We try, we refer them to the office that handles inventions. But these are people who come in and demand to be seen by someone right now!

One time when I was here there was a group of pickets outside, with signs saying that Mars Observer had been deliberately silenced. Do pickets ever come in here?

Not in here. I know what you're talking about. I guess they want to do it outside. They don't come in here, so we have been very fortunate. By the way, my interest in theater comes in handy sometimes. There are days when I'm feeling down, and I'm able to put up a good front.

But I would like to make clear to you that, even though this is a fun job, a wonderful job, it is a very dangerous job. We are in a position, my co-worker and I, where we're very vulnerable here in the front.

That is a difficult part of working here. Sometimes Deet and I feel we take our lives in our hands each day because you never know who's going to come in that front door and what's going to happen.

Even though we do have certain codes or things we can do in case of a problem, we are still up front. I have a great deal of fear about what happens if someone comes in and really wants to cause some problems. I try not to show it each day, and I continue to pray, but I do worry sometimes about the safety factor.

Another thing, we deal with a lot of people from all over the world here, and we worry about the contact with them and their illnesses. It's not uncommon, especially for the children, to come in and sneeze and cough all over us. We're faced with a lot of that. So there's some concern about that from the two of us.

After there has been a terrorist incident somewhere, does that make you more apprehensive?

Yes, it does. I'm apprehensive all the time. I try to be guarded when I come in. I'm the first one here in the morning, at 7:30. When I walk in, the first thing I do is look around and see if I notice anything different. I am aware of unusual boxes, and any box or package that was not there the night before. And my co-worker, if there is something I should know, will leave me a note.

In fact I did have a case like that one time. There was a bottle, and we didn't know what the substance in it was. We immediately called the fire department, and it turned out to be something personal to someone. But still, we are very alert to any changes, because when you work in an area you know how things should be.

We have a sign up there: it says, "parcels or purses subject to search." We want to know what's coming and going inside JPL. What's this? What's this instrument? What's this piece of equipment?

Do you, Bobbie, personally inspect the purse or the package?

Most of the visitors will let us know they have a piece of equipment. If it's in their car, they will tell the guards as they drive in. If they come in here, they will put it up on the counter and say this is so-and-so, and we will give them a property pass, and they will document what that is. When they bring their equipment back out, the employee inside, who we have notified that they have the equipment, will sign it. Most of the visitors, if they have briefcases, will ask us do we need to search their briefcase, or they will tell us they have a laptop computer. "I have this, I have that." Yes, we do look at what they're taking in; we have a need to know.

That's about equipment. Many federal installations feel they also have to look for weapons. Do you have that?

Yes, we do. When we have someone from a law enforcement agency who's come to the lab, maybe for a meeting or coming to see someone, the first thing we want to know is do they have their weapon on them. In most cases they do, and at that point we notify plant protection and they will have a conversation, and based on what plant protection tells us ....

To give you an example, recently we had a law enforcement officer that was here as a personal visitor for just a tour, but he did have his weapon on him. He was more than willing to take it back to his car and leave it in the car.

We do not have as yet any kind of weapon or metal detector, so it is difficult for us to tell who has one. Without us knowing that they're with a law enforcement agency, we don't come out and say to all of our visitors, "Do you have a weapon?"

You don't have to worry about most of the law enforcement people, but what about the odd nut? Are you concerned about some crazy coming in with a weapon? Could you spot a potential person like that?

When you have worked this long, you kind of get a feel for who might give you a problem. But usually the odd nut will not get on-lab because they don't have a contact, and they cannot go on the site without a contact.

Our guards are wonderful. We are surrounded by security here at JPL. It starts before a person even gets here. The guards at the gate are not just directing parking and traffic. They are doing much more; they are scanning. They have seen visitors walking from down at the park or whatever and they don't look quite right, and so the guards immediately will check in with us and see if everything is okay. If there is a problem here in the lobby, we do have codes and ways of getting attention, so they will come immediately.

As long as I'm here, I will try and maintain the safety and security of the people who work here, for the visitors to make them feel welcomed, and feel real good about JPL when they leave. For myself, I'll try to hold onto my job and do the best I can do for this wonderful place.

MARTY C. GREER
Automotive Technician Leadman
Born December 5, 1949 Rock Island, Illinois

Marty Greer lived in Aledo, an Illinois town of 3000 people. He went to a one-room church school in the country and worked on a neighboring farm for a year and a half while going to school.

His family moved to La Crescenta, California, in 1964, at which time he went to Glendale Academy. As a gift for graduation from 12th grade, in 1969 he and his younger brother were given the privilege of attending a "World Wide Youth Congress" as well as a tour of seven countries in Europe.

Greer attended Union College in Lincoln, Nebraska, for a two-year vocational program in auto mechanics. After completing the course he worked at a Chevrolet dealership in Montrose, California, for 10 years, and a Buick agency in Glendale another year.

His next job was at a landscape and maintenance company in Canoga Park, where he learned more about various kinds of equipment and how to do welding. He has been at JPL since October 1984.

He likes all kinds of cars and is restoring a 1937 Graham two-door business coupe. He enjoys swap meets and antiques. He collects anything automotive garage related: old tools, jacks, old oil cans, old wrenches. He also enjoys traveling and has visited 40 of the United States as well as Canada and Mexico.

> *I like to learn as much as I can if I have the opportunity.*
> *I read a lot of books, so I can learn more about engines and*
> *mechanics. I enjoy my work here at JPL very much.*
> *There are many challenges in this job*
> *as well as new things to learn.*
>   —Marty C. Greer

MARTY GREER

We're interested in the Voyager spacecraft project. We want to show that those wonderful things out in space don't just happen; there are many people we may not be even aware of, who contribute to making it happen.

Yes, you're absolutely right. There's a lot of things that are hidden in the background that people do not realize. We have standby generators to support when they're doing testing. Like if they are testing the Voyager inside the spacecraft chambers, if there was a power failure, we have standby generators that would take over and power it. We have to maintain that generator, and we do that once a week. We check the oil, we run it, we load test it, see that it does work properly, so we do have a backup.

We maintain the forklifts, so they can pick up the spacecraft and move it from building to building. Same thing with the vehicles transporting it. If they take it off the lab, if it's on one of our trailers or a truck, we have to make sure that we've maintained it, the tires, all the fluid levels; everything is ready to support the thing. If we take flight hardware to the Cape, we make sure that the trailer has all good tires on it, that the brakes are in good shape, whatever it takes to get the operation down there safely in as efficient manner as we possibly can.

Your responsibilities are not just at this lab here, they're all the way to Florida?

That's right. We have to make sure that the equipment, when we send it out on the road, is in good shape, that they won't get halfway there and something breaks down on it. We try to maintain as top notch fleet as we possibly can. If it takes new tires on it, we put new tires on it before it leaves, because when you're dealing with spacecraft, you're dealing with a lot of money. Sometimes it's a nickel or dime operation that can stop it. You don't want those kinds of things to happen. We're talking a lot of money, so we have to maintain the equipment and have it in order.

Like on our forklifts. A normal forklift doesn't have an accumulator in it, which is like a hydraulic spring; it absorbs the shock. The instrument people come, and they'll sit in the basket as it rides around here on-lab, to see what kind of movement it has because the instrumentation in the spacecraft is very delicate and you don't want to damage it while it's riding around here. So we have to make sure that we have suitable forklifts. The ones that are moving spacecraft must have this accumulator in the hydraulic system, as opposed to the shock. I'd never seen that until I came here.

Are those specially built or can you retrofit them here?

We could get it specified if we wanted; usually we just put it in ourselves. I have seen forklifts that have come from the factory with accumulators, but most forklifts don't have it.

How did you come to be working at JPL?

I was working for a landscape outfit in Canoga Park. A friend of mine was working here, and I was saying it's a long drive out there, and I would like to find something closer to my house. So my friend says if you'd be interested in coming to work at JPL, I think there is an opening. So I came and interviewed and got the job, and now I have only a 15-minute drive to work instead of 45 minutes.

For your landscape job were you a mechanic?

Yes, I've done mechanical work since 1970.

What year did you come to JPL?

I started here October 1 of 1984. I was at the tail end of the Voyager project.

It's very interesting. I never realized all the things that happen until I actually started working here, to see what goes on behind the scenes, that most people don't think about.

The computers, the electronics, how they've had to make them smaller, shrink them so they could put them in the spacecraft. I had never thought about that.

I saw where they made some kind of little filter they use in space. They're using it now for mining and stuff, and they've found out with dentists, when they're grinding the teeth, their mask is not protecting them from the dust because the particles are so small that they're still going through. They were able to detect it with this instrument they made. And I thought that's fantastic.

It's in a newspaper article that's in a magazine here at the lab. But how many people see it on the outside? That's where I wish they would do a little more public relations work, so people would know. We spend billions of dollars out there in space, but what are we getting for it here on Earth? There's a lot of things militarily, our computers and stuff, we've been able to make them smaller. And your little calculator. Tape recorder. They keep getting them smaller.

Are you also responsible for the control room in Building 230, where they have all those computers?

For that one there, they have their own generators. We don't mess with those generators. That building is kind of set off by itself. They do their own thing.

Who makes sure that their generators are working?

They have their own guy. He does their generator work. He has to know mechanical; he has to know air conditioning; he has to know electrical. He has to be all around and maintain all the equipment, just for that building itself. Those are big, big generators. There's three of them.

Is that the main exception? Are you responsible for all the other generators?

Yes. That's the only exception. All the rest of the generators on-lab we are responsible for.

What about your own experience with mechanics in running your unit? Are there procedures here that other companies or government agencies would do well to adopt? Things that you guys do?

Okay. We seem to be unique. Being in California we've had some people come out from other facilities, checking to see how we do environmentally. They were really surprised to see the things we do have. We have an oil filter crusher. It squeezes out the oil. You would be surprised how much oil is still in the oil filter, even though you turn it upside down and you think you've got it all drained out. When you squeeze it, you get all the oil out of it, and we recycle that oil. The tank we had in the ground for waste oil, we pulled that out of the ground. We've got an above-ground tank now, and the people had never seen an above-ground tank.

Now once you crush the filter we're supposed to be able to throw it in the trash, but we do not throw it in the trash. We give it to our environmental guy, and then he disposes of it. We're trying not to fill the land-fill full of this stuff. They probably take it to a recycle place, they shred it, probably burn what little residue is still in the filter, take out the metal. All of our waste, anything that is hazardous, goes through our waste department here, and they keep track of where it goes to.

So other government agencies came to see how you were doing this?

They were analyzing what we were doing to see what they could do.

How many people do you have on your staff?

There's 10 of us here that I'm in charge of: mechanics. I have a body man, I have a tire man, that details the cars, makes sure that they're all ready, that they have nice cars for them to go out on the projects and stuff.

One of the things we do, if they're taking a trailer, we make sure that the truck has the proper hitch on it, that they can tow the trailer down the road. We get it all set up for their project.

JPL is a long distance from the one-room country school you attended. What are your thoughts about the Voyager spacecraft or the project?

I know that there's some very smart people here. I don't get a chance to really go through the other buildings and see. There are a lot of buildings that are off limits as far as my clearance, that you can't get into, and know what in the world is going on. You have to know somebody or else read an article about it.

I think it is marvelous what they are doing, and how they can go out and gain the knowledge of what is out there. We're seeing medicine that they are able to do out in space, that you can't do anywhere else; they're doing it there.

I have friends who tell me, "Let me know when they're having an open house; I want to see it." I have people who go to our church come back and they'll go, "Hey man, that was fantastic to see what is going on."

Like our space simulator. It's the biggest one, from my understanding, in the world. There's only three in the world. We've had things brought over here from Europe to test because of the size of our chamber. We have a 25-ft and a 15-ft space simulator.

A friend of mine was really into space, and he was amazed to see the robotics; how they're doing welding and stuff; he's a retired machinist. Whenever I get extra pictures, I give them to him, and he's thrilled with them.

I think people need to be more aware of what's actually going on. And I'm not talking just about JPL; there's a lot of other places that are dealing with space.

JUDITH McGAVIN
Project Secretary

Born June 25, 1938   Laramie, Wyoming

Judith McGavin attended Star Valley High School in Afton, Wyoming; Ricks College in Rexburg, Idaho; and Kinman University in Spokane, Washington.

She came to JPL from the University of Idaho, where she was administrative assistant to the Dean of the College of Veterinary Medicine. After reading a *National Geographic* article about the Voyager Grand Tour, she decided she wanted to be a part of what would be a once-in-a-lifetime experience.

Since her children were through school, she left the security of a long-time position in academia and joined the JPL community in 1985 as a group secretary in the Systems Engineering Section. By the time the Voyager spacecraft was nearing Neptune she was the Voyager project secretary, fulfilling her goal of being a small part of the Grand Tour.

After the spacecraft began the interstellar phase of its mission, she moved to the flight projects assistant laboratory director's office, working under John Casani. Following the reorganization of the Flight Projects Office, she was assigned to the Telecommunications and Mission Operations Directorate, where she supports members of the director's staff.

She has been enthralled by the physical and biological sciences since her father took her to the Denver Museum of Natural History and, at the age of 10, she realized the infinite wonders of the universe. In grade school she experienced the marvelous complexity of the "inside" of things, specifically animal things. She was frequently the only girl in the class not repulsed by the fact that she could insert her entire arm into the aorta of a horse. This curiosity served her well in her future work with biomedical research graduate students.

Her fascination for finding out how things work is still a driving force, although now it takes the form of learning new software applications, trompe l'oeil painting, playing the piano, and the grandmotherly arts of knitting, crochet, needlepoint, crewel embroidery, and other things her hands can do while her eyes watch science fiction shows on television.

> *My perspective of Voyager is that of a young girl who read science fiction and saw it come true as she grew older. My perspective is one of awe and wonderment, of absolute joy to think that in this area at least we have managed to push our capabilities to the edge. Human beings are capable of almost anything if we have the time, the energy and the money to try it.*
> —Judith McGavin

JUDITH McGAVIN

How did you come to be associated with Voyager?

I became associated with Voyager on purpose. I have been enthralled with science fiction and science fact since I was 10 years old.

I was working at the University of Idaho when the Voyagers were first launched in the mid-1970s. They captivated me. Later several things happened in my life, and it became possible for me to come to Southern California. I remember thinking after Saturn, "If I don't hurry, I'm going to miss out on this whole thing." I came to Pasadena and got a job at JPL. By that time the Uranus encounter had just happened. I knew I only had one chance left. A job opened in the Voyager project office, and I was lucky enough to get it.

What year was that?

It was 1988, after the Uranus encounter, about a year before the Neptune encounter. By the Neptune encounter I was fortunate enough to be the project secretary. As I said, it was something I had set out to do. I wanted to be a part of something never done before and something that would never be done again.

Working at the office manager level gave me a good view of both ends of the process. I saw the people who worked 24 hours a day, day in and day out, getting the raw data from the Deep Space Network and translating it into the finished data we used. I saw all the departments here working together so the public could see a well-oiled organization. I also saw the near hits and near misses, the almost catastrophes and heroic saves that were part of the mission.

I was most excited by the tremendous evolution of the Voyager spacecraft from their launch in the 1970s to the last encounter more than 10 years later. The onboard computers underwent marvelous transformations in order to handle the increased expectations from the ground. When there were glitches, they were corrected without losing valuable data. This is a testimony to the care given the equipment built and flown by this Lab.

I am awed by the dedication of the people who built those two spacecraft here at JPL. They have lasted, run well, and been able to improve and gather more and better information than originally planned.

It was wonderful to see the clarity and detail of those pictures come back down from Neptune as we got closer and closer to encounter. Then came the surprises we saw at Triton! I found myself feeling the same way I did the first time I saw the flagpole in the middle of the "green" in Lexington, Massachusetts. The plaque on that pole says simply, "This is the Birthplace of Our Nation." I remember my eyes filling with tears as I thought of the sacrifices and heroics of those people. The same kinds of tears filled my eyes as I realized these images from Neptune would never happen again. It was truly a once-in-a-lifetime experience.

It was absolutely thrilling to me, but most of my family smiled at me and said, "That's nice, Mom" or "That's nice, Judy." They didn't feel the same sense of history that I did, of having been one of the pencil pushers, or keyboard pushers, that helped get things done at the Voyager project.

I had the opportunity of typing the text for the *Traveler's Guide to Neptune* for the encounter. Reading the inputs as I typed them, I enjoyed seeing what was going on in other people's minds as they went about their tasks. It is so much fun working with minds who think. I have soaked up a lot of knowledge by osmosis through the years.

So now that the encounter is over, I have moved to the Flight Projects Office. I try to keep in touch with the projects flying and getting ready to fly. Ulysses is about to be launched and Magellan is on its way. These are things I've read about since I was a young girl and I never doubted we would do them "for real".

One of my favorite magazines is *Analog*, edited, before his death, by John W. Campbell. His taste in science fiction ran toward "hard" science. There was much more engineering than fantasy in his choices. Each year I was, and I still am, delighted to see how much closer reality is becoming to what had been fiction.

I remember listening to my father as he told us about my grandfather buying the first automobile in the community in which he lived, an old Ford Model A. In just one generation look where we are! We are getting ready to go to Mars—for real. I love it! I love every second of it! I've told people here at the lab that I will still be here working when Voyager reaches the heliopause. I will probably be all hunched over and using a walker. But I will probably be using a computer, too. I will never give up. As long as things like the Voyager project are going on, I want to be in the middle of it.

I'm looking forward to the first ship landing on Mars. I hope to see it happening as a joint venture. We have the brains. Other countries have hardware and money. We ought to get together.

When people ask me what am I going to do now that the Voyager project is over, I say, "Over? Over? This is just the first quarter of the journey."

"What do you mean?" they ask, and I tell them, "The sun is a very powerful star; we don't realize how far out its influence goes. That is what Voyager is working on now. It will mark the edge of our solar system. Those little spacecraft will be going on for the next 15 to 30 years before they even reach that edge and move out into truly interstellar space."

I have tried to instill in my children this same wonderment and thirst for knowledge. I hope I can do the same for my grandchildren. I want them to look at the world and the people in it with admiration and amazement.

What has been your strongest impression of the Voyager project?

My strongest impression is the fact that it has been so successful. When I think back to the craft we started with, and when I remember what our original goals were, knowing what has been improvised since the beginning, I realize that almost every time someone said, "Let's try and see if we can do it," they were able to do it. This is still going on.

I think it is the most successful project NASA has ever undertaken. It is monumental, just because of the length of time involved and the amount of changes that were proposed and then executed by these men and women of the team. Some of these team members have devoted more than two decades of their lives to that spacecraft.

Yes, the word I would use is "successful". The things they've learned while working with those two spacecraft have made subsequent projects possible. The slingshot effect, used to shoot Voyager beyond Jupiter, has become such a well-understood concept that it pops up in the scripts of television programs. Of course, we have myriads of other things that are spin-offs of the space program.

I am a proponent of unmanned spaceflight. I know there must be persons to set up a moon base or to land on Mars, but I am fascinated by what these spacecraft can do, and by the fact that they are controlled from the ground. It is less expensive and more foolproof than manned flight.

Yes, that is my impression of Voyager: dedication and success. It is one of the monumental feats of American science.

What has been the most difficult part of your association with Voyager?

The shortsightedness of bureaucrats was very hard for me to understand. As soon as we passed Neptune, NASA lost interest. The immediate reaction was to shut down all the science platforms in order to save money. Dr. Carl Sagan rescued them for more than a year with his proposal of the Solar System Family Portrait, but the bureaucrats asked, "What is left? You've gone past the outer planets; what is left?"

Well, we don't know what is left, but we do know it is a lot! That has been the most difficult part for me. The fact that they wouldn't give the small amount of money to keep those science platforms going so we could learn as much as possible about our solar system until those little generators finally give out. As long as we have made it out there—and it is the last time we will be out there—we should get all the information and science we can. Those people just didn't see it. Yes, that was the most difficult part.

What would be an example of "science platforms"?

The experiments mounted by the different universities and agencies. There were 11 experiments on Voyager, and they were reduced to three. The platforms left are the fields, the particles, and the waves.

You referred to the way the spacecraft were built. What was your point?

My point was that when you are personally involved with a project, as the people at JPL were when they were building the Voyager spacecraft, great care is taken. When these tasks are farmed out to contractors in bits and pieces, it is very impersonal. One can't say, "this is going to be part of a wonderful project." But everyone here at JPL felt Voyager was a "wonderful project," so every part was milled and fitted and attached with great care. Everyone said, "This is going to be it," and it was.

I wish there were some way people who work for government contractors could be excited enough to go to work every day saying, "I'm really part of the future."

There is an old story about a man standing on the street corner watching two hod carriers go about their work. He asked the first man what he was doing. The man replied, "What does it look like I'm doing? I'm carrying bricks down the road." He asked the second man what he was doing and that man said, "I'm building a great cathedral."

I think that in piecemeal contracting a lot of people feel they are just carrying bricks down the road, when in fact they are really "building a great cathedral."

What would you like people in the future to know about you?

I want them to know that I looked up to the stars. I want them to know that I really believed there is more to life than what this Earth has, and that I wanted to find out what "more" was.

I want them to know that I believe there are millions and millions of planets like ours, but that the universe is so vast we will probably never encounter another one unless we break through the speed of light and are able to go at "warp eight." However, I know those planets are there.

The key is to teach the children. Something that excited my imagination 40 years ago still has me coming to work every day with a smile on my face. I would like this book to be in the hands of the children, the ones who start reading about the sky and wondering about the stars. That is where the future is. If we can lead the minds of children to think toward the stars, we will always have a way to get there.

What do other secretaries here feel about Voyager?

Their feelings run the gamut from complete disinterest to the same kind of feelings I have. I don't know of many secretaries who have been enamored of the space program since they were 10 years old, who came to work at JPL expressly because they wanted to be part of the Voyager program. I think we are all aware of what is going on, and we realize that with all the glamour comes all the paperwork. We are willing to slog through the paperwork so the glamour can take place. We are just some of the "hod carriers" saying, "I'm building a great cathedral."

JPL is part college campus and part business. I spent 14 years on a campus, so it wasn't much of a change for me to come. Many people feel privileged that they have a chance to work at a place like JPL, and they hold it in reverence, almost like a church. I don't go quite that far, but it is a nice place to work because most of the people here have the same frame of mind as I do. We feel we are doing something for the future, and we are willing to work at it.

Do you remember specifically what happened when you were 10 years old that got you thinking about space?

Yes. It was a comic book called *Planet Comics*. Years later I realized that some of the people who wrote for *Planet Comics* were Ray Bradbury, Willy Ley, and others. Some of the people who illustrated those books went on to illustrate other comics, the infamous EC comics the U.S. Senate decided were driving children to juvenile delinquency. Wally Wood was one of the illustrators and Kelly Freas. These men have gone on to make names for themselves in the science fiction community. When I went away to college my mother, bless her heart ....

Don't tell me; I've heard this before, and it happened to me, too.

Oh, isn't it awful?! Now, for a panel of pen and ink drawings from those comic books, *one* panel, you can get more than $100. I had the full sets of all four comics. Two were science fiction and two were horror fiction, *Tales from the Crypt* and *The Vault of Horror*. I loved them. They were only on the newsstand for about three years. What can you do? Your mother is cleaning house, and she had no idea what those comic books are worth! If I had them now, I could put my grandchildren through college.

But that is what got me started. To all you parents who don't want your kids to read comic books, I say, "Fie on you. Let them read!"

Speaking of children, I understand that they write to JPL.

Yes. Kids from all over the world write to us, not just the United States or western Europe, but from eastern Europe, Middle East, South America, Australia, and Indonesia. They want anything we can send them about the Voyager project and the Neptune encounter. These kids, six and seven years old, are writing thoughtful letters. You can tell they have done their homework. They know what is going on in our universe.

We send them a standard "goody package". We sent one to a young boy in Virginia. A month or so later we got a Polaroid picture of the display he had made for his local science history fair. He had done the Voyager project in general and the Neptune encounter in particular. He won first prize with it. It was extremely well done. He was just 10 years old.

These are the kids I get excited about. I know as long as there are children somewhere whose imaginations have been fired up by the Voyager project, we are going to have a space program; we are going to have people reaching out beyond the Earth to find out not just the edge of the solar system, but the edge of the universe.

DEBORAH JEAN GAG
Telephone Operator

Born March 13, 1951   Burbank, California

After graduation from Verdugo Hills High School, Sunland California, Deborah Gag joined the General Telephone Company as a toll operator in 1970. Since returning to the work force as a working mother in 1979, she has always been in telecommunications; at Walt Disney Productions, Paramount Pictures, and Warner Brothers, as well as JPL. She is on the Board of Directors of the Warner Brothers Office Employees Guild, representing some 550 clerical workers. She is also a graduate of Coro of Southern California's Neighborhood Leadership Program.

She is a member of the Bahai Faith. Her recreational interests include herb and old rose gardening, traditional folk music, nutrition, homeopathy, and phytotherapy—more commonly referred to as aromatherapy. She loves the harp, reading, and traveling.

Family and friends, relationships, enduring and eternal.
Peace, a balanced state of mind, a constructive
thought process, meditation.
Gardening and prayer nurtures the soul.
Music—waves crashing on the beach, sea gulls screeching,
dogs barking, children squealing with delight.
Vision is the key to eternity.
Trust—a stranger, a smile, a gladdened heart.
Happiness is to Be—affectionate, healthy, kind, joyful, generous,
sincere, thankful, and trustworthy.
Love, liberty and the happiness of pursuit—a wonderful life !!!!!
—Deborah Jean Gag

DEBORAH GAG

How did you happen to come to JPL and be working in this office?
I was originally working at film and television studios. I had been at Paramount Pictures for about three years, and the film industry had not been doing too well. There were a lot of layoffs, and it wasn't real secure, so I was looking for financial stability. At that time I was raising my two daughters, Camille and Erica, on my own, and I happened to see an ad in the *Los Angeles Times*. JPL had never ever placed an ad in the newspaper for a telephone operator.

Why was that?
In the early days they had a switchboard of 18 to 20 telephone operators, and as the years went on the same operators continued to work until they retired. Because we were becoming more automated, there was need for less operators. In fact, the chief operator who just retired two years ago, Bette Vandello, had been here since 1964. I think there's been a total of four or five chief operators here since 1945.

Up until the last six months those gals, all the past chief operators, would get together regularly once a month over at Kathleen's, a Pasadena restaurant, for lunch. Now some of them have moved away; some have deteriorating health. They were like an era, and we're going into a new era now. Jo Ahern, the chief operator I'm replacing, is out ill, and she's been here for about 11 years. She's seen a lot happen here.

From the time I came, in 1987, the Voyager encounter last year was the most exciting event that I'd seen at JPL. We had been building up to it for about the last year, preparing for how we would handle the situation. We worked 12-hour shifts, rotating the girls. Our influx of traffic for about one month before and one month after the encounter itself was probably 50 percent more than usual, as a result of the media coverage from around the world and people just generally interested in the space program. We'd get asked questions and transfer them to public information. Jurrie and his group were very helpful, answering the questions or calling the particular scientist that people needed to get information from.

What were some of the incidents that you remember?
What really sticks out in my mind the most was August 25, the day that Vice President Dan Quayle was here. I was scheduled to work 9:00 a.m. to 9:00 p.m., and because I was very interested in seeing him in person, I came to work at 8:00 and went up with all the thousands of other people to the mall area. It was very exciting to have that kind of media coverage due to the Voyager encounter. A government dignitary was here, and the secret police were everywhere, on all the roof tops of all the buildings. It was just generally very exciting.

I left work here that night at 9:00 and was seriously injured in a car accident which totaled my beloved 1968 California Special Mustang.

This was the evening after Quayle was here?
Yes. It was a Friday night, and I was on my way home when I was rear-ended by a drunk driver. As exciting as the day had been, it was a little bit of a letdown for me because I was scheduled to work on Saturday and Sunday as well. I had been looking forward to that, being a part of something here at JPL that had never happened before: the Voyager encounter at Neptune. But I was unable to come to work. In fact I was out for about six weeks as a result of my injuries. It could have been much worse. I was very fortunate that I had my seat belt on, but I felt that I got the short end of the stick regarding the Voyager encounter; something that culminated with a big bang for me.

It's been very slow since then. We got used to a great deal of media attention and the influx of a lot of interested parties from everywhere. It's going to take a while for anything else of that magnitude to come again. Once we start getting some good pictures from the Hubble, it will probably generate that same kind of interest. We're in a transition, I guess.

What was the most challenging thing that happened to you during that encounter period?

I would say that the additional workload it gave us was dealing with the press. There were people who would come in to von Kármán auditorium, and we would have to assist them. It was just a part of our job, except it was at different times and more of what we usually do. I don't recall any real serious challenges that were out of the ordinary, aside from it being more than usual of what we do regularly.

You were in the switchboard room?

Yes. Sometimes we would have people trying to call out on phones that were restricted, and we would have to assist them with that. For two weeks the telephone company installed a special telecommunications trailer with telephone access that the press could use without going through our switchboard, and sometimes they had problems, not understanding the system, and we would help them when they had difficulties getting out. It didn't really affect us the way it probably affected people in other technical areas.

It was just exciting to deal with more people and very intelligent people, very nice people. We were treated with a great deal of respect from everywhere around the world, and that's very nice. It's reciprocated; it goes both ways. It's a positive thing all around.

Your kids, your friends; do they ask you about what is going on here?

Yes, they do. As a matter of fact, my husband Uli and I have been married just about a year, and we went to Germany for our honeymoon for three weeks last June. His parents live in Halver, a very small town, maybe 15,000 to 20,000 people, about 60 miles northeast of Cologne. When my mother-in-law told me that the newspaper was coming to interview us, I thought she was kidding me. They did this full page article on my husband and I, and they showed pictures, some of the Public Information Office publicity shots that we had of Voyager at Saturn, Uranus, and Jupiter. They put some of this information in the newspaper, and they did this big write-up.

My in-laws in Germany are probably more impressed with the whole thing than my own family here. My own family here in the San Fernando Valley pretty much takes JPL for granted. Uli has a niece that's 14, and she got a real big kick. I gave her one of the packets with all those pictures, and her teacher had her put them all up on the board. They seem to be more interested in the scientific field over there. Here it's almost just a hobby for some people, and they're not as passionate about it as they are in Germany.

How did the papers find out about you?

Mutti, my mother-in-law, just happened to call them. Uli and I had been sweethearts in 1970, but I ended up marrying Bill Marcione, and just the fact that we had gotten back together again after a 20-year period of being apart, was of interest to the papers. Then after they found out where I worked, at JPL, they made a big story because people would enjoy reading about that. It was big fish in a little pond kind of thing.

What is your own view of Voyager?

My personal feeling is not a scientific one, but I feel intuitively that the more we learn about the cosmos the more we learn about ourselves as a society of human beings. It seems to me that the further out we extend our consciousness, the more it helps us define ourselves better internally. I think that this has been evident in the last 30 years of space investigation, because people are getting more in touch with themselves the farther out we're going. It's probably a pretty simplistic way of looking at it, but that connection is very obvious to me.

The Voyager project seemed to unify nations that ordinarily would oppose each others' success. Regardless of the current political situation, Voyager brought out the best in all parties involved, I do believe. Being aware of this fact made me appreciate being employed at JPL.

Looking ahead, one can only imagine how much more we, as a society, will benefit from global cooperation due to the fair exchange of breakthroughs in technology and science. Voyager is a classic example of what lies ahead for us if we work together.

Is there anything else you'd like readers to know about you and your work?

I think that the personnel at JPL in general are a unique group of people. The largest percent of people that work here really love what they do, and it's a pleasure to be able to work in the company of such special people. It makes it nice to come to work every day.

JACK MARTEL BLANCO
Senior Guard, Plant Protection Department

Born March 13, 1945   East Los Angeles, California

Jack Blanco, a native Californian, is the fifth of seven children in a second-generation Hispanic family. He grew up in Pico Rivera, California. After majoring in foreign language and playing baseball and football, he graduated from El Rancho High School in 1963. His ambition was to become involved in law enforcement. He followed his dream by attending Rio Hondo and East Los Angeles Colleges, majoring in police science and criminology.

In 1973 Jack was selected by the Los Angeles County Marshal's Department to serve as a peace officer. He was POST (Peace Officers Standard Training) certified after graduating from the Pasadena Police Academy. While employed by the Los Angeles County Marshal's Department from 1973 to 1986, Jack worked in a number of capacities such as Warrant Division, Prisoner Control, and Baliff Division. He was active in the Marshals' Association and Police Olympics for 10 years. In 1981 he was recognized as an outstanding officer and awarded an engraved certificate for "Unselfish Dedication" in promoting the advancement of Hispanic Americans in the field of law enforcement.

In 1988 Jack felt the need for a career change and came to work for JPL. He tries hard to make this workplace a safe and secure environment, while utilizing good communications skills with employees. He felt honored to be chosen by JPL to escort the Mars Observer Satellite from New Jersey to Cape Kennedy Space Center.

Jack enjoys spending his free time with his wife and five children. He enjoys all types of sports and is a fan of "old time" movies. He collects baseball cards and likes to travel, especially to the mountains and the beach.

JPL means a lot to me: fulfilling my dream, as a second-generation Hispanic, of getting into police work. Allowing me to work at a history-making facility. Seeing young people have a dream to shoot for. Knowing that our great nation is still, in peace, continuing to explore new horizons of our universe.
I take pride in working closely with the space program and ensuring its security. I view my job as an integral part of America's efforts in space.

—Jack Martel Blanco

JACK BLANCO

On any type of special occasions, such as the Neptune encounter, our job is mainly to make sure of the public's safety. For the public entering the Lab, to make sure that they are properly given ID, identification, to make sure that they are in the areas they should be in. There's a lot of traffic situations, where even as security-oriented as we are, we have to make sure that the safety of people will be provided for. We have dignitaries coming in, we have to check for their safety to make sure all bases are covered.

We make sure that we are in complete command at the time of any situation that we should be paying attention to, what our main control center says are the vital areas, the mainstream, the main arteries. We may have a visitor control problem in the parking area. Our business is to see that this gets rectified as soon as possible, to make people feel that they've come to something special.

When I was in law enforcement, being Los Angeles County Marshal, you watched more for the *un*lawfulness, where this is *law*fulness, with a little bit anxiety, of wanting things to go smooth, wanting the security to be upheld, but at the same time making sure that people feel comfortable, making sure that they're welcomed in the proper manner. But most of all the security part of it: to make sure that they know where they're going and where they shouldn't be.

This is our main interest, making sure that in our buildings where we're going to have any sort of activity, where news media will be, making sure the security is well taken care of there, directing visitors to where they should be, and making sure that the overall manner is calm and that they know what our job is, and that our job must be fulfilled.

Along with special events, we also maintain our normal routine security procedures—patrols, plant safety, random searches of cars leaving the lab to discourage theft of supplies or equipment—all the steps necessary to protect JPL people and property.

Were there any particular situations at the Neptune encounter that standout in your mind? It was a hot August.

It was, sir, it was. I tell you what sticks out in my mind. We had notice of a person going to make a copy of a badge to get into press conferences in von Kármán auditorium. There's so many news people from all over the world; there's hardly any room for ordinary visitors or even JPL employees.

So the regular visitor's permit, to get onto the Laboratory grounds, was not enough to get into the press conferences? You also had to have extra permission?

That's right sir. This information was fed to us early in the encounter. At that point we were going to change all badging, which meant we were talking about hundreds of people that would have to be faced with some sort of a delay period to scrutinize these badges thoroughly because we heard that a copy was being made.

But it was rectified, and I was astonished, even with my background, by the way it was handled: so professionally that we got our information out to our troops, that we did cover the area, the von Kármán auditorium, where the news media and dignitaries where checking through. We had lists of people who should be there, and once that had happened we were on top, making sure these badges were lawfully taken care of and that the visitors were listed on the sheets given to us at the initial part of the encounter.

There were exciting parts, like when the Vice President Quayle came. To hear him speak on the space program was very fulfilling. Being part of something this big and understanding that your job is, again, handling this type of situation, it makes you feel a very big part of this, it makes you feel very wonderful.

Did you have a chance to see much of what was going on during the Voyager encounter, or were you so involved in your security responsibilities that you didn't have a chance to watch?

I'm glad you mentioned that, sir. Utmost in our minds, and much as we'd like to get involved in what is happening at that time, like the space encounter, it's taking care of first things first. Excited as we were about the encounter, we first had to make sure that the job was being covered. There's a few stations where we have closed cable right through the buildings; if you're lucky enough to be stationed where there's one available, then it's okay to watch.

But any chance I got, and when I got home of course I'd read through the newspapers, just like Joe Citizen—I'm Joe Citizen, too. But being part of it here, I wish I could have got more chance to see what was going on, but I'd have to read about it, just like everybody out there and watch on TV what I'd actually seen right before my very eyes; and that was exciting.

What are your thoughts about the spacecraft? What do they mean to you?

Again, being just out there and not knowing what the space program itself is about, just on hearsay things, and all the big adventures that had taken place by the United States, I found that there's so much more to know, so much more to be involved in. The information is just mind boggling that we're getting from these things. Now I'm privy, being part of this JPL program, to know just what's involved there.

I wish I could tell people to stay on top and stay knowing what kinds of things we've learned out there from these orbiters and these Voyagers and the latest ones we've sent up, Magellan and Galileo. The information that I see now, where years ago we might have been studying the planets for vegetation, we are now heading to some place where if anybody has any doubts what's out there, they can just look at the information that we've got back. We're prepared and knowing for the future of mankind and our children's children what is going to be most valuable.

What about your family, your neighbors; do they understand what's going on?

I never realized until I got involved here just exactly what they felt. I'd get home, and I'd get calls; and they'd want to know about the encounter, and they'd want to know about Galileo and Voyager and the telescope. My nieces and nephews, any literature I get, any things we're privy to, I try to get to them. They share it in school; they call me back and say, "Uncle Jack, this was so fascinating; the teachers thought it was wonderful."

Working in the security part of it is one thing, and seeing these people that are maybe more knowledgeable, but no, I'm talking about kids that can share this kind of thing, share it at school, understand what's going on, and it's fascinating. Yes, my neighbors ask about it. I let them see the brochures I get, and they are mindful of what's going on now, and it's meaning more to me to be able to discuss it with them. It's exciting, very exciting.

Do you have a chance to discuss it with your colleagues on the force, or do they talk mostly about shop?

Mostly shop talk. We talk about special areas that we go on with this thing, but mainly we are involved with what *we* have to do and what our job is at hand, and we reflect on the Lab and what it's doing. The Lab is very good about sending out information bulletins to us. They are constantly flowing through; and we have a chance to talk about them, and we get together and chat over this. One of our men might work in a certain building; he shares with us what he saw. It's an overall interest. You've got to be interested to be able to work here and see what's going on.

Thanks very much.

My pleasure, sir, my pleasure.

CATHERINE SWIFT
Undergraduate Research Fellow

Born January 17, 1967   Honolulu, Hawaii

Catherine Swift was raised in Hawaii. After majoring in astronomy and planetary science, she graduated with a degree in geology from Caltech in 1990. During high school and college she worked on a number of different research projects in planetary science at the University of Hawaii, University of Arizona, and JPL. After graduating from Caltech, she spent several years in Hawaii working on various political and environmental projects.

In 1994 she spent seven months on Midway Atoll working with seabirds for the U.S. Fish and Wildlife Service. She is currently working on a master of science in zoology at the University of Hawaii. Her thesis research involves developing a method for broad scale distribution of a rodenticide in native ecosystems. She is conducting laboratory bioassays on wild rats at a U.S. Animal Damage Control research facility in Hilo and field trials in Hawaii Volcanoes National Park.

My career has changed from planetary science to conservation biology in Hawaii. I feel that this is a much more pressing issue because we're losing Hawaii's unique native species so rapidly. The planets seem very distant to me now. They will always be there, untouchable, at least on a human time scale, but I can see our native plants, birds, and insects disappearing all around me.
I want to be able to look back on my life and know that I have made a difference in conservation in Hawaii.
My friends and family and cat are extremely important to me as well. Their love and support have gotten me through many difficult periods in my life. Their sense of fun and different perspectives on what is important keep me from taking myself or my work too seriously.
—Catherine Swift

CATHERINE SWIFT

My involvement with Voyager centered around the encounter with Neptune, in August of 1989. I was an undergraduate in planetary science at Caltech, and I had a summer undergraduate research fellowship (SURF) at JPL with Kevin Baines and Heidi Hammel. This was an independent research project, "Ground-Based Imaging of Neptune in 1989," for which I had to write and present a paper at Caltech.

I reduced and analyzed data from near-infrared and visible ground-based images of Neptune taken by Heidi with the University of Hawaii's 2.24-m telescope on Mauna Kea Observatory. I determined latitudes and periods of rotation for cloud features in the planet's atmosphere. These results correlated well with the latitudes and periods derived from the Voyager spacecraft data set, thus validating the ground-based observations as an accurate method of monitoring Neptune's highly variable atmosphere.

My participation in the actual encounter was very small. Heidi was a member of the Voyager Imaging Team, and I was her assistant. During the 10 days before and during Voyager's closest approach to the planet, I acted as liaison between Heidi, who was observing Neptune from the 2.24-m telescope on Mauna Kea, and the other members of the Imaging Team.

I collected and conveyed relevant Voyager information to her while she was on Mauna Kea and presented preliminary ground-based results from Heidi to the Team. I also participated in the reduction and analysis of Voyager measurements of wind velocities on Neptune, so I was one of the authors on the paper that members of the Imaging Team published in *Science* on Neptune's wind speeds obtained by tracking clouds in Voyager images.

It was very exciting, both because so many discoveries were being made and because it was a major international press event. I remember going to the daily press conferences led by Ed Stone to find out what the latest results were. I was so caught up in my duties and Neptune's atmosphere was proving to be so interesting that I would only hear snippets about what else was happening unless I went to the press conferences. The conferences were packed, not just with the media, but also with people who were working on the encounter. Neptune had so many fascinating and surprising features that people weren't just interested in the specific project that they were working on.

The feeling of exploring a new place and all the discoveries that went with it made the encounter seem like quite an adventure. Neptune was so different from Earth and from what you experience in the course of your daily life that it was almost impossible to believe it really existed. The scale was so vast. Clouds larger than our planet whisk around Neptune at enormous velocities. The Voyager encounter made it seem real for a short time. I could watch images being downloaded that I knew had been taken just a few hours or days before. I can't compare it to any other experience I've had.

JURRIE VAN DER WOUDE
Public Information Officer

Born June 18, 1935   Amsterdam, the Netherlands

Jurrie van der Woude received his diploma from Josef Israels Middle School in Amsterdam in 1954. Drafted into the Royal Netherlands Air Force in 1955, he trained first as an air traffic controller and then served as fighter pilot until May 1962. He immigrated to the United States, arriving on July 4, 1962, and became a citizen in 1969.

He was employed at Caltech in August 1962 to make petrographic thin sections from geological rock samples. After a year he was hired by the leader of Caltech's Lunar Laboratory, Bruce Murray, to work in the photo lab, primarily with data from the various Ranger spacecraft to the moon.

Subsequently he handled photography for Mariners 4, 6, 7, and 9 missions to Mars and Mariner 10 to Venus and Mercury and participated in preparing NASA's atlas of Mercury.

He was assigned to JPL's photo lab in 1976, where he and R. Wichelmann shared responsibility for color quality control of all the Viking photos from Mars. In 1978 he transferred to the Public Affairs Office, where he worked with the worldwide press corps, especially during the 12-year mission of the Voyager spacecraft to the outer planets. This resulted in some of the best PBS programs about Voyager missions, both domestically and abroad.

He coproduced "Twenty-Five Years of Space Photography," an exhibit of 150 images that toured Asia and Europe and was seen by millions of people. He was also deeply involved in planning and producing the Solar System Family Portrait, made by Voyager 1 from 4 billion miles away.

He has been secretary of the Dutch-Indonesian Community Center, De Soos, for more than 20 years.

After a European upbringing, migrating to the United States seemed
the adventure of a lifetime. How could I have known that
I would embark on the greatest journey in the history of humankind:
a journey of discovery to the farthest planets,
with the greatest group of people on Earth.
Believe me, life is not a beach,
life is the Universe.

—Jurrie van der Woude

JURRIE VAN DER WOUDE

Ed killed me yesterday in Washington. There, in the eyes of the world in the IMAX theater in the Air and Space Museum, he did this marvelous presentation, as only Ed Stone can do. His voice was clear as a bell as he explained the solar system mosaic. Then the curator of the museum allowed 10 questions from the public. Ed answered them as only Ed can do. Then somebody stood up with question Number 9 and asked, "Are there going to be posters of this?" And Ed said, "Sure, there are posters." I groaned. I felt like standing up and screaming, "No, Ed, there are no posters! We didn't get the damn thing together until only last Friday—barely."

So now I wait for the influx of calls and requests because Ed Stone told them so. We'll see what we can do, of course, because this is *the* photograph, this mosaic, with that tiny dot that is us, our planet. As Carl Sagan said, "It's a pale blue dot on the stage of the cosmic arena." How we're going to produce those posters, I don't know yet, but just like we made yesterday possible, that will happen also.

Somebody asked Carl a question about planetary exploration. He was talking about the various dictionaries and encyclopedias that one still finds in the schools. Carl said, "Anything before 1975, gather it up and throw it away, as far as the planets are concerned, because that's when planetary exploration basically began in a glorious way, with Viking to Mars."

The earlier ones were flybys. We peeked. We opened the louvers a little bit while we were sneaking by somebody's living room and looked in for a short while—particularly with Venus and Mercury. Although the Voyagers were flybys also, they were much more in-depth flybys than we could do earlier. For one thing we have color capabilities in the imaging, and there are more instruments onboard. So we learned a lot.

The Voyager flights were predetermined by the orbits of the planets in their eternal dance around the sun. When we, with Voyager 1, looked back at the solar system from 4 billion miles away, that would not have been possible if we had not launched in August and early September of 1977. We had to launch at that certain time because you are dependent on the gravitational pull of all these planets that are still years away, that will have to pull you from one planet to the next one.

It's that dance where you bring your spacecraft into an orbit around the Earth. Then, with a relatively tiny rocket, you break away from the Earth's gravity. Then you coast along until Jupiter starts to pull you in and speed you up again, and slings you past, on toward Saturn. You slow down again because Jupiter keeps pulling you slightly back, but your momentum overpowers that and finally you fall within the influence of Saturn, which in the meantime has moved for several years. Saturn starts to pull you in, speeds you up, and slings you in the right direction for an encounter with Uranus five years later.

So you are dependent over a long period of time on what the planets are doing, unless you come up with a propulsion system something like nuclear, that would give you a continuous yearlong power drive. Then things could be done in a shorter term, but when we launched the Voyager spacecraft, it was a matter of a 12-year predicted, predetermined trajectory, if everything went right.

Nothing went wrong with the spacecraft, so that was the fortunate part, but if that situation had not been used—because we were not ready with the spacecraft, or for political reasons or whatever—we would have had to wait 176 years before all these planets would have made their dance around the sun and were starting it all over again. The time before the launch of the Voyagers when that happened was when Jefferson was president of the United States.

So we did it at the right time; we took advantage of the opportunity to do it. If 10,000 years from now there is still a planet with a civilization on it, there may even be a Library of Congress with books that will say that JPL and NASA and, much more important, our generation, were the first ones to explore the planets in our solar system. Who came second? Who came third? Nobody cares.

But *our* generation, with the people here at JPL, and all mankind that lived in this era—we were the travelers, we were the ones who made the greatest journey by exploring the planets in our solar system. That is something very important if you are working with a project like that. How can it be reflected in a paycheck? I would have paid for the privilege to be here.

You have been part of a fantastic voyage, and I feel fortunate to have been in on it, too.

Yes, and I'm glad that you came aboard when the time was right. It would be a very sad situation if this was a closed society and we were doing these things without making the world aware of them. The Voyager missions were journeys of discovery.

If you go back to 1976, the Bicentennial Year, when the Vikings landed on Mars, it was beyond imagination when I saw the first photographs from the surface of Mars, but basically it was not a journey of discovery, it was a journey of exploration.

We had already sailed past Mars with Mariner 4, 6, and 7, and around it with Mariner 9. We had seen it and we knew the planet, although not that well until Viking came and sent tens of thousands of photographs, of a quality which we don't have of lots of big areas on Earth. Why send man there? The American flag is on Mars; it's still standing there in two places. We have been there. We explored those foreign beaches. But Voyager took us beyond.

We knew the inner planets reasonably well, but the outer planets were something different. Jupiter? Sure, man had looked at it for more than 350 years—Saturn, basically the same way. But what did we know prior to 1986 of Uranus and Neptune? Less then you could write on that notepad on my desk. They were really mysterious. And then to see them in the way that you and I and all those other friends who came here during those encounter times saw them—you *have* to share that; it's too big to keep for yourself. Some things can be so beautiful and so immense that you have to share them.

Uranus got overshadowed by the Challenger tragedy, which really put a damper on its success. Uranus visually was not one of the most exciting objects. It was a beautiful blue, but you couldn't find a cloud there to save your life, except that one tiny white thing that fortunately popped up so we could calculate the rotation rate of the atmosphere. Of course, the magnetic field people were delighted because Uranus was laying on its side and had a magnetic axis that they couldn't have imagined.

But then, years later, to know that we were approaching the most mysterious of them all, Neptune, and remembering Uranus, saying, "We hope that this planet is a little bit better than Uranus," and then, as we served the last course that will reach your table, you came up with a planet that was indescribably beautiful and a moon that you couldn't have dreamt of—that was too much. It was really too much for me.

We knew that Neptune had a moon, Triton, but that was about the extent of our knowledge of it. Then we saw it. We sailed over the northern pole of Neptune and five hours later we were encountering Triton. That was a momentous five hours. At midnight we had a planet that we didn't know anything about, and at 5:00 in the morning we had a moon with better photography than Glendora, where I've been living as long as I've been in this country. I've never seen an aerial photograph of my own town, and here was a whole world, a glorious world!

It was about 3:00 in the morning when my eyes saw the photograph, my ears heard the people around me, and their "oohs" and the "ahs," but I was brain dead. I could not absorb it. Fortunately I heard that I was not the only one who suffered from the effects of this malady. It was a data overload. Confusing as heck. I knew exactly what photographs the spacecraft was going to make, except not in my wildest dream did I have any idea what they would turn out to be—and all these sentiments come together. I'm glad I'm not a cold scientist. You realize that this is really the last encounter. After that there are no more Voyager pictures. But my god, to go out with that planet and such a moon!

I sat there and let photo after photo wash over me. Euphoria. People were whooping and yelling. There was an emotional attachment to Voyager. It stopped being a spacecraft and became a personality, not just 1800 pounds of nuts and bolts. No other spacecraft had that feeling. The night of Neptune and Triton was comparable to the birth of my first child.

And then the amazement when you think about it: those unknown people who navigate the spacecraft over billions of miles and arrive within 10 seconds. I wish the world's airlines would run as well as we ran Voyager. You could be shopping right now in a Pasadena supermarket and be standing next to the guy who took that spacecraft over 3 billion miles and had it arrive within 10 seconds and 10 miles of its course, and you wouldn't know him. Yet he's just as much part of the Voyager family as Ed Stone or Brad Smith. They all played a part, but some got a little more exposure than others.

I don't think I will ever run into people like that again in my life, so am I entitled to get a wet nose once in awhile?

I have to add to that. In 1962 I left Holland as an immigrant to the United States. Why did I step off the train in Pasadena? I do not know. Why did I get a job at Caltech that eventually brought me right here where we now are seated? I could have stepped off the train in Albuquerque and never heard of Voyager. I don't know whether you believe in karma. I do not know myself, but sometimes I wonder why did I leave Holland, to end up doing what I did here, because nowhere else, not even if I had lived in Azusa and gotten a job there. Here, Pasadena, was the only place in the world where this would be possible, and I stumbled into it.

So you had no particular intention of going to work for aerospace?

No, none whatsoever. I was just an immigrant who had had it with Europe and its climate and its population density, although now Southern California reminds me an awful lot of western Europe as far as large numbers of people are concerned. But in general, no. I left Holland to go to a better climate, for one thing, and if you go to the United States you may as well go to California. If you watch movies in Europe, they're all in California, and so I left in '62 and ended up here, and for some unexplained reason—we had friends in town, but again, we had friends in Los Angeles as well—we left the train in Pasadena, and a couple of days later I got a job at Caltech.

Doing what?

I ground rocks into petrographic thin sections. I used to fly jet fighters. I was an air force pilot in Holland, and a couple of weeks later I was sitting in a laboratory covering myself with oil, grinding rough little pieces of rock. My only knowledge about geology when I came here was that you could do two things with a rock. You either kicked it or you threw it, and the next thing I knew I was cutting little squares out of them and grinding them. When I came in in the morning, I had to switch my brain off and think of something else while I was grinding rocks flat and glueing them to little glass slides. That was a killer job, and I did it for about a year.

At that time I earned $1.65 an hour. But those were the good old days, when a gallon of super gas cost 21 to 23 cents, depending on the gas war on the street, and so it was okay for awhile, but eventually it got to me. I couldn't handle it anymore. After about a year I took a vacation and was going to look for another job. I ran into a young doctorate at JPL with a crew cut and an inch-wide black tie like we all wore in those days, and he asked me if I was that fighter jock who was in the basement somewhere.

He turned out to be Bruce Murray. He asked, "What are you doing right now?" I said, "I'm going to the division office to see if there's a possible job for me because work downstairs is going downhill fast, and I've had it with that job." He said, "Well, take your vacation and report to my office Monday at 9:00." He had just come out of the Air Force.

I went home and told the wife I've got another job and she asked, "What is it?" and I said "I don't know." She asked, "How much does it pay?" and I said, "I don't know that either, but I'm going to work for a nice guy." And that was the very beginning of what was called, in those days, the Lunar Lab.

Bruce asked me one question: "Have you ever developed film?" Very determinedly I said "Yes"; somewhere in school I had developed one roll of black and white 35-mm. Next thing I knew we were busy with photographs of the moon from the Ranger project. It grew, and I learned to measure diameters of craters on the moon and do slope analysis, which is to determine the angle of the slope of the craters.

From there I stepped from the basement floor into the elevator and went up with everyone else. Surveyors, Lunar Orbiter, Mariner 4 to Mars in 1964—two years after I came to this country. I didn't know a heck of a lot about astronomy, but that was the very beginning.

It was 28 years ago that we went to Mars with Mariner 4 and got 21 photographs. In those days, 1964, the computers were so slow and the rate of transmission and receiving through the Deep Space Network was so slow, and the computer printers were so slow, that it was quicker to color the numbers with crayon and see what the photograph was going to be than to image process the data and get a real photograph out. I think it was 8000 bits per minute. At Jupiter Voyager transmitted 144,000 bits per second.

From then to hardware time for Voyager was just about 10 years. That was a big improvement.

Ten years. The mission that did more than any other to improve the transmission rate of data was Mariner 10 to Mercury. That was where the data bit rate went up tremendously. Mariner 10 had the capability not only of transmitting at 144 kilobits but also, on the ground, could receive the 144 kilobits. That was a huge leap forward from Mariner 4's crayon-colored photographs. But then, Orville and Wilbur Wright didn't start out with a Boeing 747 either.

About the only thing I regret sometimes is that when I go home to Europe and they ask what I'm doing for a living—friends, acquaintances, and so forth—a lot of people do not understand what I'm doing, but even less *why* I'm doing it. It's very hard to make them aware of that. "Why spend all that money?" they ask. I think I've finally found the answer: if you don't understand it, there is no way for me to explain it. If you ask a question like that, I couldn't possibly go into it.

First of all, not one penny has ever been spent in space. Not an ice cream cone or a pack of cigarettes was ever bought on the moon. That money all stays here, on this planet, and it rotates around. People seem to think that if you cancel one shuttle launch you have $450 million in the kitty. They probably think that someone else is going to get that $450 million: "Wow, look what we have left over now. We canceled a shuttle launch; now I can go buy a car and a house."

But if you can't see the value of going by planets and enriching your knowledge about your own neighborhood, why should I go through the effort of explaining it? I could not, even if I wanted to.

What percent of people do you think value and understand what you do?

It differs. Within this job you only get in touch with people who approach you, either for films or for TV productions, that are interested already in the work we're doing. For them it is 100 percent. Outside, though, it would be more like 50-50. An awful lot of people that you talk to, at a party or you talk to by accident, think sympathetically about space business, be it manned or unmanned.

Unmanned is good because it's educational, but space shuttles and astronauts are spectacular. They look good on TV. There's nothing better than to see a guy in a white suit dangling outside of a payload bay with a beautiful golden spacecraft spinning around. Great stuff. Steven Spielberg, eat your heart out. That is what the man in the street thinks. Regardless of how much it costs, what its political implications are, or whether it's too expensive, or there's not enough money in NASA, or whether we should have more or fewer space shuttles, it seems totally irrelevant to the man in the street.

What we have done with unmanned spacecraft and brought on the TV was not spacecraft and astronauts. Not even Voyager itself has ever appeared on TV. It was only the results. In this country that is very well received, even by the man in the street. In Europe you find critics, a lot of critics, even among people in the field. I have had, over dinner, amiable arguments with people from ESA, the European Space Agency. I don't know what it is: maybe it's their nationality, maybe it's their cultural background, maybe there's jealousy or envy involved.

I've been in India with a big photo exhibit of unmanned results, planetary stuff, JPL's 25-year space photography show. It has 150 photographs, the most beautiful we could select, in a really artsy exhibit. In India, with its immense population and level of poverty that is sometimes beyond our imagination, people went bananas over the show! But if you display it in *this* country, for five or six weeks, the curator is delighted with 10,000 people. He says. "My god, it was successful! We had 10,000 visitors!" In the United States, yes. In India, that number of people shuffles through the exhibit in a day. In one day!

We had a show that toured for the State Department for a year, and after the end of the year we allowed the exhibit to stay there. It is dedicated now to the people of India. Millions have seen it, and they ask questions you would not believe.

We were installing the exhibit in the Birla Museum in Calcutta, and I took a lunch break and walked outside to take some photographs. In that ocean of humanity that floated by was an extremely tall man with a lion hairdo and a beard. He was nude except for about five yards of material strategically draped around very specific parts of his anatomy, barefooted, with a staff. I saw him approach. All of a sudden I was face to face with him, on the sidewalk, and, in extremely sophisticated Oxford English, he asked me if I was an American.

Startled by his diction, I affirmed it. He said, "Are you by any chance working for NASA?" I only could nod again. He stood there, grey with dust, and said, "If you're an American with NASA you must be connected with the exhibit that will open in the museum."

I said, "Yes, I am. I'm one of the two." And I asked, "Would you like to see what we're doing?" He expressed appreciation and came along with me. I showed him the photographs, which were already installed, but we hadn't put on the captions yet so I gave him the $64 tour. He asked questions about celestial navigation, physical properties, magnetic fields, magnetospheres, bow shocks, and so forth, and finally thanked me profusely and wandered off.

And of all the personnel around me—the staff of the museum who were working there—I asked, "Who was that man?" Nobody had noticed him. They said, "Oh, he could be the director or the president of the Bank of India."

I said, "Dressed like that?"

Or he could be a shipping magnate. Maybe he was following a Hindu custom, walking the countryside for two years and living off alms. But he spoke English like I never will, and he certainly had a sharp, educated mind. He knew what he was asking.

Let's assume the man was rich and was doing a religious duty, but even the people in the street, the normally dressed people who were poor—much poorer financially than you and I—came through by the hundreds and hundreds. If I made the mistake of answering a question and said, "No, it's not quite like that. Let's go to this photograph and I'll explain to you," and I started to explain, "See, now this is the way, because of ...."

Within a second, 150 to 200 people gathered, and they wouldn't let me go. The questions came, the questions came, from every direction, and I found myself like a big ball, rolling from one photograph through the gallery to the next and explaining it all. I came back totally pooped and started all over again. I couldn't hide myself in a dhoty and look like an Indian, so as long as I stayed within the exhibit, it was repeated, all day long. But it was very gratifying.

Tell me about your routine work.

Since we're not now in an encounter, with one of the spacecraft close to a planet, I'm handling an unbelievable number of telephone calls. They're generally of the same nature; either television production companies or ad agencies. They are all busy doing something very, very, very important, and they all need photographic or videotape material, and it is always right now—"The editor wants this" or "The producer needs that," and "If I give you my Fed Express number, can I get it tomorrow?"

Well, a long time ago I gave up on that and thought, "The heck with you guys." As you can see, this stack of mail on my desk is of considerable height. I'm way behind, but they do not carry a heck of a lot of weight, in my opinion. It's all the same: it's an update of a book, of an encyclopedia, and so forth. That is one part of my life, and it is rather tiresome because it's the same violin playing over and over again, everyday, all day long.

But, to counter balance that, you can never shake the idea that what they're asking for can only be obtained in this laboratory. This is the only place in the world where what they want happened. We are the only ones with photographs of Neptune. Yes, they can go to NASA Headquarters and get the same photos, made from duplicate negatives that are sent to NASA Headquarters, because that normally happens with every encounter. But the initial telephone calls come here, and they come from all over the world.

The other part is the encounters. This laboratory flies unmanned spacecraft for NASA to explore the solar system. That is something you cannot get enough of. When I think about all the encounters I have experienced in one job or another here on-lab, I couldn't imagine prouder moments or more enjoyment, although the pressure to get the stuff out to the world is immense during the week-long encounters. You've been here during encounters; you know what a madhouse it is, but it is a fun madhouse!

The world is here. By the world I mean the press literally representing the world. They are all people who *want* to be here; they're not here because they *have* to cover it. I have been told by some press people that they fight for this type of an assignment. To a certain degree that is a feather in our cap. We must be doing something right, not only for ourselves but for the outside world as well.

I like that. I like that very much, because first of all it is another journey of discovery. You do not know, certainly in the case of Uranus and Neptune, what is awaiting you, because these planets have hardly been seen through ground-based observations. Then you see that planet get closer and closer, and in the case of Neptune it was unbelievable. The beauty was indescribable. Nobody had expected in his or her wildest dreams that a planet like that would be possible, and then a moon that was unbelievable as well, and realizing that that was the last planetary encounter of Voyager—what a way to go out! It was like Chinese New Year. There were fireworks all over the place.

So, yeah, that is the good part of life. That's in a nutshell what we do here in public information. We deal with the press, with their requests—be it for film, videotape, still photographs, slides—literally from all over the world, as you can see from that stack next to you. They're all faxes, of course, because nobody writes a letter anymore, but to me a fax is just a normal letter that arrived without the benefit of a stamp. What bothers me sometimes is they send a fax; five minutes later they call to ask if I received the fax and when can they expect the material to arrive. In most cases I haven't read the thing yet.

What is important is a letter written with pencil on sloppy paper from an eight- or nine-year-old. They draw your attention, because their request for material is generally not for financial gain or reputation or for somebody's problem with a deadline or a production time. School kids are honest in their letters. They ask you straightforward questions, some fairly tough questions. In their innocence they can nail you to the wall. They are always a challenge, and you know that they ask for stuff because they really do need it or have a genuine interest in it. So they can never be refused.

We've got lots of material for educational demands, and generally that covers it more than enough and makes these kids very happy and helps to get them an A on their school projects. We're not here to make 100,000 bucks for some publishing company. Our task, our duty is to inform the world of what we have learned.

What about you and Voyager?

Me and Voyager? Well, me and Voyager got married in 1976, and it's been one of the best marriages that I can think of. Both Voyagers, of course, but you tend to use it in the singular form: Voyager project. Voyager was one of the most exciting, most ambitious projects that one could imagine. Most spacecraft before that had *one* planet and its moons as a goal, and flew by or went into orbit, or landed, as in the case of the Vikings. But with Voyager we knew we were going to at least two planets and their moons and whatever else might be found there, with the possibility of continuing on even farther.

It was something that long ago was a wish, a dream, known as the Grand Tour, which politically ran into all kind of problems and was never allowed officially, and finally turned out to be extremely successful and exciting. I don't think anybody remained unmoved who saw the photographs of the Great Red Spot on Jupiter, and so forth. You don't have to be crazy about astronomy to look at these pictures and be deeply impressed, because we had never seen the Great Red Spot like that before.

Between 1979 and '81 our knowledge of Jupiter and Saturn and their moons and rings was improved many hundreds of times. Photographs that we had taken from Earth, you could tear up and throw away; they were no longer of importance.

But then, when Voyager 2's trajectory enabled us to continue on, to do that Grand Tour of Uranus and Neptune, it was very hard to describe what our feelings were. Its trajectory, which took it by these two planets, had been very carefully calculated. We knew it was possible. We knew how close we could go by these last two planets, which were the most mysterious ones in the solar system. We knew little about Uranus and hardly anything about Neptune.

You were here for these encounters, Dave, and you know what the results were. Who was more excited during those days? Whether it was the press, the staff, or the scientists, I think we were all on a high. So how can you knock that? How can you say this or that was a much more important mission? They were all fantastic in their own right. They taught us a tremendous amount about the planets, and at the same time they made us a heck of a lot smarter in flying spacecraft to far-away goals. And we did that in a grand style. If you take a spacecraft that has been under way for many years and then finally reaches a planet 3 billion miles away from Earth and is only 10 seconds early, somebody somewhere along the line did something very, very right, and there's no way you can feel but proud.

My boss keeps giving me a paycheck every two weeks. I don't dare tell him he doesn't need to do that; I would gladly pay to work here. JPL is a unique situation. It's a place that grows on you because you find the best of people here, people who *want* to work at JPL, not because it's just another job. Everybody has his own reasons, but in general it boils down that those people always had JPL somewhere in their mind, and finally saw an opportunity to work here, in whatever capacity.

I think you will find a deeper and stronger feeling among those who are more or less directly involved with flights. But even people who are not directly involved with the actual project—for instance, guards at the gate, or the people who move your desk from one office to another or pick up the big boxes in the hallway and carry them for you to another department, or the telephone operators—those people are just as essential here as anybody else, although their job doesn't bring them in direct touch. Yet Voyager and the other flights did rub off on them. They shared in the glory. And that makes this a very pleasant place to work. I wish we would have more parking lots, but that's about the only problem I can think of, the only shortcoming at JPL.

Of course these marvelous things didn't just happen. Hundreds of people worked thousands of hours to bring them about, and I suppose there were personal sacrifices, too.

Yes, in my case it cost me a family. You could say, "What a heck of a price to pay." That is relative. I happened to have a wife who was a good lady, of course; otherwise, I wouldn't have married her, long before this all started. But she worked in another part of industry in Pasadena and was not interested in the space business.

While I was riding high about the success and the photographs—and I'm talking about a long time ago—I'm talking about black and white photographs of Mars, be they Mariner 4, 6, or 7 pictures, it doesn't make any difference. It was the first time we saw another planet in all its glory, and you come home and you cannot explain to your wife why you are walking on the ceiling in euphoria. She had her problems in the office; they were mundane compared to what I had just gone through that day. I was floating around like a balloon because of those pictures.

Then you say, "Look, if you don't understand this and I cannot talk to you about it, nor do you understand why I spend so many hours at my work because when these pictures are coming in ...." You don't want to go home. If you even go to the bathroom, you miss 10 or 15 photographs that come down; how can you afford to go home?

So there's nobody to blame for that. It was the time, the opportunity, and the mission kept going on, and you grow farther and farther apart until finally the rubber band somewhere along the line snaps. In this case it was her, and I have to admit that she may have been right, that it must have been a pretty dumb life, on her part, to be married to me. The stress comes along, and you're unsettled, and in in my case it happened in the middle of Viking.

There was a funny twist connected to that. At that time I was running the production end of the photo lab here; I was not in the public affairs office. I did not like a lot of things about the photo lab at the time. At that time it was, to me, not one of the most pleasant places to work. Not the job; I was in charge, I was involved in the color quality control of the Viking lander photographs, the pictures that Viking sent back from the surface of Mars, and nothing could be more interesting than that. But it was upwards, in the direction of management, that I had my problems.

Then one fateful day, early in the morning, she said, "We better separate." That came as a cold shower. By the time it was 11:00 a.m. we'd had several phone calls. There was no more talk about separation but about filing papers for divorce. So in the afternoon I talked to one of my bosses, and I said, "Look, if I act kind of erratic in the next couple of weeks, please bear with me; my wife was just talking divorce."

He said, "Let's go to my office, because I have to talk to you as well." In his office I elaborated a little more on the pending breakup of my marriage, and then he said, "What I wanted to talk to you about is I have to let you go."

[Spontaneous surprised laugh]

Dave, your reaction was exactly like mine. I thought I'd seen a lot of Chevy Chase movies, but I've never seen a movie where a guy's wife screamed divorce and he got fired in the same day. I started to laugh, because it dawned on me that I had literally hit rock bottom. I didn't know anyone else who had run into such a situation. It was more funny than tragic. I almost collapsed with laughter. My boss thought that I had gone over the edge and that I was about to kill him. But at that moment, and still today, I thought it was one of the most humorous situations that I'd ever been in, and of course there was only one way to go, if you're on rock bottom, and that is up.

So I talked to the manager here at the Public Affairs Office. At that time it was Frank Collela, who already had his eyes on me because of my capability with language with the foreign press and my knowledge of the photo lab and my Caltech background. He asked me to start Monday, but I said no. I took a long time off to handle the kids.

My wife left the house. I spent the whole summer with my four kids. I took them to school and had a marvelous time with them. I made the splitup bearable for them, and as a result I still have the kids at home. They're well into their 20s now. I thought that after I got the divorce there would be an era of peace and tranquility, but they never left. They all came back, like boomerangs, so somewhere I must have done something right.

That was the summer of Viking, '76. So encounters with planets are sometimes marked by other events in your personal life as well, and I've heard more of these stories here on-lab. When you get people who are that dedicated to what they are doing—because it deserves to be done, because time dictates you to do it, because if you don't do it *this* year, the planets will not allow you to do it again for *x* number of years—that is another drive. And if you're bit by the bug and you listen to that inner voice, and if your better half does not understand that, you have to make a choice. In many cases the choice is a relatively easy one; an easy one to make, not necessarily an easy one to live. But, it happens.

While we are talking about personal lives, what events in your youth might have prepared you for the Voyager experience?

I lived in the eastern part of Amsterdam. Just east of us, within a three-minute walk, was Zuider Zee Park and the Amsterdam Rhine Canal. Behind that is the Zuider Zee, and there is a little triangular island. That whole part, from the canal off towards the northeast, was all German terrain. You were not allowed to cross the bridges. Germans were guarding them. They had a flying boat base there, and on that little island was a V-1 ski slope, a launch pad. At the far end of the park, they had a pit, a natural depression, which they launched their stuff out of.

As an eight-year-old kid you wanted to know what the heck is going on there. And like a rat, you can rustle your way in through the bushes, no matter how many Germans were guarding that base. We always made it. One day there was a V-2 standing not far from us, 300 or 400 yards away at the most. Activity around it drew our attention: "Aha, something must be going to happen." Then they launched it. It went up, never turned around, came back down tail first, fell over, and exploded. We were too close; the heat wave that went over us singed us, and the sound of the explosion made me deaf for awhile.

You come home with the most innocent of faces, and you see your mom doing this lip speaking bit on you: "What happened?" You do that balancing act: if I tell her where I've been, she's going to kill me, so I better shut up, but you couldn't hear what she was saying, so it was a no-win situation.

V-1s coming off that island were a familiar situation. We couldn't get close to the actual launch site because we were separated by the canal, but they flew directly over us on their climb out. They were maybe 150 feet up. We could see the steam come off the launch ramp, and the little cart fell off. We could see all that so clearly, what happened directly after launch.

There were several launches a day. We heard their throbbing sound. We knew they were flying bombs, but they became so routine that we didn't even look up anymore. The only thing that we reacted to was if the engine stops the thing is crashing. One day we were playing soccer in the park, and we heard one come off the ramp—and then it stopped. Without looking we started running. We even left the ball. We were running for our life.

Then all of a sudden this thing came sliding past us on the ground. It had bellied down on the grass field where we were playing, and there is this swishing sound and smoke coming out of its pipe. We make a 90-degree turn without losing speed and run off in another direction. Then we stop and wonder, "Why am I running? It ain't exploding." So we went over to it, looking at it from closeby. Ten minutes later trucks came with the German soldiers, and we got a swift kick in the tail.

It was fun. That was an exciting time to be a kid.

The V-1s flew over regularly. They were not that exciting. But V-2s went really high, up in a big ballistic curve. One day I saw one explode very high. I don't know whether it had re-entered or not. There were a couple of very short flashes off a vapor trail, and then there was a huge explosion way, way up. I never heard the explosion, the sound of which never reached the ground, but there was this huge flash, then trails of smoke, pieces, plumbing, whatever was left came drifting down. The smoke disappeared long before the stuff hit the ground.

Another time I saw a complete missile way up in the sky come down at a tremendous speed and explode in the fields or meadows south of Amsterdam. It was kilometers away. I assumed they were fired at Antwerp but never made it. By that time, late 1944, the Germans had been pushed way back.

Maybe that's where I got my first taste for planetary exploration, for rocketry, except it was on the wrong side of the fence.

It was an exciting time to be a kid. One day five of us were playing by the railroad and we snuck onto a freight train. Then the train began to move. We saw our whole neighborhood go by, we recognized streets but from an angle we hadn't seen before. It was really exciting, but by that time the train had speeded up so fast we couldn't get off.

We visited all sorts of weird places we had never seen. Whenever the train stopped we jumped off to see if there was food to be stolen somewhere. The one place we knew was the train, so we always got back to it, but it kept going in the wrong direction. After a day or so we no longer heard Dutch; everyone was speaking German.

When we finally left the train a week later we were in Leipzig, East Germany, 350 miles away. I realized later that the train may have been going to the Eastern Front; if we had stayed on it a few days longer we might have ended up in Stalingrad!

We decided that if we were ever going to get back to Amsterdam we had better start walking. So we walked back all the way. We stayed on farms, walked along with German soldiers, hitched rides on German trucks, as long as we were going west. We finally made it back, in March '44. We had left in August '43.

Everybody went crazy that we had returned safely, but it was not a big deal for us kids. I was worried about a possible beating I would get for staying away that long.

Then the police came. Dutch law was maintained under German occupation, so there was Dutch police, and higher up there was German police. Everybody had been concerned because five young kids in one block disappeared in a puff. One moment we were there and then all of a sudden we were gone. No witnesses, nobody to come home and say, "They stepped on the train." Of course, we had made sure that nobody was watching us, because that train was in the German territory.

So the Dutch police had been looking for us, and wanted to know what had happened. The Germans came also, to try to figure it out. They were not happy that we had walked all over Germany without papers, without ever having been stopped or checked. It showed that their security system was leaky as a sieve.

From that experience as a kid I always have this feeling in the back of my mind that there's nothing in this world, if I want to have it, that I cannot steal or finagle or organize it. That train experience happened at the right age. You develop the basic instincts of a rat. It all depends on the times, and those were remarkable times.

What role did you play during Voyager's encounters?

In an encounter I was the link between the Imaging Team and the image processing lab and the photo lab. I generally run about 100 to 120 hours a week, and you see your bed for a couple of hours every other night if you're lucky. I say, "if you're lucky," but that's not the truth either because you do it yourself that way because you don't want to miss out on it. You don't want to drive home and go to bed and sleep when all these glorious things, when all that history is happening.

But also I have to see that everything keeps moving and the production keeps going, that the colors are right, that the images are cropped correctly, and that the captions get written and reproduced, and that the photographs and the captions do get together. We hand them out to the press, and then it starts all over again the next day. We go home after an encounter week with baggy eyes and totally drained physically and mentally, but God, are we happy!

Nobody ever asked me to work 100 hours a week, in whatever form: as a request or as an order. That's nonsense; you cannot ask, and it wasn't done. But there are tasks that have to be done, and I was the only one who could do some of them. I coordinated the flow of data between the Imaging Team, image processing lab where the data got computer processed, and the photo lab. I'm working now in the Public Affairs Office, but I know the capabilities of the photo lab because at one time I was one of the supervisors there, so I knew the people; I knew their capabilities, so it was a matter of coordinating.

At the same time, I knew what was required of every product that comes each step down the way to the photo lab and even in the photo lab, so I also had control. I signed off on the right color balance. I signed off that this photo, out of a whole series of photographs of the same subject, was the one that would be press released the following day; the Imaging Team had that trust in me.

I started to feel rather uncomfortable. The approval procedure had gradually changed. It had started years before as an everyday meeting with scientists of the Imaging Team and all these other people: photo lab and image processing lab and public information. Each afternoon they decided what was going to be shown and talked about the following day at the press briefing, and the products to support that. During the night when the photographs started to come out of the photo lab as a finished product, and the captions had to be written, there was always an Imaging Team member or some other scientist present.

But that, over the years, became less and less the custom, until I finally started finding myself signing off on things, and so one day I stepped up to Brad Smith, the Imaging Team leader, and said, "Brad, I'm feeling rather uncomfortable about having the responsibility of determining what is the right product for tomorrow." He said, "Look, if I knew anybody on the team who could do a better job than you, that person would be there, so shut up."

To me that was like getting a medal; in fact it was a lot better than getting a medal, because of the faith that somebody you greatly respect has in you. Since you are regarded with their complete trust, you do the task, and if that means a 100 or 120 hours a week, you do it—no questions asked. I would do it again; except when you get older, it don't come easier.

And it didn't always go smoothly. There were snags along the way. One happened on Black Thursday, in the week of the very first encounter with Jupiter. Black Thursday was when we lost people. It's normal for computers to crash and processors to go down or chemicals get spoiled because of a mistake under the pressure. But none of that happened. It was just people that got wiped out.

There are always totally unexpected things that can happen in spite of the fact that everything has already been planned years before. In our case the planets were predetermined, probably billions of years before, and when you finally reach the point where you are about to encounter, with a new team, a planet that you've never been to before, you know exactly down to one-tenth of a second when the photographs and data will start coming in, and where your functions start. There's the Imaging Team, the computer people who handle the data, the photo lab, the Public Information Office—and yet in spite of all that organization and timeline stuff, there are still things that happen that are unexpected. They're based on inexperience, and in this case they affected the photo lab.

The photo lab people have to produce all those things to be handed out to the press. When you do that for the first time, you don't know each other's capabilities, nor your own.

Black Thursday was the day when not only were we at the closest point to the planet Jupiter itself, but we discovered also that Io had volcanoes, and so the scientists were going bananas because of all the goodies that came down during that day and the night before. It was decided that the following day we were going to release 29 individual photographs times 600 prints, because there were would be 600 people in the auditorium, representing press organizations from around the world.

Well, that may have sounded like a reasonable number—29 times 600 8 × 10s plus all the slides, plus all the viewgraphs, plus all the stuff that had to be projected onto the wall—but there were only so many people in the photo lab. Yet they did it; they did it, but at great expense to the following press briefings, because I have seen people laying in the hallways of the photo lab sound asleep. They literally worked until they fell at their posts, but the world never knew about it. The press got their photographs, and the newspapers and magazines and television news were full of the prints. The Lab looked like a million roses or a million bucks, and in the photo lab they were oblivious to it because they were snoring in the hallways.

At the following press briefing 24 hours later, we had only four press releases; not that there wasn't much more to be shown to the world, but we simply couldn't produce them because we did not have the people.

The press does appreciate us. At encounters they come here from all over the world: the TV networks and newspapers and magazines and wire services. I think, of all the press media, the wire services are the hardest nosed business, for the reason they are guys who are not under a simple deadline: every second of their life is a deadline. If they get something, they move it on the wires *now*. It may still make the evening press in Europe, and if you get your news on the wire in time, you may make the early morning press in the Far East, so the wires are the most hurried of all the press units.

My rules were always, when the photographs come out of the photo lab, go to my photo trailer. You have to wait until the press briefing is over and the question and answer session starts. That is the moment when I handed out the envelope with the photographs to the wire services, because they run with it. They instantly disappear to their bureaus to get those things out into the world. The rest of the media wait until the press briefing is over, and then the newspaper people, the TV guys get it first, and in the second wave the magazines, who are not on a tight deadline. Those were ironclad rules. There was no bending those rules. That's the way my world ran, and I stuck to that.

Last August, when we went by Neptune, I got some bottles of wine and thank you cards from the wire people. That's the first time I've heard that ever happened. So somewhere along they must have gotten an extremely fair shake at JPL. They got rather sentimental, realizing that it was over. This was the last encounter. There were no more new planets that we would see. That was basically the way that party happened in von Kármán auditorium, when we turned around and had a buffet set up for the press, thanking them for carrying our word into the world. It was rather emotional.

What about your relationship with scientists on the project?

The planetary community is still a relatively small group of scientists worldwide. It was much smaller in the 1960s and 1970s; you could almost count the planetary scientists on the fingers of your hands. I knew many of them when they were students. When you have known them that long, it is not an ordinary relationship. We don't see each other often, but you have a very special feeling for them because you helped them with the photographs for their thesis, sat in when they defended their thesis to give them moral support, and socialized with them on weekends. I knew their fathers and in some cases their mothers.

Larry Soderblom is an example. There's more to him than just being deputy team leader of the Imaging Team, much more. I still remember Larry coming in as a new graduate student in Dr. Murray's group. I barbecued hamburgers with him in the backyard. I knew his two huge red Irish setters, Cayenne and Fuego, who long since have gone to the happy hunting grounds, where they're chasing those 60-pound jackrabbits in the sky.

We all had kids, and changing diapers is one thing I became very proficient at. Whether its yours or anyone else's, it doesn't make any difference; changing a diaper is changing a diaper on a little squirmy kid. Then he disappears out of your life, and the next time you see him he's a six-feet-four giant, and I ask, "Did I change *that* thing's diaper?"

That's part of JPL and Caltech. You feel protective towards these people, and they are always slightly different from the other scientists who are members of a team because you did not know them as students. That makes a difference. It is a special relationship. You know each other's capabilities, you know what you can count on.

That is very, very important. In a case like that, no words need to be said. You'd go through hell for these people because they are more than just friends, and because of what you are doing together. It's hard to explain that to an outsider if you haven't lived it. There were thousands of people here during these encounters, but their tasks were different from ours. So you get these old relationships that date back to the 1960s through student times.

And there are the newer generations, like Rich Terrile, one of the Saturn ring scientists—and your daughter; I was so glad to see her working here last summer.

So it is not just taking a spacecraft and going by a planet and sending back pictures; it's much more than that. It's very deep feelings and personal relationships. Admiration, though it is not a mutual admiration society; it's simply people working together and working together very well and having a great goal.

Yesterday in Washington all the project managers were there for the presentation of the solar system mosaic: Ray Heacock, who took the birds by Jupiter, Ek Davis who took both the Voyagers by Saturn, then Dick Laeser by Uranus, and Norm Haynes by Neptune, and George Textor who now is the project manager and takes the spacecraft out of the solar system.

I had five envelopes with the mosaic photos under my arm, and after the briefing was over I took one envelope out and said, "Captain Ray, it's been a privilege serving on your ship, sir." "Captain George, you better get those things out of the solar system on the right course." I went by all of them, because they were literally captains of a ship. They were in command of the Voyagers. They had the ultimate responsibility, and they had their officers—the principal investigators, the guys responsible for the various science instruments—onboard below them, just like a captain of a ship. Being part of that team taking it by the planets is like indeed being on a ship of discovery. So I gave them their envelope with the solar system photographs.

Tell me about those photos.

When you are really involved, you know exactly what the spacecraft is going to do during an encounter. People like you who come in for the encounter, and the press people, are given a time line. The time line is like a timetable or a departure schedule of an airline, where you can see exactly what happens on what hour of the day. It lists the actions that the spacecraft has to perform at given times on given days, particularly during the approach to a planet and in the subsequent encounter. In that week of the close flyby, the time line is narrowed down to one-tenth of a second.

Well, I have that in my head. When you work with a spacecraft like Voyager for well over 12 years, the timeline gets imprinted in your brain. So, by the time we go into an encounter, I can wake up at 3:00 in the morning—if I do get an opportunity to sleep—and look at my watch and I know that Voyager is taking a sequence of 18 photographs at that moment.

I don't know how I do that or what makes me do that; I gave up a long time ago trying to figure out what it is, but you have that whole imaging sequence somewhere in your noodle, and you know it 24 hours a day when you're in the encounter, so it is easy to visualize in the mind's eye how Voyager would turn around and look back at our sun and where these planets are. You can "see" that, and you know it would look great. Finally everything falls into place and somebody says, "Okay," maybe just to get rid of us, to get us off his back, "Do it"; and the permission was given.

Wasn't this part of the mission plan from the beginning?

No, it wasn't. But all that time, all those years, Carl Sagan had talked about it. Already when we were just past Jupiter, he wanted to look back at the Earth. Of course that was shot down because you're not going to jeopardize the optical system, run the risk of having it burn out before you get to the secondary target, which was Saturn. So that was never allowed.

Then after Saturn he tried again: "We *have* to look back at our sun and our planets." That didn't happen either, because by now Uranus and then Neptune became possible. So it was not until after the whole solar system was behind us that they said, "Okay, there's nothing to do anymore with the cameras," that the decision was made to do that.

But in order to get that decision made and the permission to go ahead and do it, an awful lot of convincing had to happen, because very few people really believed in it—you could count them on the fingers of both hands and still have fingers left over.

It took some convincing to get permission?

Yeah, because you have to send commands to the spacecraft, you have to plan where the camera has to aim, in what sequence do the photographs have to fall in order to get all planets in the solar system that are visible. That requires people to make these commands possible; to do the computer language. Then you need the Deep Space Network to send these commands to that spacecraft. Then the Deep Space Network is needed again, to receive the return data. There is a lot of work involved.

I was not part of that chain, or doing the convincing, or even begging on my knees or kissing somebody's ring to make that possible. Candy Hansen was much more involved than I was. But I knew all along that we had to do it.

Voyager 1 took the photographs on February 14, 1990. Thirty-nine wide-angle shots and seven times three narrow-angle shots, or telephoto shots, of the whole solar system, starting at Neptune and walking the camera footprint inward towards the inner solar system: over Uranus, Saturn, Jupiter, and so forth, and then with a loop around, covering the sun, and all the little guys around it, which are us: Venus, Earth, Mercury, Mars.

They were stored onboard Voyager on the tape recorder, for the simple reason that the Deep Space Network—those big antennas in Australia, Madrid, and Goldstone, California—were too busy with Magellan and Galileo to listen to poor old Voyager. At the end of March, we played that sequence of photographs back, and the last series of photographs, five or six, were being received by the Madrid antenna when it rained. That blocked out the data, so in April we had to give it another shot. This time the antenna in use was the one at Goldstone, which turned out to have a hardware problem. As a result we still did not get those six photographs, so we had to do it again in May, when we finally got them.

Now we had all the photographs down on Earth, and then followed an intense period of computer enhancement, because if you are almost 4 billion miles away and you look back at the solar system, even with 1500-mm telephoto lens, the planets are not very big. The largest planet, in the photographs and in reality, is Jupiter. It is almost four pixels across. The inner planets are a totally different ball game. Earth and Venus were about 37 microns, an awfully small dot, smaller than a pixel. They had to be enhanced in the computer to make them a little more visible than they are in real life.

The ones that fell through the cracks were, starting from the outside and going in, Pluto, which was literally all the way over in left field and also is extremely small, so we didn't even attempt to photograph it. Another one that got lost was Mercury; it is very small and is so close to the sun that it disappeared in the glare of the sun. But Mars also was not visible, and there was a reason for that, too.

When the photographs were taken, Mars, in its orbit around the sun, was almost directly between the sun and the spacecraft. That, combined with the fact that it is only one-half the size of Earth, makes for a very small dot, so poor Mars disappeared in the shuffle as well. But there are Earth, Venus, Jupiter, Saturn, Uranus, and Neptune, and of course our sun is clearly visible.

That is our neighborhood, and that was the first time—hopefully not the last time, but certainly the last time in *our* lifetimes—that a manmade object was in a position to take a photograph from that vantage point. So although the planets won't dazzle you with their beauty, as most planets do when you are close by, historically it's a mosaic that will blow your socks off. My God, we are small!

There's another effect added to that. You get refraction of light, scattering off irregularities within the optical train of the camera—like off the edge of the lens, the lens tube. So pictures showing spikes apparently radiating from the sun in all directions; that's not true. The sun is an evenly distributed source of light. But photographing it with all the objects in the way, you get a lot of scattering, which makes the sun look more like we *think* the sun ought to look, with beams going out of it, but that's not correct. But it happened, and there was nothing we could do to avoid these wide beams, scattering in every direction.

As it happened, by luck, Earth is in one of these "sun beams," if you want to call them that, which makes for a beautiful piece of art, with the dark background and the golden orange beam going straight across the picture. Within that beam, that tiny light dot is us, the planet that is us. People who are religiously inclined would say, "See, we are blessed." I would say, "Is the IRS living on this tiny dot?"

If you would give that negative to a photo lab to be printed, without telling what it was about, a dedicated photographer would print it and then touch that "dust spot" away, not realizing that that spot which looks like a stray speck of dust on the negative is actually us: the whole planet, with you, me, and everybody else on it.

I always had that solar system picture in my mind. My mind's eye could see it. I knew somewhere, not exactly what day, even what year, but there was going to be a point in time where Voyager could turn around and look back and see the whole solar system; our neighborhood. And when it became a reality that in fact it could be physically done, I already knew what it was going to look like. I didn't know precisely the positions of the planets in their orbits, or how brilliant they were going to be, but I knew in general what it was going to look like.

When finally all the data was computer enhanced and put together, it was impressive! It looked better than I had envisioned it would be. In these 60 individual photographs that were transmitted back, there is the whole mosaic, the whole solar system. But that one piece of information, that photograph with the golden sunbeam with the tiny speck that is us—that did it! That was worth all the effort. That was far prettier than I had expected.

If you look at the full globe photographs of Earth that were done stunningly by the Apollo astronauts—Earth is glorious. You see all the cloud banks, the oceans and continents, and you can still see where everybody lives. You can see Africa, you can see Europe. The big land masses are all definitely recognizable. You can almost find where Gorbachev lives or where George Bush is playing golf. But when you see it from the edge of the solar system, the way we shot it, Earth is only part of an entourage around a relatively small star.

Look at it another way. Here you see a star from 4 billion miles away, with a full arrangement of planets. We live there; it's our neighborhood. But what kind of a chance do we have to look at other stars and find the planets around them? We're only 4 billion miles away, and look how small our planets are.

So that is the story of this final picture. Some people call it the picture of the century. I do not know whether that would be the right term for it, because many great photographs were shot in this century, and this was the first time for mankind, so how can we limit it to a century? The Solar System Mosaic is the official title for it; Family Portrait is for insiders. Either way it is fantastic—and it really is the last picture.

Jurrie, the Dutch were great explorers and wonderful artists. You've combined both of these in the Solar System Mosaic.

CONCLUSIONS

CONCLUSIONS

The mission was far from flawless. A series of equipment failures, some of them significant, began soon after launch. In 1980, before the second Saturn encounter, mission controllers estimated that Voyage 2 had a 70 percent chance of remaining functional at Uranus, and only a 30 percent chance at Neptune.*

How did Voyager beat the odds? Probably because the spacecraft was more reliable than the statisticians had forecast, but there was also more to the project than engineering and science. A number of other factors were present.

Clearly Defined Goals

Goals were formulated in great detail, years in advance, in innumerable discussions between the many individuals and groups involved. For example, the detailed time line for Uranus indicated that on January 24, 1986, at 9:21 a.m. Pacific Standard Time, Voyager 2 would take four pictures of Ariel. Between 9:30 and 9:42, the spacecraft would roll −65 deg from Canopus to execute a plasma-wave observation between 9:42 and 9:44 for detection of a possible plasma sheet near the equatorial plane of Uranus. And from 9:43 to 9:51, a photopolarimeter observation of Miranda would study variations in surface brightness to determine structure and composition.†

Concern with Reality

Achieving these goals required decisions based on reality rather than on wishful thinking. Goals could be imaginative, even daring, but ultimately had to lie within immutable limits imposed by resources actually available and by the laws of physics. For example, the spacecraft could only carry so much weight (258 pounds of instruments) and would make only one pass by the planet (at 61,148 mph at Neptune). Every ounce, every second was precious.

System Perspective

These conditions forced Voyager personnel to view the project as an interrelated whole. They had to take each other's functions into account, because of the inescapable constraints dictated by the weight of the spacecraft and by the time available during encounters. Each ounce added by one instrument either would have to reduce the weight margin allocation the designers had set aside or could mean less weight was available for the others. Each second obtaining data for one science team, as the craft sped past the planet, reduced the time available for the other 10 teams.

Criteria for Evaluating Achievement

Arrivals of the spacecraft at the selected planets were obvious accomplishments, as were the spectacular images of these distant worlds. The quality and quantity of science data provided additional criteria for evaluating success. Goals were defined in such detail that success could be stated in numerical terms. Virtually all the tens of thousands of observations planned were actually received, so despite the loss of some data from the scan platform problem at Saturn, the success rate for the Voyager project was very close to 100 percent.

---

* Richard Berry, "Voyager Science at Saturn," *Astronomy*, Feb. 1981, p. 22.
† Uranus Encounter Time Line, pp. 17, 18.

Open Internal Communication

Questions, criticism, and suggestions from even the youngest rank-and-file workers were taken seriously. Review sessions were built into the planning process, and crucial decisions were mercilessly scrutinized. Nobody wanted a spectacular public failure for which they were in any way responsible.

Competent, Dedicated Personnel

Competent personnel were recruited and nurtured. Challenges were met, problems were solved or circumvented by the diligence and ingenuity of the technical staff, most of whom are known only to their immediate colleagues and co-workers. Everyone I interviewed had a sense of their involvement with the project. Even the people at the lowest levels responded enthusiastically when I mentioned Voyager.

Continuity

Voyager was a model of a long-duration project, demonstrating what could be done by maintaining an effective organization, a work group, over many years. In addition to building on their observations from the past, Voyager veterans praised the cohesion, camaraderie, and teamwork that evolved and contrasted it with their experiences on other projects, more typical and shorter lived.

These attributes are not unique to Voyager; they also may be found in other successful organizations. Nonetheless Voyager embodied them to an unusually high degree and therefore could serve as a model that other organizations might strive to emulate, or as a yardstick with which to assess their own performance.

We must acknowledge that luck played a part: the timing was right. The Grand Tour depended on a crucial element not present in most technological triumphs. It was enabled by an external process over which humans have no control: the inexorable motions of the planets in their orbits, unlocking the easy path to the outer solar system only once every two centuries. If we had missed that three-year launch opportunity that opened in the late 1970s, the journey would have become far more difficult.

While it is unwise to underestimate the rate of future technological advance, the planets are rapidly diverging from their optimal path, vastly increasing the distances between them and complicating the trajectories. It is possible that, had we missed the recent opportunity, a grand tour by a single spacecraft would not have been feasible until the middle of the 22nd century. But by that time the tour probably would not be needed, because the remaining outer planets will have long since been visited by single planet missions already designed, obviating the prime reason for a grand tour.

So the Voyagers' tour may be even more than a once-in-a-lifetime event. Unless our descendants in future centuries retrace the Voyagers' routes for sentimental, historical reasons, this may well be the only Grand Tour ever.

What will be the Voyagers' long-term significance? Barely 20 years after launch, they have already faded from the public spotlight and are not even mentioned on the large scoreboard in JPL's courtyard. I think again of the question posed by *Astronomy*'s editor, Richard Berry, during the Voyager 2 encounter with Saturn: "What's happening here?"

My thoughts go back to the excitement and record heat of Southern California in the summer of 1981. He asked the question while we were sitting in the press room at JPL two evenings after Voyager 2's closest encounter with Saturn. The room was almost empty; the feverish activity of the preceding days had died down.

CONCLUSIONS

We watched the closed-circuit television screen count off the hours, minutes, and seconds until the closest approach to the ringed planet that night. Hundreds of cars had clogged the parking lot and the street leading to JPL as people waited expectantly, watching the digital clock and the succession of photographs—waiting for what, precisely, they did not entirely know. It resembled New Year's Eve: a moment in time designated as important, although it was actually indistinguishable in most ways from those that preceded and those that followed.

Yet we gave it meaning.

The long-awaited moment of closest encounter was an instant that fell not at an even hour, such as midnight or noon, but rather at the messier moment of 8:25 p.m. Pacific Daylight Saving Time, Tuesday, August 25, 1981.

It would be an hour and a half before actual word from Voyager could reach us across the billion miles of space to confirm that it had indeed arrived at the point of closest approach. The message would be further delayed because that point was hidden behind the planet, so the radio messages could not even be started on their way until after Voyager had emerged from the occultation.

What was happening here? Thousands of people had focused their interest on an inanimate object more than a billion miles away, an object they would never see again—neither they nor any human, nor probably any other sentient beings in our family of planets. Only a fraction of them had actually seen the small craft before it left Earth four years earlier, so they had not had an opportunity to develop a first-hand awareness of this ungainly looking Voyager.

Yet there they were, hundreds of technicians, scientists, reporters, secretaries, and security personnel gathered at JPL. A thousand press passes had been distributed to media representatives who ranged from reporters for obscure journals in obscure towns to the famous faces seen nightly on network television news.

A few miles away, in the Pasadena municipal auditorium, thousands of others attended the Planet Fest, saw exhibits, and heard reports on the past, present, and future of space exploration. Across the nation and around the rest of the world, millions more followed Voyager in the mass media.

Why were they interested?

Voyager did not carry people or creatures of another species, like ET, with whom we could identify and interact. Voyager is not alive and does not even look like a living being. It lacks the anthropomorphic appeal of R2D2 and other personable humanoid machines. It is an inanimate object, and a spindly one at that, without the smooth aerodynamic features found in fictional and real spacecraft from Flash Gordon to the space shuttle.

Voyager, in fact, lacks intrinsic value of practically any sort. It is a conglomeration of rods and gears and computers, of plastic, metal, and glass. There is little about it that would bring wealth to a celestial scavenger interested in precious metals, little to bring its finder riches or power. In centuries past, merchants had been concerned about the fate of distant craft that they hoped would bring back wealth such as spices and silks from faraway Cathay and Cipangu—but this Voyager will never return.

Instead, it serves other functions. The most obvious is that it sends us information about new worlds and wonders never before seen at close range.

But there is another, less obvious aspect of Voyager. Voyager is, in some ways, a totem. A totem is an object, animate or inanimate, that is regarded by a group as having a special relationship to that group. Primitive peoples attributed extraordinary powers to their totem and believed that it would assist them. While today we are more sophisticated, there are several totemic aspects of Voyager.

The totem can be a central element in the social organization of a group, providing a focal point around which people can gather. The totems of primitive religions were an important factor in developing cohesion within early societies. In our time, Voyager, particularly at the planetary encounters, was a center, a common ground around which many groups and thousands of people gathered.

Some of these groups have lasted for years—the Voyager science teams, for example. Others, like the crowds at JPL and the Pasadena Planet Fest, were momentary. Regardless of the duration of the group, whether long lasting or short lived, people were brought together by their shared interest in Voyager.

Their interest is reminiscent of the veneration primitive people had for their totems. We do not worship Voyager, but we do have positive feelings about it and concern for its well-being. This was particularly evident in the apprehension felt as Voyager flew behind Saturn out of radio contact for several hours, in the relief when it emerged on the other side, and in our subsequent dismay when we learned that something had happened to the scan platform.

Why were people concerned about Voyager?

Voyager affected people vocationally and recreationally. For some people, Voyager provided jobs, and data from it has affected careers. Many others with no occupational connection were fascinated by it and by the images it sent to us from distant worlds.

Voyager also had some less favorable consequences, as people devoted so much time to it that they neglected their private lives. And the loss of data from the jammed scan platform meant that some planned reports could not be published.

We will never know all the effects of Voyager, but the point is that Voyager did have an impact on people while it was on Earth, and still does, now that it is billions of miles away and hurtling ever farther into space. Its impact is due not to any magical power but rather to the fact that it was involved in some way with people's lives.

Another totemic element of Voyager is that it stands for the positive aspects of our civilization, particularly our science and technology.

While we devote billions of dollars to weapons, Voyager flies on a mission of peace.

Instead of polluting the environment and consuming Earth's limited resources, Voyager glides through space using the gravitational forces of other planets.

In a world rife with greed and sensual desire, Voyager provides intellectual challenges by sending us fascinating data about distant planets.

In a world full of deceit, corruption, and trickery, it breaks no campaign promises; it makes no surreptitious political deals. Few heroes are evident today, and Voyager particularly compensates for this dearth.

Voyager represents some of the finest aspects of our civilization, with admirable qualities too often missing in prominent humans.

Voyager has powers that not long ago would have seemed like magic. One of these is remote viewing—the ability to look beyond the normal limits of human vision, to see events in distant places. This was a cherished dream of primitive sorcerers and persists as a fact of present-day parapsychology. Yet the reality of Voyager surpasses the wildest hopes of past magicians and present mystics. Voyager sends us closeup pictures of objects billions of miles away, of rings and worlds and other wonders whose very existence was unknown to the ancients—or even to scientists a few years ago.

A final totemic aspect of Voyager is immortality—immortality for the totem itself and, in a sense, for us, too. Voyager will not die. It is exempt from the degenerative aging processes of living matter, and it is also spared the corrosive effects of planetary atmospheres. All terrestrial creatures eventually expire, and even inanimate monuments as vast as the pyramids of Egypt slowly weather into dust.

Outer space is free of these hazards, although it does have others, including radiation and space debris. Apart from a collision, however, there is little to diminish Voyager's life expectancy of a million years or more.

CONCLUSIONS

In addition to an extremely long life for itself, there is immortality in another sense. Voyager may bring us immortality, not for individuals or even for our species, but for our existence and accomplishments on Earth.

The spacecraft will leave the solar system to travel perhaps for eons in interstellar space. If intelligent beings eventually find it, they may wonder where it came from and who made it. In this respect Voyager may be viewed as an extension of ourselves, carrying with it our hopes for immortality, a wish that our labors and our existence will someday be recognized, perhaps long after we and the Earth have vanished.

We do not relate to Voyager in the same naive manner as our ancestors would, but our actions retain traces of ancient social processes. This is not surprising; in spite of the changes wrought during the intervening millennia, we still face some of the fundamental problems that confronted early hominids. Although our dangers take a different form, they are very real.

Now, in the space age, we may be experiencing some of the same forces that impelled our remote ancestors to leave their familiar habitats and venture into the unknown. Voyager encounters, while celebrating a marvel of modern technology, also included elements going back thousands of years into the past.

Although participants were aware that these were important events, few realized the full significance of our distant technological totems. Civilizations are judged by their capacity to transcend mundane human problems. On the eve of the 21st century, our planet is threatened by an exploding population and serious environmental degradation. The Voyagers offer inspiring reminders that we can respond effectively to challenges.

Have we really reached the Red Limit, as Bruce Murray fears—the boundary beyond which we never will be able to explore, because distant galaxies are receding from us at speeds approaching the velocity of light? Perhaps we should settle for Moustafa Chahine's question: "What's next?"

APPENDICES

APPENDIX A
Voyager Chronology

Grand Tour alignment discovered		1965
Project officially begins		July 1, 1972
Launch	Voyager 2	Aug. 20, 1977
	Voyager 1	Sept. 5, 1977
Jupiter encounter	Voyager 1	Mar. 5, 1979
	Voyager 2	July 9, 1979
Saturn encounter	Voyager 1	Nov. 13, 1980
	Voyager 2	Aug. 26, 1981
Uranus encounter	Voyager 2	Jan. 24, 1986
Neptune encounter	Voyager 2	Aug. 25, 1989
Solar system mosaic	Voyager 1	Feb. 14, 1990
Heliopause		2012 AD
Communication ends		2017 AD
Reach Oort Cloud		26,000 AD
First time a Voyager is closer to a star (Ross 248) other than our sun		40,176 AD
Closest approach to Sirius		296,036 AD
Past 12 nearest stars		1,000,000 AD

APPENDIX B

Voyager Phases Discussed

Phase	Arden Albee	Charles Avis	Jack Blanco	Roger Bourke	John Brickett	John Casani	Robert Cesarone	Moustafa Chahine	Henry Cox	Esker Davis	Bobbie Fishman	Gary Flandro	Deborah Gag	Thomas Gavin	Marty Greer	Candice Hansen	Norman Haynes	Raymond Heacock	Charles Kohlbase	Richard Laeser	Peter Lyman	Howard Marderness	Judith McGavin	Ellis Miner	Bruce Murray	Robert Parks	William Pickering	Robert Post	Bud Schumeier	William Shipley	Anita Sohus	Glen Southworth	Edward Stone	Catherine Swift	George Textor	Jurrie van der Woude	Peter Wells
History — Before July 1, 1972	X	X		X	X	X	X	X		X		X		X			X	X	X	X	X	X		X	X	X	X	X	X	X		X	X		X	X	
Development / Project start through launch — July 1, 1972–Sept. 5, 1977	X	X			X	X		X		X		X		X			X	X	X	X	X	X			X	X		X	X	X	X	X	X				
Postlaunch cruise through Jupiter encounter — Sept. 6, 1977–Aug. 8, 1979	X	X				X		X		X		X		X			X	X	X	X	X	X	X	X	X	X	X	X	X	X	X	X	X		X	X	
Post-Jupiter cruise through Saturn encounter — Aug. 8, 1979–Sept. 25, 1981	X	X				X					X	X				X	X	X	X	X	X	X		X				X			X		X		X		
Post-Saturn cruise through Uranus encounter — Sept. 26, 1981–Feb. 24, 1986	X							X	X	X	X	X				X	X	X	X	X	X	X		X				X					X		X	X	
Post-Uranus cruise through Neptune encounter — Feb. 25, 1986–Sept. 25, 1989	X	X						X	X	X	X	X	X			X	X	X	X	X	X	X	X	X				X					X	X	X	X	X
Post-Neptune interstellar phase — After Sept. 26, 1989							X	X	X							X	X											X					X		X		

419

APPENDIX C
Science Investigations at Saturn

Cosmic Ray Particles
Purpose: To measure the distribution, composition, and flow of high energy trapped nuclei and to examine energetic electron spectra

Imaging Science
Purpose: To measure atmospheric dynamics, to determine geologic structures of satellites, to search for new rings and satellites, and to determine the structure and properties for rings

Infrared Radiation
Purpose: To determine atmospheric composition, thermal structure, and dynamics of Saturn and Titan

Low Energy Charged Particles
Purpose: To measure distribution, composition, and flow of energetic ions and electrons and to examine satellite–energetic particle interactions

Magnetic Fields
Purpose: To measure magnetic fields of Saturn and Titan and magnetospheric structure and interactions with Titan

Photopolarimetry
Purpose: To measure brightness and polarization of light from Saturn and Titan and to conduct occultation measurement of small-scale ring structure

Planetary Radio Astronomy
Purpose: To determine polarization and spectra of radio frequency emissions, to measure plasma densities, and to determine the rotation period of Saturn

Plasma Particles
Purpose: To measure magnetospheric ion and electron distribution, solar wind interaction with Saturn, and magnetospheric interaction with Titan

Plasma Waves
Purpose: To measure plasma electron densities, wave–particle interactions, and low-frequency wave emission

Radio Science
Purpose: To measure Saturn's atmospheric and ionospheric structure and composition, to determine surface temperature and pressure on Titan, and to measure satellite mass

Ultraviolet Spectroscopy
Purpose: To measure the upper atmospheric composition and structure of Saturn and Titan, to examine auroral processes, and to determine distribution of ions and neutral atoms

APPENDIX D
JPL Grand Tour Memo 1966

Interoffice Memorandum
312.5–201
October 10, 1966

TO: T. A. Barber/P. N. Haurlan
FROM: R. D. Bourke/G. A. Flandro
SUBJECT: Comments on proposed grand tour mission study.

In view of the opportunity in 1977 to fly a single spacecraft to Jupiter, Saturn, Uranus, and Neptune some comments pertinent to a possible mission study are in order.

1. Scientific requirements for close up investigation of the planets beyond Jupiter have not been generated in house. This may be due to a general feeling that we could never get a spacecraft to that part of the solar system rather than a lack of scientific interest in the outer planets. IIT Research has recently generated some science objectives in this area.

2. Guidance requirements for this mission are a major question, but initial work by ITTRI indicates the problems may not be as severe as they at first appear. Based on the Earth-Venus-Mercury experience, it would seem likely that at least two maneuvers would be required between each planet with the possible exception of the Uranus-Neptune leg where one might be sufficient. Thus a minimum of 7 maneuvers is probably required. The related area of navigation using Earth-based radio tracking with or without onboard measurement requires attention.

3. Communication and solar distance are extreme. Neptune encounter occurs at 30 AU from the sun so that solar illumination is down by a factor of 900 from that at the Earth with associated attitude and thermal control implications. Communication space loss is roughly 30 db worse than at one AU. One way light times to the Earth is about 4 hours which may have a marked effect on encounter operations. Completely automatic encounter sequences triggered by planet acquisition will probably be required.

4. Flight times are also extreme. A reasonable launch window for a C_3 of 130 km²/sec² leads to a flight time of 8.4 years, and for 100 km²/sec² (about the minimum launch energy in any opportunity) flight time is 12 years. Reliability implications are obvious. In addition, consumables (attitude control gas, nuclear fuel, midcourse fuel) may dictate the size of the spacecraft. Continuous ground operations are almost out of the question; consideration of hibernating spacecraft definitely in order. (It is interesting to speculate on what can happen during the course of a

mission of this duration. The government's administration will turn over at least once and the Congress several times. The space budget and interest could do anything. The Laboratory's whole direction may change: we could launch while devoted to space exploration and finish while interested in oceanography. At the end of the mission, it's unlikely that any significant fraction of the personnel on the project will have been associated with it at the outset. In the event of an anomaly, it may be very difficult to find someone who is familiar with the spacecraft first hand.)

5. Launch energies are high. The Saturn V is theoretically capable of achieving these energies with Voyager class payloads but structural modifications may be required. An opportunity with comparable energy requirements occurs in 1978 but flyby distance at Saturn is rather low.

6. It would seem that considerable overall development of multiple planet technology would be in order to accomplish this mission. This might be best done by flying a multiple planet mission prior to the grand tour. Unfortunately no Earth-Jupiter-Saturn opportunities (the only multiple outer planet missions with reasonable flight times) occur in the early 1970s. Thus accomplishing a 1973 Earth-Venus-Mercury mission may be a necessary prerequisite for the grand tour. A Jupiter flyby with a "paper" Saturn could be used to check out the entire concept prior to the 1977 opportunity. A 1973 Jupiter flight would be ideal; a 1975 mission would be of marginal use due to the fight time involved.

INDEX

AACS, *see* Attitude and Articulation Control Subsystem (AACS)
AC+79 3888, 270
Acord, Don, 191
Actors and actresses, 361, 364
Actuators, 15, 276, 277–279
Aim points, 118
Air Force
   business dealings, 164
   First Aerospace Squadron, 190
   power source development, 157
   shuttle use, 110
Albee, Arden, 291–297, 303
Alexander, George, 73
Algorithms, 140, 255, 317, 354
Allocation of resources, 117
Alpha Ophiuchi, 270
American Rocket Society, 26
Analytical chemistry, 250–251
Andromeda, 30
Antennas, 226, 240
   arrangement of, 242
   data transmission and, 70
   DSN, 12
   Galileo program, 204–205
   spacecraft communication and, 27–28
   use of, 243
Antigravity, 26
Apollo missions, 27, 204
Approach guidance, 110
Armstrong, Neil, 364
Asteroids, 29–30, 70
Astronautica Acta, 65, 66
Astronomy, 74, 314
Asymptotic velocity, 270
AT&T, 260
Atmospheric motion studies software, 283
Attitude and Articulation Control Subsystem (AACS), 154–155
Attitude control system, 109, 110, 116, 148, 150
   memory refreshing, 127, 128
   slew command, 278–279
Australia, 243, 245
Autographs, 195

Automation, 288
Automotive technicians, 371–374
Aviation Week, 81
Avis, Charles, 281–288
Awards and medals, 166–167, 195
Azimuth actuator, 276, 277–279

Backup mission load, 354–355
Bacon, Roger, 26
Baines, Kevin, 390
Barlow, Richard, 140
Beerer, Joe, 72, 85
Belcher, John, 311
Berry, Richard, 410
Beta Pictoris, 306
Binder, Eando, 28
Black Thursday, 402–403
Blanco, Jack Martel, 385–387
The Blazing World, 25
Blinn, Jim, 19
Book of War Machines, 26
Boriakoff, Valentin, 258–260
Bourke, Roger D., 75–82, 183
Bradbury, Ray, 13, 212
Brickett, John E., 357–362
Briden, Jim, 251
Bridge, Herb, 311
Briggs, Geoff, 45
British Interplanetary Society, 26, 68
Brown, Jerry, 293, 296
Bryden, Jim, 156
Budget issues, 52–53, 124–125
   contributions to, 186
   Davis and, 161–164
   estimates, 234
   Grand Tour, 202, 225–227
   manned program, 179
   mission completion and, 169
   mission problems and, 168
   photographs and images, 286
   staff reductions and, 195
Burrows, William E., 72–73
Business management, 161–162

Calibration plaque, 283
Camelopardalis, 30
Camera engineers, 287

Cameras, 186–187, 217
　capability of, 305–306
　Ranger 6, 105
　see also Scan platform
Capacitors, 251–252, 255
Carter, Jimmy, 259
Casani, John R., 56, 85, 101, 113–120, 151
　cost pressures, 227
　decision making, 115–117
　Kohlhase and, 91
　management style, 234
　responsibilities, 210, 218
Cassini system, 158, 236–237
CATS, see Computer Accessed Telemetry System (CATS)
Cavendish, Margaret, 25
CBS Laboratories, 259
CCS, see Command Computer Subsystem (CCS)
Centaur, 88, 108, 147
Cesarone, Robert John, 21, 261–273
Chahine, Moustafa T., 299–306, 413
Challenger, 96, 109, 110, 205, 293–294, 312
Chauffeur service, 340
Children
　involvement in missions, 308, 365, 383
　JPL correspondence, 379
　public relations, 397–398
Chronology of missions, 417
Circuit breakers, 245
Clarke, Arthur C., 68, 74, 212, 343
CMOS, see Composite Metal Oxide Semiconductor (CMOS)
Coding devices, 116, 117
Cole, Patricia, 19
Collela, Frank, 400
Collins, Andy, 349
Colorado Video, 11, 258–260
Comic books, 379
Command Computer Subsystem (CCS), 148, 149, 154–155
Command loads, 126
Command Moratoria, 94–95
Communication systems, 139–140, 148
Communications, 410
　importance of, 336–337
　results of, 129–130
　scientific discoveries, 38–41
　signal-to-noise ratio, 116
　see also Deep Space Network (DSN)

Composite Metal Oxide Semiconductor (CMOS), 149, 253
Computer Accessed Telemetry System (CATS), 146
Computer Command Subsystem (CCS), 279
Computers, 28
　Galileo, 236
　generators for, 373
　integrated circuits, 251, 253
　memory, 214, 226, 280
　new system capabilities, 285
　reprogrammed, 205
　security issues, 352
　STAR, 146, 225
　use of, 187–188
Conflict resolution, 321–322, 348
Congress
　funding, 233
　Grand Tour and, 200–201, 203
　space exploration support, 215
Congreve, William, 26
Conrath, Barney, 311
Constant launch energy, 66
Constellations, 30
Consumables, monitoring of, 169–170
Contingency planning, 92–93
Contractor's management meetings, 231
Convolutional coding, 247
Corporal vehicle, 122
Cortright, Ed, 200
Cosmic ray particles, 421
Costs, 51, 124–125, 163–164, 186, 295
　budget decisions, 128
　design requirements, 206
　Earth–sun images, 305–306
　funding profile, 148
　Grand Tour, 225–227
　growth in, 233
　hardware size and, 236
　OPM, 147
　procurement, 226
　radiation environment, 149
　see also Budget issues; Funding
Cox, Henry G. (Hank), 239–248, 355
Crocco, G. A., 69
Cruise periods, 192–193, 242, 278
Cunningham, Glenn, 149, 156
Curiel, Robert, 343
Cutting, Elliott (Joe), 62–65, 68, 69, 77

Danielson, Ed, 217
Dash-pot damper, 150
Data compression devices, 116, 117
Data processing systems, 177
Data transmission, 55, 69–70
Davies, Mert, 217
Davis, Esker K., 15, 151, 159–171
The Day the Earth Stood Still, 264
Decision making
   Albee, Arden, 294
   Avis, Charles, 284–286
   Bourke, Roger D., 81–82
   Casani, John R., 115–117
   Cesarone, Robert John, 267–268
   Chahine, Moustafa T., 304–305
   Cox, Henry G. (Hank), 243–244
   Davis, Esker K., 169–170
   Hansen, Candice J. (Candy), 321
   Haynes, Norman R., 184, 268
   Laeser, Richard P. (Dick), 177–178, 243
   Lyman, Peter, 140–141
   Miner, Ellis D., 310–311
   Murray, Bruce, 218–219
   Parks, Robert J. (Bob), 126–128
   Pickering, William H., 200–202
   Post, Robert, 335–336
   Schurmeier, Harris M. (Bud), 110–111, 129
   scientists, 45–47, 304
   Shipley, William S., 230–234
   Sohus, Anita M., 351
   Stone, Edward C., 41–42, 45–46, 53, 129, 271, 309–310, 318, 321
   Textor, George, 193–194
   Wells, Peter W., 342
Deep Space Network (DSN), 95–96, 153, 160–161, 243–244, 252
   antenna, 12
   capability of, 292
   changes to, 188
   description, 240
   Laeser and, 174
   Mariner mission, 395
   role, 55, 135
   stations, 27–28
   Voyager relationship, 143–144
Defense Nuclear Agency, 227
Deflections, 8, 87
de Fontana, Joanes, 26
Delivery error, 272

Delta V, 97–98
Diogenes, Antonius, 25
Discoverer satellites, 38
Documentation, 346–347
Documentation representatives, 345–355
Draper, Ron, 230, 252
Drew, Russell, 217
DSN, *see* Deep Space Network (DSN)
DynaSoar, 110
Dyson, Freeman, 212

Earl, Jim, 38
Earth, 349
   phonograph record, 11, 258–260
   role in exploration, 302
Earthquakes, 92
Ecology issues, 297, 373
Edison, Thomas Alva, 26
Editors, 345–355
Education, 90, 397–398
Ehricke, Krafft, 62–63, 66, 74
Electric propulsion, 71
Energy issues, 218
Engineers
   Avis, Charles, 281–288
   Cesarone, Robert John, 21, 261–273
   characteristics of, 312
   Cox, Henry G. (Hank), 239–248, 353
   Gavin, Thomas R., 156, 249–255
   Marderness, Howard P., 275–280, 348
   overview, 221
   scientists and, 42–45, 201, 211
   Shipley, William S., 77, 156, 223–237, 252
   Southworth, Glen R., 11, 257–260
Environmental risks, 86, 92–93
Equipment, security of, 358–362
Ethical issues, 232
Europa, 45–46, 88
European Space Agency, 396
Exciters, 139
Explorer 1, 8, 27, 228
Exploring Space, 72

Family relationships, 88–90, 344, 360, 399–400
   mission impact on, 267, 335, 383, 387
   mission impact on children, 308, 365, 383
   visits by, 365–366

Fault protection routines, 226, 229–230, 253–254
   algorithms, 317, 354
   corrective action, 150, 155
   logic of, 98–99
Fawcett, Bill, 156
Felberg, Fred, 118
Ferris, Timothy, 213
Fishman, Bobbie, 363–369
Flandro, Gary A., 8–9, 61–74, 76, 80, 183, 187
Flight Data Subsystem (FDS), 93, 148, 149, 153–155, 277, 279
Flight Data System, 225, 253
Flight operations, 56, 150–151
Flight software, 149–150
Flight status bulletins, 345
Flight team, 143
Flying boats, Germany, 400–401
Forman, Edward, 7
Forward, Bob, 212
Fowler, Willy, 212
Francis, Karl, 21
From the Earth to the Moon, 25
Funding, 54–55, 148, 206, 241
   solar system mosaic, 324, 325
   see also Budget issues; Costs

Gag, Deborah Jean, 381–384
Gagarin, Yuri, 27
Galileo, 25
Galileo mission, 109, 112, 177, 204–205, 236–237, 254
Ganymede, 45
Gavin, Thomas R., 156, 249–255
Generators, 373
Geology, 216–217
Geysers, 39, 49
Gindorf, Tom, 156, 251
Goddard, Robert, 26–27
Golay encoded data, 153
Gold, Elliot, 258
Golovine Award, 68
Goodwin, Francis, 25
Government Accounting Office, 305
Grand Tour, *see* Outer Planets Grand Tour; Views of missions
Gravity assist, 62–66, 146, 151, 157
   Bourke and, 73
   flight opportunities, 85–87
   Grand Tour, 200–201
   Mariner missions, 160

Gray, Don, 268
Great Dark Spot, 49
Great Red Spot, 49–50
Greer, Marty C., 341, 371–374
Groom, Don, 72
Ground testing, 277
Guards, 385–387
Guggenheim Aeronautical Laboratory, California Institute of Technology, 7
Gulliver Joi, 26
Gunpowder, 26
Gurnett, Don, 311
Gyro error alarm, 253–254
Gyro saturation, 119

Haas, Conrad, 26
Haddock, Gordon, 156
Hagar, Tony, 15
Halley's Comet, 294
Hammel, Heidi, 321, 390
Hanel, Rudy, 107, 300, 311
Hannaway, Wyndham, 259
Hansen, Candice J. (Candy), 88, 315–325, 349, 350, 405
Hardening of spacecraft, 51, 107, 225, 227, 250–251
Hardware
   changes to, 120
   contingency planning and, 92–93
   control of, 265
   problem plaque, 15–16
   problems with, 152–154
   quality of, 280
   reuse of, 254
   uses of, 235
   see also Spacecraft
Haynes, Norman R., 181–188, 268
Heacock, Raymond L., 123, 145–158, 190
   Davis and, 161–162, 164
   leadership style, 166
   Parks and, 126
   personality, 230
   problem plaque, 15–16
   responsibilities, 126, 134, 137, 141, 210, 218–219, 225
Health issues, 88–89, 191, 286, 344
Heliocentric positions, 63–66, 73
Heliopause, 54, 55, 195, 269
Heliosphere, 354–355, 417
Herschel, William, 314
Heyser, Richard, 258

Hibbs, Al, 293
Himmler, Heinrich, 27
Hine, Maynard, 349
Historical perspectives, 213–214, 393, 410–413
    early Voyager development, 30, 85–87, 115–117, 139
    photographs, 334–335
    planetary exploration, 273
    space travel, 25–26
    spacecraft, 354
Howard, Don, 156
Hoyle, Fred, 26
Hubble Space Telescope, 51, 176, 317
Human exploration, 301
Hunter, John, 156, 229
Hydrazine, 97

Icaromenippus, 25
Icy planets, 296
Illustrations
    Neptune time line, 17
    Voyager, 10
    Voyager deflections, 8
    see also Photographs and images
Image motion compensation, 153–154
Image processing laboratory (IPL), 331–332, 401–403
Images, *see* Photographs and images
Imaging science, 421
Imaging sequence design, 316
Imaging Team, 319–320, 390, 401–403
India, photography exhibit, 396
Inflation, 152
In-flight problems, 150–151, 253
Informal documentation, 347
Information sharing, 130
Infrared investigation, 311–312
Infrared radiation, 421
Infrared spectrometer, 107
Integrated circuits, 149–150, 251, 253
Interferometer, 300
International cooperation, 243
Interplanetary ballistic trajectory program, 64–65
Interstellar space, 54, 269
Io, 296
    photographs, 88, 171, 318
    torus, 50, 97
    volcanoes, 49, 91–92
IPL, *see* Image processing laboratory (IPL)

Jacobson, Bob, 21
James, Jack, 228
Jet Propulsion Laboratory (JPL)
    Albee employment, 292
    Avis employment, 281
    Blanco employment, 385, 386
    Brickett employment, 357, 358
    Bruce employment, 114
    Casani employment, 113
    Cesarone employment, 261, 262
    Chahine employment, 299, 300, 303
    characteristics of, 80–81
    Cox employment, 239, 241
    Davis employment, 159, 160
    environment of, 235
    Fishman employment, 363
    Flandro employment, 62
    Gag employment, 381, 382, 384
    Gavin employment, 249, 250, 252
    Greer employment, 372
    Hansen employment, 315, 316
    Haynes employment, 181, 182–183
    Heacock employment, 145, 146
    Kohlhase employment, 83, 84
    Laeser employment, 173, 174
    Lyman employment, 133, 136
    Marderness employment, 275, 276
    McGavin employment, 375–376
    mechanical operations, 371–374
    Miner employment, 307, 308, 309, 310–311
    MJS 77 project, 147
    Murray employment, 209, 210, 215–216
    NASA relationship, 71–73, 77, 81, 182
    objectivity of, 123–124
    operation of, 352
    organization of, 24–25, 114, 156, 227
    Parks employment, 121, 122
    perception of, 218
    photographs, 5–6
    photography lab, 330
    Pickering employment, 199, 200
    Post employment, 329, 330
    security issues, 358–362
    Shipley employment, 223, 224
    Sohus employment, 345, 346
    Southworth employment, 257, 258
    success of, 130–131
    support services, 327–406
    Swift employment, 389–390
    Textor employment, 189, 190

Jet Propulsion Laboratory (JPL)
(continued)
   tour groups, 366–367
   van der Woude employment, 391, 398
   visitors to, 358–359, 361, 363–369
Johnson, Torrence, 20, 293
Joint space ventures, 179
Jones, Chris, 156
Joseph, Al, 71
Journal of Spacecraft, 8–9
Journey into Space, 213–215
JPL, *see* Jet Propulsion Laboratory (JPL)
JST, *see* Jupiter-Saturn-Titan (JST) mission
JUN, *see* Jupiter-Uranus-Neptune (JUN) mission
Jupiter, 39
   flight obstacles to, 29–30
   flight revisions, 150–151
   radiation environment, 97, 149
   turbulence, 49–50
"Jupiter Effect," 63
Jupiter-Saturn-Pluto mission, 105
Jupiter-Saturn-Titan (JST) mission, 78
Jupiter-Uranus-Neptune (JUN) mission, 105–108

Keller, Warren, 78–79
Knowledge error, 272
Kohlhase, Charles E., 56, 59, 68, 72, 83–100
   characteristics, 349
   consumables monitoring, 170
   photograph, 15, 19
   public relations, 353

Laboratory directors
   Murray, Bruce, 13, 70–72, 108, 109, 114, 115, 122–123, 126, 134, 155, 166, 212, 209–220, 294–295, 394–395, 413
   overview, 197
   Pickering, William H., 8, 68, 84–85, 104, 122, 199–207, 227, 240
Laeser, Richard P. (Dick), 141, 173–180, 183, 349
   Davis and, 164
   decision making, 177–178, 243
   JPL operations, 352
   leadership style, 166–167
   photograph, 15
   public relations, 353
   responsibilities, 348
   Textor and, 190
Lakia (dog), 27
Lane, Jerry, 253
Lane, Lonne, 350
Laser sails, 212
Launch energy, 66–67
Launch vehicles, 119
Launch windows, 66–67, 87, 323
Law enforcement agency personnel, 369
Leadership style, 138
   differences in, 166
   Lyman, Peter, 171
   Parks, Robert J. (Bob), 166
Lee, Gentry, 134
Liebman, Hans, 300
Lightning proofing, 232, 234
Ling, Helen, 64
Linick, Sue, 350
Literature, space travel, 25–26
Littman, Mark, 68–70
Long, Jim, 191
Low, David, 364
Low energy charged particles, 421
Lukian of Samasata, 25
Lunar landings, 122–123, 301
   see also Ranger project
Lyman, Peter, 123, 133–144
   decision making, 140–141
   Parks and, 126
   responsibilities, 126, 218, 219

MacDonald, Rob Roy, 224
Madge, A. V., 26
Magnetic fields, 49, 50, 70, 421
Magnetosphere, 50
Mail department, 341
Malina, Frank, 7
The Man in the Moone, 25
Management relations, 111, 123
Management style, 231
   Marderness, Howard P., 276
   Miner, Ells D., 348
   Parks, Robert J. (Bob), 123–124, 126–128
   Shipley, William S., 231–232
   Stone, Edward C., 42, 310, 321, 348
Maneuver accuracy model, 271
Maneuver analysis, 205, 264

Marble, Frank, 65
Marderness, Howard P., 275–280, 348
Margraf, Harry, 156
Mariner Jupiter-Saturn (MJS) mission, 51, 71–72, 81, 85–86, 106–108
  early development of, 76–78
  prelaunch, 148–150
  proposal for, 147, 200
  transition to, 202–203
Mariner-Jupiter-Saturn-Uranus mission, 116
Mariner missions, 105–107, 204, 228–229, 395
  history, 29–30
  see also name of specific mission
Mariner Venus-Mercury mission, 160, 225
Mark, Hans, 72–73
Mars, 29, 194
Mars Observer, 205
Martin Marietta, 84
Massachusetts Institute of Technology, Uranus photographs, 260
Masursky, Harold, 14
Matrix organization, 114, 156, 227
McClelland, Larry, 259
McGavin, Judith, 375–379
McKay, Chris, 213
McKinley, Ed, 15
Mechanic's shop, 341
Media
  advances in, 292
  amount of, 186
  negative press, 178
  Voyager coverage, 164–166
  workload and, 383
Mercury, 29, 215
Merrill, Grayson, 63
Meyer, Jerome, 26
Meyer, Peter, 38
Miles, Ralph, 64
Military Sea Transport Service, 135
Miller, Ron, 25
Miner, Ellis D., 13, 241, 303, 307–314, 348
Minovitch, Michael, 62–65, 76, 80, 183
Miranda, 49, 178
Mission control team, 265
Mission design, 79, 205, 263
"Mission Design Guidelines and Constraints," 95
Mission design managers
  Kohlhase, Charles E., 15, 19, 56, 59, 68, 72, 83–100, 170, 349, 353
Mission planning, 96–97, 117, 234–235
MJS, see Mariner Jupiter-Saturn (MJS) mission
Modeling the atmosphere, 244–245
Momentum cancellation, 276
Monitoring systems, 245
Montgomery, Larry, 349
Morabito, Linda, 92
Morrison, David, 29
Murakami, Shozo, 346
Murray, Bruce, 108, 109, 114, 134, 209–220, 394–395, 413
  appointments, 134
  Casani and, 115
  data transmission and, 70
  decision making, 218–219
  leadership style, 166
  moons of outer planets, 71
  Parks and, 122–123
  photograph, 13
  responsibilities, 126, 294–295
  TCM problem, 155

Nakata, Albert, 15
National Aeronautics and Space Administration (NASA), 151–152, 293–294, 312–313
  awards and medals, 167
  budget, 152, 162–164, 202
  funding, 206, 241
  Grand Tour and, 71–73, 77, 106, 200–201, 202
  Haynes relationship with, 185
  JPL relationship, 71–73, 77, 81, 182
  Neptune mission, 52–53
  power source development, 157
  Ranger project, 203
  Saturn support, 165
  science experiments selection, 201, 202
  shuttle development, 109
  space exploration support, 215
National Geographic, 89
National Geographic Society, 259
Naugle, John, 200, 210
Navaho project, 115
Naval architects, 135–136
Navigation team, 262–264

Neal, Roy, 165
Negator spring, 150
Neptune mission, 49, 183–184
   communication delays, 28–29
   encounter time line, 17
   magnetic field, 50
   NASA and, 52–53
   photographs, 185–186
   research, 390
   scientific discoveries vs engineering capability, 44–45
Ness, Norm, 350
New Zealand, 206
Newton, Isaac, 26
Nixon, Richard, 71
Nonstandard commands, 127
North American, 115

Occultation zone, 47, 266–267, 272
Odetics, 93
Of the Wonderful Things Beyond Thule, 25
Operations, 142, 174, 219
OPM, *see* Outer Planets Missions (OPM)
Optical navigation, 263–264
Orbit determination, 263–264, 272
Orbital energy, 62–64
Outer Planets Grand Tour, 30, 56
   cancellation of, 125
   continuation of, 158
   costs, 225–227
   funding, 300
   initial reaction to, 76–78
   name adoption, 69
   NASA reaction to, 77
   photographs, 73
   proposal for, 105, 134, 200–201
Outer Planets Missions (OPM) project, 65–66, 71, 74, 76–78, 146–147

Pancreatic cancer, 88–89
Parks, Robert J. (Bob), 121–131, 170, 190
   decision making, 126–128
   leadership style, 166
   responsibilities, 134–135, 137, 141, 151, 218, 219
Parsons, John, 7
Parts selection, 130, 225–226
Patrick Air Force Base, Cape Canaveral, 160
Peer review, 245

Penzo, Paul, 72, 85
Perce, Elbert, 26
Personnel, 114
   age, 316
   appointments, 218
   as spokespersons, 292
   changes in, 311, 318
   characteristics of, 52, 99–100, 138–139, 207, 241, 255, 273, 280, 306, 312–313, 314, 348–350, 365–366, 384, 410
   commitment to project, 193, 194
   competition among, 211
   conflict resolution, 322, 348
   contributions of, 246–247
   counseling for, 309
   dynamics of, 191–192
   family relations, 88–90, 267, 308, 335, 344, 360, 365–366, 383, 387, 399–400
   holiday schedules, 95
   hours worked, 125, 286–287, 303, 308, 322, 343, 358, 360, 376, 382, 401–403
   incompatibility of, 175–176
   mechanics, 341
   morale, 123, 350
   motivation of, 120, 302
   NASA, 226
   number of, 185, 190–191, 202, 236, 309, 317, 374
   office politics, 124
   output of, 286
   peer review, 245
   personal sacrifices, 248
   problems of, 140–141, 309–310
   reduction in, 195, 218
   relationships, 311, 317, 322, 348
   responsibilities, 316–317
   security of, 358–359
   selection of, 110, 128–129, 193
   staff meetings, 141
   staffing level, 55
   stress, 88, 241, 309, 344
   transition to Galileo, 177
   turnover, 347
Peter the Great, 26
Phases of missions, 419
Phobos spacecraft, 138, 211, 220
Phonograph record, 11, 258–260
Photo lab supervisors, 329–337

Photographs and images, 292
  availability of, 285–286
  camera system, 217
  collection, 5–23
  demand for, 330–331
  down link process, 320
  Europa, 88
  exposure timetables, 319
  Grand Tour, 73
  historical significance, 194
  image motion compensation, 153–154
  image processing, 282, 331–332, 401–403
  imaging sequence design, 316
  impressions of, 376, 393–394
  Io, 88, 171, 318
  JPL lab, 329–337, 401–403
  Jupiter, 142, 156–157
  Mars, 194
  Miranda, 178
  missing data, 312
  Neptune, 185–186, 193, 390
  phonograph record, 11, 258–260
  photo lab supervisors, 329–337
  photography show, 396
  printing requirements, 332–333
  production of, 331–332, 401–403
  reaction to, 246
  recognition for, 335–336
  Saturn, 165
  significance of, 333–334
  smears on, 277–278
  solar system, 305, 333–334
  uplink process, 320
  Uranus, 193
  visualization of, 404
Photopolarimetry, 421
Pickering, William H., 68, 84–85, 104, 122, 199–207, 240
  cost pressures, 227
  decision making, 200–202
  photograph, 8
Pickets, 368
Pioneer missions, 29–30, 38–39, 54, 98, 149, 211, 250–251
Planet Comics, 379
Planetary exploration
  attitude about, 73
  challenges to, 302
  discoveries, 49–50
  historical overview, 29–30
  personnel characteristics, 52
Planetary radio astronomy, 282, 421
Planets Beyond, 68
Plant protection, 357–362, 385–387
Plasma particles, 421
Plasma wave subsystem (PWS), 282
Plasma waves, 421
Political issues, 179, 214–215, 217, 297
Posen, Dan Q., 262
Post, Robert, 20, 329–337
Potts, Chris, 21
Power source development, 157
Prelaunch period, 134, 176
Press conferences
  computer memory, 187
  graphics and, 347, 350–351
  photographs and images, 14, 285, 286–287, 330
  planning for, 337
  Saturn mission, 164–165, 178
  scientific discoveries, 39–41
  security issues, 386–387
  Uranus mission, 21, 178
Press releases, 282
Principal investigators, 211, 221, 304, 310
Principia, 26
Problem/failure reports, 250, 252
Problem-solving techniques, 305
Procurement, 226, 231–232
Project Astro, 314
Project managers
  Casani, John R., 56, 85, 91, 101, 113–120, 151, 210, 218, 227, 234
  Davis, Esker K., 15, 151, 159–171
  Haynes, Norman R., 181–188, 268
  Heacock, Raymond L., 15, 123, 145–158, 166, 190, 210, 218–219, 225, 230
  Laeser, Richard P. (Dick), 15, 141, 166, 173–180, 183, 348, 349, 352, 353
  Lyman, Peter, 123, 133–144, 218
  overview, 101
  Parks, Robert J. (Bob), 121–131, 166, 170, 190, 218, 219
  Schurmeier, Harris M. (Bud), 56, 84, 101, 103–112, 114, 123, 146, 151, 174, 210, 215, 218, 225, 230
  Textor, George, 15, 101, 189–196
Proof test model, 118, 235

Propellant tanks, 96–97
Propellants, 97–98
Propulsion, 26–27, 252–253
Prospector, 233
Public Information Office, 383
Public information officers, 391–406
Public relations, 89–90, 195–196, 296, 353, 391–406
   children, 379, 397–398
   discussion group, 13, 212
   increase in, 294
   international, 396–398
   outreach, 314
   photograph distribution, 334
   quality of, 292
Public safety, 385–387
Publicity, Uranus flights, 118

Quality assurance, 340
Quayle, Dan, 361, 382

Radar, 122
Radiation design margin (RDM), 149
Radiation environment, 97, 107, 232, 250–251, 271, 276–277
   bremsstrahlung, 219
   effect on spacecraft, 70
   effects of, 149
   information sharing and, 130
   intensity of, 45, 51
   magnitude of, 227
   size of, 115
   study of, 38
Radiation hardening, 51, 107, 225, 227, 250–251
Radio beams, 46
Radio frequencies, 95–96, 244
Radio Frequency Subsystem (RFS), 139–140, 250, 251–252
Radio messages, 27–28
Radio occultation, 44
Radio science, 240, 242–245, 277, 421
Radioisotope thermoelectric generators (RTGs), 146, 150, 157–158
Radios
   development, 227–228
   TOPS and, 224–225
   see also Receivers
Random noise, 247
Ranger project, 104–105, 111, 203, 228–229

RDM, *see* Radiation design margin (RDM)
Readers' Scope, 26
Real-time frames, 287
Receivers, 91, 94, 151–153, 228, 254–255, 354
   failure of, 140, 205
   problems with, 155–156
Receiving services, 341
Recordkeeping, transportation, 342–343
Recycling, 373
Reed-Solomon coding, 153, 247
Reiber, Duke, 346
Remote sensing, 110
Research fellows, 389–390
Rest frequency, 94
Review boards, 292
Ring occultation, 46–47
Risk management, 183–184
Robotic missions, 27–29, 179, 214, 300–301
Rocket propulsion, 26–27
Rockets, historical overview, 26–27
Roll maneuvers, 119
Roth, Duane, 21
Royal Astronomical Society, 26
RTG, *see* Radioisotope thermoelectric generators (RTGs)
Rudd, Dick, 352
Ryne, Mark, 21

Safety issues, 46–47
Sagan, Carl, 13, 19, 212, 213, 335, 364, 377, 405
Sander, Mike, 156
Satellites, 216–217
Saturn
   rings, 46–47, 49, 78, 98
   science investigations, 421
   secondary flight objectives, 117–118
"Saturn and the Mind of Man," 13, 212
Saunders, Steve, 293
Savino, Joe, 156
Scan platform, 43, 91, 99, 151, 177, 178, 191
   actuator and, 139
   booms, 150
   conflicts and, 321
   images and, 283
   images lost and, 266–267
   photograph, 15

position change and, 313
press conference and, 164–165
problem cause, 220
resource allocation and, 117
slewing of, 277
spacecraft maneuvers and, 205
Scarf, Fred, 311, 348
Schurmeier, Harris M. (Bud), 56, 101, 103–112, 114, 123, 151
   decision making, 110–111, 129
   Heacock and, 146
   Kohlhase and, 84
   Laeser and, 174
   personality of, 230
   responsibilities, 210, 215, 218, 225
Science enhancements, 147–149
Science experiments, 201, 202, 318, 377–378, 421
Science fiction, 262
Science payload, 148–149
Science Steering Committee, 110–111
Science Steering Group, 41, 271, 310, 351
Science teams, 312, 321
Scientists
   Albee, Arden, 291–297, 303
   Chahine, Moustafa T., 299–306, 413
   challenges to, 52
   decision making, 41–43, 45–47, 53, 304
   engineers and, 42–45, 201, 211
   Hansen, Candice J. (Candy), 88, 315–325, 349, 350, 405
   Miner, Ellis D., 13, 241, 303, 307–314, 348
   overview, 289
   public information office and, 403–404
   role of, 48
   Stone, Edward C., 13, 21, 35, 37–57, 111, 126, 129, 141, 194, 211, 215, 248, 271, 293, 303, 309–310, 311, 318, 321, 325, 347, 348, 351, 353, 392
   time commitment, 48
Scott, Rudolph, 7
Scout, 319
Secretarial staff, 375–379
Security issues, 351–352, 368, 386–387
   JPL, 358–362

terrorism, 369
weapons, 369
Self Test And Repair (STAR), 146, 225
Semiconductors, 149, 253
Sequence planning, 41–42, 94–95, 127, 192–193, 265, 292, 313
Sequence teams, 194–195
Sergeant missions, 122, 228–229
Shaeffer, Joe, 156
Shatner, William, 344
Shatz, Bill, 156
Shell Development, 136
Shipley, William S., 77, 156, 223–237, 252
Shipping services, 341
Shuttle development, 71
Signal frequency, 247, 263, 266
Simpson, John, 38
Sirius, 269, 270
Slewing, 277–279
Smith, Brad, 217, 316
Smith, M. O., 7
Snyder, Don, 64
Social issues, 124
   planetary exploration, 273
   solutions to, 169
Society for Spaceship Travel, 25
Soderblom, Lawrence, 20, 217, 403
Software, 149–150, 188, 241–242, 354–355
   atmospheric motion studies, 283
   attitude control patch, 278
   corrections to, 219
   modifications to, 192–193
   problems with, 190, 219
   Scout, 319
Sohus, Anita M., 345–355
Solar electric propulsion, 71
Solar system
   formation, 300
   mosaic, 305, 324–325, 404–406, 417
   photographic tour of, 284
Solar System Exploration Committee, 294, 295
Solar wind, 54
Sonic shock, 54
Sorel, Charles, 25
Southworth, Glen R., 11, 257–260
Space Flight, 62–63
Space Flight Operations Plan, 346

Space Science Board, 106
Spacecraft
　course corrections, 270–271
　design, 50–51, 136, 137, 139–140, 196
　effectiveness of, 314
　evolution of, 206
　future operation of, 195
　historical perspective, 354
　impressions of, 387
　longevity of, 196, 269
　moving of, 372
　navigation problems, 272–273
　perceptions of, 196
　sequence planning and, 196
　speed of, 269–270
　views of, 210–212
　see also Hardware
Spacecraft convoy, 342
Spacecraft team, 276
Spehalski, Dick, 156
Sputnik, 27
Staff meetings, 141
State Department, 96
Stembridge, Charles, 191, 349
Stevens, Bob, 143–144
Stewart, Homer Joe, 69, 71, 80–81
Stone, Edward C., 35, 37–57, 111, 126, 141, 194, 248
　decision making, 41–42, 45–46, 53, 129, 271, 309–310, 318, 321
　management style, 321
　photograph, 13, 21
　press conferences, 347, 351
　public relations, 353, 392
　responsibilities, 52–53, 55, 126, 211, 215, 303, 311
　role of, 293
　solar system mosaic, 325
Stress, 241
Struts, 97
Sturms, Fran, 62, 77, 156
Subsystems
　Attitude an Articulation Control Subsystem, 154–155, 252
　Command Computer Subsystem, 148, 149, 154–155, 279
　Computer Command Subsystem, 279
　development of, 108–109
　Flight Data Subsystem, 93, 148, 149, 153–155, 277, 279

Flight Data System, 225, 253
　memory, 280
　photograph, 14
　plasma wave subsystem, 282
　Radio Frequency Subsystem, 139–140, 250, 251–252
　reliability of, 237
　TOPS and, 224–225
Sullivan, Walter, 13
Summerfield, Martin, 27, 65
Sun sensors, 116
Support services
　automotive, 371–374
　documentation representative, 345–355
　overview, 327
　photography lab, 329–337
　plant protection, 357–362, 385–387
　public information office, 391–406
　research fellows, 389–390
　secretarial staff, 375–379
　telephone service, 381–384
　transportation dispatcher, 339–344
　visitor control, 363–369
Surface barrier detectors, 119
Surveyer project, 122–123
Swift, Catherine, 389–390
System design, 43–44, 226
Systems engineering, 43, 136–137
Systems engineers, 231–232

Tape recorders, 117
TCMs, *see* Trajectory Correction Maneuvers (TCMs)
Teamwork, 99–100, 124, 156–157, 195, 207, 348
　characteristics of, 175–176, 323–324
　decision making, 53
　lack of, 134
　operations team, 137–138, 142
　personalities and, 230
　robotic missions, 179
　spirit of, 123
　strategy for, 177
　trajectory programs, 79–80
　trust and, 230–231
　views of, 211
Technical documentation, 346–347
Technical support, 156
Telecommunications, 28, 86
Telemetry systems, 146, 242

Telephone numbers, 116
Telephone operators, 381–384
Television
    coverage, 292
    graphics, 347
    production, 395–397
Terrorism, 369
Textor, George, 56, 101, 189–196
    decision making, 193–194
    Haynes and, 187
    Laeser and, 190
    photograph, 15
Thermoelectric Outer Planet Spacecraft (TOPS), 76, 105, 106, 146–149, 157–158, 224–225
Thrusters, 266
Time-lapse movies, 284, 287
Time lines
    Neptune encounter, 17
    Uranus encounter, 18
Timing errors, 272
Titan, 46–47, 49, 69, 82, 118, 155
Titan IV launch vehicle, 108, 109, 110
Titan-Centaur launch vehicle, 147
TOPS, see Thermoelectric Outer Planet Spacecraft (TOPS)
Tracking and data systems, 160–161
Trajectories, 44–45, 62–68, 150, 155, 200
    accuracy of, 271
    choice of, 69–70
    corrections in, 264, 271–272
    decisions involving, 304
    graphic display of, 79–80
    planning for, 267–268
    profiles of, 66–68
    selection of, 86, 270–271
Trajectory Correction Maneuvers (TCMs), 154–155, 244–246
Trajectory designers
    Bourke, Roger D., 75–82
    Flandro, Gary A., 61–74
Transportation services, 339–344
Traveling wave tube amplifier (TWTA), 148, 227–228
Traxler, Marv, 241
Trip to Mars, 26
Triton, 47, 272, 393
    geologic surface, 49
    scientific discoveries vs engineering capability, 44–45
Tucana, 270

Turbulence, 49–50
TWTA, see Traveling wave tube amplifier (TWTA)
Tyler, Len, 47

Ultraviolet spectrometer, 45, 47
Ultraviolet spectroscopy, 421
Ulysses missions, 54
University of Chicago, 38
University of Hawaii, telescope, 390
Unmanned exploration, 25, 27
Uranus, 47
    description of, 185
    photograph, 20
    time line, 18

van Allen, James, 71, 216
van der Woude, Jurrie, 292, 330, 391–406
Velocity, asymptotic, 270
Venus Orbiting Imaging Radar, 294
Vera Historia, 25
Verne, Jules, 25, 26
Video converters, 258
Video modulation test system, 258
Views of missions, 246, 295, 301, 353–354
    Bourke, 78–79
    challenges to, 117
    characteristics of, 235–236, 351
    conclusions, 409–413
    continuity of, 410
    contributions of, 3–4, 124, 297, 312, 322
    disappointments, 50, 124–125, 297
    discoveries of, 320
    evaluation criteria, 409
    expectations of, 320–321
    failures, 90–91, 112
    future of, 53–54, 322–323
    impressions of, 293, 316–319, 362, 374, 376–379, 383–384, 392–394, 398–399
    Kohlhase, 87–88, 98–100
    lessons learned
        Albee, Arden, 295–296
        Avis, Charles, 288
        Casani, John R., 120
        Chahine, Moustafa, 300–301
        Cox, Henry G. (Hank), 247, 248
        Davis, Esker D., 167, 169

lessons learned (continued)
    Flandro, Gary A., 74
    Haynes, Norman R., 187–188
    Kohlhase, Charles E., 87, 98, 100
    Lyman, Peter, 142–143
    Miner, Ellis D., 313–314
    Parks, Robert J. (Bob), 124
    Pickering, William H., 204–206
    Schurmeier, Harris M. (Bud), 112
    Sohus, Anita M., 352–353
    Stone, Edward C., 53
    Textor, George, 196
  memories of, 330–331
  obstacles to, 4
  positive aspects of, 144, 266
  problems with, 214–215, 220, 301, 377–378
  scientific impact, 38
  Stone, 53, 57
  see also Historical perspectives
Viking missions, 147–148, 225
Visitor control greeters, 363–369
Visitors, 358–359, 361
  control of, 363–369, 386
  parking, 365

Vogt, Rochus (Robbie), 38, 111, 311
Volcanic caldera, 171
Volcanoes, Io, 296, 318
von Braun, Werner, 8, 71
Voyager Bulletins, 346
Voyager Interstellar Mission, 54–55

Waff, Craig, 76
Wave tube amplifiers, 227–228
Webb, Jim, 233
Weightlessness, 26
Wells, Peter W., 19, 339–344
Wilkins, John, 25
Williams, Larry, 341
Williams, Max, 68
Wilson, Jim, 346
Winds, 49
Wooley, Richard, 26
Work environment, 365, 366–367
Wright, Frank, 156

X-band system, 93, 148

Yardumian, Louie, 114–115